BEAMS, PLATES, AND SHELLS

KV-638-595

STRATHCLYDE UNIVERSITY LIBRARY
30125 00089942 6

ML

Books are to be returned on or before
the last date below.

07 JUN 1999
08 DEC 1999
4 DEC 2000
16 SEP 2003

LIBREX

2 0 AUG 1997

McGRAW-HILL
BOOK COMPANY
New York
St. Louis
San Francisco
Auckland
Bogotá
Düsseldorf
Johannesburg
London
Madrid
Mexico
Montreal
New Delhi
Panama
Paris
São Paulo
Singapore
Sydney
Tokyo
Toronto

ANDERSONIAN LIBRARY
★
WITHDRAWN
FROM
LIBRARY
STOCK
★
UNIVERSITY OF STRATHCLYDE

LLOYD HAMILTON DONNELL

Professor Emeritus
Illinois Institute of Technology

Adjunct Professor
University of Houston

ENGINEERING SOCIETIES MONOGRAPHS

Beams, Plates, and Shells

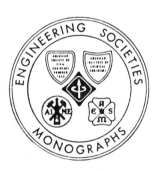

Library of Congress Cataloging in Publication Data

Donnell, Lloyd Hamilton, 1895–
Beams, plates and shells.

(Engineering societies monographs)
Includes index.
1. Girders. 2. Plates (Engineering) 3. Shells (Engineering) I. Title II. Series.
TA660.B4D66 624′.177 76-20665
ISBN 0-07-017593-4

BEAMS, PLATES, AND SHELLS

Copyright © 1976 by McGraw-Hill, Inc. All rights reserved.
Printed in the United States of America. No part of this publication may be reproduced, stored in a retrieval system, or transmitted, in any form or by any means, electronic, mechanical, photocopying, recording, or otherwise, without the prior written permission of the publisher.

1 2 3 4 5 MPMP 7 9 8 7 6

D
624.1772
DON

To

SYLVIA

and a friendship and romance

still strong after more than half a century

ENGINEERING SOCIETIES MONOGRAPHS

Bleich: *Buckling Strength of Metal Structures*
Crandall: *Engineering Analysis*
Donnell: *Beams, Plates, and Shells*
Ippen: *Estuary for Coastline Hydrodynamics*
Knapp, Daily, and Hammitt: *Cavitation*
Nachtrieb: *Principles and Practice of Spectrochemical Analysis*
Timoshenko and Gere: *Theory of Elastic Stability,* Second Edition
Timoshenko and Goodier: *Theory of Elasticity,* Third Edition
Timoshenko and Woinowsky-Krieger: *Theory of Plates and Shells,* Second Edition

ENGINEERING SOCIETIES MONOGRAPHS COMMITTEE
A. S. C. E.
Marcel P. Aillery
Paul Parisi
A. I. M. E.
John V. Beall
Lawrence G. Kuhn
A. S. M. E.
Philip G. Hodge, Jr.
Allan R. Catheron
I. E. E. E.
E. K. Gannett
Edward E. Grazda
A. I. Ch. E.
Harold I. Abramson
Oliver Axtell

Five national engineering societies, the American Society of Civil Engineers, the American Institute of Mining, Metallurgical, and Petroleum Engineers, the American Society of Mechanical Engineers, the Institute of Electrical and Electronics Engineers, and the American Institute of Chemical Engineers, have an arrangement with the McGraw-Hill Book Company, Inc., for the production of a series of selected books adjudged to possess usefulness for engineers and industry.

The purposes of this arrangement are: to provide monographs of high technical quality within the field of engineering; to rescue from obscurity important technical manuscripts which might not be published commercially because of too limited sale without special introduction; to develop manuscripts to fill gaps in existing literature; to collect into one volume scattered information of especial timeliness on a given subject.

The societies assume no responsibility for any statements made in these books. Each book before publication has, however, been examined by one or more representatives of the societies competent to express an opinion on the merits of the manúscript.

S. K. Cabeen, CHAIRMAN
Engineering Societies Library
New York

CONTENTS

PREFACE

The three subjects covered in this book, beams (particularly beams of rectangular cross section, which serve as an introduction to and special case of the other two subjects), plates and shells, each have a large literature and would justify books of their own. The present treatment represents at best an introduction to these subjects, with unusual emphasis upon practical aspects such as the errors made in the various approximations commonly used and methods for getting more accurate results when this proves necessary.

The book makes no pretense of covering modern world developments in its subjects, particularly in the most advanced and active field, that of shells. The author has never been a 'scholar' in the sense of being a diligent student of the world literature in any field, and since the book is being written a number of years after his retirement from full time activity he is poorly equipped to present a review of recent developments.

Instead, this is a very personal book, presenting its subjects from the author's own point of view. Comparatively few references are given, and those which are given are predominantly and unabashedly to the author's own work. This is certainly not because he considers his own work more important than that of others; rather it is because this is the work with which he is most familiar, and because it is his belief or at least his hope that this work is worth presenting even where the work of others may have gone far beyond it.

Among many questions which may be raised regarding the book, a rather obvious one is why, having decided to use scalar mathematics in presenting the subject, some mathematical shorthand at least was not adopted, such as using subscripts for differentiation. The answer for this particular possibility, and similar ones could be cited for others, is that subscripts were needed for other purposes. And, in the author's not very humble opinion, 'shorthand' such as $\partial^4 F_x / \partial x^4 = F_{x,xxxx}$ or $F_{x,4x}$ is mathematical jargon which saves little space and adds one more distraction, at least for the less mathematically sophisticated reader, to divert his attention from the very real difficulties of a difficult field. On the other hand symbols such as the Laplacian operator ∇^2 and its multiples save so much space and are used so often that it pays to initiate any readers who may not be familiar with them.

It is not easy to give proper credit. In the first place it is necessary to apologize for the many developments which are presented without proper attribution to those who have made them before. The lack of acknowledgement in some cases may be due to ignorance of recent prior publication, the work presented having been done independently by the author (there is a great deal of material in the book which is new as far as the author is concerned); such an error is at least pardonable in the present era of science explosion. Less pardonable are omissions of attribution due to thoughtlessness or ignorance of the originator of work which has long been so familiar that one tends to think of it as part of the public domain or as a routine application of well known principles.

Only a few people have read parts of the manuscript during its preparation. The author acknowledges with thanks the comments of 'Bo' Almroth and Elfriede Tungl, who read the early chapters, and particularly of Eli Sternberg, who read the first and most of the third chapter and made many valuable comments, particularly on the history of elasticity theory. Of course only the author can be blamed for the book's no doubt numerous errors of typography as well as substance, and he will be very grateful to anyone who takes the trouble to tell him of either errors or omissions of attribution, which he will try to correct if and when it becomes possible.

By far the greatest help which the author has received has been from Douglas Muster, and words of thanks to him are quite inadequate. He is the author's former student and colleague and was the host for the lecture series at the University of Houston which gave the impetus for writing the book. Not only was he largely responsible for initiating this project, but he has provided great help and encouragement throughout the years of its writing, and has taken it upon himself to arrange for its publication; without him the book would not have been written.

LLOYD DONNELL

CONSISTENTLY USED SYMBOLS

E, G, ν, τ—material properties, namely modulus of elasticity, shear modulus = $E/2(1+\nu)$, Poisson's ratio, strength property (τ_y = yield point, etc.)

σ, ϵ, F, M—stress, strain, and force and moment per unit length of section, all used with subscripts indicating directions

u, v, w and u_x, u_y, u_z—displacements of middle surface point and of general point, in x, y, z directions.

x, y, z—rectangular, and cylindrical (x being the axial, y the circumferential directions) coordinates.

c, $h = 2c$—half thickness, thickness of beam, plate, or shell.

$\mathscr{E}$—strain energy

∇^{2m}—$(\partial^2/\partial x^2 + \partial^2/\partial y^2 + \partial^2/\partial z^2)$ applied m times

1
GENERAL PRINCIPLES

1.1 INTRODUCTION

This book is based in large part upon the lectures given for a class in shell theory which was presented at the University of Houston in 1966. Flat plates and beams are considered as an introduction to and as special cases of shells. The discussion is principally of elastic, homogeneous, isotropic constructions with constant thickness or sections, and these conditions will be assumed unless otherwise indicated.

While the theoretical coverage of the fundamentals of the subject is quite complete, the emphasis is upon engineering aspects rather than upon the more mathematical approach which is becoming increasingly prevalent. The book is thus directed particularly to the engineer who is interested in applications to real situations. However, it should also be of interest to those making a career in shell theory, as a supplement to the more mathematical studies, which sometimes give minimum attention to the physical meanings of their mathematical statements or their relevance to the actual problems with which engineers and scientists must deal.

Besides stressing physical meaning, considerable attention is given to the neglections and approximations which even so-called 'exact' solutions involve. It is sometimes said that since real problems are too complex to study in their entirety, we must replace them with simplified, ideal 'mathematical models' which are more practical to study. But this is a rather dangerous way of looking at things, since it tempts us to forget the approximations which we are actually making and to be satisfied with our idealized solutions. As engineers we do not deal with idealized shells but with complex and messy structures like airplane wings. We have to decide what are the outstanding characteristics which can not be neglected in any given study and take these into account, at the same time considering how good an approximation to the real problem we are getting in neglecting the other things.

In general, any worth-while simplifications are valid only in certain ranges of applications, and thus we have various kinds of shell theories for different applications. Much of the difficulty in shell theory lies in deciding what simplifications are valid in certain types of applications, and the range of their validity. The range and accuracy of the various theories presented are discussed at the ends of Chaps. 3, 5, and 7. An understanding of the physical meaning of the approximations and of the terms which we are actually neglecting is obviously the first step in trying to reach reasonable conclusions.

Some space is also devoted in this first chapter to discussion of such things as material properties, failure theories, and philosophy of design (factors of safety, etc.). Such questions are evidently not peculiar to shell theory, but they are highly relevant to a rational approach to any field of design and are too often ignored completely.

All derivations are carried out in ordinary 'number mathematics' (or, strictly speaking, scalar mathematics, since most of the numbers involved represent the magnitudes of physical characteristics). Vector and tensor notations and mathematics provide powerful shorthands as well as shortcuts to certain mathematical operations, but it is the author's opinion that they are not inherently useful to engineers except insofar as familiarity with them may be necessary to read literature written in their language; however, this is only a matter of language, like learning German or Russian in order to read that literature.

Of course these forms of mathematics have important uses for deriving complex theories, but theories so set up can seldom be used directly for solving specific numerical problems. In general, in order to solve numerical problems, vector or tensor relations must first be translated into the corresponding number relations. When a subject as complex as shell theory is set up in tensor form approximations are usually introduced in translating it into number form, and the physical meaning and justification of these approximations may not be very clear. If no approximations are introduced the total task is likely to be greater than it is if a more direct approach is used.

The symbols used in all equations will of course represent the magnitudes of the physical quantities referred to. Since the meaning will always be obvious we can use the same symbols in referring to the physical quantities themselves in the text.

While the problems which we study will chiefly be statics problems, we will generalize the methods which we develop and make them applicable to dynamics problems by the use of D'Alembert's principle, that is the inclusion among the actual forces on bodies, which are exerted upon the bodies by *other* bodies by contact or by action at a distance, of so-called 'inertia forces', treating these as if they were real forces, which of course they are not. Thus, discussion of equilibrium equations will hereafter be understood to include equations of motion, and forces will be understood to include 'inertia forces' by the use of the D'Alembert principle.

As is customary in scientific books (as distinct from many handbooks) no units will be specified for the physical quantities involved in the relations which we derive. All correct physical relations must be 'dimensionally consistent', that is all terms which are added or subtracted or equated must represent the same kind of physical quantity or combination of physical quantities. In statics problems, which involve only the physical quantities *length* and *force*, any units can be used for these quantities provided that they are used consistently wherever the quantities are involved.

Dynamics problems involve also the physical quantities *time* and *mass*, since the 'inertia force' on a body is calculated as the mass of the body times its acceleration, directed opposite to the acceleration. The 'inertia forces' will then be measured in the same units as the real forces only if we use a 'consistent' system of units for length, force, time, and mass; this means that the units are chosen so that unit force would give unit acceleration to unit mass.

Table 1.1 shows the four consistent systems of units in common use, the first

Table 1.1

System of units	Length	Force	Time	Mass
British–American engineering	foot	pound	second	slug
Metric engineering (MKS force)	meter	kilogram	second	metric slug
CGS metric	centimeter	dyne	second	gram
SI (Système Internat. d'Unités)	meter	newton	second	kilogram

two in engineering and the last two in pure science. The same results will also be achieved if any units for length, force, and time are used consistently throughout, and the common practice is followed of taking the mass of a body as W/g, where W is the weight of the body (the force acting on it due to gravity) and g is its acceleration due to gravity; W and g must be measured at the same location and this will be achieved if their nominal values are used, which are based on sea level at the latitude of Greenwich.

The effects of curvature, either initially present or produced by flexure, are extremely complex. One method of dealing with the consequent difficulties, which the author used in teaching for many years, is to introduce approximations from the very beginning, approximations which can be justified only on intuitive physical grounds. This is unsatisfying intellectually, and its consequences are also unsatisfactory if we are to judge from the differing results which have been arrived at with this process by different authors.

To avoid this difficulty as well as the purely mathematical approach discussed above, a general shell theory with minimal approximations is introduced at the beginning of the discussion of shells in Chap. 6, which can then be applied to any particular cases, making whatever neglections seem to be appropriate. Thus, instead of merely having a vague notion of the approximations made, the reader has all the terms in front of him and can see exactly what he is leaving out, with the physical meaning of both the terms retained and the terms omitted. Perhaps surprisingly, there is no particular difficulty in setting up such a general theory in number mathematics form if no attempt is made to simplify it by reducing the number of unknowns while it is in its general form; the task of doing it is actually not much greater than that of setting up a limited theory for a particular range.

The beginner is prepared for such a general treatment by preliminary studies of the simpler special cases of beams and flat plates in Chaps. 2 to 5. These are important in themselves and illustrate most of the factors involved, the methods of setting up problems and the effects of the approximations which are customarily made in the more complex shell theory, where the effects of approximations are much harder to evaluate. Use of consistent coordinate systems and symbols makes it easy to demonstrate how the simpler and more easily understood cases are special cases of the general relations, and thus to clarify the meaning of the general relations.

Many examples are included but no problems. Readers who wish to spend some time to increase their understanding are advised, after each development, to carry out the development independently without reference to the text, in which the development is often merely outlined. It is surprising how something which seems very simple when it is done for you proves to be not so easy when you try to do it yourself. In any case 'doing it yourself' is the best way to really understand it and to retain that understanding.

1.2 SOME FUNDAMENTALS

Shell theory and its special cases of the theories of flat plates and beams are branches of mechanics, which is one of the basic divisions of physics. Mechanics might be defined as that branch of science which deals with the relationship between the forces acting upon bodies and their motions. The general concept of motion includes displacement, as well as the time rate of change of displacement or velocity, the time rate of change of velocity or acceleration, etc. Relative displacements of different parts of a body in general produce strains, which are related to the displacements by geometrical considerations.

The forces exerted upon a body such as a shell by other bodies, by contact or by action at a distance, are the 'loads'; these include the reactions, which we distinguish from other loads only for convenience. Besides the forces on the body as a whole we are interested in the forces which act on all parts of the body, and for a particular part these include the forces exerted on that part by neighboring parts across the surface which separates them, that is stresses.

Essentially we are interested in stresses because experience has shown that their intensity is related to the possibility of failure of the material, as well as to the strains (and hence the displacements) by relationships characteristic of the materials involved. The forces on any body are related to each other by the conditions of equilibrium of the body; thus we can relate together the whole gamut of things in which we are interested.

In shell theory we are concerned with the loading and motion of bodies which have one dimension, the 'thickness', small compared to the other dimensions. Such bodies have much less resistance to displacement in the transverse direction (the direction of the small thickness) than in the other directions. Hence we are usually principally concerned with these transverse displacements or 'deflections', the loads which produce them, and the accompanying strains and stresses.

To summarize, in shell theory as in all continuum problems of solid mechanics (that is problems involving bodies whose material is considered to be distributed continuously throughout their volume) we are interested primarily in the relations between the *loads*, *stresses*, *strains*, and *displacements*. Other physical quantities may of course be involved, for example temperature in thermal stress problems and time and mass in the 'inertia loads' in dynamics problems, but it is convenient to concentrate our attention upon the above four basic quantities, and to take other physical quantities into consideration in the definitions of these four or in the relations between them. It is convenient to tabulate these quantities and the relations between them as shown in Table 1.2.

Table 1.2

Quantities	Relation	Type of relation
Loads		
	Equations of equilibrium	Physical law
Stresses		
	Stress–strain relations	Empirical—based on experiments
Strains		
	Strain–displacement relations	Geometric—based on logic
Displacements		

The equations of equilibrium apply equally to problems of fluid mechanics, but the unlimited relative displacements possible in fluids makes a somewhat different approach advisable for the other relations. Temperature effects will not be considered in this book, but thermal stress problems could be included by adding αT to the normal strains due to the loads, where α is the coefficient of thermal expansion of the material and T is the rise of temperature above that of the unloaded state.

It might be argued that the ultimate proof for the equations of equilibrium is as much experimental as it is for the stress–strain relations. However, the equations of equilibrium are philosophically simple and logically appealing relations which all experiments confirm to be exact, and they thus deserve the dignity of being considered to be physical laws. On the other hand, in spite of the historical importance of the discovery of Hooke's 'Law', this relation is in practice almost always, perhaps always, an approximation.

For example, although deviation from strictly elastic behavior, or 'elastic hysteresis', may not be measurable in simple tests, its presence is implied by fatigue phenomena or heat formation in rapidly repeated cycles of stress. Outside the elastic range the stress–strain relations can usually only be expressed by empirical formulas or curves, and the approximate nature of such relations is even more pronounced. Some day we may be able to derive such relations from the laws governing intermolecular forces, but it remains to be seen if it will be practical to use such derivations in routine calculations.

1.3 LOADS, STRESSES, STRAINS, DISPLACEMENTS

There are many important points which should be discussed about these basic elements of our subject, and it is convenient to take up these in some cases somewhat unrelated points in one place, before going into specifics concerning their relations to each other. Many of these points are familiar and are consequently usually taken for granted, but there are some aspects of them which are not so familiar, and all of them are important enough to make a review worth while.

Loads

As a first discussion of the approximations which we make, it is instructive and illuminating to list some of the loads which we ignore or approximate because they are unimportant in most problems; we must be aware of what we are doing because they may be important in some special problems.

In many cases we neglect weight and the parts of the reactions which are due to weight, either because these are unimportant compared to other loads or because it is easier to consider the effects of other loads and the reactions due to them to be superposed upon an initial loading condition consisting of the weight and the reactions due to weight; the conditions under which such loading conditions can be superposed will be discussed later in Sec. 1.4. When we do consider weight, that is the gravitational attractive force of the earth's mass on the mass of the body studied, we use the nominal weight and assume that it acts uniformly over all of the body, neglecting its variation with altitude, and its variation with location due to the earth's nonhomogeneity and deviation from a true spherical shape, as well as the slight variation in magnitude and direction over the body due to the different locations of different parts of the body.

Of course we usually neglect such very small forces as the gravitational attraction of neighboring small bodies and of the sun, moon, etc., stray magnetic or electrostatic forces, radiation pressure, etc., but we also commonly neglect the very large force of the atmospheric pressure acting on the surface of each body, except in applications where a vacuum is maintained on part of it. We can neglect atmospheric pressure in most cases because its effect nearly cancels out on a body as a whole. However, it causes a small compressive stress and an upward buoyant force (due to the different altitudes and hence different pressures on the top and bottom of the body) which is equal to the weight of the air displaced (roughly a thirteenth of a pound per cubic foot at sea level) and which is of course a major force for lighter-than-air craft. We also often ignore or approximate the variations in magnitude and direction of air pressure forces which are due to the motion of a body relative to the atmosphere and which add up to air resistance, lift, etc.

In general engineers are concerned with 'macro' phenomena, that is with average conditions over areas and volumes of the dimensions which are most important in our machines and structures; the most striking example of this is our treatment of an actual body as a continuum, which is possible because the only thing which is ordinarily important is the average over such dimensions of the enormous number of separate molecular forces, masses, and motions. Thus, among the motions producing 'inertia loads' to which our engineering bodies are actually subjected, we can ignore the violent thermal molecular motions; this is because their effects practically average out for any body larger than a colloidal particle, for which thermal agitation in the fluid and particle results in the 'Brownian' motion visible under a microscope.

At the other extreme we ignore cosmic motions, of the earth about its axis and about the sun, the motions of the solar system and the galaxy, etc. The velocities are very large but they are nearly constant in magnitude and the radii of curvature of the paths are enormous. The most important of these cosmic motions is the rotation of the earth about its axis, which produces a centrifugal 'inertia force' equal to about a third of one percent of the weight at the equator; we customarily lump this in with the nominal weight, although it does not have the same direction except at the equator. In many problems we can also ignore motions such as those of ships and other vehicles, and even those of moving parts of machines when the motions are slow enough and other loads are large enough.

Stresses

As has been discussed before, to consider all facets of the behavior of deformable bodies (and all bodies are deformable) we must study not only the whole body but every part of it. The forces which act upon a typical *part* of a body consist of that

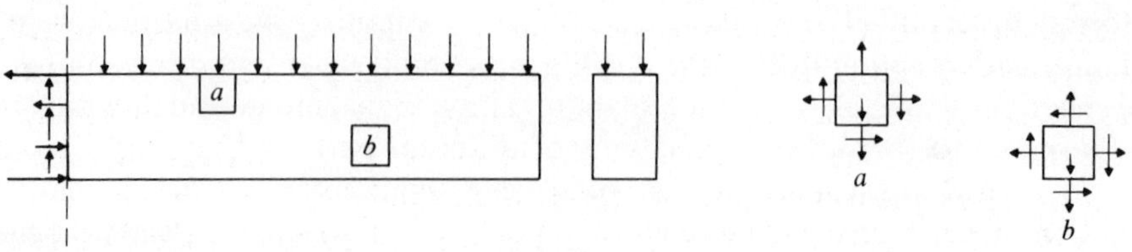

FIGURE 1.1

portion of the loads on the body as a whole which happen to be acting upon that particular part, and also the forces which are exerted upon that part by the adjacent parts of the body (that is, stresses). Thus the equations of equilibrium, applied to such parts as free bodies, give us relations between the loads and the stresses.

Figure 1.1 shows a cantilever beam loaded over its top surface and by its weight, and two parts or 'elements' of this beam *a* and *b*. The element *a* has acting on it the part of the weight acting on this element, the part of the top load acting on the top side of the element, and the forces on the other sides of the element which are exerted by surrounding elements (stresses). The element *b* has acting on it its weight and the stresses on all its sides.

The stress which acts on each side usually acts at some oblique angle and for convenience we separate it into 'normal' and 'shear' components, normal and tangential to the face; we will designate the strains corresponding to these stresses as 'normal' and 'shear' strains respectively. The stresses do not necessarily have the sense and magnitude indicated on these free body diagrams; we will present a complete discussion of the relation between shear stresses on different sides and other such matters in Sec. 3.1. All the forces shown, in fact *all* forces, are 'external forces' *with respect to the bodies on which they act.* The sometimes used concept of 'internal forces' is a superfluous and somewhat misleading one and we will not use it. The important thing to keep in mind is the body on which the forces act.

There is no necessity for making any particular distinction between the distributed force on the top face of element *a* and the distributed forces on the other faces. They are all essentially molecular forces, which are chiefly electrostatic (in this sense even 'contact' forces are really due to 'action at a distance', but our ordinary, common-sense concept of contact forces is nevertheless a useful and valid one) and are exerted by molecules outside the boundary of the element on molecules inside that boundary. The only real distinction between the forces on the top face and those on the other faces is that the molecules outside the top boundary of the element are also outside the whole body, while those outside the other boundaries are in the whole body; however there need be no distinction at all as far as the element is concerned.

As a matter of fact the compressive stress on the upper surface of the cantilever beam, produced by the load acting on the beam, differs only slightly from the compressive stress on a horizontal surface a little below the top surface, and the stresses on horizontal surfaces still lower change gradually and continuously as we go from the top surface downward. In other words, the distributed 'load' on the top surface is actually a part of —a continuation of—the stress condition or 'stress field' which exists throughout the body.

This is true for all loads exerted by contact on all bodies. In the case of a concentrated load on a point of the external surface, the stress field simply becomes more concentrated the closer one gets to the point, that is the stresses become greater on elements closer to the point, until *at* the point we have theoretically a stress of infinite intensity on an infinitesimal area (or in practice a high intensity on a small surface, since, due to deformations around it, there is actually no such thing as a true concentrated load as discussed on P. 144).

Stresses and Stains

The word 'stress' has been used somewhat loosely above. Hereafter we will use the word to signify only unit stress or force per unit area, which must be multiplied by the area on which it acts (an infinitesimal area if the stress is a varying one) to obtain the force on the area for use in the equations of equilibrium. The magnitude of a uniform stress on a certain area is thus defined as the force on the area divided by the area, while the magnitude at a certain point of a variable stress is defined as the limit of this ratio as the area around the point approaches zero. These definitions and the corresponding definitions of strains are very familiar, but a less familiar question arises as to just what we mean by 'area' or 'length', since all the dimensions of a deformable body change during loading.

We will consider the boundaries of the elements of a body to be *fixed in the body* and thus to move and deform as the body deforms; for example we can mark with ink the outline of the outer face of such an element on the surface of a sheet of soft rubber, and observe its deformation as we stretch the sheet. Each face of the element has an 'initial' area when the element is in an unstressed state, that is when there are no stresses on it (in practice, no stresses *due to the loads* on the body of which the element is a part—most bodies have some initial stresses not due to loads, but we usually have to ignore these in our calculations since they are likely to vary greatly from specimen to specimen and their magnitude can not easily be determined; we will discuss such stresses and their importance later in Sec. 1.7).

Besides the 'initial' area of the element face there will be a different 'final' area in the stressed and consequently strained state which we are studying. The magnitude of the average normal or shear stress in this stressed state could be defined as the total normal or shear force on the face divided by either the initial area or the final area of the face. Either definition would be valid since it can be used to correctly describe actual conditions. Similar questions arise in defining strains; for example small normal strains can be defined as the ratio of the changes in certain dimensions to either the initial or the final value of these dimensions.

Whatever definitions are used all the relations involving stresses or strains (and this includes all the relations mentioned in Table 1.2) must be *consistent with these definitions*, that is theoretical relations must be set up and experimental data must be interpreted to accord with the definitions which we have chosen. In general we will use definitions based upon *initial* areas and dimensions (as is common practice) since this results in the simplest relations and interpretations. Thus we define a stress as a force divided by the initial area of the surface on which the force acts, and a normal strain as a change in length divided by the initial length. Of course this question is unimportant when the strains are small, as they are in the elastic deformation of hard materials, but it is important and can be a source of confusion in studies of rubber-like materials, plastic deformation of ductile materials, etc.

Displacements

The displacements of points in a body often have practical importance in themselves, and, as has been pointed out, relative displacements of different points produce strain. Points in a body are considered to be fixed in the body, and are specified by their initial coordinates in a coordinate system which is usually fixed relative to the earth, whose motion is generally neglected. In dynamics problems it may be desirable to use a moving coordinate system which is fixed relative to certain points of the body; in this case the displacements can be divided into two parts, a 'rigid body' displacement corresponding to the motion of the unstressed body with the coordinate system, and the displacement of points relative to the coordinate system; the latter determines the strains, and in general both parts cause 'inertia forces' which must be included in the loads.

The displacement of a point relative to the coordinate system can be described in various ways, but we will describe it by the components of the displacement tangential to the coordinate lines at the initial position of the point. Thus in Fig. 1.2 if a polar coordinate system is used and a point of the body moves from the initial position P to the final position P', we define the displacement by the components u_r and u_θ drawn tangential to the radial and circumferential coordinate lines at the initial position P. It might seem simpler to define the displacement by the difference between the coordinates of P' and of P, but this

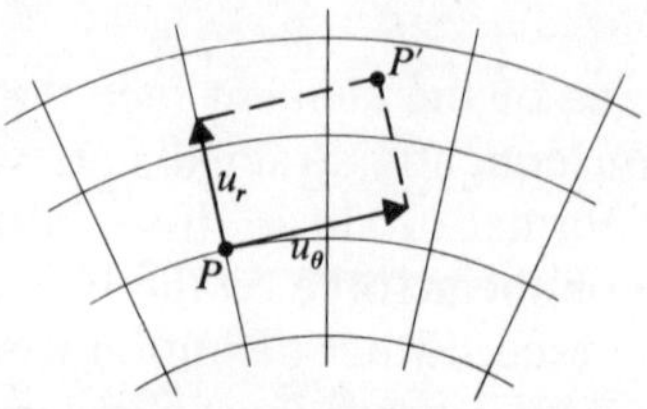

FIGURE 1.2

involves greater difficulties in the general case. It of course makes no difference which definition is used if rectangular coordinates are employed, but other coordinate systems are more suitable in many cases. We will use coordinate systems lying in and normal to the unstressed middle surface of the bar or plate, and in the case of shells following the 'lines of curvature' of this surface, as discussed in Chap. 6.

1.4 ENERGY METHODS

Let us consider now the relations between the basic quantities. The principle of equilibrium is well understood and neither this nor the geometrical strain–displacement relation requires discussion at this time. This is a convenient place, however, to discuss the fact that for the elastic case, that is for bodies whose material can be assumed to follow Hooke's Law and in which the stresses do not exceed the elastic limit, the equations of equilibrium can be replaced in whole or in part by considerations involving the elastic strain energy, that is the potential energy stored up during the elastic straining of the body (for example the energy stored up in the winding of a watch spring), which can be calculated as the sum of the work done in deforming each part of the body.

Superposition

Of the various energy principles which have been formulated we will use only the 'law of virtual work', since it will serve all our purposes and does not require that the condition of 'superposition' applies, as other energy methods do. The condition that superposition applies is that the effects of a combination of load systems acting on the same body must be a simple superposition, that is addition, of the effects of each load system acting separately; for example, if a load system A on the body produces a tensile stress σ_A in a certain direction at a certain location, while another load system B produces a similar stress σ_B, then loads A and B acting together must produce a similar stress of magnitude $\sigma_A + \sigma_B$, and likewise for strains and relative displacements. The condition of superposition applies in general only if there is a linear relation between the loads and the stresses, strains, and displacements; this in turn requires not only that the material follows Hooke's Law but that relative displacements be infinitesimal, so that the changes in the geometry of the problem due to the displacements can be ignored. Beams, plates, and shells are relatively stiff in the middle surface directions and change of geometry in these directions is seldom important, but change in geometry due to flexure in the weak transverse direction is often important and in such cases superposition can not be assumed to apply.

Law of Virtual Work

Since this law does not depend upon superposition it can be used for large as well as small displacements. The law states simply that during an infinitesimal and possible change in the displacements the work done by the loads, that is by all the external forces on the body, equals the change in the elastic strain energy. A 'possible' change in displacement is a displacement varying continuously with the coordinates which does not violate the boundary conditions, for instance by involving displacements or rotations at points where the constraints prevent them. It should be noted that the actual displacements may be large, and only the *change* in them must be small. Such small, possible displacements are called *virtual* displacements, whence the name of the law; the word 'virtual' is a traditional one which has no other meaning to us now in this context.

If the change in displacement violated the boundary conditions we would be studying a different system than the one we started with. If it were not very small we would be studying the system under loading which varied between the real one and the different loading which would be required to produce the changed displacements. The law of virtual work can be considered to be a simple application of the general law of the conservation of energy, limited in this case to mechanical work, and it is consequently easy to remember and to apply without danger of confusion. It would theoretically be possible to apply comparable methods to non-elastic problems although this would involve far greater and probably impractical complications.

The law of virtual work can also be derived from the equations of equilibrium and vice versa, which demonstrates their interchangeability. Imagine a body to be replaced by an equivalent system of particles interconnected by massless elastic springs. On each particle act part of the loads, and also the forces due to the springs. Let the displacement of a typical particle in x, y, z directions be u, v, w, and the change in these displacements du, dv, dw. Then apply the equations of equilibrium $\Sigma f_x = 0$, $\Sigma f_y = 0$, $\Sigma f_z = 0$ to each particle, multiply these equations respectively by du, dv, dw for each particle, and add all the equations together. In the resulting expression the products of the load components by the displacement components (which will be in their own direction and at their points of application) add up to the work done by the loads, while corresponding products involving the forces due to the springs add up to the negative of the change in elastic strain energy, the sum of both being zero.

Similarly, the equations of equilibrium can be derived from the relation between the elastic strain energy and the work done by the loads which is given by the law of virtual work, by the application of variational calculus. While this is not difficult when simple expressions for the strain energy can be used, it would not be easy to do it with indirect expressions for the strain energy such as those derived from the more exact general strain–displacement relations given in Chap. 6, which involve many equations and numerous intermediate quantities. In any case, it seems more meaningful to derive the equilibrium equations, as we will, directly from the physical problem, in accordance with the simple concept of equilibrium.

Some approximate relations derived directly from equilibrium considerations do not check with those derived from energy considerations using similar approximations in the expression for strain energy; the mathematically minded have advanced this as a 'test' which such relations fail to pass. It can readily be shown that exact relations do pass such a test, but this fact does not mean that the test in any way measures the degree of exactness of approximate relations, although the test is no doubt an indication of consistency. Some of the more approximate and limited relations pass the test, while admittedly more exact and widely applicable relations do not.

The law of virtual work applies also to our generalization of statics problems to dynamics by using D'Alembert's principle and including 'inertia forces' among the loads. In this case the 'work done' by 'inertia forces' really represents a change in kinetic energy, but the law can be applied in the same manner as in statics problems.

In practice, applications of energy methods can involve assuming from general experience or experiments a shape or 'mode' of displacement which approximates the actual shape, and using the law of virtual work to determine its magnitude; this is commonly called the 'Rayleigh method' from its originator. Or the displacement can be described by a series of component displacements with unknown magnitudes which, taken together, can approach the exact displacement, the component magnitudes being determined by the law of virtual work; this is commonly called the 'Rayleigh–Ritz method' and can theoretically give exact series solutions. Since strain energies involve averages of conditions all over a body these series usually converge rapidly for such things as deflections of beams, or buckling strength or frequency of vibration, which depend upon conditions all over the body. Energy methods are much less useful for calculating local conditions such as the stress at a particular point, since the convergence in such cases is likely to be slow. These points will be illustrated in later chapters.

As is shown by example at the end of Chap. 2, solutions obtained by the energy method in general exaggerate the stiffness and hence the buckling strength and vibration frequency of the bodies studied, but at least give a limit for these values. This is because the inexact shape assumed by cutting off the series used (at one term in the Rayleigh method or at several terms in the Rayleigh–Ritz method) can be considered to be the exact shape for a case in which additional constraints have been applied to force the body to take the assumed shape, and constraints of any kind reduce deflections, that is increase apparent stiffness.

A method of solution which was developed by Galerkin is somewhat analogous to the energy method, but we will not use it here since it seems to have no advantage over the energy method in the cases we shall consider and lacks the latter's simple, universal physical appeal.

The availability of high speed computers has made practical the development of approximate methods of numerical computation such as that of 'finite elements', in which the structure is broken down into a finite number of parts and the many relations representing the equilibrium of each element and the conditions for continuity between the elements are solved directly or by an energy approach. Such methods are extremely valuable because of their great flexibility. They are more suitable for solving individual problems than for developing broader solutions for whole classes of problems, and they must of course utilize the basic principles with which we will chiefly concern ourselves here. For details regarding their use readers are referred to the many specialized works on such subjects.

1.5 STRESS–STRAIN RELATIONS

The applications discussed in this book will be for the elastic case, for which the stress–strain relations are expressed by the familiar and relatively simple Hooke's Law, which will be formally stated in Sec. 3.1 in the discussion of elasticity theory. Actual materials do not follow this law exactly. Some, such as cast iron, have a slightly nonlinear stress–strain relation. Even those which seem to have a linear relation up to the elastic limit show a minute difference between loading and unloading ('elastic hysteresis', probably significant in connection with fatigue), and temperature effects, manifested by difference between constant temperature or isothermal straining (with very slow changes in loading) and adiabatic straining (with very rapid changes in loading), and somewhat similar electrostatic effects. Such deviations from Hooke's Law are usually of negligible importance in practical problems, and will not be considered here.

The equations of equilibrium and the strain–displacement relations, such as those developed for the general shell in Chap. 6, are not limited to the elastic case and apply to all materials and conditions. To partially match this generality we will discuss in the remainder of this chapter some of the aspects of more general stress–strain relations, as well as closely related subjects such as failure theories, factors of safety, etc., which are basic to all design.

The most familiar stress–strain relation, and one which illustrates many important points, is that obtained in a tension test of a mild steel specimen. If the test is run so as to avoid the effect of inertia of the testing machine, the results usually look something like the full line in Fig. 1.3. The 'proportional limit', τ_p,

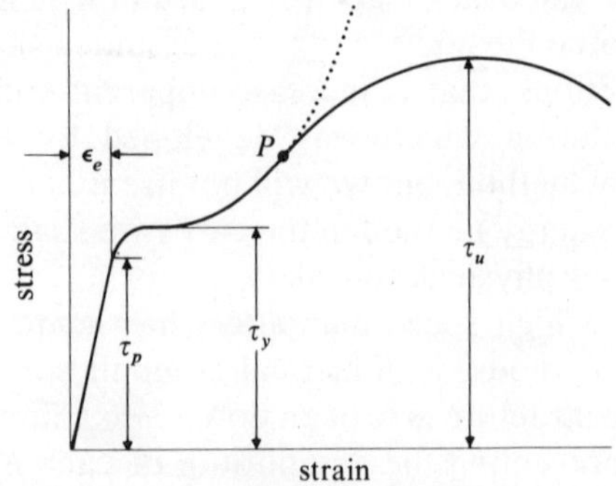

FIGURE 1.3

which is defined as the limit of the initial straight portion of the diagram, can be assumed to coincide (as closely as either of them can be measured) with the 'elastic limit' τ_e, defined as the highest stress for which the specimen will spring back to its original dimensions when the stress is removed. The 'yield point' τ_y is the stress at which straining takes place at constant stress, while the 'ultimate strength', τ_u, is self explanatory. It should be noted immediately that such a diagram is grossly distorted in order to make these features clear. The vertical coordinates are about to scale, but the elastic strain ϵ_e is greatly exaggerated in comparison with the plastic strain in the remainder of the diagram, which should actually be several hundred times as large as ϵ_e.

The stress plotted is the total tensile load divided by the initial cross-sectional area. This is of course only the average tensile stress over the cross section, but up to the onset of 'necking' the stress will be nearly uniform in the part of the specimen used for measurements, so that it will also be nearly the actual stress at each point of a cross section. Necking, which involves localization of the yielding over a short length of the specimen (due to instability in the process of uniform yielding) starts at about the point of inflection *P*, where the curvature of the stress–strain line changes sign from concave upward to concave downward. From there on, the diagram reflects more the change in shape of the specimen due to necking than the material behavior, and the actual resistance of the material as it would be measured by the resisting force per unit of final area keeps on rising after point *P* until failure occurs, in the manner suggested by the dotted line.

At the middle of the neck, Fig. 1.4(a), where most of the action is taking place during this period, conditions are becoming more complex. The tensile stress is not only nonuniform, with a concentration of stress at the outside, as suggested in Fig. 1.4(b), but due to the curvature of the outer surface at the neck the tensions in the outer fibers on each side of the center of the neck have an outward radial component as indicated in Fig. 1.4(a). This puts the material under a strong radial tension which is zero at the surface and increases toward the center, as suggested in Fig. 1.4(c), and which is superposed on the longitudinal tension. All this results in the complex failure shape shown in Fig. 1.4(d), which will be discussed in Sec. 1.6 under theories of failure.

The proportional limit and yield point are not affected by these complications due to necking, and this is fortunate since they are the most important properties for design. In early days it was customary to base design upon the

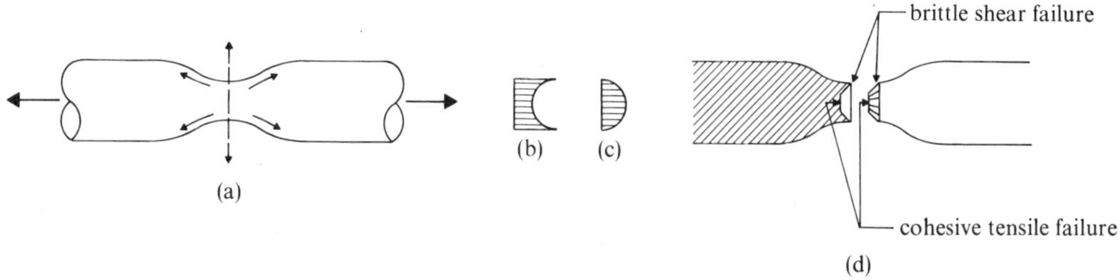

FIGURE 1.4

ultimate strength, but this is now considered old fashioned, and design is usually based upon the yield point. It is still logical, however, to use ultimate strength as a basis for the design of bodies such as hoisting ropes, which are subjected to uniaxial tension and can be allowed to stretch permanently somewhat as long as they do not break; for such cases we could make the most effective use of materials with different τ_u/τ_y ratios by basing design on ultimate strength rather than upon yield point. For bodies which are subjected to many cycles of reversed stress or of changes in tensile stress, the endurance limit of the material, usually defined as the largest reversed stress which the material can withstand for a very large number of cycles, is a logical basis for design, while for bodies whose stiffness is a paramount factor the modulus of elasticity E may be as important in design as any measure of strength.

For most design purposes it is logical to use as a basis for design either the yield stress, the stress at which large and usually prohibitive plastic deformation can occur, or the elastic limit, the limit for no plastic deformation whatever. Unfortunately these two stresses are somewhat indefinite and difficult to define and to determine in many cases. It is common experience that the value of τ_p or τ_e determined from experiments depends a great deal upon the sensitivity of the instruments used to detect the first deviation from linearity or the first evidence of permanent change in dimensions; the more sensitive the instruments the lower the values obtained, and it has been suggested that if our instruments were sensitive enough we would find these values to be zero. The yield point is definite enough and easy enough to measure for mild steel, but most of the materials which are now in use have no true yield point as it was defined originally, their stress–strain diagrams looking something like Fig. 1.5(a).

Artificial Elastic Limit and Yield Point

The concept of an elastic limit as the stress at which yielding first starts, or of a yield point as the stress at which yielding becomes very large, is so useful in design that numerous artificial 'elastic limits' and 'yield points' have been suggested for use with materials for which these properties are indefinite or nonexistent. These are defined as the stresses at which the permanent strain, the total strain or the slope of the stress–strain curve have certain rather arbitrarily chosen values; such values may have been reasonable for the purpose for which they were first chosen, but they are likely to be less reasonable for other conditions.

A more rational approach is to use for this purpose the proportional limit and yield point of an idealized material which really has these properties as they were traditionally defined, and which differs as little as possible in behavior from the real material, so that the difference would not be important in engineering applications. Figure 1.5(b) shows how this can be done.* If, instead of plotting

*L. H. Donnell, 'Suggested New Definitions for Proportional Limit and Yield Point', *Mech. Engng*, November, 1938, p. 837.

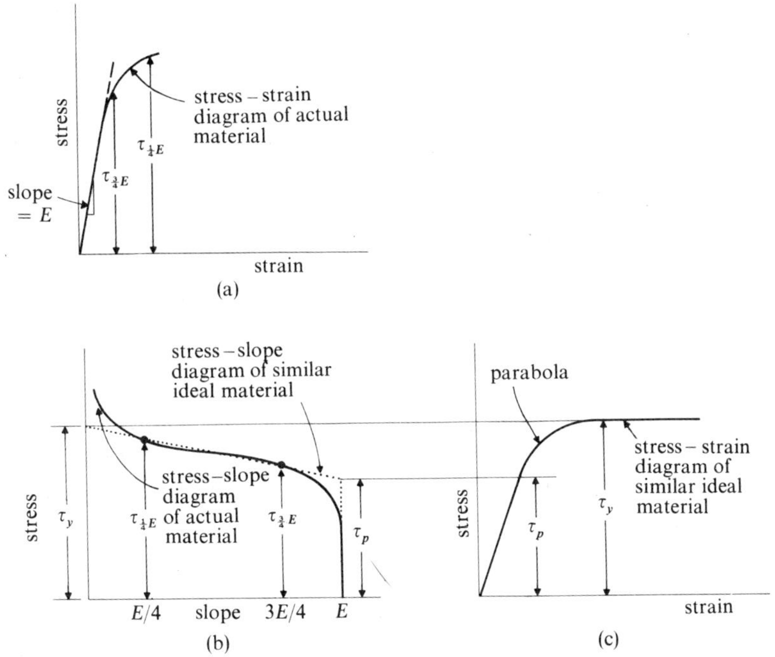

FIGURE 1.5 (from L. H. Donnell, 'Suggested new definition for proportional limit and yield point', *Mech. Eng.* Nov. 1938, p. 837, Courtesy of ASME)

stress versus strain as in Fig. 1.5(a), we plot stress versus the slope of the stress–strain diagram, we obtain a curve such as is shown by the full line in Fig. 1.5(b). All materials seem to give stress–slope curves of this type. Such curves can be closely approximated by the two dotted straight lines shown, a vertical line at the slope E and a diagonal line such as one which intersects the full line at the slopes $E/4$ and $3E/4$. These dotted lines would constitute the stress–slope curve of the idealized material whose stress–strain curve is shown in Fig. 1.5(c), and whose true proportional limit and yield point would be the values of τ_p and τ_y shown, that is the stresses at the intersections of the dotted diagonal line with the vertical dotted line and with the vertical axis, Fig. 1.5(b). Using these as the artificial proportional limit and yield point of the real material we can define these as:

$$\tau_p = 1{\cdot}5\tau_{3E/4} - 0{\cdot}5\tau_{E/4}, \qquad \tau_y = 1{\cdot}5\tau_{E/4} - 0{\cdot}5\tau_{3E/4}$$

where $\tau_{E/4}$ and $\tau_{3E/4}$ are the stresses at which the slopes of the stress–strain diagram of the real material are $E/4$ and $3E/4$ respectively. Similarly, it might be useful in plasticity studies to approximate the stress–strain behavior of the material by using the analytically simple stress–strain lines of Fig. 1.5(c) or the dotted stress–slope lines of Fig. 1.5(b).

1.6 COMBINED STRESSES, FAILURE THEORIES

The complexity of the subject of mechanical behavior of materials is indicated by the fact that so far we have considered only some aspects of the behavior of materials under static uniaxial tension. While elastic behavior under the most complex stress condition is relatively simple (that is, the general Hooke's Law given in Sec. 3.1), a good deal of the voluminous literature on plasticity is concerned with the determination from uniaxial test data of how plastic flow will take place under bi- or triaxial stress conditions in certain applications. Further complications would be introduced if we considered the behavior of nonhomogeneous or nonisotropic materials or those with time-dependent (viscous) material properties, the effects of fatigue, temperature, corrosion, etc.

In a discussion of this sort it can only be hoped to call attention to some of the more important matters which often tend to be overlooked. Some information about behavior under different conditions can be obtained from other 'static' tests such as compression and torsion tests, or 'dynamic' tests such as fatigue and Izod tests; there are difficulties in performance and interpretation for these, as for the tension test. It is not difficult to obtain the most complex triaxial stress conditions (that is, involving stresses in three directions) in tests, but it is often difficult or impossible to interpret tests quantitatively because of unknown stress distributions, especially after even small amounts of plastic flow occur.

By considering the equilibrium of a small element which is bounded by the surfaces on which we know the stresses and the surfaces on which we want to know them, we can determine the normal and shear stresses on planes at any angle at the point considered, in terms of the known normal and shear stresses on three (usually perpendicular) planes. In this way it can be shown that the description of any triaxial stress condition can be simplified by selecting three mutually perpendicular 'principal' planes, on which there is no shear stress but only the 'principal' normal stresses; let us call these principal stresses $\sigma_1, \sigma_2, \sigma_3$. They include the algebraically maximum and minimum normal stresses (considering tension as positive, and compression as negative) on any plane at the point. There are also 'principal' shear stresses $\sigma_{12}, \sigma_{23}, \sigma_{31}$, having magnitudes $(\sigma_1 - \sigma_2)/2$, $(\sigma_2 - \sigma_3)/2$, $(\sigma_3 - \sigma_1)/2$ on planes at 45° to the planes on which act σ_1 and σ_2, σ_2 and σ_3, and σ_3 and σ_1 respectively; these include the maximum shear stress on any plane.

Thus in a bar under a uniaxial tensile stress σ_x, the principal stresses are $\sigma_x, 0, 0$ and the principal shear stresses are $\sigma_x/2, 0, \sigma_x/2$. In a plate under equal biaxial stresses $\sigma_x = \sigma_y = \sigma$, the principal stresses are $\sigma, \sigma, 0$ and the maximum shear stress is not $(\sigma - \sigma)/2 = 0$, as is sometimes mistakenly assumed, but $(\sigma - 0)/2 = \sigma/2$. In both cases, when yielding occurs it is due to the shear stress, not to the larger normal stress.

In problems of beams, plates, and shells the important known stresses are usually limited to one plane ('plane stress', discussed in Sec. 3.2), say normal stresses σ_x, σ_y on surfaces perpendicular to the x, y directions and shear stresses σ_{xy} on both of these surfaces in the xy plane. In such cases it is shown in elementary strength of materials that the principal normal and shear stresses are:

$$\sigma_1 = (\sigma_x + \sigma_y)/2 + \sigma_{12},\ \sigma_2 = (\sigma_x + \sigma_y)/2 - \sigma_{12},\ \sigma_3 = 0$$

where:

$$\sigma_{12} = \sqrt{\{(\sigma_x - \sigma_y)^2/4 + \sigma_{xy}^2\}} \qquad (1.1)$$

Theoretically it is possible to obtain experimentally any triaxial stress condition except three tensions, by loading a thin walled tube with a combination of axial tension or compression and both internal and external pressures. Such a test can be interpreted quantitatively even after plastic flow has taken place, because the stresses are nearly uniform across the thickness and can be determined from simple static considerations. Some of these and simpler tests have been made, and a good deal is known, or can be conjectured with reasonable certainty, about the way in which flow and failure of material takes place under general stress conditions.

It would seem that in general three distinct types of 'failure', that is deviation from elastic behavior, can occur. In the first, plastic flow or yielding, the flow of molecules is complex but is presumably in large part a movement of molecules on one side of shear planes relative to those on the other side, in the direction of the planes; in this movement a typical individual molecule would break its intermolecular force bond with the molecule opposite it as it slides past, but not before it has established an intermolecular bond with the next molecule; thus permanent deformation occurs but the material still hangs together. In the second, 'cohesive' tensile failure, the action is also complex but again must consist in large part of a direct pulling apart of the molecules on one side of a plane on which there is a tension stress, from those on the other side of the plane, until the molecular bonds are ruptured for a large number of the molecules; we will call the tensile stress at which this can occur τ_c.

The third type, less well recognized, is brittle shear failure, which is like the first except that molecules on one side of the shear planes break their bonds with the molecules on the other side before establishing bonds with the next molecule, and the body 'comes apart' on a certain shear surface. A weak molecular bond, indicated by a relatively low value of τ_c, as well as presence of a tensile stress on the shear plane would obviously favor such a brittle rather than plastic shear failure, while a high value of τ_c and compressive stress on the plane would favor plastic failure; there is evidence that these factors do enter the picture.

All three of these types of failure are illustrated in the tension test of a mild steel specimen, Fig. 1.4. Large amounts of plastic flow of course occur before final breaking. The instant before the break occurs the material near the axis at the neck is subjected not only to a high axial tension σ_1, but also to somewhat smaller radial tensions $\sigma_2 = \sigma_3$. The maximum shear stresses are therefore much lower relative to the maximum tensile stress σ_1 than they are in uniaxial tension, and, due to the progressive diminution of the cross-sectional area, σ_1 eventually reaches a value in the neighborhood of the cohesive strength τ_c, and cohesive failure occurs near the axis at the neck. At the outside surface of the neck there is no radial tension, so the shear stresses have their full value relative to those in uniaxial tension. Shear failure can therefore occur, and, due in part at least to the high tensile stress on the shear surface, it is a brittle shear failure at about 45° to the axial direction. Of course the action is complex, and before the break occurs the material has been 'strain hardened' by the yielding, as exemplified by the rising resistance after the flat yielding part of the curve, Fig. 1.3 (this strain hardening is chiefly in the longitudinal direction, so that the material is no longer isotropic in its strength properties); this strain hardening means an increased resistance to ductile shear flow but not necessarily to the other two types of failure.

To describe the yielding and cohesive failure behavior of any material we would evidently need to know the value of the strength characteristics τ_y and τ_c (the uniaxial tension stresses at which yielding and cohesive failure respectively occur if nothing else happens first) together with the laws relating these to the values of $\sigma_1, \sigma_2, \sigma_3$ at which these types of failure would occur under more complex stress conditions. An additional law would be required to give the values of $\sigma_1, \sigma_2, \sigma_3$ at which brittle shear failure would occur, with perhaps a third strength characteristic; or perhaps the law for this involves only the other two strength characteristics together with the shear and normal stresses acting on possible failure surfaces. Not much is known about this, but something can be reasonably conjectured.

There is no doubt that the ratio τ_c/τ_y is a measure of the 'ductility' of the material. Under simple stress conditions, materials with a high value of this ratio compared to unity, that is high resistance to cohesive failure and relatively low resistance to yielding, fail first by yielding and are therefore called 'ductile', while those with a low ratio fail first in a brittle manner and are called 'brittle'. However, any material can fail in either way under suitable stress conditions. Thus, von Kármán showed that a marble compression specimen would yield like a specimen of soft copper if it was subjected to lateral compressive stresses of the order of magnitude of the longitudinal compression, thus increasing the compressive stresses on shear planes. The simplest way to produce cohesive failure in a ductile material is by subjecting a deeply notched specimen such as shown in Fig. 1.6 to tension. This produces shear stresses on diagonal sections such as that shown dotted, but it can be seen that the area which would have to be sheared simultaneously is very large compared to the cross-sectional area at the notch which is subjected to tension; consequently a true tensile (that is, cohesive) failure occurs before appreciable yielding (however, very localized yielding will occur at the corners of the bottom of the notch).

Unfortunately determination of the yield condition for a brittle material requires elaborate equipment for subjecting the sides of a compression specimen to hydraulic pressure, and accurate direct determination of τ_c for a ductile material has proved to be impractical. We cannot simply divide the load at failure

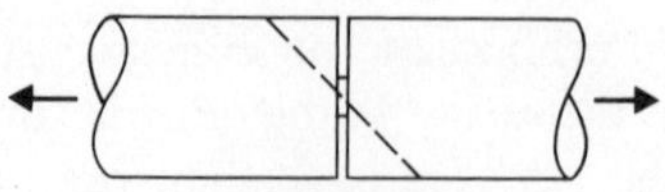

FIGURE 1.6

in Fig. 1.6 by the cross-sectional area at the notch. This gives the average stress, but, unlike ductile failure, brittle failure can start as a crack in a very small area which is highly stressed due to stress concentration (stress concentration is very high in this case); because the specimen is thereby weakened and there is a high stress concentration at the end of the crack the failure will proceed into the originally less stressed areas. The small yielding at the corners of the notch, which relieves some of the stress concentration, makes it impossible to calculate accurately the stress concentration which existed at the instant the crack started. There are similar difficulties with other types of tests which might be devised for measuring τ_c in ductile materials.

All brittle failure tests are subject to large scatter, because a crack can start at any of the defects which are unavoidable in all specimens. Tests involving ductile flow, on the other hand, are relatively consistent, because slight yielding can occur around defects, or any other source of stress concentration, without weakening the specimen, in fact with the effect of relieving the stress concentration, as will be discussed in the next section under the importance of ductility.

General Conditions for Failure

In spite of the difficulties of determining strength characteristics such as τ_c a good deal is known about the laws governing failure. The failure under any stress condition of an isotropic material, whose properties are independent of time, can be completely described by surfaces in a three-dimensional plot of the principal stresses σ_1, σ_2, σ_3, such as that shown in Fig. 1.7.* Starting at the origin, that is with zero stresses, if the magnitude of a certain stress combination is gradually increased (that is if we go away from the origin in a certain direction) eventually at a certain point some kind of failure will occur; the locus of such points is a failure surface. There will be failure surfaces for each type of failure; these will intersect, and only those parts of the surfaces closest to the origin are of interest to us if we are considering *primary* failure, that is the failure which occurs first. Anything beyond the brittle failure surfaces is of no interest at all, since the specimen will break and is finished as soon as these surfaces are reached; beyond the yielding surface the specimen is still hanging together but conditions become more complex, as discussed later. For the present we will restrict the discussion to primary failure.

It is fairly well established that yielding failure depends upon all the principal shear stresses and not merely upon the largest of them, and occurs when the elastic energy stored up by shear straining reaches a limiting value. This condition describes a yielding failure surface in the shape of a circular cylinder whose axis *OC*, Fig. 1.7, passes through the origin *O* and makes equal angles with all three axes. The radius of the cylinder must be such that the cylinder surface cuts each σ_1, σ_2, σ_3 axis at a distance τ_y from the origin, since this is the condition

*L. H. Donnell, *A Study of Triaxial Stresses and Failure of Material*, Office of Naval Research Report, 1955.

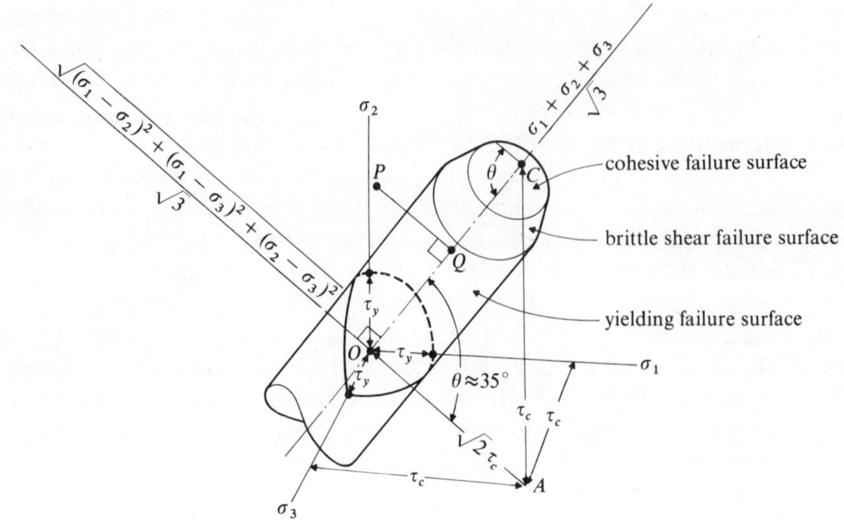

FIGURE 1.7 (from L. H. Donnell, 'A study of biaxial stresses and failure of material, ONR Report, 1955, courtesy of Office of Naval Research, Arlington, Va.)

for yielding under uniaxial tension. A more conservative assumption that yielding depends only upon the maximum shearing stress would give a slightly smaller surface in the shape of a cylinder with the cross section of a regular hexagon inscribed in the circular cylinder, and having the same intercepts with the $\sigma_1, \sigma_2, \sigma_3$ axes.

An assumption that cohesive failure will occur when the maximum principal stress reaches τ_c would give a cohesive failure surface consisting of three planes parallel to the σ_1, σ_2, the σ_2, σ_3, and the σ_3, σ_1 planes, at distances τ_c from each plane, producing a sort of three cornered roof on the end of the yielding failure cylinder. It is possible that, if the truth could be determined, cohesive failure would also be found, as in the case of yielding failure, to be dependent upon all three principal stresses and not just upon the largest of them, and that its failure surface would also be symmetrical with respect to the axis *OC*. In any case such an assumption gives a reasonable approximation to the more usual assumption described above. By taking the cohesive failure surface as a cone which is tangent to the three planes mentioned above, as indicated in Fig. 1.7, the values predicted by it will be on the safe side, which is desirable for a failure which is subject to so much scatter.

The brittle shear surface, if it exists (it could conceivably lie outside the intersection of the other surfaces for some materials) can be conjectured to form another conical surface between the surfaces for yielding and cohesive failure, where both the shearing stresses and tensile stresses are high, as suggested in Fig. 1.7. The composite failure surface is thus a tube closed at the tension end but open at the compression end, consistent with the assumption that materials can withstand an unlimited uniform triaxial compression.

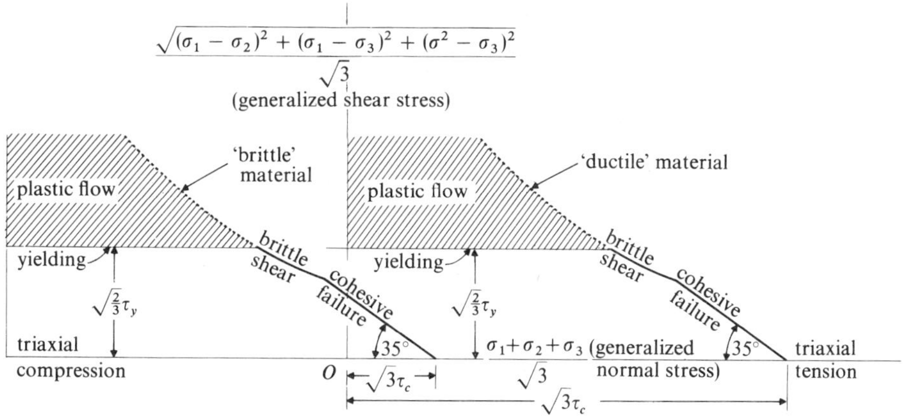

FIGURE 1.8 (from L. H. Donnell, 'A study of biaxial stresses and failure of material', ONR Report, 1955, courtesy of Office of Naval Research, Arlington, Va.)

Since the conjectured composite surface is symmetrical about OC, it can be completely described by a two-dimensional (and therefore more convenient) plot consisting of the intersection with the composite surface of any plane containing OC. Such a plot is shown in Fig. 1.8. The horizontal and vertical coordinates of this plot are found as follows: Let P be a general point of the three-dimensional plot, with coordinates $\sigma_1, \sigma_2, \sigma_3$; let Q, with coordinates a, a, a (to be determined) be the projection of P upon the axis OC, whose ends have coordinates 0, 0, 0 and τ_c, τ_c, τ_c, as shown in Fig. 1.7. Then by an extension of the theorem of Pythagoras (as shown in Fig. 4.6 and the accompanying text, Sec. 4.2) the square of the distance between any two points equals the sum of the squares of the differences between the coordinates of the points. Hence $\overline{OC}^2 = \tau_c^2 + \tau_c^2 + \tau_c^2 = 3\tau_c^2$, $\overline{OP}^2 = \sigma_1^2 + \sigma_2^2 + \sigma_3^2$, $\overline{OQ}^2 = 3a^2$, $\overline{PQ}^2 = (\sigma_1 - a)^2 + (\sigma_2 - a)^2 + (\sigma_3 - a)^2$. But, since PQ is perpendicular to OQ, $\overline{OP}^2 = \overline{PQ}^2 + \overline{OQ}^2$, from which, using the above relations, we find $a = (\sigma_1 + \sigma_2 + \sigma_3)/3$. Putting this value into the expression for $\overline{OQ}$, we find that the distance parallel to OC from the origin to any point $\sigma_1, \sigma_2, \sigma_3$ is equal to $(\sigma_1 + \sigma_2 + \sigma_3)/\sqrt{3}$, which can be considered to be a generalized normal stress. Putting the value of a into the expression for $\overline{PQ}$ we find that the distance perpendicular to OC from the origin to any point $\sigma_1, \sigma_2, \sigma_3$ is equal to $\sqrt{\{(\sigma_1 - \sigma_2)^2 + (\sigma_1 - \sigma_3)^2 + (\sigma_2 - \sigma_3)^2\}}/\sqrt{3}$, which is proportional to the well-known 'octahedral shear stress' and to the square root of the elastic shear strain energy, and can be considered to be a generalized shear stress.

Since yielding occurs when, say, $\sigma_1 = \tau_y$, $\sigma_2 = \sigma_3 = 0$, the radius of the cylindrical yielding failure surface is $\sqrt{\{(\tau_y - 0)^2 + (\tau_y - 0)^2 + (0 - 0)^2\}}/\sqrt{3} = \sqrt{2/3}\tau_y$, and the condition for yielding is therefore:

$$(\sigma_1 - \sigma_2)^2 + (\sigma_1 - \sigma_3)^2 + (\sigma_2 - \sigma_3)^2 = 2\tau_y^2 \tag{1.2}$$

For the plane stress condition $\sigma_x, \sigma_y, \sigma_{xy}$, using Eq. (1.1) this becomes:

$$\sigma_x^2 - \sigma_x\sigma_y + \sigma_y^2 + 3\sigma_{xy}^2 = \tau_y^2 \tag{1.3}$$

The equation of the assumed conical cohesive failure surface, with apex at C and tangent to the planes through C parallel to the $\sigma_1\sigma_2$ planes, etc., is obtained by setting $\overline{PQ}/\overline{QC} = \overline{PQ}/(\overline{OC} - \overline{OQ}) = \tan\theta = 1/\sqrt{2}$. This gives the equation $\sqrt{2}\sqrt{\{(\sigma_1-\sigma_2)^2+(\sigma_1-\sigma_3)^2+(\sigma_2-\sigma_3)^2\}}+\sigma_1+\sigma_2+\sigma_3 = 3\tau_c$.

Possible typical plots are drawn in Fig. 1.8 for 'ductile' and for 'brittle' materials, characterized by large and small τ_c/τ_y ratios respectively.

1.7 PLASTIC FLOW, DUCTILITY

The areas inside the full lines in Fig. 1.8 are the regions of elastic deformation, where the relations between loads, stresses, strains, and displacement are studied by the theory of elasticity, or approximations to this theory such as classical shell theory. The region above the brittle failure lines is of no practical interest, as discussed before. The shaded regions between the horizontal yielding line and the dotted brittle failure lines are regions of plastic flow, where the relations between loads, stresses, strains, and displacement are studied by plasticity theory. As stated before, we will not discuss any applications of this theory, or theories for more complex time-dependent materials, but the general shell equilibrium and strain–displacement relations which will be developed later apply to all such cases. As to the stress–strain relations, which distinguish this region from the elastic region, we will make only a few general observations. If there is no reversal of the direction of plastic straining then it may be sufficiently accurate to assume that there is a one-to-one relation between stress and strain ('deformation theory'), and there is some experimental evidence for generalizations such as that the ratio between each principal shear stress and its corresponding shear strain is equal to this ratio for each of the other principal shear stresses, and that if the octahedral shear stress or the generalized shear stress of Fig. 1.8 is plotted against a corresponding shear strain, the test points for all kinds of stress conditions will lie close to a common line, which can be regarded as a generalized stress–strain curve.

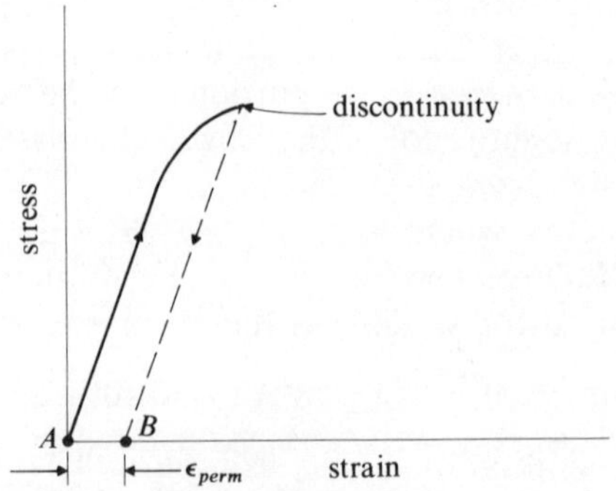

FIGURE 1.9

If there is any reversal of the direction of straining, then stress and strain will not be unique functions of each other. For instance, if a tensile test is carried into the yielding region and the specimen is unloaded, then there will be a discontinuity in the stress–strain relation at the point of reversal of the direction of loading and straining, Fig. 1.9, the material following the dotted stress–strain line parallel to the original elastic line during unloading and being left with a permanent strain $\epsilon_{perm.}$. Thus the material can exist in two strain states A and B, both with the same zero stress. In such cases a point-to-point study of the straining must be made, in infinitesimal steps ('incremental' theory) or at least in steps each of which involves no reversal of the direction of straining.

Importance of Ductility

Some generalizations important for design can be made about ductility and plastic flow without going into detailed calculations of this flow. Ductility under the stress conditions present is well known to be one of the most important properties of engineering materials. It is chiefly important because, by a small and harmless localized plastic flowing which produces negligible general changes in shape it tends to even out stress distributions, and thus to relieve stress concentrations, irregular distribution of contact forces due to imperfect fit between parts, initial stresses due to forming (which can be of the order of magnitude of the yield point), etc. Hence it is generally safe to design machines and structures which are made of ductile materials and which are under more or less steady load conditions by using elementary strength-of-material analytical methods (including the classical beam, plate, and shell theories) which ignore the above factors and assume the simplest possible distribution of stresses, whose magnitudes can then usually be determined from equilibrium considerations only.

An exception to this rule would occur if a nearly equal triaxial tensile stress condition exists somewhere in the structure; as discussed above this could lead to brittle behavior of materials normally considered ductile. However, such a condition is very unlikely in any ordinary structure. To be sure, a triaxial tensile stress is known to exist at the center of the end of a crack in a sheet under tension, and at one time rather widespread tearing failures occurred in welded ship plating and tanks, probably due to rapid spreading of very localized and otherwise harmless fatigue cracks. However, the three tensions in this case are probably very far from equal and the failures were traced to impurities in the mild steel used which greatly reduced its ductility at the low temperatures at which the failures occurred.

For brittle materials and, as discussed later in this section, for ductile materials subjected to fatigue conditions, such strength-of-materials methods must be replaced by a consideration of the initial stresses which may be present and a more accurate analysis of the stresses due to the loads, based upon the theory of elasticity (Sec. 3.1) or methods of experimental stress analysis. Initial stresses in brittle materials are produced by casting, quenching, welding, etc., and can also be high. It is not easy to determine the magnitude of initial stresses in an individual specimen by a non-destructive test, but such stresses can be reduced by partial or complete annealing, or sometimes by modification of the manufacturing process. Fatigue failure, as well as static brittle failure are usually precipitated by the defects always present. These types of failure are associated primarily with tensile, as opposed to compressive, stresses, partly at least because initial or developing cracks close up under compression. Due to surface oxidation and other effects, initial defects and cracks are more likely at the surface of bodies, and this is also where tensile stresses due to flexure are the highest; hence it is sometimes advantageous to produce initial compressive stresses at the surface intentionally, by peening operations in ductile or quenching in brittle materials.

Limit Design

The reliance of engineers upon ductility of their materials to save them from the consequences of the crudities of their constructions and methods of analysis, really represents an application of the principle of 'limit design', although it is seldom recognized as such. This principle, applicable only to statically loaded structures made of ductile materials, sets the ultimate load-carrying capacity of a structure as the minimum loading at which the loads are resisted on some cross section entirely by material which has reached the yield point, instead of the loading at which the maximum stress reaches some value. Below this loading some of the material resisting the loads must be elastic and therefore strained by no more than the small elastic strains, and hence the general displacements of the structure should still be of the order of magnitude of elastic displacements; above this loading, on the other hand, the displacements can increase without limit until breakage occurs. While this theoretical assumption is reasonable, it is obvious that the actual magnitudes of the displacements will depend upon the geometry of the structure. Whether they represent a reasonable limit to the usefulness of the structure depends upon the structure's purpose; for most structures they will, but for machine parts they often will not. In using limit design methods some consideration should also be given to the possibility that the plastic straining which is relied upon before ultimate capacity is reached may become localized and so result in necking and breakage before the theoretical ultimate capacity is reached.

For a material like mild steel which can be assumed to have a constant resistance after yielding, the method of limit design may enormously simplify analysis, by substituting for the complex elastic analysis of highly redundant structures the assumption of simply calculated (by assuming bending stresses all equal to $\pm\tau_y$) bending resistance at points of maximum bending moment in beams ('plastic hinges'), of fixed tension in tie rods (calculated by assuming uniform tension stress equal to τ_y), and of the buckling compressive stress in struts. These assumptions generally reduce the structure to a simple statically determinate one. By giving a more realistic appraisal of the real limiting loading of a structure than the elastic analysis does, the method may make it safe to allow higher loadings in some cases, although this may mean designing things like tie-rods so that localized yielding at the end connections will not cause premature failure before the contemplated general strains are reached.

Amount of Ductility Required for Various Purposes

The various methods which are used to increase the strength of materials—heat treatment, cold working, and in some cases alloying—also usually decrease the ductility; the combination of high strength and high ductility ('toughness') is hard to get and must usually be paid for in some way. The increasing use of high-strength materials therefore makes important the question of *how much* ductility is needed for various purposes. A simple answer to this can of course not be given and in many particular cases is not known. However, it can be said that the large amount of ductility present in mild steel and other very ductile materials is needed only for such applications as forming operations, while a relatively small amount of ductility usually suffices for the purposes discussed above. This is true in spite of the fact that engineering specifications, based on nothing more than previous successful use of a highly ductile material, often require a much larger amount of ductility.

This point is dramatically illustrated by the difference between ordinary cast iron and ordinary glass, which have rather similar strength properties. Both have zero ductility according to the standard measure of this property, the elongation and reduction in area in a tensile test. Nevertheless one can make a prick-punch mark in cast iron without difficulty, but not in glass. This indicates that ordinary cast iron has enough ductility for many purposes even though the ductility is too small to be measured by the crude standard measures; the successful use of cast iron in many engineering applications (where ordinary glass would certainly be almost useless) confirms this. The difference between cast iron and glass shows up in a test like the Izod test, which measures the energy absorbed in breaking a notched bending specimen; this suggests that the real significance of this test may be as a measure of amounts of ductility which are too small to be measured by the standard static tests, but which are nevertheless very important.

Stress cycles (particularly their tensile part, as discussed earlier) which are repeated a large number of times produce a 'fatigue' failure, even in ductile materials, which is much like a static failure in brittle materials, failure starting with a crack and proceeding without any measurable plastic flow. In such cases, therefore, the ductility of the material cannot be relied upon, and the design methods for brittle materials must be used, as discussed previously. Steels generally have threshold values called 'endurance limits' for each type of stress cycle, which can be repeated indefinitely without failure, but most other materials can withstand only a certain number of cycles of a given type. This critical number becomes very large for small stress cycles but is always finite. Machines such as transportation equipment are subjected to many different types of stress cycle, and the simplest assumption, that failure may occur when the fractions of the fatigue life incurred under each type of cycle (equal to the number of cycles incurred divided by the corresponding critical number) add up to unity, is supported by some experimental evidence.*

At the other extreme, viscous material properties, while usually undesirable and making an analysis involving time effects necessary, are favorable insofar as they may permit some plastic adjustments to take place, given enough time. All materials probably have some viscosity in addition to their other properties even under simple stressing, as illustrated by the fact that thin stone slabs used as beams (and according to some accounts even steel bridges) have been found to sag measurable amounts over a period of many years.

1.8 FACTORS OF SAFETY, PHILOSOPHY OF DESIGN**

Finally, something should be said about factors of safety and related questions in design. Learned tomes and research papers are devoted to questions which may affect design costs, weights, etc., by a more 10 or 20 percent, while a general discussion of factors of safety, which may have many times as much effect on the design, is usually ignored or relegated to an unsatisfactory paragraph or two in books on strength of materials.

We might start by asking 'Where do the factors of safety, which we use so confidently, come from?'. Did someone just pick out of the air a factor of two, or four? Sometimes it seems that way, and sometimes it is to be feared it approaches it in practice. However, the history of most factors of safety in widespread use is probably more like the following:

A long time ago somebody had the job of building an important structure. He didn't have much to go on, but he did the best he could and 'guesstimated' a factor of safety, a large one to be on the safe side, using his experience, common sense, and 'engineering judgement'. (The author believes that there is an important place for 'engineering judgement', which in its best form involves taking a mental weighted average of many factors gleaned from broad experience; in the hands of experienced individuals such mental integration may be surprisingly accurate, but it can hardly be recommended for general use when more rational means are available.)

Let us say that the design worked out all right. The structure performed its function and no serious troubles appeared, so the next man and the next who had to design a similar structure used the same methods of analysis and the same factor of safety. This went on until someone, thinking about the costs of materials, the advantages of weight-saving, etc., said 'I think this factor of safety is bigger than necessary', and designed his structure with a smaller factor. Maybe *it* turned

*See discussion of Goodyear Stress Change Recorder, *Goodyear–Zeppelin Report M 102.7*, 1939, p. 8, by L. H. Donnell; also quoted in *SESA Handbook of Experimental Stress Analysis*, 1949, p. 242.

**This discussion is taken from a talk given some years ago by the author.

out all right, and so his fellow engineers followed suit and used it too. Perhaps this happened several times, and the factor of safety in general use became smaller and smaller, until finally a lot of failures began to occur, and engineers said 'We've gone too far', and boosted the factor up. So, by a process of trial and error, in the way that folklore is developed, a compromise was reached which was not too wasteful of material and did not result in too many failures.

Somewhere along the line some of the steps of this process may be carried out by a group of eminent and dignified authorities, gathered around a conference table, and the result may be called part of an engineering 'code', but the mental processes are likely to be much the same. There is no reason for ridiculing this process. Each step is logical, and the final result is sometimes the most ideal one which could be achieved, taking into consideration some of the manifold economic and other practical factors which are actually involved in a better and more complete way than any theoretical analysis could hope to do.

However, we need not look far to see the limitations to such a process. First of all how do we know that such a number is a constant, rather than a function of the length, height, material, speed, location—all the variables which enter into all problems even when they are in the same 'field'? Experience results in some average factor which gives results which do not work too badly with usual values of all these variables. However, we might have achieved better results by using different factors in different cases; perhaps we could have eliminated most of the failures which did occur by using a larger factor in certain cases, and safely saved material by using a smaller factor in other cases. Secondly, what effect should it have on a traditional factor of safety if we change our methods of analysis, the materials we use, etc.? In general it should permit a reduction in the factor of safety if we improve our analysis and so reduce some of the inaccuracies and uncertainties which a factor of safety allows for. Finally such a method of arriving at a factor of safety is inapplicable in a new field, and almost everything we do involves starting from scratch in some respects.

Other difficulties and complications can arise. For instance, when stresses are proportional to loads it makes no difference whether the factor of safety is applied to the stresses, giving us a so called 'working stress', or whether it is applied to the loads. But there are many cases in which the stress is not proportional to the load, and then it does make a difference where the factor of safety is put. In some fields, such as aeronautical structures, it is customary to put the factor of safety on the load, while in most others the usual practice is to put it on the stress. If we don't know which to do we may use whichever method is most conservative, but this is no solution; it is easy to be on the safe side by using a big factor of safety (except for certain well known cases where adding material reduces rather than increases safety), and as a last resort this is better than nothing. But it is a confession of a failure to understand what we are doing, and of course is incompatible with maximum efficiency.

What then *is* the correct philosophy to use in approaching this subject? To answer this question we must ask ourselves what is the reason for a factor of safety, and what we wish to accomplish by it. The reason for using a factor of safety is obviously that our methods for analyzing a problem to determine a proper design, are far from complete. We do not consider all the factors involved and our treatment of those we do consider is far from being 100 percent accurate. This does not necessarily mean that the solution is to improve our methods of analysis. Of course we should be as accurate as we can be, considering the effort we put on the problem, but there are always compromises to be made involving economic and time factors, which determine how much effort we can afford to spend for an analysis. If we are making only one machine and it does no more than cost a little extra money for materials if we make a rough analysis and use a large factor of safety to be on the safe side, then a proposal to spend more than could be saved by making an exhaustive analysis is plainly impractical. But when weight saved has a large effect on efficiency, or enormous production makes a small economy on each unit mount up, it pays to spend more time and money on the analysis. If it paid to spend the time and money, we might be able to track down and make accurate allowance for all the many factors which enter into an actual design problem. In such a case the need for a factor of safety would vanish.

There are two basic ways in which we can improve our usual approach to this question and our understanding of it. First we should make a conscious effort to separate and analyze and take care of the various uncertainties which are involved in any given problem. And, second, we should make an effort to analyze in our own minds what it is that we want to accomplish by the factor of safety.

Let us consider the first of these propositions—what are the various uncertainties which enter a problem? Can we afford to lump them all together in allowing for them? We could make a far better attempt at allowing for them properly if we tried to separate them and to estimate what allowance is needed for each. Consider the question of putting the factor of safety on the stress or on the load. Where does the uncertainty really lie? Is it in the magnitude of the load which we have estimated, or is it in the magnitude of the stress which we estimate will cause failure? It is more likely that there is some uncertainty about both, and that there are also uncertainties in all the other quantities which we take into consideration in our calculations. For example, all the dimensions which enter any calculation are subject to some variation due to manufacturing tolerances and errors, and to reductions in service due to wear, corrosion, etc. The speeds, densities, coefficient of friction, elastic constants, and all the other data which enter into the calculations are likewise subject to some variation or uncertainty, which can be estimated at least a good deal more accurately than by the process of drawing an overall factor of safety out of the hat. Moreover, the assumptions which we make regarding the relations between the various data (the formula or theory we rely on) and the assumption that this represents the critical condition which will determine failure are subject to some question. While we may not want to take the trouble of developing or using more complete or accurate formulas, or of examining all the possible relationships which *might* be critical, if we make a conscious review of these uncertainties in our calculations we should be able to make a better estimate of the allowance which should be made for each than if we make no such attempt.

This proposal might be carried out by using the most unfavorable value which could conceivably occur for each of the data entering the problem, and the most unfavorable relations between them that might apply, and then using no factor of safety as such. Or we could use the nominal values for each element in the calculation, but multiply each element by an 'elemental factor of safety', which would be the ratio between the most unfavorable value and the nominal value for that element, with a similar factor to allow for inaccuracies in the relation.

Such individual or 'elemental' factors of safety would have a more permanent, universal application than the usual 'shotgun' factor of safety. Estimating the value of such an elemental factor of safety is something which at least can be grasped by the mind, with the expectation that, with the aid of experience and common sense, a reasonable value can be arrived at. Or groups of engineers, gathered in code-making committees, or by discussions in the technical publications or at technical meetings could pool their experience and set up general guides, rules, recommended limiting values, etc., for these elemental factors of safety, to help the designer in making his choice. But this would immediately bring up the need to consider the second of the two propositions discussed—the need to understand what it is that we want to accomplish by the factor of safety.

If we examine our desires we will find that practically never do we really want to completely eliminate the possibility of failure. Of course we would like to do this if it did not cost us anything, but we don't want to pay the penalties which it would actually involve. Even in the case of machines or structures whose failure jeopardizes human life, do we, for instance, really want to design ships so that they can outlive the worst combination of storms and bad luck that could ever occur in all human experience? At least we never have done this, and it is to be doubted if we will ever want to attempt it; man is an adventuresome animal, and he prefers small risks to the almost infinite conservatism, the sacrifice in efficiency and the shackling of his instinct to create which would be required to completely eliminate failures. What we really try to do is to design so as to reduce the probability of failure down to a certain level. What that level is depends upon the application, and even for the same application our ideas about it change with our standards of social and individual values. If we could do everything rationally, our object would be to select our unfavorable values or our elemental factors of safety so as to bring the probability of failure down to some preconceived level. Usually the possibility of several kinds of failure can be envisaged and the level of probability to which we wish to reduce each may be different. Partial failures which only involve a little expense and loss of time can obviously be allowed to happen more frequently than failures which might involve loss of life or limb. In doing all this we would have to use theories of probability, and this would involve considering that the probability that many independent happenings (such as actually having the most unfavorable value of each element) will occur concurrently is much less than the probability of each happening separately.

2

CLASSICAL BEAM THEORY

2.1 THE LOVE–KIRCHHOFF APPROXIMATION

Classical plate and shell theories, like the classical (elementary) beam theory, are based upon the approximate assumption, initiated for beams by Bernoulli but first applied to plates and shells by Love and Kirchhoff, that straight lines normal to the middle surface before deformation remain straight, normal to the middle surface and unchanged in length after deformation. This means that if the initial and final position of points on the *middle surface* are known, the initial and final positions of *all* points of the shell wall will also be known; hence the strains everywhere can be calculated in terms of the displacements of the middle surface alone. This gives the enormous simplification of in effect converting the plate or shell problem from a three- to a two-dimensional problem, and the beam problem to a one-dimensional one.

The Love–Kirchhoff approximation involves neglecting the effects on deformation of both the transverse strains and the transverse stresses. In the case of a beam these effects include those of transverse shear strains or changes in angles caused by the transverse shear stresses, the transverse normal strains caused by transverse normal stresses (that is, tensile or compressive stresses in the direction of the thickness, for example the compressive stress under the load in Fig. 1.1a), transverse normal strains caused by longitudinal stresses (calculated by using Poisson's ratio) and, similarly, longitudinal normal strains caused by transverse normal stresses. The longitudinal strains caused by the longitudinal stresses are the only strains considered in the classical beam theory. Similar considerations apply in plate and shell theory. After deformation, the lines mentioned above are actually no longer exactly normal to the middle surface because of transverse shear strain, and no longer straight because this strain varies with the distance from the middle surface; and they no longer have the same length because of transverse normal strains.

As we will prove later, in Chap. 3 and elsewhere, the errors due to this approximation are negligible for slender beams and thin plates and shells made of homogeneous materials under most loading conditions of practical interest, and we will use this approximation in setting up most of our general theories. However, we will also make studies of the magnitude of the errors made under various conditions, and develop second-approximation corrections for the neglections involved; in many cases these corrections do not greatly complicate computations, and so a good deal of the advantage of the Love–Kirchhoff approach is retained. It is important to know the conditions under which the Love–Kirchhoff approximation is adequate, and when correction is needed. This can most easily be done in applications of beam theory; moreover, the elementary beam theory can be compared to the more exact beam theory which is readily obtainable from two-dimensional elasticity theory.

The conclusions arrived at with beams apply also in general to plate and shell theory and these cases will be further discussed in later chapters. We will find that corrections are needed in general only for built-up structures (such as truss-type beams, or plates and shells made of 'sandwich' material), in which the central parts are lightened and have a relatively low resistance to transverse shear, or for homogeneous structures in which the wave length of the deflection is of the order of magnitude of the thickness (for instance for thick, stocky structures, or for the higher modes of vibration which are characterized by short wave lengths).

2.2 CLASSICAL BEAM THEORY

In the present chapter we will consider some applications of beam theory illustrating important aspects of plate and shell theory which are more difficult to explain and evaluate in these more general cases. For this purpose it is sufficient to consider straight, uniform beams of rectangular cross section of height $h = 2c$, loaded symmetrically with respect to a vertical plane of symmetry zx, Fig. 2.1(a), so that displacements will be principally in this plane. For convenience we will

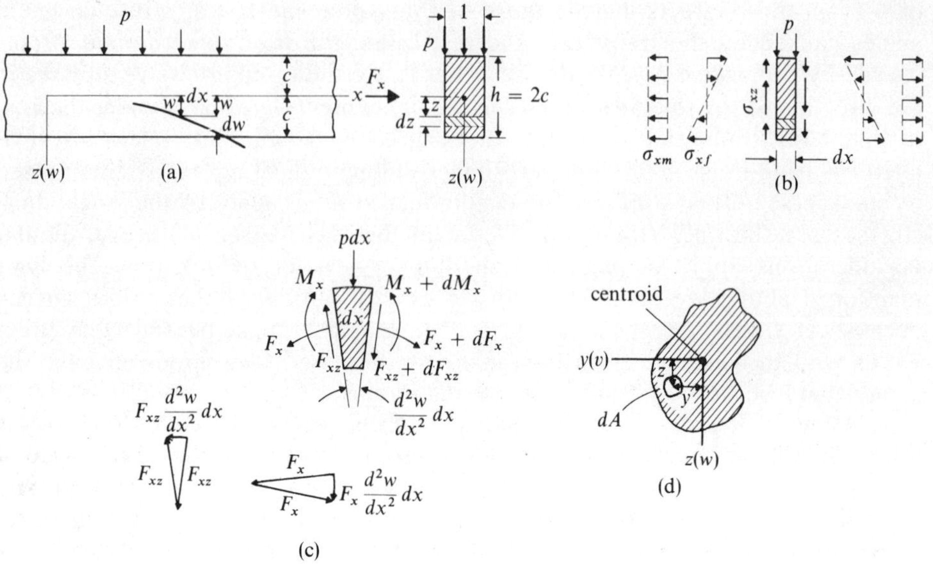

FIGURE 2.1

assume the width of the beam to be unity, as indicated in the figure, so that the bending moment M_x, the transverse shear F_{xz}, and axial force F_x represent moment and forces per unit width, to match the application of the same symbols used later in studying plates and shells. For rectangular sectioned beams having a width different from unity, M_x, F_{xz}, and F_x must be multiplied by this width to obtain the total moment and forces acting on cross sections or ends of the beam.

Much of the theory developed will be applicable also to beams of non-rectangular section, as will be discussed in Sec. 2.3. To serve the double purpose of an introduction to plates and shells, and also of application to more general beams, basic equations which are applicable to beams of general cross section will be given in two forms, involving h or c for the first purpose, and involving the moment of inertia I and area A of the cross section for the latter purpose. In applying the equations in the form involving I and A, M_x, F_{xz}, F_x, and p must be taken to represent the *total* bending moment and transverse and axial forces and load per unit length respectively.

The loads on the beam involve the transverse downward load shown, a unit pressure $p = p(x)$ (meaning that p is a function of x and only of x) which is uniformly distributed over the width while varying in the x direction. There may also be an axial tensile load F_x applied at the centroids of the ends. Under this loading the beam undergoes small transverse deflection $w(x)$ in the z direction.

An elementary length dx of the beam between adjacent cross sections is acted on and deformed by stresses such as those shown in Fig. 2.1(b). Integrated over the cross section, the shear stresses result in the transverse shear force F_{xz}; the normal stresses result in the normal force F_x acting at the center and a couple

M_x; all of these forces in general change with x, as indicated in Fig. 2.1(c). F_{xz} and M_x are of course the 'transverse shear' and 'bending moment' of elementary strength of materials, which can be determined from the transverse and moment equilibrium of the portion of the beam on one side of the section; the axial force F_x can similarly be determined from the axial equilibrium of this portion of the beam. The system of coordinates and the symbols and the positive directions shown are those most commonly used, and they are consistent with those used later for plates and shells so as to clarify the connection between the more general theories and the simpler theories for special cases.

We assume that deflection *slopes* are small compared to unity. In most practical problems this is true even when deflections rise to the magnitude which we will call the 'large deflection' range. Slopes of the order of unity are unlikely to occur except in extreme cases which involve thin wires (the 'elastica' problem) or thin plates or shells which are bent into shapes developable into their original shape, or which are made of rubber-like materials or deformed far into the plastic range; the general relations developed in Chap. 6 apply to such cases, but otherwise we will not consider them. For the present, therefore we make no distinction between a slope such as dw/dx, which is by definition the tangent of the angle of rotation of the middle surface at the point x, and the sine of this angle, or the angle itself measured in radians, or between the cosine of such an angle and unity. The angle between the two cross sections of Fig. 2.1(c) after deformation can therefore be taken as the rate at which the slope dw/dx changes as we go in the x direction, d^2w/dx^2, times the distance we go in this direction, dx, as indicated.

Besides the error made in using the tangent of the angle for the angle, there may actually be displacements in the x direction to be considered as well as the transverse displacement w; hence dx should strictly be replaced by $dx(1+\epsilon_{xm})$, where ϵ_{xm} is the axial strain at the middle surface discussed below. The exact mathematical expression for the curvature of a w vs. x curve, $(d^2w/dx^2)/[1+(dw/dx)^2]^{3/2}$, may not apply here because it does not consider the displacements in the x direction. The errors caused by these approximations are negligible in beam problems such as we will consider.

Due to the relative rotation $(d^2w/dx^2)\,dx$ of the two cross sections, there will be total longitudinal tensile strains $\epsilon_{xm}\,dx - z(d^2w/dx^2)\,dx$, where ϵ_{xm} is the unit tensile strain at the middle surface and z is the distance from the middle surface. Dividing this by the initial length of all the longitudinal fibers, dx, we obtain the longitudinal tensile strains:

$$\epsilon_x = \epsilon_{xm} - z\frac{d^2w}{dx^2} \tag{2.1a}$$

Multiplying this by the modulus of elasticity of the material, E, we have the longitudinal tensile stresses:

$$\sigma_x = \sigma_{xm} + \sigma_{xf} = E\epsilon_{xm} - Ez\frac{d^2w}{dx^2} \tag{2.1b}$$

where σ_{xm} and σ_{xf} are the uniform 'membrane' and the linearly varying 'flexural' parts of the longitudinal stress respectively, as shown in Fig. 2.1(b). Multiplying the stress by the area dz of the element on which it acts and integrating this elemental force over the cross section, that is as z goes from $-h/2$ to $h/2$, we obtain the longitudinal tensile force $F_x = Eh\epsilon_{xm} = \sigma_{xm}h$. Multiplying the elemental force by z we obtain its moment about the middle surface, and integrating these elemental moments over the cross section we obtain the bending moment $M_x = \int_{-h/2}^{h/2} \sigma_x z\, dz$:

$$M_x = -\frac{Eh^3}{12}\frac{d^2w}{dx^2} = -EI\frac{d^2w}{dx^2} \tag{2.2}$$

We next consider the equilibrium of the moments of all the forces on the element of Fig. 2.1(c) about an axis perpendicular to the paper through the center of the element. We ignore the moment of dF_{xz}, as this disappears in the limit compared to that of F_{xz} as dx approaches zero. Using Eq. (2.2) and remembering that we have assumed the beam to have a constant cross section:

$$F_{xz}\, dx - (M_x + dM_x) + M_x = 0, \qquad F_{xz} = \frac{dM_x}{dx} \tag{2.3}$$

$$F_{xz} = -\frac{Eh^3}{12}\frac{d^3w}{dx^3} = -EI\frac{d^3w}{dx^3} \tag{2.3a}$$

It is important to note that when the main part of a quantity, such as M_x, disappears from the equation because its values on opposite sides of the element cancel, then the *change* in the quantity (in this case dM_x) becomes the important part and must be considered. On the other hand when the main parts of a quantity, such as F_{xz}, remain in the equation, then the change in this quantity (in this case dF_{xz}) can be neglected because, as noted above, in the limit its effect approaches zero compared to the effect of the main parts, as dx approaches zero; this involves no approximation. There will be many applications of this principle in the following pages.

From the equilibrium of forces on the element in the x direction, using Eq. (2.3a):

$$dF_x - F_{xz}(d^2w/dx^2)\, dx = 0, \qquad dF_x/dx = F_{xz}\, d^2w/dx^2$$
$$= -EI(d^3w/dx^3)(d^2w/dx^2) \approx 0 \tag{2.3b}$$

From this result it can be seen that, for the case when there is no axial loading except at the ends, any change of F_x along the axis will be a 'nonlinear' quantity, proportional to a higher power or product of w or its derivatives. Even for the 'large deflection' case considered later this will be small, and F_x can be taken without serious error to be constant along the length and equal to the applied axial loads at the ends. If there are no axial loads or constraints against axial motion at the ends, ϵ_{xm} and σ_{xm} or F_x can therefore be taken as zero everywhere without serious error. Absence of axial constraint at the ends implies a roller or similar support at one end, but in practice even without such provision the axial stresses and strains are not likely to be important unless the beam is held firmly at both ends and the deflections are of the order of the thickness (a case which is considered later in Sec. 2.6).

For the case of a beam which is acted upon by distributed axial forces f_x per unit length in the x direction, the equation of equilibrium in the x direction can be taken as $dF_x + f_x\,dx = 0$, $dF_x/dx = -f_x$, ignoring the nonlinear term $F_{xz}\,d^2w/dx^2$. We will consider some problems of this type in Sec. 3.3, Eqs. (3.31a,b).

Transverse Equilibrium Equation

Finally, from equilibrium of forces on the element in the z direction, using Eq. (2.3):

$$dF_{xz} + p\,dx + F_x\frac{d^2w}{dx^2}\,dx = 0, \qquad -\frac{dF_{xz}}{dx} = -\frac{d^2M_x}{dx^2} = p + F_x\frac{d^2w}{dx^2} \tag{2.3c}$$

and, using Eqs. (2.2) or (2.3a) and $F_x = \sigma_{xm}h$:

$$\frac{Eh^3}{12}\frac{d^4w}{dx^4} = p + \sigma_{xm}h\frac{d^2w}{dx^2}, \qquad \text{or} \qquad EI\frac{d^4w}{dx^4} = p + F_x\frac{d^2w}{dx^2} \tag{2.4}$$

For the case of transverse loading only, F_x, $\sigma_{xm} = 0$ and Eq. (2.4) becomes:

$$\frac{Eh^3}{12}\frac{d^4w}{dx^4} = p, \qquad \text{or} \qquad EI\frac{d^4w}{dx^4} = p \tag{2.4a}$$

It should be noted that Eqs. (2.3), (2.3b), (2.3c) are exact equilibrium equations while Eqs. (2.2), (2.3a) (2.4), (2.4a) are Love–Kirchhoff approximations.

Equation (2.4) represents the balance of forces, per unit length of beam, which tend to *produce* and to *resist* deflections in the weak transverse direction. This and the corresponding equations for plates and shells are therefore the paramount relations in all the theories. Consider the physical meaning of the different terms in Eq. (2.4). Their signs depend upon the sign conventions used and have no special physical significance.

The first term, $EI\,d^4w/dx^4$, represents a *resistance* to deflection, calculated as the variation in the transverse shear F_{xz}, the moment of which balances the variation in bending moment M_x which is caused by the change in curvature; this is the 'flexural' resistance to deflection, proportional to the flexural stiffness of the beam, EI. The second term, p, is the transverse loading tending to produce deflection, or, if it represents a distributed reaction, tending to prevent it; in the first case it is usually independent of the deflection w while in the second case it may be proportional to w (the case of 'elastic support').

The third term represents the transverse component, due to the curvature d^2w/dx^2, of any axial force F_x which may be present. The first two terms are 'first degree', 'linear' terms with respect to the deflection, that is they are independent of or vary with the first power of w or its derivatives. The d^2w/dx^2 part of the third term is also first degree. If F_x is a force independent of the deflection, due to an axial end load which is so applied that it remains constant during the deflection, then the third term is also first degree and the equation is linear in w. If F_x is negative, that is compression, then this force tends to *produce* deflection; if it is big enough (the Euler stability limit) it keeps the beam deflected without the help of any transverse load; if it is less than the Euler load it increases the deflection due to any transverse load which may be present. If F_x is positive, that is tension, it acts with the flexural resistance to decrease deflection.

The axial force F_x can also be produced *by* the deflection. This is the case if the beam supports prevent the ends of the beam from moving toward each other, as in the case considered later in Sec. 2.6. Then if the beam is bent by lateral forces, its middle surface will be stretched, since it is now curved and longer than it was originally, and the supports will exert a tensile force F_x on the beam which will increase with the square of the deflection. Since F_x is multiplied by d^2w/dx^2, the third term in this case will increase with the cube of the deflection, and the equation will be nonlinear in w. When deflections are small, such a higher degree term is negligible in magnitude compared to the first degree terms, but it becomes very important as the deflection increases (long before the error due to using the tangent of the angle for the angle becomes important). If the beam does not have the same length initially as the distance between the end attachments, so that there is an initial force F_x which changes as the deflection increases, then we have a combination of the above cases, the initial force giving a first degree term and the change in it a higher degree term.

The three types of terms in the transverse equilibrium equation of beams discussed above are typical of terms found in the transverse equilibrium equations of plates and shells, and the observations made about them above apply equally to those more general cases.

Filament and Membrane Theories

In the cases where F_x is tension and so the third term acts to resist deflection, the flexural resistance EI may be zero and the equation then applies to flexible filaments, such as chains, ropes or cables, under tension. Stresses like σ_{xm} which are uniformly distributed across the thickness are called 'membrane' stresses, because membranes such as soap film and balloon or blimp fabric have little or no flexural stiffness and must rely on such stresses to carry their loads, like the chains, ropes, and cables mentioned above. In curved beams and shells such membrane stresses on the ends of an element have transverse components which are due to both the curvature produced by the deflection and to the initial curvature.

For structures made of membranous materials we can set up simplified 'membrane theories' by omitting the flexural terms. Membranous materials would collapse under compressive stress, so they can only be used where internal pressure (inflated structures) or other forces keep the wall in tension, that is where the principal stresses in the plane of the wall are tensile. Membrane theories may also be applicable to shells with walls having appreciable flexural stiffness if they have such a shape and loading that the flexural part of the resistance is negligible.

Thus for shells having axially symmetrical shape and loading and hence also deflections, large membrane resistance builds up with very little change in curvature, except around discontinuities in shape or in loading and on surfaces which are nearly flat and normal to the axis of symmetry, where there may be important flexural stresses. Steel tanks made in the shape that a rubber bag would assume under similar loading are familiar applications.

Stresses in Beams

As was mentioned in the beginning, the stresses considered in the classical beam theory are the longitudinal stresses σ_x corresponding to F_x and M_x and the transverse shear stresses σ_{xz} corresponding to F_{xz}, Fig. 2.1(b). As in elementary strength of materials the former can be found from the derivation of F_x and M_x, and the latter from the longitudinal equilibrium of an element bounded by adjacent cross sections and a horizontal plane a distance z from the middle surface. The longitudinal stress σ_x is, using Eqs. (2.1b), (2.2):

$$\sigma_x = \sigma_{xm} - Ez\frac{d^2w}{dx^2} = \frac{F_x}{h} + \frac{12M_xz}{h^3} = \frac{F_x}{A} + \frac{M_xz}{I} \tag{2.5}$$

The transverse shear stress σ_{xz}, derived in elementary strength of materials and for the more general case of a plate in Sec. 4.3, is, for a rectangular cross section:

$$\sigma_{xz} = -\frac{E}{2}(c^2 - z^2)\frac{d^3w}{dx^3} = \frac{6F_{xz}}{h^3}(c^2 - z^2) \tag{2.5a}$$

The transverse shear stress has a parabolic distribution, with a maximum value $3F_{xz}/2h$ which is 1·5 times the average value F_{xz}/h. It comes to zero at the top and bottom of the cross section, as it must if the top and bottom surfaces of the beam are free from shear loads. It is shown in Sec. 3.1 that the shear stresses on all four sides of an infinitesimal rectangular element must be equal in magnitude; hence for an infinitesimal element at the top or bottom of a beam, such as element a, Fig. 1.1, the shear stress must be zero on the vertical sides because it is zero on the top or bottom sides.

The stresses given by Eqs. (2.5), (2.5a) are the stresses on sections normal to the axis (and in the case of shear stresses also on sections parallel to the axis). There are in general stresses on sections at *all* angles, and the possibility suggests itself that some of these may be greater than those on sections normal to the axis. At any given point the largest normal stress on sections at various angles is the 'principal stress', which can be found as was outlined in Sec. 1.6 Eq. (1.1) and has a value for this case (since we ignore transverse normal stresses) of $\sigma_1 = \sigma_x/2 + \sqrt{\{(\sigma_x/2)^2 + (\sigma_{xz})^2\}}$. Confining the discussion to beams of rectangular cross section in which F_x, $\sigma_{xm} = 0$, and using expressions (2.5), (2.5a) for σ_x and σ_{xz}, we find the principal stress at a point a distance z from the axis:

$$\sigma_1 = \frac{6M_x}{h^2}\left[\frac{z}{h} + \left\{\frac{z^2}{h^2} + \left(\frac{F_{xz}h}{4M_x}\right)^2\left(1 - \frac{4z^2}{h^2}\right)^2\right\}^{1/2}\right] \tag{2.5b}$$

In this equation $6M_x/h^2$ is the maximum value of σ_x, which is the value ordinarily used in design. The quantity in the brackets is a factor by which this must be multiplied to obtain the maximum stress at any angle at any point on the cross section; it has a value of unity at the top and bottom ($z/h = \pm 1/2$), and the question is whether it may have a value greater than unity at some other point.

Calculation with various values of $F_{xz}h/4M_x$ and values of z/h from zero to $\pm 1/2$ show that the quantity in the brackets never exceeds unity unless $F_{xz}h/4M_x = \sigma_{xz_{max}}/\sigma_{x_{max}} > 0{\cdot}92$, that is unless the transverse shear stresses are of the same order of magnitude as the longitudinal bending stresses, in which case the Love–Kirchhoff assumption is not a good approximation. When membrane stresses are present there is a somewhat greater likelihood that principal stresses at an oblique angle may be greater than the stresses on normal sections, because on these sections the transverse shear stresses and the normal stresses may both be nearly a maximum at the same point, whereas when there are no membrane stresses the transverse shear stress is zero where the normal stress is a maximum and vice versa.

2.3 SOLUTIONS OF CLASSICAL BEAM THEORY

Boundary Conditions for Beams

The solution of the equilibrium equations (2.4) or (2.4a) must satisfy not only these equations but also the 'boundary conditions' at the points where the beams are held, usually the ends. These equations are of the fourth order, that is they contain as their highest derivative the fourth derivative of w with respect to x, d^4w/dx^4. The most general solution of such an equation, that is the most general expression for w as a function of x which will satisfy the equation, will contain four 'arbitrary' constants of integration (arbitrary as far as the equation is concerned but of course not arbitrary with respect to the physical problem), which can be used to satisfy two conditions at each of two locations along the beam where there are constraints. If these locations are at the ends of the beam, as is usually the case, the most common 'end conditions' are that at the ends, using Eqs. (2.2), (2.3):

$w,\ M_x = 0$ (that is $w = 0$ and $M_x = 0$) and therefore $w, \dfrac{d^2w}{dx^2} = 0$, for a 'hinged' (simply supported) end

$$w, \frac{dw}{dx} = 0, \text{ for a 'fixed' (built-in) end} \tag{2.6}$$

$M_x,\ F_{xz} = 0$, whence $\dfrac{d^2w}{dx^2}, \dfrac{d^3w}{dx^3} = 0$, for a 'free' end

For an elastic instead of a rigid vertical support, the condition $w = 0$ can be replaced by $\beta w = \pm F_{xz} = \mp (Eh^3/12)\, d^3w/dx^3 = \mp EI\, d^3w/dx^3$, where β is a measure of the stiffness of the support (force per unit vertical displacement) and the signs are for the left and right ends of the beam respectively. Similarly, for an elastic resistance to rotation, $M_x = 0$ can be replaced by $\beta' \, dw/dx = \mp M_x = \pm(Eh^3/12)\, d^2w/dx^2 = \pm EI\, d^2w/dx^2$, where again the signs are for the left and right ends, and β' is a measure of stiffness against rotation (moment per radian rotation). β' would have a negative value in applications where an end rotation invokes a moment which tends to *increase* the rotation, as may occur in a member of a truss or frame when other members attached to the same joint tend to buckle.*

The conditions $w = 0$ and $dw/dx = 0$ can apply also to constraints applied at intermediate locations on the beam. However, conditions involving F_{xz} or M_x become more complex for an intermediate support, since the magnitude of the force or moment exerted by a support is equal to F_{xz} or M_x respectively only at the end of a beam. For such cases it is usually simplest to study the parts of the beam on each side of the support separately, satisfying the continuity conditions that the displacement and slope, w and dw/dx, are the same for the two parts at the support, and that the forces and moments exerted by the two parts and by the support are in equilibrium.

Before leaving this subject it should be pointed out that in actual physical problems support conditions are seldom as clear-cut as the above discussion implies, and in some cases a detailed study of them may be called for, resulting in more complex boundary conditions (but not a greater number of conditions) then those mentioned. For example, a hinge always involves some friction, even if very slight. Hence the moment at a 'hinged' support will equal not zero but the friction moment, which may be a constant or proportional to the reaction exerted by the support, or something more complex. If a 'hinged' support involves simply resting one side of a beam or plate against a rigid support, then, since the support is off-center, rotation will result in a tangential friction force which produces both a resisting moment and an axial load; when the deflections are large the ends will tend to 'pull in' and the tangential friction force will reverse in direction.

A soft rubber beam or plate cemented at the end or edge to a steel wall would approach the simple theoretical condition of a fixed or built-in boundary. However, if a steel beam or plate is welded to a steel wall the bending stresses extend into the wall, almost unchanged at first but becoming spread out and small at a distance of the order of the thickness, h. The same is true of a clamped end, but in this case in addition to the strains in the clamp there must be some slippage at the edge of the clamp, where the stress concentration would theoretically be infinite if there were no slippage.

The effect of friction moments at hinged supports is similar to that of *decreasing* the length of the beam or the dimensions of a plate or shell, while the effect of support strains or slippage in fixed supports is similar to that of *increasing* the length or dimensions, by an amount ranging from close to zero for

*L. H. Donnell, 'The critical axial compression or tension on a bar, for all possible positive and negative end fixities', Reissner Anniversary Volume, *Contributions to Applied Mechanics*, J. W. Edwards, Ann Arbor, Mich., 1949, p. 183.

'frictionless' hinges or the rubber beam on the steel wall to probably a maximum of about one half the thickness (or one times the thickness to allow for supports at two ends).

Boundary conditions such as those of Eqs. (2.6) which consider only the *resultant* forces or moments on the boundary, or the slopes or displacements of the *middle surface* (or some other particular surface) might be called 'gross boundary conditions'. To satisfy completely the actual conditions at each boundary would in general mean to satisfy not only somewhat different conditions than those of Eqs. (2.6), but far more conditions than two. Complete boundary conditions in a beam problem would involve the specification of the stresses or displacements, or a relation between these, at *every point* of the cross section, and this represents theoretically an infinite number of conditions. Some of these conditions may be satisfied accidentally by the solution of Eqs. (2.4) or (2.4a) which we obtain in a given case, because any solution describes *some* stress and displacement at every point of the cross section, and these may happen to be the stresses and displacements desired. But this is in general unlikely and if we confine ourselves to solutions of the fourth order equations obtained by using the Love–Kirchhoff approximation we can be sure of satisfying only two conditions (that is we can only vary two conditions at each boundary arbitrarily, to fit the requirements). We of course use these two conditions to get the best approximation we can, by satisfying conditions on the resultants of the stresses on all points of the cross section or on the displacements of the middle surface, and this satisfies conditions at the boundaries to about the same degree of approximation as we satisfy the conditions at other points along the beam. This is sufficient for most practical problems, and there would be little point in satisfying conditions more exactly at the boundaries than elsewhere, although this could be done by superposing other solutions. Practical ways of obtaining better approximations where these are required are discussed in Chap. 3.

Mathematical Solutions

Equations (2.4) and (2.4a) take different forms requiring different types of solutions when they are applied to various physical problems, We will take up these solutions as we discuss the corresponding problems, but some general remarks may be helpful here. We will confine these remarks to the linear case, that is to cases where higher powers than one or products of w or its derivatives are not involved; while direct solutions of nonlinear equations are sometimes possible, as in some cases discussed in Sec. 5.1, more often other methods must be resorted to, such as the energy method.

For the purpose of discussing solutions, the terms in Eqs. (2.4) or (2.4a) can be classified into two types, those proportional to w and those independent of w. The first category consists of the first and third terms in Eq. (2.4) and any part of p which varies with w, such as the distributed reactions which act on a beam which rests on an 'elastic foundation', or the 'inertia forces' in lateral vibration problems, which are proportional to the second derivative of w with respect to time. The second category consists of that part of p which represents transverse loads so applied that their magnitude can be taken to be independent of the deflection.

Then in the most general case the solution can be found as the sum of two solutions: (1) a 'particular solution' which satisfies the equation as a whole, that is any expression for w which, when substituted into the terms involving w, balances the term which does not contain w; and (2) the solution of the 'homogeneous equation' obtained by omitting the term which is independent of w, that is an expression for w which when substituted into the terms involving w balances itself.

The general solution of the homogeneous equation will contain as many constants of integration as the order of the equation, and this should permit satisfaction of the boundary conditions. The sum of the particular and homogeneous solutions is the general solution of the equilibrium equation, that is the most general expression for w (which may take various forms all of which are equivalent) which will satisfy the equilibrium equation. This could of course be sought directly; the purpose of finding a separate 'homogeneous' part of the solution is one of efficiency, to solve an important part of the problem, applicable to many kinds of loads, only once, instead of having to repeat the process for each load.

As to the particular solution, which must satisfy the equation as a whole, it is generally impossible to find such a solution in 'closed' (non-series) form when the transverse loading is discontinuous, for example when it involves concentrated loads or sudden changes in distributed loads. That is, no single function for w when substituted in the terms containing w will yield a discontinuous function which balances the discontinuous loading function for all values of x, that is at all points along the beam. In such cases it is possible to study parts of the beam between discontinuities separately, satisfying continuity conditions where these parts are joined, as discussed above for intermediate elastic supports. However this is often not very practical in plate and shell problems, and for beams it is usually as easy or easier to take the particular solution as an infinite series, individual terms of which do not balance the loading term but whose sum can balance this term. We will find that it is generally possible to find such series which converge rapidly for most purposes, that is which give excellent approximations when only the first few terms are used. Because of these facts we will usually find it convenient to use such series solutions for all transverse loadings, even continuous loading such as a uniform loading for which a closed solution is possible.

For the linear case the homogeneous equation usually has a well known closed solution, and systematic methods for finding such solutions are discussed in mathematics books. If the coefficients of the terms are constants with respect to x (which would not be the case for the $F_x\, d^2w/dx^2$ term if there are axial forces distributed along the length, instead of being applied only as end loads) then a systematic way to find the solution is to take w as proportional to the exponential function $w = e^{cx}$. Since this has the same general form when it is differentiated any number of times, the factor e^{cx} can then be cancelled from the equation, leaving an algebraic 'characteristic equation' in c. This algebraic equation will have the same degree as the order of the differential equation (in our case four) so that a corresponding number of values of c ('roots) will satisfy it. Then the exponential function formed by using each of these values of c, multiplied by an arbitrary constant, must be a solution of the differential equation, and their sum will also be a solution. If a number m of the roots have the same value, then it can be shown that exponential functions formed by using this common value and multiplied by $x, x^2, x^3, \ldots, x^{m-1}$ are also solutions. If some of the values of c are complex, the corresponding exponential functions can be converted into trigonometric and hyperbolic functions by use of the definitions $\sin cx = (e^{icx} - e^{-icx})/2i$, $\cos cx = (e^{icx} + e^{-icx})/2$, $\sinh cx = (e^{cx} - e^{-cx})/2$, $\cosh cx = (e^{cx} + e^{-cx})/2$. If the derivatives are all odd or all even, this process can be shortened by starting with trigonometric or hyperbolic functions, since they have the same form when differentiated an even number of times.

If w is a function of more than one independent variable, for example of x and t in the dynamic beam problem, or of x and y in the static plate problem, etc., then a solution of the resulting partial differential equation can generally be found by 'separation of variables', for example by taking w as a product of exponential or trigonometric or hyperbolic functions of the different variables. For instance, in the dynamic beam problem w can be taken as a sine or cosine function of t multiplied by an unknown function $X(x)$ of x alone. Then, since only even derivatives of t are involved, all the terms will contain this sine or cosine function as a common factor, and when this is cancelled out an ordinary differential equation in X remains, which can be solved as above.

Beams of General Cross Section

It has been mentioned that the present study of beams of rectangular cross section applies in certain cases to other cross sections. For simplicity we will confine the discussion to transverse loading only, with $\epsilon_{xm} = 0$, but the conclusions apply also to the general case. Figure 2.1(d) shows the cross section of a uniform bar of general cross section, with the x axis passing through the centroids of the cross sections before displacement. Let the bar undergo displacements v in the y direction as well as displacements w in the z direction. A typical element of the cross section has the area dA and is at the point y, z as shown. Then by the same reasoning and using the same approximations as were used in deriving Eq. (2.1b), there is a longitudinal stress on dA of:

$$\sigma_x = -E\left(z\frac{d^2w}{dx^2} + y\frac{d^2v}{dx^2}\right) \tag{2.7}$$

Multiplying this stress by dA and integrating this elemental force over the whole cross-sectional area A, we obtain

$$F_x = -E\, d^2w/dx^2 \int_A z\, dA - E\, d^2v/dx^2 \int_A y\, dA = 0, \text{ since} \int_A z\, dA, \int_A y\, dA = 0$$

for y, z axes through the centroid.

Integrating the moments of this elemental force about the y and the z axes, we obtain the moments of the internal forces about these axes, M_y (the moment which is called M_x elsewhere in this book) and M_z, which equal the moments of the external forces on the portion of the bar from the cross section to one end, that is the 'bending moments' about the y and the z axes:

$$\begin{aligned} M_y &= -E\frac{d^2w}{dx^2}\int_A z^2\, dA - E\frac{d^2v}{dx^2}\int_A yz\, dA = -EI_y\frac{d^2w}{dx^2} - EI_{yz}\frac{d^2v}{dx^2} \\ M_z &= -E\frac{d^2w}{dx^2}\int_A yz\, dA - E\frac{d^2v}{dx^2}\int_A y^2\, dA = -EI_{yz}\frac{d^2w}{dx^2} - EI_z\frac{d^2v}{dx^2} \end{aligned} \tag{2.7a}$$

where I_y (called I elsewhere), I_z, I_{yz} are respectively the moments of inertia about the y and z axes passing through the centroid of the cross section, and the product of inertia about these axes. Solving Eqs. (2.7a) simultaneously for d^2w/dx^2 and d^2v/dx^2 we obtain:

$$\frac{d^2w}{dx^2} = \frac{M_yI_z - M_zI_{yz}}{E(I_{yz}^2 - I_yI_z)}, \qquad \frac{d^2v}{dx^2} = \frac{M_zI_y - M_yI_{yz}}{E(I_{yz}^2 - I_yI_z)} \tag{2.7b}$$

The bending stresses can then be obtained in terms of the bending moments M_y, M_z from Eq. (2.7). The slope of the 'neutral axis' is found by setting σ_x in Eq. (2.7) equal to zero, from which its slope is $z/y = -(d^2v/dx^2)/(d^2w/dx^2) = (M_zI_y - M_yI_{yz})/(M_zI_{yz} - M_yI_z)$.

If the product of inertia is zero, which is always the case for at least one set of perpendicular axes called the 'principal axes', then Eqs. (2.7b) reduce to $M_y = -EI_y\, d^2w/dx^2$ and $M_z = -EI_z\, d^2v/dx^2$. Comparing these with Eq. (2.2), it can be seen that Eqs. (2.2), (2.3), (2.4) and the equations which will be derived later from (2.4), (2.5) (but in general not (2.5a)) can all be used (with the proper subscripts) in their forms involving I for any cross-sectional shape, provided that the bending moments act in the planes of the principal axes of the cross section. If the bending moments do not act in these planes they can be separated into components which do act in these planes.

Axes of symmetry are always principal axes, since products of inertia with respect to them can easily be shown to cancel by symmetry. For symmetrical cross sections (such as the rectangular cross sections which we have been studying) the transverse shear stresses will from symmetry have zero net moment about the x axis; if the external loads also have zero moment about the x axis, as in the cases which we will consider, there will be no twisting of the bar. If the cross section is not symmetrical or the loads do have a moment about the x axis, there may be twisting as well as bending of the bar; the study of such cases is outside the scope of this book.

We will not discuss here the methods for studying beams which involve 'bending moment diagrams' and the areas and 'area moments' of these diagrams, since these very useful methods are covered in elementary strength of materials and are not suitable for the study of two-dimensional structures such as plates and shells. Instead we will use mathematical solutions of the differential equations, using series solutions when they are necessary or convenient. Such methods are not only of interest for application to beams, but they are of special interest to us here because they are simpler versions of the methods which are usually most useful for plates and shells.

2.4 LATERALLY LOADED BEAMS WITH HINGED ENDS

For a beam of length l, taking the origin at the left end, as shown in Fig. 2.2, the hinged end conditions $x = 0, l$: w, $d^2w/dx^2 = 0$ will be satisfied if we take w in the form of the sine series:

$$w = \sum_m w_m \sin\frac{m\pi x}{l}, \qquad m = 1, 2, 3, \ldots \tag{2.8}$$

since the expression for d^2w/dx^2 will be a similar sine series (with each term multiplied by $-m^2\pi^2/l^2$) and $\sin m\pi 0/l$, $\sin m\pi l/l = 0$. Equation (2.4a) can also be satisfied if p is a similar function of x:

$$p = \sum_m p_m \sin\frac{m\pi x}{l} \tag{2.8a}$$

Substituting expressions (2.8) and (2.8a) into Eq. (2.4a), we obtain:

$$\frac{\pi^4 Eh^3}{12l^4}\sum_m m^4 w_m \sin\frac{m\pi x}{l} = \sum_m p_m \sin\frac{m\pi x}{l} \tag{2.9}$$

This equation can evidently be satisfied if each coefficient of the series on the left equals the corresponding coefficient of the series on the right (and in general this is the only way it can be satisfied for all values of x), that is if:

$$\frac{\pi^4 Eh^3}{12l^4} m^4 w_m = p_m, \qquad w_m = \frac{12l^4}{\pi^4 Eh^3}\frac{p_m}{m^4} \tag{2.9a}$$

Using this value of w_m in expression (2.8), we obtain:

$$w = \frac{12l^4}{\pi^4 Eh^3}\sum_m \frac{p_m}{m^4}\sin\frac{m\pi x}{l} \tag{2.10}$$

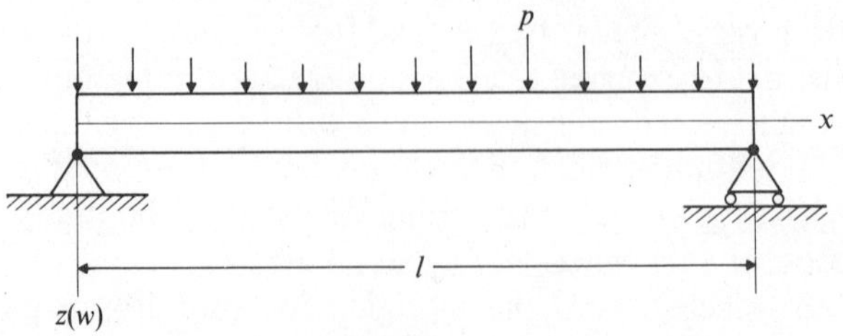

FIGURE 2.2

As discussed in mathematical treatises, the function $p(x)$ given by the sine series (2.8a) can be made to have any practical shape, that is it can represent any desired distribution of lateral load, by choosing proper values for the coefficients p_m. Figure 2.3(a) shows the first three components of such a sine series. At (b) it is shown that if the first two terms which are symmetrical about the center (namely those for odd values of m, such as $1, 3, \ldots$) are added together in the proper proportion we obtain the full line curve shown, which begins to approximate a uniform distribution; on the other hand if we subtract these terms in a suitable proportion as shown at (c) we begin to approximate a load concentrated at the middle. To obtain unsymmetrical distributions we would have to also use terms with even values of m, which have antisymmetrical shapes about the center (that is, they are symmetrical on either side of the center except for opposite signs). It can be shown that we can approach any shape we like as closely as we like by combining enough terms.

Unlike a full Fourier series, which contains both sine and cosine terms, such a simple sine series cannot have a value other than zero exactly *at* the ends, although if enough terms are used it can approach the correct value a very short distance from the ends. Actually the convergence is little affected even when the loads have large values near the ends, because the portion of the loading close to the ends has little effect on deflections and bending stresses.

We can determine the proper values of the coefficients p_{mn} by the process of 'harmonic analysis'. This depends upon the fact that the terms of such a sine series (and of certain other series such as a similar cosine series, a combination of sine and cosine series or Fourier series, etc.) are 'orthogonal' or 'normal' to each other in the interval $x = 0$ to $x = l$. This means that the integral over this interval of the product of two different terms is zero. On the other hand the square of any term has in this case an average ordinate of exactly one-half, so that its integral from 0 to l is $l/2$.

These facts can easily be checked by reference to integral tables, and their

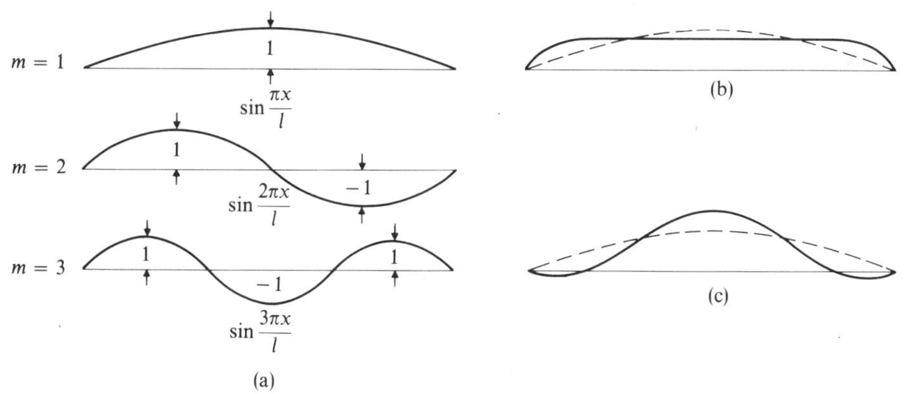

FIGURE 2.3

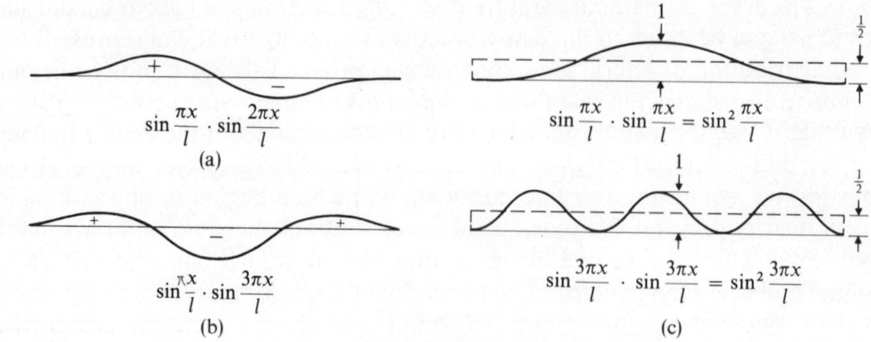

FIGURE 2.4

reasonableness is shown by Fig. 2.4. It is evident that the product of any symmetrical (odd value of m) term by an antisymmetrical (even value of m) term will give an antisymmetrical shape, as shown at (a), and the positive and negative areas under this will therefore always cancel. It is not so evident that the positive and negative areas will cancel for a product of two different symmetrical or antisymmetrical terms, but if we sketch them roughly it can be seen that this is reasonable; at (b), for instance, there are two positive areas to one negative area, but the positive areas are obviously smaller than the negative one. On the other hand the square of any term will have all positive ordinates, with values ranging from zero to unity, and we can see in (c) that the ordinates obviously do average one-half.

To determine the values of a particular coefficient p_m in Eq. (2.8a) we multiply both sides of this equation by $dx \sin m\pi x/l$ and integrate over the length l. Then on the right side of the equation only the product of this by the similar term in the series gives an integral which is different from zero:

$$\int_0^l dx\, p(x) \sin\frac{m\pi x}{l} = \int_0^l dx \left(p_1 \sin\frac{\pi x}{l} + \cdots p_m \sin\frac{m\pi x}{l} \cdots\right) \sin\frac{m\pi x}{l} = p_m \frac{l}{2}$$

$$p_m = \frac{2}{l}\int_0^l dx\, p(x) \sin m\pi x/l \tag{2.11}$$

Using this in Eq. (2.10):

$$w = \frac{24l^3}{\pi^4 Eh^3} \sum_m \frac{1}{m^4} \sin\frac{m\pi x}{l} \int_0^l dx\, p(x) \sin\frac{m\pi x}{l} \tag{2.12}$$

For any given distribution of load $p(x)$, we can find the deflection $w(x)$ from Eq. (2.12), and then the longitudinal bending and shear stresses from Eqs. (2.5) and (2.5a).

Uniform Load on Hinged End Beam

If $p = p_0$, a constant,

$$\int_0^l dx\, p_0 \sin m\pi x/l = p_0 \int_0^l dx \sin m\pi x/l = -\{p_0 l/(m\pi)\}(\cos m\pi - \cos 0),$$

which equals $2p_0l/(m\pi)$ for odd values of m and zero for even values of m. Then Eq. (2.12) becomes:

$$w = \frac{48l^4p_0}{\pi^5Eh^3}\sum_m \frac{1}{m^5}\sin\frac{m\pi x}{l}\quad (m = 1, 3, 5, \ldots) \tag{2.13}$$

For such a symmetrical load on a uniform beam the maximum deflection is at the center of the beam, that is at $x = l/2$, and is:

$$\begin{aligned} w_{max.} &= \frac{48l^4p_0}{\pi^5Eh^3}\left(\frac{1}{1^5}\sin\frac{\pi}{2}+\frac{1}{3^5}\sin\frac{3\pi}{2}+\cdots\right) \\ &= \frac{48l^4p_0}{\pi^5Eh^3}\left(1-\frac{1}{3^5}+\frac{1}{5^5}-\cdots\right) \end{aligned} \tag{2.14}$$

Using only the first term of the series we find $w_{max.} = (l^4p_0/Eh^3)/6{\cdot}38$. The correct value for the number in the denominator is 6·4, so the value obtained with one term of the series is only 0·4 percent too high; with two terms it is as close to the correct value as can be calculated with a slide rule.

The above series $1-1/3^5+1/5^5-\cdots$ has a known sum, $5\pi^5/1536$, which yields the exact value for the maximum deflection immediately. Many of the other solutions in series form which will be obtained or can be obtained by the methods which will be described, have known sums (which can be found in mathematical treatises on function theory, or handbooks) which convert these solutions to a 'closed', or non-series, form. However, it is the author's experience that for numerical work with most of the series which we will consider, the required accuracy can be obtained by evaluating a few terms of the series, cutting the series off when the terms become negligible, in less time than by attempting to calculate or look up the sum of the series.

Concentrated Load on Hinged End Beam

Consider the case of a concentrated load P acting at the point $x = x_0$. For the purposes of calculation this can be considered to be a distributed load of $p = P/\Delta$ acting uniformly over an infinitesimal length Δ of the beam, centered about the point of application of P, with zero loading everywhere else, as indicated in Fig. 2.5.

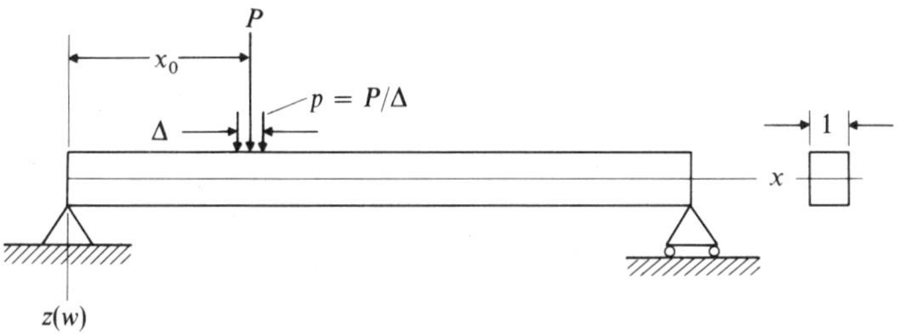

FIGURE 2.5

Then the integral $\int_0^l dx\, p(x) \sin m\pi x/l$ will evidently be zero everywhere except in the length Δ. Since this length is infinitesimal, $\sin m\pi x/l$ can be taken to have the constant value $\sin m\pi x_0/l$ over this length, and the integral is therefore $\Delta(P/\Delta) \sin m\pi x_0/l - P \sin m\pi x_0/l$. Equation (2.12) therefore becomes:

$$w = \frac{24Pl^3}{\pi^4 Eh^3} \sum_m \frac{\sin m\pi x_0/l}{m^4} \sin \frac{m\pi x}{l} \tag{2.15}$$

For the case of a concentrated load in the center of the beam $x_0 = l/2$. For such a symmetrical load the maximum deflection will also be at the center, $x = l/2$, and will be:

$$w_{max.} = \frac{24Pl^3}{\pi^4 Eh^3} \sum_m \frac{\sin^2\left(\frac{m\pi}{2}\right)}{m^4} = \frac{24Pl^3}{\pi^4 Eh^3} \sum_m \frac{1}{m^4}, \qquad (m = 1, 3, 5, \ldots) \tag{2.16}$$

Using only one term of this series, we find $w_{max} = Pl^3/48{\cdot}7EI$, which is 1·5 percent lower than the correct value; with two terms it is 0·3 percent low, etc.

Solutions can be obtained in a similar manner for several concentrated loads, or concentrated loads combined with distributed loads. Or, since we are considering here solutions for elastic beams with small deflections, for which displacements, strains, and stresses vary linearly with the loads, solutions for parts of the loading can be superposed; that is, the displacement or stress at each point for the entire loading can be calculated as the algebraic sum of the displacements or stresses at that point for the different parts of the load.

Free Vibration of Hinged End Beam

In the case of a beam undergoing lateral acceleration, the loading p is provided by the 'inertia forces' due to the acceleration $\partial^2 w/\partial t^2$ (we must of course use partial derivatives here, since the deflection w will be a function of both x and the time t). Since w represents a downward displacement, $\partial w/\partial t$ represents a downward velocity and $\partial w^2/\partial t^2$ a downward acceleration, giving rise to an upward 'inertia force' (which can be imagined as a pseudo-'resistance' to acceleration). The 'load' on an element of the beam of length dx is therefore $p\, dx = -\rho h\, dx\, \partial^2 w/\partial t^2$, where ρ is the mass of the material per unit volume, $h\, dx$ is the volume, and $\rho h\, dx$ is the mass of the element. Equation (2.4a) can therefore be written for this case as:

$$\frac{Eh^3}{12}\frac{\partial^4 w}{\partial x^4} = -\rho h \frac{\partial^2 w}{\partial t^2}, \qquad \text{or} \qquad \frac{Eh^2}{12\rho}\frac{\partial^4 w}{\partial x^4} = \frac{Ek^2}{\rho}\frac{\partial^4 w}{\partial x^4} = -\frac{\partial^2 w}{\partial t^2} \tag{2.17}$$

where k is the radius of gyration of the cross section.

Equation (2.17) and the boundary conditions can be satisfied if we take w in the form:

$$w = w_m \sin \frac{m\pi x}{l} \sin 2\pi N_m t \tag{2.18}$$

where m is any integer. This deflection satisfies the boundary conditions: $x = 0, l$: $w, \partial^2 w/\partial x^2 = 0$, and describes a vibration whose 'amplitude' (defined as the maximum displacement at any place or time) is w_m, and whose shape or 'mode' of vibration is described by $\sin m\pi x/l$. The factor $\sin 2\pi N_m t$ changes from 0 to $+1$ to 0 to -1 and back to 0, ready to start the whole cycle over again, as the time t changes from 0 to $1/N_m$. Hence there will be N_m cycles or 'vibrations' per unit of time, and N_m represents the 'frequency' of vibration in the mth mode, that is in the shape $\sin m\pi x/l$.

Expression (2.18) is an exact closed solution of Eq. (2.17), because if we substitute it into this equation we obtain:

$$\frac{m^4\pi^4 Ek^2}{\rho l^4} w_m \sin\frac{m\pi x}{l} \sin 2\pi N_m t = 4\pi^2 N_m^2 w_m \sin\frac{m\pi x}{l} \sin 2\pi N_m t \tag{2.19}$$

which can be satisfied for all values of x and t if:

$$\frac{m^4\pi^4}{l^4}\frac{Ek^2}{\rho} = 4\pi^2 N_m^2, \qquad N_m = \frac{m^2\pi k}{2l^2}\sqrt{\frac{E}{\rho}} \tag{2.20}$$

Note that the frequency is independent of the amplitude w_m, but depends very much upon m, that is upon the mode. The lowest or 'fundamental' frequency is for $m = 1$, that is for vibration in one half sine wave. The next 'harmonic', for $m = 2$, has four times the fundamental frequency, etc.

For small displacements such vibrations of different frequency can be superposed upon each other, and this could have been described mathematically by using a series (that is a summation sign) in Eq. (2.18), and setting each coefficient of the series on the left in Eq. (2.19) equal to the corresponding coefficient of the series on the right, as we did with Eq. (2.9). However this would represent physically a superposition of many separate closed solutions, rather than a series solution of the type which we have been discussing, in which all the terms together are theoretically required to provide one solution.

In using the above solution for higher harmonics, it should be remembered that it is based upon the Love–Kirchhoff neglection of transverse strains and stresses, and that it considers only the acceleration in the transverse direction. It is a good approximation as long as the half wavelength of the deflection l/m is large compared to the thickness h. For shorter wave lengths the effects of transverse strains and stresses and of longitudinal accelerations ('rotatory inertia') must be included. This is done in Sec. 3.5, giving Eq. (3.65) for N_m.

2.5 AXIALLY LOADED BEAMS WITH HINGED ENDS, STRUTS

Strut with Initial Crookedness

All actual beams have a certain amount of initial crookedness, that is crookedness in the unloaded condition. Carefully made beams, for example bars machined from larger bars or subjected to a straightening operation, will have less crookedness than those not so carefully manufactured. Even if there were no measurable crookedness of the outside surfaces of the beam, however, there would always be an 'equivalent' crookedness (deviation from a straight line of the 'elastic axis', the effective axis of the beam) due to elastic nonhomogeneity and anisotropy of the material (due for instance to the multicrystalline structure of metals, inclusions, blow holes, etc.). Initial crookedness is usually of no importance for beams having lateral loading only, but in buckling problems it may be very important.

In the following discussion it will be assumed that the z axis is taken in the weakest direction; buckling under axial compression will generally occur in this direction. Taking the y, z axes, Fig. 2.1(d), as the principal axes of the cross section (that is the axes for which the product of inertia of the cross section is zero, as discussed in Sec. 2.3), the smallest moment of inertia about any axis through the centroid will be about one of these axes, say the y axis. If the end constraints are the same for bending in either direction, then buckling will occur at the lowest load by bending about this axis, that is in the z direction. If it is not obvious which direction is the weakest direction then both directions should be studied.

Consider a hinged-end bar with an initial deviation $w_0(x)$ of its center line from the straight x-axis passing through the centroids of its ends, Fig. 2.6, where:

$$w_0 = \sum_m w_{0m} \sin \frac{m\pi x}{l} \tag{2.21}$$

Such a series, as we have shown, can represent any shape (including eccentric application of the load) and each of the coefficients w_{0m} of the different terms will have some specific value for any individual beam.

Suppose now that the beam, which in such a case is usually called a strut or a column, is subjected to a compressive force $P = -F_x$ acting along the x axis as shown in Fig. 2.6. Under this loading there will be a lateral *movement* $w(x)$, producing additional displacement of the center line from the x axis, as indicated in Fig. 2.6. This movement can be expressed by a similar series:

$$w = \sum_m w_m \sin \frac{m\pi x}{l} \tag{2.22}$$

We will assume that there is no lateral loading p in this case; if there were, it could be included and expressed by a similar series, and its effect would be found to be the same as if the deflection produced by it alone represented an initial deviation from straightness, superposed on that given by Eq. (2.21).

For this case the flexural resisting moments will be zero when the movement

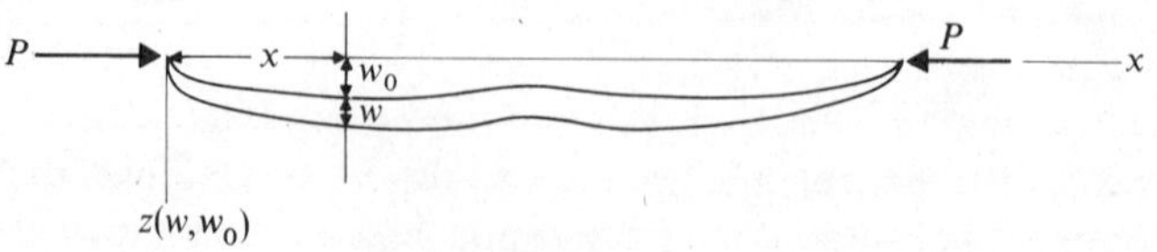

FIGURE 2.6

w is zero and they will depend upon w alone. On the other hand the transverse force due to the axial load P will be P times the *total* curvature, which is due to both w_0 and w. The equation corresponding to Eq. (2.4) for this case is therefore:

$$EI\frac{d^4w}{dx^4} = -P\frac{d^2(w+w_0)}{dx^2} = -P\left(\frac{d^2w}{dx^2} + \frac{d^2w_0}{dx^2}\right) \tag{2.23}$$

Substituting expressions (2.21) and (2.22) into Eq. (2.23), and equating the coefficients of the corresponding sine terms on both sides of the equation we obtain:

$$\frac{m^4\pi^4EI}{l^4}w_m = \frac{m^2\pi^2}{l^2}P(w_m + w_{0m}), \qquad w_m = \frac{w_{0m}}{\dfrac{m^2\pi^2EI}{Pl^2} - 1} \tag{2.24}$$

The movement w of the initially crooked strut under the axial compression P will therefore be

$$\begin{aligned} w &= \sum_m \frac{w_{0m}}{\dfrac{m^2\pi^2EI}{Pl^2} - 1} \sin\frac{m\pi x}{l} \\ &= \frac{w_{01}}{\dfrac{\pi^2EI}{Pl^2} - 1}\sin\frac{\pi x}{l} + \frac{w_{02}}{\dfrac{4\pi^2EI}{Pl^2} - 1}\sin\frac{2\pi x}{l} + \cdots \end{aligned} \tag{2.25}$$

Consider, now, how the movement w will vary with the load P. When $P = 0$ the denominators in expression (2.25) will all be infinite, and the movement w will be zero. When P approaches the value π^2EI/l^2 (the 'Euler buckling load') the coefficient of the first term will approach $w_{01}/0$ and so will 'blow up' and approach infinity, while the coefficient of the second term will only approach $w_{02}/3$, of the third term $w_{03}/8$, etc. Since in a practical strut the coefficients of the initial deviation w_{02}, w_{03}, etc., will not be likely to be larger than w_{01}, (they are usually smaller, the higher the value of m) it can be seen that only the first term of the initial deviation (the term having the shape in which the bar actually buckles) is important. Figure 2.7(a) shows very clearly how the different terms in expression (2.25) increase as P goes from 0 to π^2EI/l^2, and how unimportant the higher harmonics of the deflection are in this case.

Perfect Strut

For the purely theoretical case of a 'perfect' strut, w_0 and its coefficients w_{01}, w_{02}, etc., would all be zero. When P reaches π^2EI/l^2 the first term in (2.25) then has the indeterminate coefficient 0/0. Theoretically it might remain zero, (represented mathematically, as in all stability studies, by the 'trivial' solution $w = 0$ of the transverse equilibrium equation) but in practice some accidental transverse force, a component of the weight, a slight movement of air, etc., is sure to be acting upon it. As stated previously, such a continuing force produces deflections similar in effect to an initial deviation from straightness, and w_1 will immediately blow up as

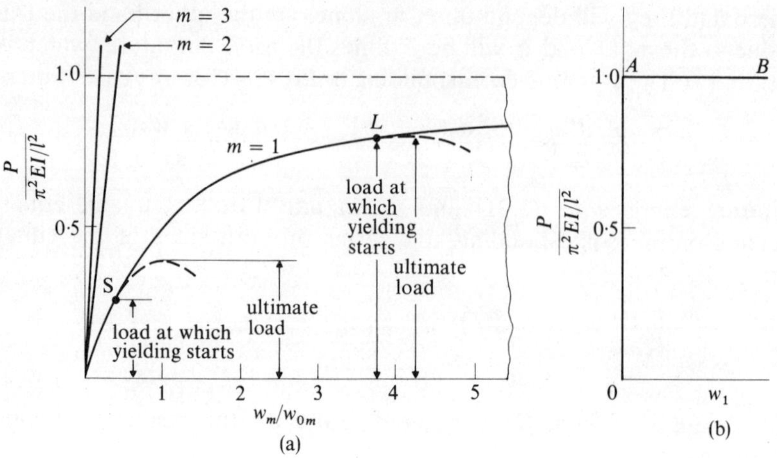

FIGURE 2.7

indicated by the horizontal line AB in Fig. 2.7(b). If P goes slightly above the value π^2EI/l^2, then even transitory transverse impulses, due to unavoidable vibrations or sound waves, are sufficient to produce the same effect. Theoretically, the same thing could also happen to w_2 when P reaches $4\pi^2EI/l^2$, but of course in practice the strut would have buckled long before this load could be reached.

Such large results triggered by very small causes are quite common. Resonance of a frictionless vibrating system produced by a very small vibratory disturbance is another example (there are very close analogies between resonance and buckling phenomena). The dependence of the lift of a wing upon the viscosity of the fluid moving past it is another example; theoretically, for a 'perfect fluid' without viscosity, there would be no circulation around the wing and hence no lift, but the least amount of viscous friction at the trailing edge, where the streams flowing over and under the wing meet with initially different velocities, is enough to unify these velocities locally and thus cause a circulation and a lift.

In a study of the buckling of a perfect strut it is unnecessary and meaningless to use a summation sign in the expression (2.22) for w, since all of its terms would remain zero except for the one term for which the load required to cause buckling is a minimum; however it is necessary to use the subscript and factor m in expressions such as (2.22) unless we are sure beforehand which buckling mode will require the smallest load. The above discussion can be regarded as a physical interpretation of the mathematical 'eigenvalue' problem.

A very important difference between buckling problems and problems involving lateral loading is that the buckling loads such as P are finite when the deflections are still zero or only infinitesimal, or in the case where small initial deviations are considered, P is large compared to the forces and moments produced by the deflections. This means that approximations and neglections which may be legitimate in dealing with the forces and moments produced by the deflections may not be good enough in dealing with the much larger force P. We will have examples of this later.

For the perfect elastic strut, if the compressive force P is below the critical Euler value the strut will return to its initially straight form if it is given a small displacement such as $w_1 \sin \pi x/l$ and then released; we describe its condition before it was displaced as a condition of 'stable equilibrium'. If the force P could be raised a little above the critical value without precipitating buckling and the strut were then displaced and released, it would deflect further without limit, that is buckle; we say that it was in a condition of 'unstable equilibrium'. If the load P is exactly at the critical value and the strut is displaced a small amount and then released it would remain in equilibrium in this displaced position; we say that it is in a condition of 'neutral equilibrium' under small displacements.

We will call a study by linear (small-displacement) theory of the conditions under which a structure or structural element with perfect shape and elasticity, under a load tending to buckle it, can be in a state of neutral equilibrium (that is can be in equilibrium in a displaced position) a 'classical stability' study. Until comparatively recent times theoretical studies of buckling problems were confined to such idealized solutions. The engineers who had to use such elements in their machines and structures early found that these solutions sometimes showed little relationship to actual behavior. Thus classical stability studies are satisfactory for very slender struts, but because of the limits to the elastic behavior of actual materials, empirical results are still relied on for most applications. When classical stability theories became available for more complex elements, it was found that nonelastic behavior was only one of the causes for serious discrepancies between such theories and experiments. For example, classical stability theory predicts many times the actual strengths of very thin shells under axial compression; on the other hand, the classical theory predicts only a fraction of the actual ultimate strengths of very thin hinged or fixed edge plates under compression or shear (although it does predict when buckling will commence). These discrepancies become greater the thinner the material and hence the smaller the average stress, so they are not due to inelastic effects.

Adequate theories are needed even more for the more complex cases than for the case of simple struts, because the greater number of variables makes it less practical to cover all combinations of these variables by empirical methods. The rise of the aeronautics industry, with its need for weight efficiency, has necessitated more realistic studies of many buckling problems, with a consideration of not only the factors considered in the classical stability study but of many other factors which sometimes have an important effect upon the behavior of actual specimens—nonelastic as well as elastic behavior of materials, finite displacements, initial imperfections which are always present, inertia, etc. There is an obvious parallel in this with the history of fluid mechanics—the early preoccupation of theoreticians with idealized 'perfect fluids', the failure to answer many practical questions and the consequent reliance upon empirical methods, until aeronautical problems forced the consideration of important additional factors which affect real phenomena.*

Strut on an Elastic Foundation

Unlike a simple strut, plates and shells characteristically buckle in many waves. This type of buckling and many of the consequences of it are illustrated by the simple case of a strut on an 'elastic foundation'. Consider the case of a hinged-end strut, with an elastic support along its length such as could be provided by a row of many small springs between the strut and a rigid foundation. We assume that the foundation conforms to the initial crookedness, as suggested in Fig. 2.8 and as would be the case for a railroad track on a tamped gravel bed. When the strut is deflected, the elastic support exerts a distributed supporting force per unit length, proportional to the movement w and equal to βw, where β is the supporting force per unit length per unit deflection and is a measure of the stiffness of the elastic support. Adding $p = -\beta w$ to Eq. (2.23), Eq. (2.4) for this case becomes:

$$EI\frac{d^4w}{dx^4} = -\beta w - P\left(\frac{d^2w_0}{dx^2} + \frac{d^2w}{dx^2}\right) \tag{2.26}$$

Using expressions (2.21) and (2.22) for w_0 and w and again equating the coefficients of the corresponding terms, solving for w_m and using this in (2.22), we find

$$w_m = \frac{w_{0m}}{\dfrac{m^2\pi^2EI}{Pl^2} + \dfrac{\beta l^2}{Pm^2\pi^2} - 1},$$

$$w = \sum_m \frac{w_{0m}}{\dfrac{m^2\pi^2EI}{Pl^2} + \dfrac{\beta l^2}{Pm^2\pi^2} - 1} \sin\frac{m\pi x}{l} \tag{2.27}$$

A typical term of this series will 'blow up' when P reaches the value:

$$P = \frac{m^2\pi^2EI}{l^2} + \frac{\beta l^2}{m^2\pi^2} = \left(\frac{m^2\pi^2}{l^2}\sqrt{\frac{EI}{\beta}} + \frac{l^2}{m^2\pi^2}\sqrt{\frac{\beta}{EI}}\right)\sqrt{\beta EI} \tag{2.28}$$

The term which makes this expression a minimum will blow up first. We assume for the moment that m can have any value (actually it must be an integer to satisfy the boundary conditions) and set the derivative of the expression (2.28) with

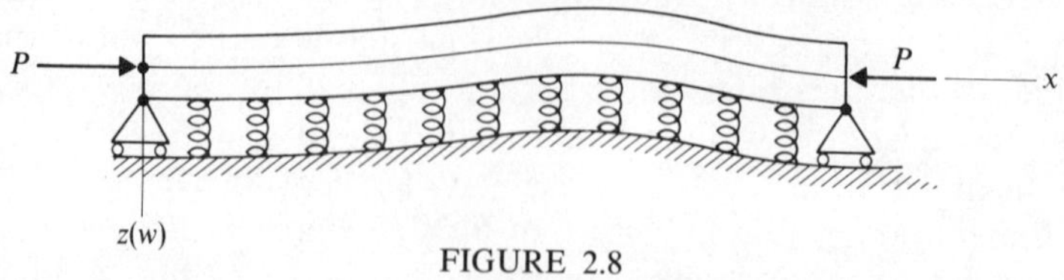

FIGURE 2.8

*L. H. Donnell, 'Recent developments in the study of buckling problems', *Applied Mechanics Reviews*, vol. 5, no. 7, 1952, p. 289.

respect to m or $(m^2\pi^2/l^2)\sqrt{EI/\beta}$ equal to zero. Solving for the critical m, corresponding to the term which will blow up first, we find:

$$m_{cr} = \frac{l}{\pi}\left(\frac{\beta}{EI}\right)^{1/4} \tag{2.29}$$

Substituting this back into Eq. (2.28) we obtain P_{cr}:

$$P_{cr} = 2\sqrt{\beta EI} \tag{2.30}$$

Equation (2.27) can then be written

$$w = \sum_m \frac{w_{0m}}{\dfrac{P_{cr}}{2P}\left(\dfrac{m^2}{m_{cr}^2} + \dfrac{m_{cr}^2}{m^2}\right) - 1} \sin\frac{m\pi x}{l} \tag{2.31}$$

Figure 2.9(a) shows plots of P/P_{cr} against w_m/w_{0m} for $m_{cr} = 2$, while a similar plot is shown at (b) for $m_{cr} = 8$. It will be seen that (unlike the case of the simple strut for which m_{cr} is unity) when m_{cr} is large compared to unity, the components of the initial deviation from straightness which have values of m close to m_{cr} also become quite large as the load P increases, and these can not be neglected in considering the stresses and the question of when yielding of the material will occur.

When m is large there is little error in ignoring the condition that it must be an integer. When m is small the problem can be studied with the aid of a diagram like Fig. 4.17, Sec. 4.4, for the buckling of plates.

Load on Strut at Which Yielding Starts

Going back to Fig. 2.7(a) for the case of a simple strut, it can be seen that if the strut remained elastic the load P on an actual crooked strut would approach

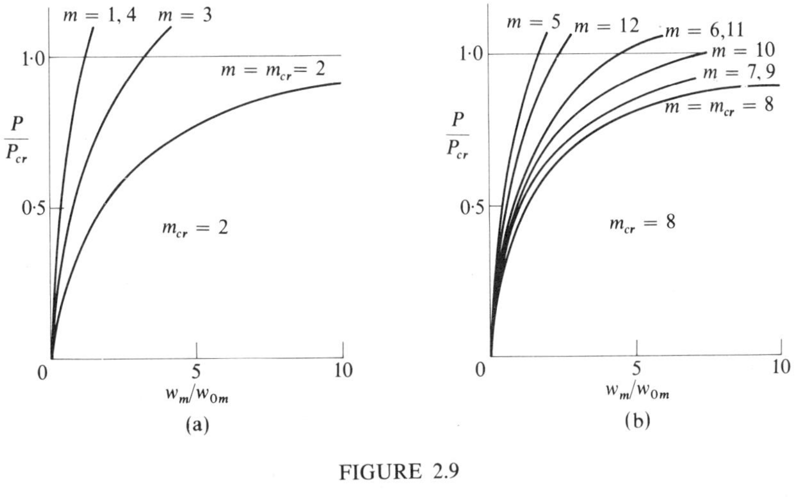

FIGURE 2.9

asymptotically the Euler buckling load for a perfect strut, $\pi^2 EI/l^2$, but would never quite reach it. Actually as the load P and consequently the movement w increase, the average compressive stress on a cross section will increase with P (and thus with the vertical coordinate, Fig. 2.7(a)) while bending stresses will increase with w (and thus with the horizontal coordinate). The maximum stress due to the sum of these will be at the middle of the strut on the concave boundary, where the compressive bending stress is a maximum. It will be a uniaxial stress, and appreciable yielding will therefore start when it reaches the yield point of the material τ_y. Neglecting the higher harmonics of the displacement, $w_2 \sin 2\pi x/l$, etc., which are very small in this case, the condition for yielding can be taken, using Eq. (2.24), as:

$$\begin{aligned} \tau_y &= \frac{P}{h} + \frac{M_x h}{2I} = \frac{P}{h} - \frac{Eh}{2}\left(\frac{d^2 w}{dx^2}\right)_{x=l/2} = \frac{P}{h} + \frac{\pi^2 Eh}{2l^2} w_1 \\ &= \frac{P}{h} + \frac{\pi^2 Eh}{2l^2} \frac{w_{01}}{\dfrac{\pi^2 EI}{Pl^2} - 1} \end{aligned} \tag{2.32}$$

The value of w_{01}, the amplitude of the harmonic component of the initial deviation which has the shape of a half sine wave, will depend upon the method of fabricating the strut, and for struts made by the same process it will also vary on the average with the length l and thickness h, being likely to be larger for a slender strut. If it were cut from a bar with a constant small curvature it would vary with l^2, and for the same l it would be likely to be larger for a thin than for a thick bar. It seems reasonable, therefore, to assume that for the average strut w_{01} will have a value

$$w_{01} = \frac{Ul^2}{\pi^2 h} \tag{2.33}$$

where h represents twice the distance from the neutral axis to the furthest fiber and π^2 is included to simplify the results. U is a dimensionless number, an 'unevenness' factor, which should on the average depend only upon the manufacturing process. It will be subject to scatter, but the final resulting scatter will be much less than it would be if we assumed values for w_{01} which were independent of the dimensions. It should be noted that variations in U will produce much smaller variations in the failure load. In any case, variations in failure loads due to variations in initial defects represent a fact of life with which engineers must deal. By allowing for the systematic part of these variations we can eliminate a considerable part of the uncertainties, which otherwise would have to be allowed for as unpredictable scatter unless we subject each individual specimen to measurements or tests.

We have treated w_{01} above as a geometric defect, but other defects which have been mentioned, due to anisotropy, nonhomogeneity, etc., as well as accidental disturbances (which there is reason to believe are much less important in most practical problems than defects) all have similar effects upon buckling, and w_{01} and hence U can be regarded as referring to an 'equivalent geometric deviation' which would have the same total effect as all these causes taken together. The possible maximum value of such a total effect is of course the arithmetical sum of the values of the different effects, but its actual value is unlikely to be much greater than the geometric effect alone, since the different effects are as likely to subtract from each other as they are to add to each other. Such questions are important in all buckling problems, particularly in the case of axial compression of a cylinder, which will be discussed in Chap. 7. The relatively simple analysis of them for a strut throws light on all these cases.

Using Eq. (2.33) in (2.32) and solving for P, we find the load P at which yielding starts to be:*

$$\frac{P}{A}=\frac{P}{h}=\tau_y\left(N-\sqrt{N^2-\frac{\sigma_{cl}}{\tau_y}}\right), \qquad \text{where} \qquad 2N=1+\frac{\sigma_{cl}}{\tau_y}+\frac{UE}{2\tau_y} \tag{2.34}$$

where A is the cross-sectional area, and σ_{cl} is the 'classical stability stress', which in the present case is the Euler stress $=(\pi^2EI/l^2)/A=\pi^2Eh^2/12l^2=\pi^2Ek^2/l^2$ where k is the radius of gyration. From comparison of this formula with actual buckling tests and from measurements of the crookedness of actual struts, the unevenness factor U can be expected to vary from about 5×10^{-5} (minimum values for struts made with care for straightness) to 100×10^{-5} (maximum values for ordinary construction). Figure 2.10 shows results obtained from Eq. (2.34) (full lines) compared to some traditional empirically derived formulas (the dotted lines).

It can be shown** that Eq. (2.34) applies to struts of all types if σ_{cl} is taken

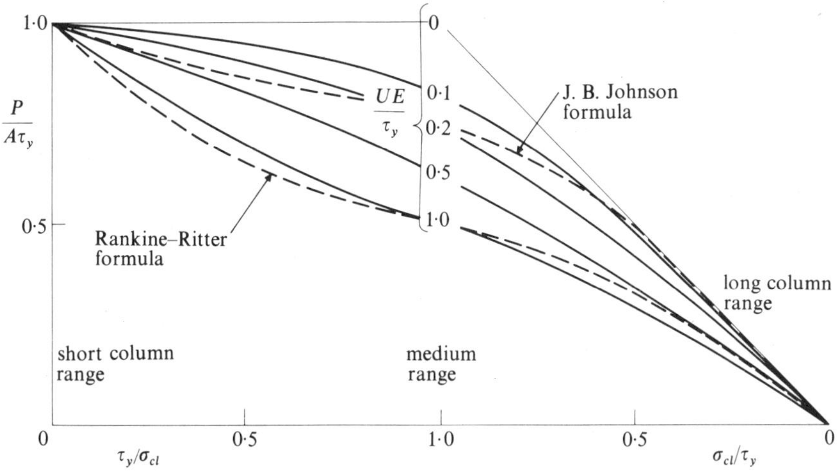

FIGURE 2.10 (from L. H. Donnell and C. C. Wan, 'Effect of imperfections on buckling of thin cyl. and columns under axial compression', *Trans. ASME*, vol. 72, p. 73, 1950, courtesy of ASME)

*L. H. Donnell and C. C. Wan, 'Effect of imperfections on buckling of thin cylinders and columns under axial compression', *J. Appl. Mech., Trans. ASME*, vol. 72, 1950, p. 73.

**L. H. Donnell and V. C. Tsien, 'A universal column formula for load at which yielding starts', *N.A.C.A.* Tech. Note 3415, 1955. This paper also presents some experimental confirmation of Eq. (2.33).

as the 'classical stability stress' for the strut considered (that is, the average axial stress producing buckling as calculated by classical bending theory under the assumption of a perfectly concentric load and perfect shape and elasticity) and if in some cases somewhat larger values of U are used than the values given above for the case of a simple hinged end strut. Equation (2.34) can thus be considered to be a 'universal column formula' for the load on a strut at which yielding commences. In the case of a fixed-end strut the above values of U should be multiplied by from 1·0 to 1·5. For the case of a strut on an elastic foundation, calculations of the condition for yielding similar to the above but considering the effects on the stresses of all the components of the deflection, particularly those with values of m near the critical value m_{cr}, indicate that the average value of U should be multiplied by a factor of $2-(1/m_{cr}^2)$ and the maximum value of U by a factor of $3-(2/m_{cr}^2)$.

Ultimate Strength of Struts

After yielding starts, experience shows that the ultimate resistance of a strut will rise only a little for slender struts of strong materials (the 'long column' range, characterized by large values of τ_y/σ_{cl} compared to unity). However, for stockier struts of lower yielding materials (the 'short column' range, characterized by small values of τ_y/σ_{cl}) the ultimate resistance may be considerably higher than the value given by Eq. (2.34), and in some applications it may be safe to base design upon this larger ultimate resistance.

These general conclusions from experience can also be deduced from the following reasoning. After yielding, the plastic region will grow gradually from the point at which it starts. The load deflection curve will be below that calculated on the assumption of elastic action, but since the plastic region grows only gradually, the load deflection curve after yielding will be tangent to the corresponding elastic curve at the load at which yielding started, as suggested by the dotted lines in Fig. 2.7(a). Due to this fact it is evident that for the short column range, for which yielding starts at a low value of $P(\pi^2 EI/l^2)$ where the elastic curve is nearly vertical (for example, the point S, Fig. 2.7(a)) the ultimate load, reached at the point where the plastic curve turns over and becomes horizontal, must be considerably higher than the load at which yielding starts. On the other hand in the long column range (illustrated by the point L) where the elastic curve is already nearly horizontal, there will be little difference between these two values.

2.6 AXIALLY LOADED BEAM WITH HINGED ENDS, NONLINEAR CASE

This is not a case of great practical importance in itself, but it illustrates very important principles in 'finite deflection' or 'large deflection', nonlinear problems of plates and shells, for which solutions are very much more difficult. For the purpose of illustrating principles it is sufficient to take the following very simple case.

Figure 2.11(a) shows a beam of length l, with ends which can rotate freely but cannot move horizontally, under a vertical load $p = p_1 \sin \pi x/l$. The hinged end conditions and Eq. (2.4) can be satisfied if $w = w_1 \sin \pi x/l$, where w_1 is the center deflection. Let Δl be the difference between the length of the deflected and the undeflected middle surface. Then, using Fig. 2.11(b) and the binomial theorem $(1+a)^n = 1 + na + n(n-1)a^2/2 + \cdots \approx 1 + na$ when a is small compared to unity, the uniform part of the axial stress, σ_{xm}, will be:

$$
\begin{aligned}
\sigma_{xm} &= E\frac{\Delta l}{l} = \frac{E}{l}\int_0^l (ds - dx) = \frac{E}{l}\int_0^l \left(\sqrt{dx^2 + dw^2} - dx\right) \\
&= \frac{E}{l}\int_0^l \left\{\left[1 + \left(\frac{dw}{dx}\right)^2\right]^{1/2} - 1\right\} dx = \frac{E}{l}\int_0^l \left\{\left[1 + \frac{1}{2}\left(\frac{dw}{dx}\right)^2 \cdots\right] - 1\right\} dx \qquad (2.35) \\
&\approx \frac{E}{l}\int_0^l \frac{1}{2}\left(\frac{dw}{dx}\right)^2 dx = \frac{E}{2l}\frac{\pi^2 w_1^{\,2}}{l^2}\int_0^l \sin^2\frac{\pi x}{l}\, dx = \frac{\pi^2 E w_1^{\,2}}{4l^2}
\end{aligned}
$$

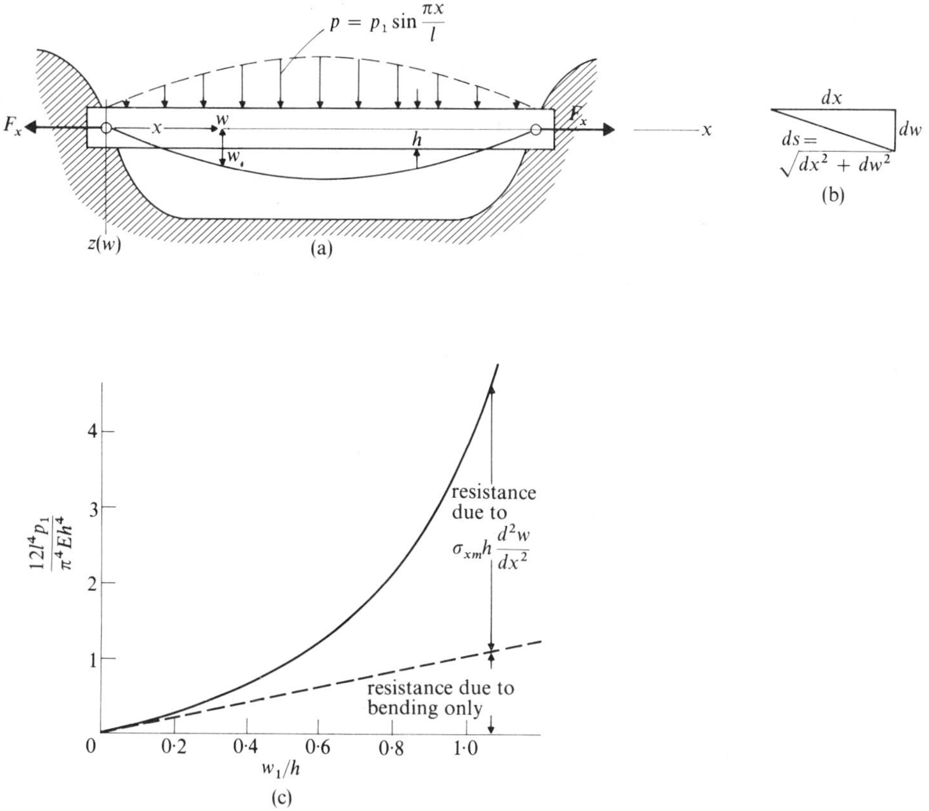

FIGURE 2.11

Substituting w, p, and σ_{xm} into Eq. (2.4), dividing through by $\sin \pi x/l$, and rearranging, we obtain

$$\frac{Eh^3}{12}\frac{\pi^4}{l^4}w_1 = p_1 - \frac{\pi^2 E w_1^2}{4l^2}\frac{\pi^2}{l^2}hw_1, \qquad \frac{12l^4 p_1}{\pi^4 Eh^4} = \frac{w_1}{h} + 3\left(\frac{w_1}{h}\right)^3 \tag{2.36}$$

The first term on the right-hand side of Eq. (2.36) represents the flexural resistance to p_1. With it alone we would have the elementary linear formula for beam deflection. In Fig. 2.11(c) the dotted line shows this elementary relation while the solid line shows the complete relation (2.36); it can be seen that the elementary, linear theory gives a good approximation as long as the deflection is small compared to the thickness, say for $w_1 \leqslant 0{\cdot}1h$ (for $w_1 = 0{\cdot}1h$, the error in omitting the nonlinear term is 3 percent). For larger deflections the part of the load taken by $\sigma_{xm} h\, d^2w/dx^2$, which is represented by the second term on the right-hand side of Eq. (2.36), increases rapidly and must be taken into consideration. For large enough deflections, say for $w_1 \geqslant 3h$, the *flexural* resistance, represented by the first term on the right-hand side of Eq. (2.36), can be ignored in comparison to the second term, that is the beam can be treated as a filament like a cable or rope (for $w_1 = 3{\cdot}3h$ the error in omitting the flexural term is about 3 percent, always assuming, of course, that the material remains elastic). These findings apply qualitatively to large deflections of plates and shells.

2.7 BEAMS UNDER END CONDITIONS OTHER THAN HINGED

Method of Superposition

Small deflection solutions of the hinged beam case, such as those discussed in the last article, can be converted to other end conditions by superposing on them the case shown in Fig. 2.12 of a beam under zero transverse loading on the longitudinal sides and under end moments and transverse shear $x = 0, l$: $M_x = M_1, M_2$, $F_{xz} = F = (M_2 - M_1)/l$. The deflection w and end rotations θ_1, θ_2 for this case are well known:

$$\begin{aligned} w &= [(M_2 - M_1)x(l^2 - x^2) + 3M_1 lx(l - x)]/6lEI \\ \theta_1 &= (2M_1 + M_2)l/6EI, \qquad \theta_2 = (M_1 + 2M_2)l/6EI \end{aligned} \tag{2.37}$$

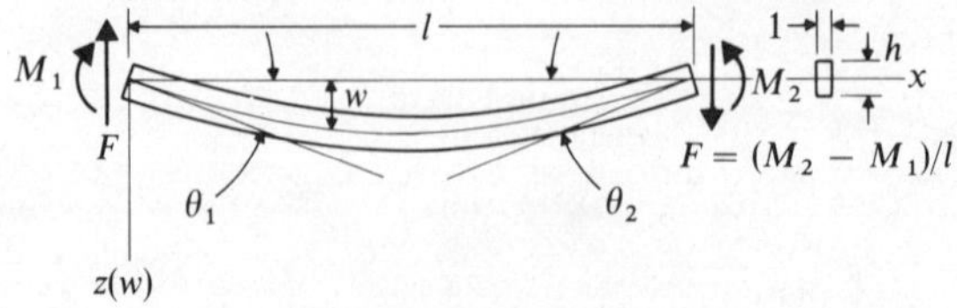

FIGURE 2.12

These expressions can be derived by substituting the w_h part of expression (2.47) given below into the boundary conditions $x = 0$: $w = 0$, $d^2w/dx^2 = -M_1/EI$; $x = l$: $w = 0$, $d^2w/dx^2 = -M_2/EI$, solving for C_0, C_1, C_2, C_3, and using $\theta_1 = (dw/dx)_{x=0}$, $\theta_2 = -(dw/dx)_{x=l}$. The exact solution for this case is presented later in Sec. 3.3 and is found to coincide with this classical solution. For $M_1 = M_2$, $F = 0$; this is called the case of 'pure bending'. For M_1 or M_2 zero it is the solution for a cantilever beam fixed at one end and loaded by a force F at the other end, applied by shear stresses parabolically distributed over the end.

For example, a case of transverse loading of a hinged-end beam can be converted to that of a fixed-end beam under the same loading by superposing the above solution, with M_1 and M_2 chosen to give end rotations equal and opposite to those found for the hinged-end case. Similarly such a transverse-load, hinged-end case can be converted to the case of the same loading on a beam fixed at the left end and free on the right end, by superposing the above solution with $M_2 = 0$ and F equal and opposite to the right-hand reaction found for the hinged end case; the whole beam must then be given a rigid body rotation sufficient to make the net rotation of the left end equal to zero.

As an application, consider the case of a uniform loading $p = p_0$ on a hinged-end beam, Eq. (2.13). The following calculations can be made using Eq. (2.13) in its series form, but for simplicity we will use instead the elementary strength-of-materials closed solution: $w = p_0(x^4 - 2lx^3 + l^3x)/24EI$ to which Eq. (2.13) converges. Then $dw/dx = p_0(4x^3 - 6lx^2 + l^3)/24EI$. To obtain the case of uniform loading on a fixed-end beam we take $M_1 = M_2 = M$, $\theta_1 = \theta_2 = Ml/2EI = -(dw/dx)_{x=0} = (dw/dx)_{(x=l)} = -l^3p_0/24EI$, from which $M = -l^2p_0/12$. Superposing the two cases, we find the total deflection, $w = p_0(x^4 - 2lx^3 + l^3x - l^3x + l^2x^2)/24EI = p_0(x^4 - 2lx^3 + l^2x^2)/24EI$, and the maximum deflection at $x = l/2$, $w_{max} = l^4p_0/384EI$.

Similarly, to obtain the case of uniform loading of a beam fixed at the left end and free at the right end, we take $M_2 = 0$, $F = -M_1/l = p_0l/2$ from which $M_1 = -l^2p_0/2$. To make the slope at the left end zero, we add a rigid body rotation described by $w = Cx$. Superposing the three cases we have

$$\begin{aligned} w &= Cx + p_0(x^4 - 2lx^3 + l^3x - 4l^3x + 6l^2x^2 - 2lx^3)/24EI \\ &= Cx + p_0(x^4 - 4lx^3 + 6l^2x^2 - 3l^3x)/24EI, \\ dw/dx &= C + p_0(4x^3 - 12lx^2 + 12l^2x - 3l^3)/24EI \end{aligned}$$

Taking $C = 3p_0l^3/24EI$ to make the slope zero at $x = 0$, we finally obtain $w = p_0(x^4 - 4lx^3 + 6l^2x^2)/24EI$, giving the maximum deflection at $x = l$, $w_{max} = l^4p_0/8EI$.

This method is probably the simplest way of dealing with various end conditions in beams. However, it is not usually applicable for plates and shells in the form presented above, so we will take up other methods which are useful in the latter cases as well.

'Normal Modes' of Vibration

For end conditions other than hinged it is sometimes more convenient to take the origin of coordinates at the center of the beam, as shown in Fig. 2.13. This simplifies the study of beams having symmetrical end conditions, that is the same conditions at both ends, under symmetrical loading (in which case symmetrical functions such as even power functions, or cosine or hyperbolic cosine functions can be used) or under antisymmetrical loading (in which case antisymmetrical functions such as odd power functions, or sine or hyperbolic sine functions can be used); for other cases it is usually no disadvantage.

Equation (2.17) applies to any uniform beam undergoing lateral acceleration. The general solution of this equation can be found as discussed in Sec. 2.3, and can be written in the following form for application to the beam shown in Fig. 2.13:

$$w = \sum_m w_m (A_m \sin c_m x/a + B_m \sinh c_m x/a + C_m \cos c_m x/a$$
$$+ D_m \cosh c_m x/a) \sin 2\pi N_m t, \qquad m = 1, 2, 3, \ldots \tag{2.38}$$

where c_m can have a succession of distinct values (not necessarily integers) and N_m is the frequency of free vibration in the mode corresponding to one of these values. Substituting expression (2.38) into Eq. (2.17) and equating each coefficient of the series on the left to the corresponding coefficient of the similar series on the right, we obtain the frequency of free vibration in this mode:

$$\frac{Ek^2}{\rho}\frac{c_m^{\ 4}}{a^4} = 4\pi^2 N_m^{\ 2}, \qquad N_m = \frac{c_m^{\ 2}k}{2\pi a^2}\sqrt{\frac{E}{\rho}} \tag{2.38a}$$

We will leave the value of w_m in expression (2.38) arbitrary, corresponding to the arbitrary amplitude of free vibration. This leaves five undetermined constants, $c_m, A_m, \ldots, D_m$. Since Eq. (2.17) is fourth order, only four of these constants can be determined from it. We can then assume a value for one of them so as to give convenient expressions for the rest. For example, if desired we might make the maximum value of the parenthesis equal to unity and thus make w_m the amplitude of the vibration.

We will determine corresponding values of c_m and of three of the coefficients from the four boundary conditions, such as those given in Eqs. (2.6). The value of c_m then determines the frequency N_m from Eq. (2.38a), and, together with the corresponding values of the coefficients, the mode shape given by the parenthesis of Eq. (2.38). The value of w_m remains arbitrary, as explained above.

As an example, consider the free vibration of a beam fixed at both ends. For simplicity consider first the symmetrical modes of vibration. For such vibrations

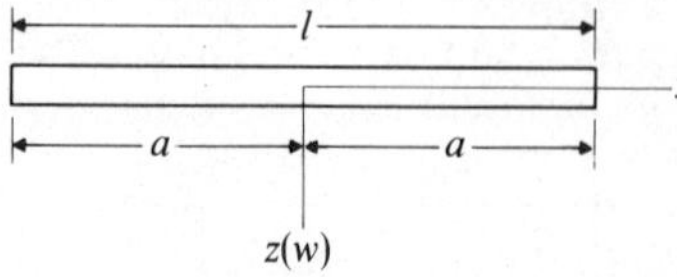

FIGURE 2.13

the coefficients A_m, B_m of the antisymmetrical terms in expression (2.38) can be taken as zero. To satisfy the fixed-end conditions we substitute $x = \pm a$ into the expression (2.38) for w and into the expression for $\partial w / \partial x$ obtained from (2.38), and set the resulting expressions equal to zero. We thus obtain:

$$C_m \cos c_m = -D_m \cosh c_m, \qquad C_m \sin c_m = D_m \sinh c_m \tag{2.39}$$

Assuming that $C_m = 1/\cos c_m$, and solving Eqs. (2.39) for D_m and c_m, we obtain $D_m = -1/\cosh c_m$ and:

$$\tan c_m = -\tanh c_m \tag{2.40}$$

Equation (2.40) can be solved for c_m by such methods as plotting $\tan c_m + \tanh c_m$ against c_m and measuring the values of c_m where the curve intersects the axis; as great accuracy as desired can be obtained by plotting to larger and larger scale around these points of intersection. In this way we find that $c_1, c_2, c_3, \ldots = 2{\cdot}365, 5{\cdot}498, 8{\cdot}639, \ldots (m - \frac{1}{4})\pi$ (in addition to the 'trivial' solution $c_m = 0$ for no motion at all). Each of these values, substituted into Eq. (2.38a) and the above expressions for C_m and D_m and Eq. (2.38), gives the value of the frequency and shape of the corresponding mode.

The antisymmetric modes can be studied in the same manner by taking C_m and D_m equal to zero, and similarly for other end conditions. Table 2.1 shows the

Table 2.1 CONSTANTS FOR MODE AND FREQUENCY OF FREE VIBRATION OF BEAM WITH VARIOUS END CONDITIONS

End condition at $x=-a$	$x=0$	$x=a$	A_m	B_m	C_m	D_m	c_m
fixed (symmetric modes)		fixed	0	0	$\dfrac{1}{\cos c_m}$	$\dfrac{-1}{\cosh c_m}$	$\tan c_m = -\tanh c_m$, $c_m = 2{\cdot}365, 5{\cdot}498, \ldots (m-\frac{1}{4})\pi$
fixed (antisymmetric modes)		fixed	$\dfrac{1}{\sin c_m}$	$\dfrac{-1}{\sinh c_m}$	0	0	$\tan c_m = \tanh c_m$, $c_m = 3{\cdot}927, 7{\cdot}069, \ldots (m+\frac{1}{4})\pi$
	hinged	hinged	1	0	0	0	$\sin c_m = 0$, $c_m = m\pi = \pi, 2\pi, 3\pi, \ldots$
	hinged	fixed	$\dfrac{1}{\sin c_m}$	$\dfrac{-1}{\sinh c_m}$	0	0	$\tan c_m = \tanh c_m$, $c_m = 3{\cdot}927, 7{\cdot}069, \ldots (m+\frac{1}{4})\pi$
	hinged	free	$\dfrac{1}{\sin c_m}$	$\dfrac{1}{\sinh c_m}$	0	0	same as for hinged-fixed
	fixed	fixed	$A = -B = 1/(\sin c_m - \sinh c_m)$ $-C = D = 1/(\cos c_m - \cosh c_m)$				$\cos c_m \cosh c_m = 1$, $c_m = 4{\cdot}730, 7{\cdot}853, \ldots (m+\frac{1}{2})\pi$
	fixed	free	$A = -B = 1/(\sin c_m + \sinh c_m)$ $-C = D = 1/(\cos c_m + \cosh c_m)$				$\cos c_m \cosh c_m = -1$, $c_m = 1{\cdot}875, 4{\cdot}694, 7{\cdot}855, \ldots (m+\frac{1}{2})\pi$
	free	free	$A = B = 1/(\sin c_m - \sinh c_m)$ $C = D = -1/(\cos c_m - \cosh c_m)$				same as for fixed-fixed

values of the coefficients and of c_m obtained in this way for all combinations of hinged, fixed, and free ends.

It can be checked by the use of tables of integrals that the series given by expression (2.38) for any of the cases shown in Table 2.1 are orthogonal for the interval represented by the length of the beam. That is, the integral over this length of the product of any term of the series, represented by the entire quantity in the parenthesis in expression (2.38) by any other term (corresponding to a different value of m) is equal to zero. Equation (2.17) and its solution (2.38) can also be used for problems involving other boundary conditions, such as the elastic constraints discussed in connection with Eq. (2.6); however, the series obtained for such cases are in general not orthogonal.

Transverse Load Problems Using Normal Modes

For obtaining solutions of Eq. (2.4a) for beams under transverse loading with end conditions like those listed in Table 2.1, it is possible to use the normal modes of vibration of a beam with the same end conditions in the same manner as the sine functions used previously in Sec. 2.4 for a beam with both ends hinged. However, the functions for other than hinged conditions are not as simple to work with as sine functions.

As an example, consider the case of a beam fixed at both ends and under a loading symmetrical about the center, taking the ends of the beam at $x = \pm a$, Fig. 2.13. In this case the deflection will be symmetrical and can be described by the symmetric vibration modes for fixed ends, which satisfy the boundary conditions. Using a series made up of these modes in the same way that we used sine series in Sec. 2.4, we take the deflection $w(x)$ and loading $p(x)$ as:

$$w = \sum_m w_m \left(\frac{\cos c_m x/a}{\cos c_m} - \frac{\cosh c_m x/a}{\cosh c_m} \right) \tag{2.41}$$

$$p = \sum_m p_m \left(\frac{\cos c_m x/a}{\cos c_m} - \frac{\cosh c_m x/a}{\cosh c_m} \right) \tag{2.42}$$

where $c_m = 2{\cdot}365,\ 5{\cdot}498, \ldots (m - \tfrac{1}{4})\pi$. Substituting these expressions into Eq. (2.4a) and equating the coefficients of the corresponding terms of the series, which in this case have the same form, we find:

$$\frac{Eh^3}{12} \frac{c_m^4}{a^4} w_m = p_m, \qquad w_m = \frac{12a^4}{Eh^3 c_m^4} p_m$$
$$w = \frac{12a^4}{Eh^3} \sum_m \frac{p_m}{c_m^4} \left(\frac{\cos c_m x/a}{\cos c_m} - \frac{\cosh c_m x/a}{\cosh c_m} \right) \tag{2.43}$$

Since the terms of these series are orthogonal to each other in the interval $x = -a$ to $x = a$, we can determine the coefficients p_m in the same way as in harmonic analysis by multiplying both sides of Eq. (2.42) by $dx[(\cos c_m x/a)/(\cos c_m) - (\cosh c_m x/a)/(\cosh c_m)]$ and integrating over this interval. Using integral tables and trigonometric and hyperbolic transformation formulas, and the relation $\tan c_m = -\tanh c_m$, Eq. (2.40), we find:

$$\int_{-a}^{a} dx\, p(x) \left(\frac{\cos c_m x/a}{\cos c_m} - \frac{\cosh c_m x/a}{\cosh c_m} \right)$$
$$= \int_{-a}^{a} dx \left[\cdots p_m \left(\frac{\cos c_m x/a}{\cos c_m} - \frac{\cosh c_m x/a}{\cosh c_m} \right) \cdots \right] \left(\frac{\cos c_m x/a}{\cos c_m} - \frac{\cosh c_m x/a}{\cosh c_m} \right) \tag{2.44}$$
$$= \left(\frac{1}{\cos^2 c_m} + \frac{1}{\cosh^2 c_m} \right) a p_m = [(1 + \tan^2 c_m) + (1 - \tanh^2 c_m)] a p_m = 2 a p_m$$

Solving for p_m and inserting in Eq. (2.43), we obtain:

$$w = \frac{6a^3}{Eh^3} \sum_m \frac{1}{c_m^4} \left(\frac{\cos c_m x/a}{\cos c_m} - \frac{\cosh c_m x/a}{\cosh c_m} \right) \int_{-a}^{a} dx\, p(x) \left(\frac{\cos c_m x/a}{\cos c_m} - \frac{\cosh c_m x/a}{\cosh c_m} \right) \tag{2.45}$$

Applying Eq. (2.45) to the case of a concentrated load P located at the center, $x = 0$, of the beam, the integral in this equation is $P(1/\cos c_m - 1/\cosh c_m)$ and the maximum deflection at the center $x = 0$ of the beam is, using $I = h^3/12$:

$$w_{max} = \frac{Pa^3}{2EI} \sum_m \frac{1}{c_m^4} \left(\frac{1}{\cos c_m} - \frac{1}{\cosh c_m} \right)^2 = \frac{Pa^3}{2EI} \left(\frac{(-1{\cdot}403 - 0{\cdot}186)^2}{2{\cdot}365^4} \cdots \right) \tag{2.46}$$

Using the one term of this series shown, we obtain $w_{max} = Pa^3/(24{\cdot}8EI)$. The correct value for the number in the denominator is 24, so the value of w_{max} obtained with one term of the series is 3·3 percent too small; with two terms it is 0·5 percent too small, and with three terms 0·2 percent small.

Such normal mode functions can not in general be used in this direct manner for solutions of Eq. (2.4) involving axial loading. This is because, unlike the fourth derivative, the second derivative of a function such as (2.38) or (2.41) does not have the same form as the original function, because the trigonometric terms in the function change sign while the hyperbolic terms do not change sign. However, solutions can be obtained, at the expense of much greater complications, by making use of the orthogonality of the original series to replace each altered function by an infinite series of terms like the original ones. This method will be explained and illustrated in Sec. 4.5 in application to similar but more difficult problems of plates.

Transverse Load Problems Using Particular and Homogeneous Solutions

The method of studying transverse load problems by use of the normal mode functions is of interest as an application of orthogonal series other than the simple trigonometric series, and has some usefulness in plate and shell problems. As was discussed in Sec. 2.3, the more general method is to take the solution of Eq. (2.4a) in the form $w = w_p + w_h$. Here the part w_p of w is a particular solution, that is any function which satisfies Eq. (2.4a) irrespective of boundary conditions. The part w_h is the general solution of the homogeneous equation obtained from Eq. (2.4a) by retaining only the terms involving w, which, in the case when p represents a simple transverse load independent of w, is $(Eh^3/12)d^4w/dx^4 = 0$, or $d^4w/dx^4 = 0$. It is evident that such a combined solution will also satisfy Eq. (2.4a), and the general solution of the fourth-order homogeneous equation will contain just enough arbitrary constants of integration to satisfy the boundary conditions. The combination $w_p + w_h$ is of course the general solution of Eq. (2.4a), that is a solution which contains, not necessarily in a unique form, all possible solutions of this equation. This method is essentially the same as the 'method of superposition' for the case of beams, but the following presentation is more general and applicable to plates and shells, so it is worth going over the same ground again from a different viewpoint.

It is convenient to take the sine series function used before with hinged ends as the particular solution. If the origin of coordinates is taken at the left end, the solution of Eq. (2.4a) can then be taken as:

$$w = \sum_m w_m \sin\frac{m\pi x}{l} + C_0 + C_1x + C_2x^2 + C_3x^3 \tag{2.47}$$

Using Eq. (2.8a) for $p(x)$, Eq. (2.4a) will be satisfied and Eqs. (2.9) to (2.11) will still apply, so that w_m can be found for any loading $p(x)$ from Eqs. (2.9a) and (2.11), and will have the same values as were found before for the hinged-end case. C_0, C_1, C_2, C_3 can then be used to satisfy any boundary conditions.

For the case of the left end fixed and the right end hinged, for example, using expression (2.47) in the boundary conditions $x = 0$: w, $dw/dx = 0$; $x = l$: w, $d^2w/dx^2 = 0$, we obtain:

$$\begin{gathered} C_0 = 0, \qquad C_1l + C_2l^2 + C_3l^3 = 0 \\ (\pi/l)\sum_m mw_m + C_1 = 0, \qquad 2C_2 + 6C_3l = 0 \end{gathered} \tag{2.48}$$

Using values of w_m from Eqs. (2.9a) and (2.11), C_1 can then be found from the third equation of (2.48), and the second and fourth equations can then be solved simultaneously for C_2 and C_3.

For symmetrical end conditions and loading it is simplest to take the origin in the center, Fig. 2.13, as discussed before. Using only the odd values of m in (2.47), substituting $(x + a)$ for x in the sine series and using trigonometric transformations, and using the symmetrical (even-powered) power terms, Eq. (2.47) can then be written:

$$w = \sum_m w_m \sin\frac{m\pi}{2}\cos\frac{m\pi x}{l} + C_0 + C_2x^2 \qquad (m = 1, 3, 5, \ldots) \tag{2.49}$$

The coefficients w_m of the harmonic components are not affected by the shift in origin, and the values of w_m can still be obtained from Eqs. (2.9a) and (2.11).

Lateral Load on Fixed End Beam Using Particular and Homogeneous Solutions

For fixed ends, the conditions $x = \pm a$: w, $dw/dx = 0$ give:

$$C_2 = \frac{\pi}{4a^2} \sum_m m\, w_m, \qquad C_0 = -a^2 C_2 = -\frac{\pi}{4} \sum_m m\, w_m \tag{2.50}$$

Putting these values back into Eq. (2.49) we obtain

$$w = \sum_m w_m \left(\sin \frac{m\pi}{2} \cos \frac{m\pi x}{l} - \frac{m\pi}{4} \frac{a^2 - x^2}{a^2} \right) \tag{2.51}$$

The maximum deflection, which is at the center, $x = 0$, is then:

$$w_{max} = \sum_m w_m \sin \frac{m\pi}{2} + C_0 = \sum_m w_m \left(\sin \frac{m\pi}{2} - \frac{m\pi}{4} \right) \tag{2.52}$$

For the case of a central load P, using the values of w_m given by Eq. (2.15) for $x_0 = l/2$, Eq. (2.52) becomes:

$$\begin{aligned} w_{max} &= \frac{2Pl^3}{\pi^4 EI} \sum_m \frac{\sin \frac{m\pi}{2}}{m^4} \left(\sin \frac{m\pi}{2} - \frac{m\pi}{4} \right) \\ &= \frac{Pl^3}{EI} \frac{2}{\pi^4} \left[\left(1 - \frac{\pi}{4}\right) + \frac{1}{81} \left(1 + \frac{3\pi}{4}\right) + \frac{1}{625} \left(1 - \frac{5\pi}{4}\right) + \cdots \right] \end{aligned} \tag{2.52a}$$

Using one term of the series, we obtain $w_{max} = (Pl^3/EI)/227$; with two terms the denominator is 190·3, with three terms 193·8, and with four terms 191·6, etc., as compared to the correct value of 192.

Several illustrations of series methods of solution have now been given, using familiar cases so that the answers obtained can be compared to known results. In application to practical problems the correct answer will of course usually be unknown, but it can be seen that by comparing the results of using successively larger numbers of terms, a good idea can be obtained of when it is safe to cut the series off. To be sure, there are cases of 'mathematically convergent' series where such conclusions can not safely be drawn by studying the first few terms; however, the series solutions considered here are what might be called 'practically convergent' series, for which such difficulties should not arise.

Axially Loaded Beams

For problems involving axial loading with or without transverse loading, the homogeneous equation obtained from Eq. (2.4) can be written as $d^4w/dx^4 - (12\sigma_{xm}/Eh^2)\, d^2w/dx^2 = 0$. If there is no transverse loading, this is the entire equation and no particular solution is needed. If F_x and hence σ_{xm} is constant, as in the classical stability problem, then this is a linear, homogeneous equation with constant coefficients.

The general solution of this homogeneous equation for the case when F_x and $\sigma_x m$ are constant, found as discussed in Sec. 2.3, can be written as $A + Bx + C \sin(-12\sigma_{xm}/Eh^2)^{\frac{1}{2}}x + D\cos(-12\sigma_{xm}/Eh^2)^{\frac{1}{2}}x$ when σ_{xm} is negative (compression), or $A + Bx + C\sinh(12\sigma_{xm}/Eh^2)^{\frac{1}{2}}x + D\cosh(12\sigma_{xm}/Eh^2)^{\frac{1}{2}}x$ when σ_{xm} is positive (tension). This solution is explored over a wide range in the reference cited in the discussion following Eq. (2.6). The reader is referred to mathematical treatises for a broader coverage of this subject.

2.8 ENERGY METHOD APPLIED TO BEAMS

Elastic Strain Energy in a Deformed Beam

As was discussed in Sec. 1.4, the elastic strain energy stored up during the elastic straining of a body is the sum of the work done in deforming each part of the body. No work is done during a 'rigid body' motion of a body in equilibrium, that is motion of the body as a whole which does not involve any change in shape of the body, because during such a motion the work done by the balanced forces acting upon the body cancels and there is no work done in deforming the body. We can assume without error, then, that a beam undergoes a rigid body motion such that one side of an element remains stationary and no work is done by the forces acting on that side, and all the work is done by the forces on the other sides on their motion relative to the stationary side.

An element of the beam of unit width, height dz, and length dx, Figs. 2.1(a), (b), undergoes a change in length $\epsilon_x\, dx$, and its right end will move in the x direction by this amount relative to the left end. The force acting in the x direction upon the right end is the stress $\sigma_x + d\sigma_x$ times the area dz upon which it acts. During the change in length $\epsilon_x\, dx$ this force changes linearly from zero to its final value $(\sigma_x + d\sigma_x)\, dz$, as indicated by the dashed line in Fig. 2.14. The work done is the area under this line, shown shaded, or $(\sigma_x + d\sigma_x)\, dz\epsilon_x\, dx/2$. The work done by the change in stress $d\sigma_x$ disappears in the limit compared to the work done by the main part of the stress σ_x as dx approaches zero, and can be ignored.

Using expressions (2.1a, b) for ϵ_x and σ_x and integrating over the whole beam, the elastic strain energy in the beam is therefore:

$$\mathscr{E} = \frac{1}{2}\int dx \int_{-h/2}^{h/2} dz\sigma_x\epsilon_x = \frac{E}{2}\int dx \int_{-h/2}^{h/2} dz\left[\epsilon_{xm}^2 - 2\epsilon_{xm} z\frac{d^2w}{dx^2} + z^2\left(\frac{d^2w}{dx^2}\right)^2\right]$$

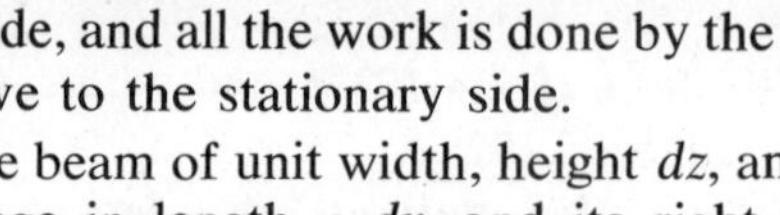

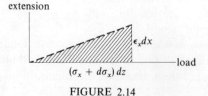

FIGURE 2.14

Since d^2w/dx^2 and ϵ_{xm} do not vary with z, and it was shown in Eq. (2.3b) that $F_x = Eh\epsilon_{xm}$ is practically constant with x, we then obtain:

$$\mathscr{E} = \frac{l}{2Eh} F_x^2 + \frac{Eh^3}{24} \int \left(\frac{d^2w}{dx^2}\right)^2 dx = \frac{l}{2EA} F_x^2 + \frac{EI}{2} \int \left(\frac{d^2w}{dx^2}\right)^2 dx \qquad (2.53)$$

where l, A, I are the length, area, and moment of inertia of the cross section of the beam, and the integration is over the length of the beam.

The first term of this expression, $lF_x^2/2EA$, is the membrane strain energy stored up by the axial force F_x, if there is one, doing work on the elastic change in length of the axis. This would have to be taken into consideration in a case such as that studied in Sec. 2.6, of a beam which is prevented from axial movement at the ends, if we studied it by the energy method instead of by the equilibrium method which we found it convenient to use. In cases of axial load where the ends are free to move axially, as in the buckling case, the work done by the external axial load due to the elastic change in length of the axis just cancels the above axial elastic strain energy in the virtual work equations, and it is therefore customary to omit both of these terms. However, the work done by the external axial load on the decrease in the distance between the ends which is due to the curving or change in curvature of the center line must be considered, as it will be in the case considered later in this section.

There will also be elastic strain energy stored up by the transverse shear force F_{xz} acting upon the shear strain which it causes, as suggested in the picture of a small element shown in Fig. 2.15. In accordance with the Love–Kirchhoff approximation we neglect this shear strain and therefore the strain energy corresponding to it. However this strain energy would have to be included if we wished to study the effect of this approximation by the energy method, as we do later by other methods in Sec. 3.5.

The elastic strain energy of bending for an entire uniform beam under simple transverse loading can therefore be taken as:

$$\mathscr{E} = \frac{EI}{2} \int \left(\frac{d^2w}{dx^2}\right)^2 dx \qquad (2.54)$$

where the integration is carried out over the length of the beam.

Exact (within the limitations of the Love–Kirchhoff approximation) solutions of beam problems can be obtained by expressing the deflection w as an

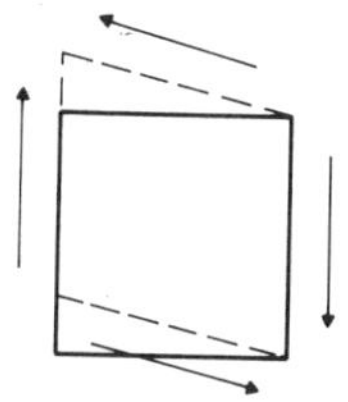

FIGURE 2.15

infinite series with unknown coefficients, each term of which satisfies the end conditions and which can converge to the correct shape, and determining the coefficients by application of the law of virtual work (the 'Rayleigh–Ritz' method). In applying the law of virtual work, as will be demonstrated later, we take as virtual displacements the displacements which would be produced by a small change in one of the unknown coefficients. In this way we can obtain as many equations as there are unknowns.

If the second derivative d^2w/dx^2 of the series used is orthogonal over the interval represented by the length of the beam, then the expression for elastic strain energy (2.54) will contain only terms involving squares of the unknown coefficients, because all cross product terms will drop out in the integration. Since the change in strain energy is the rate at which it varies with the particular coefficient (that is its partial derivative with respect to the coefficient) times the change in the coefficient, each virtual work equation will contain only one unknown which can then be solved for explicitly.

If a non-orthogonal series is involved, the strain energy expression will contain products as well as squares of the unknowns, and the virtual work equations will in general contain all, or at least more than one, of the unknowns, and must be solved simultaneously. This greatly increases the labor involved, and limits the number of terms it is practical to employ. In applying similar methods to plates and shells, especially when the edges are not hinged, or deflections are not small compared to the thickness so that nonlinear theory must be used, the virtual work equations (which often represent the only practical way to obtain any solution at all) may be so difficult to solve that it is only practical, with the time and equipment available, to use one term (the 'Rayleigh' method) or at most a few terms, and it may be difficult to tell how accurate an approximation is being obtained. The closeness of the approximation in this case of course depends upon how closely the function or the few functions chosen represent the correct shape, which may be known only roughly from experiments. Although in the case of beam problems such cases seldom arise or they have other more satisfactory solutions, these questions can be considerably clarified by simple illustrations from beam problems.

The external loads can be considered to remain constant during 'virtual' infinitesimal changes in the displacements, and each will of course do work equal to its product by the displacement at and in the direction of the force. Reactive forces at rigid reactions thus will do no work, and neither will reactive moments at hinged or fixed reactions, since either the moment or the angle through which it acts will be zero. On the other hand, elastic reactive forces or moments will in general do work, because the force or moment will act over some distance or angle of rotation.

We will consider first 'exact' series solutions. If the same series are used as were used with the equilibrium method the results obtained by the energy method are usually also the same, but such cases are useful for illustrating the application of the energy method.

Laterally Loaded Beams with Hinged Ends

For the case of a beam with hinged ends, taking the origin at the left end, the end conditions are satisfied if we take w as the sine series Eq. (2.8). Using this in Eq. (2.54) the elastic strain energy is:

$$\mathscr{E} = \frac{\pi^4 EI}{2l^4} \int_0^l dx \left(m^2 w_m \sin \frac{m\pi x}{l} \right)^2 = \frac{\pi^4 EI}{4l^3} \sum_m m^4 w_m^2 \tag{2.55}$$

The displacement $dw_m \sin m\pi x/l$ which would be produced by a small change dw_m in a typical coefficient w_m is a virtual (small, not violating boundary conditions) displacement. During this virtual displacement the change which will occur in the elastic strain energy $\mathscr{E}$ (considered as a function of all the w_m's, as given by Eq. (2.55)) is the rate at which $\mathscr{E}$ changes with w_m times the amount dw_m which w_m changes, or $(\partial \mathscr{E}/\partial w_m)\, dw_m$.

Since the change in displacement is infinitesimal, a lateral load $p(x)$ can be considered to remain constant during this change in displacement (p and w are proportional, but we are now considering only a small *change* in w, as illustrated in Fig. 2.16). Hence the work done by p in a length dx of the beam is $p\, dx\, dw_m \sin m\pi x/l$. (When p and w are proportional, the total work done by p during the whole displacement can be calculated, and the work done during the virtual displacement could be found as the partial derivative of this work with respect to w_m times dw_m; however, this method is not recommended, as it is no easier and involves some pitfalls.)

By the law of virtual work, for the whole beam, then:

$$\frac{\partial \mathscr{E}}{\partial w_m} dw_m = \frac{\pi^4 EI}{2l^3} m^4 w_m\, dw_m = dw_m \int_0^l p(x) \sin \frac{m\pi x}{l} dx \tag{2.56}$$

Note that the summation sign is dropped, because $\partial \mathscr{E}/\partial w_m$ is the partial derivative of $\mathscr{E}$ with respect to one particular coefficient, and is therefore zero for all the other coefficients. Also note that dw_m can be placed outside the integral sign, since it is not a function of x. Dividing Eq. (2.56) by dw_m, solving for w_m and putting its value back into Eq. (2.8) (and using $I = h^3/12$), we obtain Eq. (2.12), which can then be applied as before to any distribution of $p(x)$.

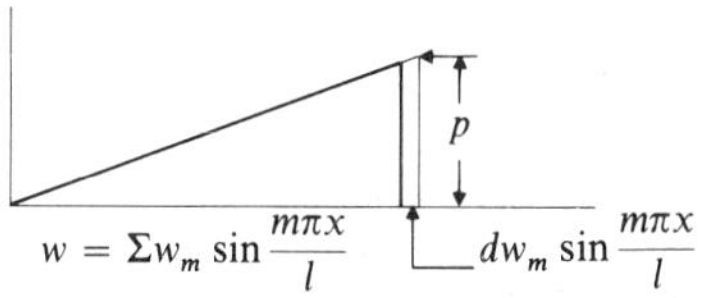

FIGURE 2.16

Free Vibration of Hinged-End Beam

Assume that the deflection w varies with x and with the time t as given by Eq. (2.18), $w = w_m \sin m\pi x/l \sin 2\pi N_m t$. The 'inertia load' is, as in Eq. (217), Sec. 2.4, $p(x, t) = -\rho h \partial^2 w/\partial t^2 = 4\pi^2 N_m^2 \rho h w_m \sin m\pi x/l \sin 2\pi N_m t$. Treating this 'inertia load' the same as an ordinary load, as discussed in Sec. 1.4, the virtual work equation for a small virtual displacement $dw_m \sin m\pi x/l \sin 2\pi N_m t$ will be

$$\begin{aligned}\frac{\partial \mathscr{E}}{\partial w_m}\, dw_m &= \frac{\partial}{\partial w_m}\left[\int_0^l \left(\frac{\partial^2 w}{\partial x^2}\right)^2 dx\right] dw_m \\ &= \int_0^l p(x, t)\, dw_m \sin\frac{m\pi x}{l} \sin 2\pi N_m t\, dx\end{aligned} \tag{2.57}$$

Substituting the values of w and p given above, carrying out the integrations and dividing through by $\sin^2 2\pi N_m t$, we obtain the same result as before, given in Eq. (2.20).

Strut with Hinged Ends

Consider first the idealized case of an initially perfectly straight strut. If we do not know beforehand that the strut will buckle in a simple sine wave shape, we can take w in the general form (2.8), which can represent any possible buckling shape. As in the case discussed in Sec. 2.6, the strut will shorten due to the lateral deflection w by the amount Δl which, referring to Fig. 2.11(b), is:

$$\begin{aligned}\Delta l &= \int_0^l (ds - dx) = \int_0^l [(dx^2 + dw^2)^{1/2} - dx] \\ &= \int_0^l \left\{\left[1 + \left(\frac{dw}{dx}\right)^2\right]^{1/2} - 1\right\} dx \approx \frac{1}{2}\int_0^l \left(\frac{dw}{dx}\right)^2 dx \\ &= \frac{\pi^2}{2l^2}\int_0^l \left(\sum_m m w_m \cos\frac{m\pi x}{l}\right)^2 dx = \frac{\pi^2}{4l}\sum_m m^2 w_m^2\end{aligned} \tag{2.58a}$$

The work done by P during the virtual displacement $dw_m \sin m\pi x/l$ is then $P(\partial \Delta l/\partial w_m)\, dw_m$, and by the law of virtual work,

$$\begin{aligned}&\frac{\partial \mathscr{E}}{\partial w_m}\, dw_m = P\frac{\partial(\Delta l)}{\partial w_m}\, dw_m \\ &\frac{\pi^4 EI}{2l^3} m^4 w_m = \frac{\pi^2 P}{2l} m^2 w_m, \qquad P = \frac{m^2\pi^2 EI}{l^2}\end{aligned} \tag{2.58b}$$

As discussed previously, P can have various distinct values corresponding to values of m of $1, 2, 3, \ldots$, that is theoretically the strut can be in equilibrium in various shapes, each with a different value of P, but only the lowest value of P, for $m = 1$, is of practical importance.

For the strut with an initial crookedness w_0, given say by Eq. (2.21), undergoing a movement w given by (2.22), the bending stresses will be zero when w is zero and will depend on w alone, so that Eqs. (2.54) and (2.55) still apply for the strain energy. The shortening due to the initial curvature (that is the difference between the length of the crooked center line and the straight distance between the ends) is the same as found in Eq. (2.58a) except for substitution of w_0 for w, while the total shortening after the deflection w is the same as in Eq. (2.58a) except for substitution of $(w + w_0)$ for w. The net shortening during the loading is the difference between these, or:

$$\Delta l = \frac{\pi^2}{4l} \sum_m m^2[(w_m + w_{0m})^2 - w_{0m}^2] = \frac{\pi^2}{4l} \sum_m m^2(w_m^2 + 2w_{0m}w_m) \qquad (2.59a)$$

The virtual work equation in this case is then:

$$\frac{\partial \mathscr{E}}{\partial w_m} dw_m = P\frac{\partial(\Delta l)}{\partial w_m} dw_m, \qquad \frac{m^4\pi^4 EI}{2l^3} w_m = \frac{m^2\pi^2}{2l} P(w_m + w_{0m}) \qquad (2.59b)$$

This is the same as relation (2.24), obtained by the equilibrium method.

Energy Method Used to Obtain Approximate Solutions

Enough examples have now been given to illustrate the application of the energy method to different types of loading, and it is apparent that for use with exact series solutions (that is solutions which can converge to the exact solution within the limitations of the classical beam theory) the energy method represents merely an alternative method of setting up the problem. However, as will be seen later, the energy method sometimes gives different results than the equilibrium method for individual terms even when used with the same series, although both methods converge to the same exact result.

When used with only one or a few terms, the energy method often provides a useful approximation in cases where an equilibrium solution would not be practical. It has been argued that use with the energy method of any displacement functions (in the beam case the function $w(x)$) which satisfy the boundary conditions and which 'look right', that is which have the general form which experience or experiments predict, will give a reasonably good approximation; if the functions contain a few unknown numbers which modify the shape and which, when determined by energy principles (using as virtual displacements the displacements produced by a small change in each) can bring the shape closer to the correct shape, the result should be even better.

While there is some truth in this statement, it is very vague in many ways. Some perspective on the degree of approximation which is likely to be achieved can best be obtained by comparing the maximum deflections found by the energy method by using different deflection functions, for a simple case such as that of a fixed-end beam under a load at the center. Table 2.2 shows the functions used for $w(x)$ for this case, together with the corresponding values of the 'deflection coefficient' K in the formula $w_{max} = Pl^3/KEI$, which were obtained by applying the law of virtual work in the same manner as in the examples which have been given above, $l = 2a$ being the length.

The first function $w(x)$ is the exact one, found in elementary strength of materials. It applies only to the part of the beam from the origin to the concentrated load in the center; from symmetry the total strain energy can be obtained as twice that found for this part of the beam. As was discussed in Sec. 2.3 no non-series ('closed') solution can be expected to describe the displacement over the whole beam when there is a discontinuity such as that which is produced by a concentrated load; to be sure, the displacement w and the slope dw/dx must be continuous over the whole beam, but the bending moment and hence the curvature d^2w/dx^2 have a discontinuity at the concentrated load. Since the function chosen was the exact one, the energy solution gives the exact result.

Table 2.2 COMPARISON OF DEFLECTION COEFFICIENTS FOR CENTRALLY LOADED, FIXED-END BEAM, OBTAINED BY ENERGY METHOD WITH DIFFERENT DEFLECTION FUNCTIONS $w(x)$

No.	Origin of coord.	$w(x)$	K one term	K two terms
1	left end	$w_0(3ax^2 - 2x^3)$ (for $x = 0$ to $x = a$ only)	192 (exact)	
2	center	$\sum_m w_m (\cos c_m x/a - A_m \cosh c_m x/a)$ $(c_m = 2{\cdot}365, 5{\cdot}498, \ldots)$ $(A_m = \cos c_m/\cosh c_m = -0{\cdot}1328, 0{\cdot}0058, \ldots)$	198·7	193·5
3	center	$\sum_m w_m [\sin m\pi/2 \cos m\pi x/l - (m\pi/4a^2)(a^2 - x^2)]$ $(m = 1, 3, 5, \ldots)$	201·3	193·5
4	center	$\sum_m w_m (a^2 - x^2)^2 x^m$ $(m = 0, 2, 4, \ldots)$	204·8	195·1
5	left end	$\sum_m w_m (1 - \cos m\pi x/a)$ $(m = 1, 2, 3, \ldots)$	194·8	192·6
6	left end	$\sum_m [\sin m\pi x/l - m/(m+p) \sin (m+p)\pi x/l]$ $(p = 2,$ $m = 1, 3, 5, \ldots)$	274	249

The second and third functions are the same as used before in Eqs. (2.41) and (2.51) respectively with the equilibrium method, and for the second function the results are the same as before. In the case of the third function, however, a comparison of the results shown in the table with those of Eq. (2.52a) shows that the energy method in this case gives different results from the equilibrium method, and converges considerably better. The difference in this case is due to the x^2 term in $w(x)$, which disappears from d^4w/dx^4 in the equilibrium equation but does not disappear from d^2w/dx^2 in the strain energy expression. The better convergence with the energy method represents perhaps a partial confirmation of the optimistic view of the energy method mentioned before.

The fourth function is typical of power functions, which can take various forms and involve simple integrals, although the convergence is not particularly good in this case. The fifth function, composed of one-minus-cosine terms, is perhaps the most commonly used and the most useful for two fixed ends (or two fixed edges in the case of plates and shells), being easy to use and converging well for this case; however, unlike most of the other functions it can be used only for such fixed-end cases. The $m = 2$ term happens to be zero for this particular load (it would not be zero in general) so the two-term solution was obtained for $m = 1, 3$.

The sixth and last function would be very useful if it gave more accurate results, since it is versatile (with $p = 2$ it satisfies the end conditions for two fixed ends, while with $p = 1$ it satisfies them for one end fixed and the other hinged), and it can be 'normalized' or made orthogonal by grouping similar sine terms together (with, however, two unknowns in the coefficients of most terms). The reason that it gives such poor results with fixed boundary conditions is not accidental, and could have been predicted without this evidence. Although it satisfies formally the fixed-end condition of zero deflection and slope, the sine functions of which it is composed all have zero curvature at the ends; hence any combination of them will still have zero curvature there, whereas end fixity always requires a fixing moment to prevent the rotation which would otherwise occur at the end. All the other functions given have some curvature of the right sign at the ends, which will converge to the right curvature when many terms are used. While this type of function is disappointing for this purpose, it has some uses and is important for illustrating some of the pitfalls which await the unwary.

In general, all series solutions, whether used with energy or equilibrium methods, converge better (or give more accurate results when only a few terms are used) for properties like deflections or buckling loads, which depend on some kind of average of conditions over the entire length (or over the entire area of a plate or shell), than they do for local properties such as the bending moment at a particular point, which depends only upon conditions at that point. This is because the integration used with the energy method and the harmonic analysis or its equivalent used with the equilibrium method, is essentially an averaging process. The series used in the sixth function of Table 2.2 could never converge to the correct bending moment at the ends, although it would converge (very slowly) to the correct bending moment a short distance from the ends.

It may be noticed that the approximate deflection coefficients given in Table 2.2 are all larger than the exact value, that is the predicted deflections are all smaller than they should be. The energy method gives an exact solution when the exact deflection shape is assumed, and an inexact shape can be considered to be the exact shape for a case in which additional constraints (for example, suitably distributed elastic reactions) force the body to take the assumed shape. Since constraints of any kind tend to reduce deflections, an approximate solution, satisfying the boundary conditions, obtained by the energy method always indicates the body to be generally stiffer or to have higher buckling loads or vibration frequencies than is actually the case.

3

IMPROVEMENTS ON CLASSICAL BEAM THEORY

3.1 ELASTICITY THEORY, GENERAL

In 'classical' three-dimensional elasticity theory we use Hooke's Law (we will confine ourselves here to isotropic materials), and satisfy the other relations of Table 1.2, Sec. 1.2 (the equations of equilibrium and the geometrical strain–displacement relations) exactly, without approximations such as the Love–Kirchhoff assumption.

However, we assume that relative displacements are small enough to ignore the effect of the change in the geometry of the problem due to them, that is of the change in the shape of the body and in the geometric relationship between the loads. As discussed in Sec. 1.4, in beam, plate, and shell theory only those changes in geometry which are due to flexure in the weak transverse direction are likely to be important, and these are likely to be important only for 'slender' beams and 'thin' plates and shells, for which the Love–Kirchhoff assumption is an excellent approximation so that the more accurate methods of elasticity theory are not needed; such 'finite deformations' involve nonlinear equations and are considered in Secs. 2.6, 5.1, and more completely in Chaps. 6 and 7.

Coordinates, Stress Symbols

We will use 'right handed' rectangular coordinates, which can be interpreted to mean that if the thumb of the right hand is pointed in the direction of the x axis the fingers point in the direction of rotation required to bring the y axis into the z axis (the letters being used in their alphabetical order x, y, z) as shown in Fig. 3.1.

A typical element of the body studied, with edges parallel to the coordinate axes, Fig. 3.1, may have stresses on all six of its sides, exerted by neighboring elements (or in the case of a side which forms part of the external surface of the body, by external loading). We separate the stress on each side into components parallel to the coordinate axes as indicated in the figure. The components normal to the sides are called 'normal' stresses and designated by subscripts x, y, z corresponding to their directions, which are of course the same as the directions of the normal to the sides on which they act, for example σ_x as shown. The tangential components are called 'shear' stresses and are designated by two subscripts, the first of which indicates the direction of the normal to the side on which the stress acts, and the second the direction of the stress, for example σ_{xy} and σ_{xz} as shown.

The normal stresses are taken as positive when they are away from the element, that is tensions. If this positive normal stress on a side is in the direction of the corresponding coordinate axis then we take the positive directions of the two shear stresses on the same side as the directions of the other coordinate axes; if the positive normal stress is opposite in direction to the corresponding coordinate axis then we take the positive directions of the shear stresses also

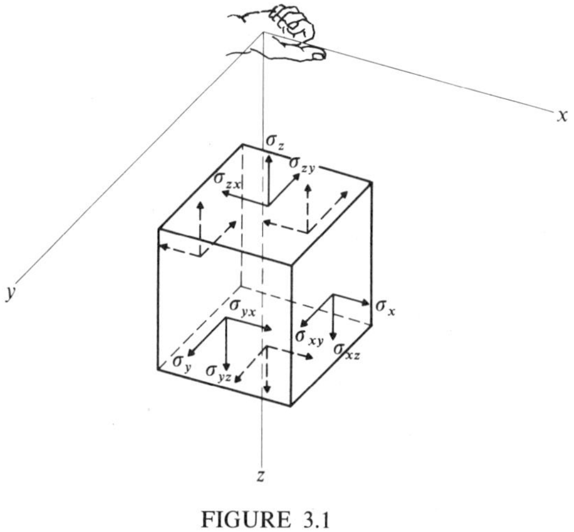

FIGURE 3.1

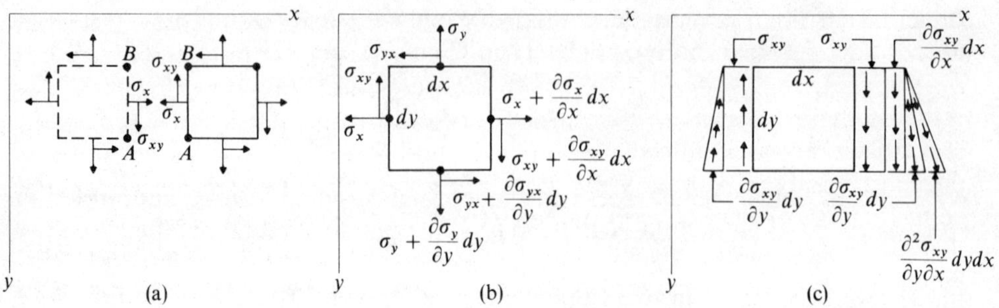

(a) (b) (c)

FIGURE 3.2

opposite in direction to the coordinate axes. Thus if stresses are positive they have the directions shown in Fig. 3.1; if negative they of course would have the opposite directions.

Figure 3.2 shows views of elements as seen when looking in the direction of the z axis. Figure 3.2(a) shows two neighboring elements having a common side AB (they are shown separated so as to provide room for the arrows and symbols shown). It can be seen that if the above sign conventions are used for the stresses on all elements, the stresses acting on sides AB of the neighboring elements will be action and reaction, as they of course must be. The sign conventions are consistent with those used previously, as in Fig. 2.1(b).

Symmetry

If AB is a plane of symmetry of the stresses, then shear stresses on this plane such as σ_{xy} must be zero, because otherwise they cannot be the mirror images which symmetry requires. The stresses will be symmetrical about a plane if the body and its loading are symmetrical about that plane, except possibly in cases of instability or vibration.

Equations of Equilibrium

In the above we have not distinguished between the stress components on opposite sides of the element. However in general the stresses vary from point to point, that is they are functions of the coordinates, as designated by the symbolism $\sigma_x = \sigma_x(x, y, z)$, etc. If σ_x is the normal tensile stress at a point a distance x from the origin, then the normal tensile stress a small distance dx further on will be σ_x plus the rate $\partial\sigma_x/\partial x$ at which σ_x changes with x times the amount dx which x changes, that is $\sigma_x + \partial\sigma_x/\partial x\, dx$, as shown in Fig. 3.2(b).

We can ignore the effects on equilibrium of changes in stresses *along* infinitesimal sides of elements. To show that this is reasonable, Fig. 3.2(c) shows the σ_{xy} stresses on two opposite sides together with the changes which take place in them, both along the sides and from one side of the element to the opposite side.

It can be seen that in the equation of moments the forces due to the σ_{xy} stresses form a couple, with the moment arm dx. The moments due to all the *changes* in σ_{xy}, say about the center of the element, are 'small quantities of higher order', that is they contain one or more factors of dx or dy more than the moment of σ_{xy} does, and therefore (since all the moment arms and areas are of the same order of magnitude and σ_{xy} and its derivatives are in general finite) they tend to zero compared to the moment of σ_{xy} as the dimensions of the element approach zero.

In the force equilibrium equation $\Sigma f_y = 0$, the σ_{xy} stresses cancel each other, but the change $(\partial\sigma_{xy}/\partial x)\,dx$ from one side to the opposite side remains and must be considered. The changes *along* the sides cancel each other and $(\partial^2\sigma_{xy}/\partial y\,\partial x)\,dy\,dx$ is a small quantity of higher order than $(\partial\sigma_{xy}/\partial x)\,dx$. Similar arguments can be cited for the normal stresses. Consequently we only need to consider the stresses proper and their changes from one side of the element to the opposite side, all of which can be considered to be uniformly distributed, and whose resultants can therefore be taken to act at the centers of the sides as suggested in Fig. 3.2(b).

The neglection of small quantities of higher order involves no error, in contrast to 'finite deformation' effects caused by changes in geometry due to deformation, whose neglection generally involves an approximation which may or may not be important in particular cases. In the present case multiplication of the stress by the original rather than the final area involves no error, since we define stress in terms of original area; however, deformation causes changes in directions and moment arms, neglection of which involves some approximation.

From the above discussion it is evident that moment equilibrium about an axis in the z direction requires that:

$$(\sigma_{xy}\,dy\,dz)\,dx - (\sigma_{yx}\,dx\,dz)\,dy = 0 \tag{3.1}$$

where $\sigma_{xy}\,dy\,dz$ is a *force* (the unit stress σ_{xy} times the area $dy\,dz$ of the side on which it acts) and dx is the moment arm of the couple of which this force is a part, etc. From Eq. (3.1) and similar equations for moments about the x and y directions it can be seen that:

$$\sigma_{xy} = \sigma_{yx}, \qquad \sigma_{yz} = \sigma_{zy}, \qquad \sigma_{zx} = \sigma_{xz} \tag{3.2}$$

This means that the order in which the subscripts for shear stresses are written is immaterial, and for simplicity we will hereafter use only the symbols σ_{xy}, σ_{yz}, σ_{xz}. But we will need the above definition of double subscripts for the forces and moments produced by such stresses when integrated over the thickness of a shell, because for that case the order of the subscripts will not in general be immaterial.

Besides the three equations of moment equilibrium of the element, three force equilibrium equations must be satisfied. In these the main part of the stresses cancel each other on opposite sides of the element (σ_x on the left side cancels the σ_x part of the stress on the right side, etc., Fig. 3.2(b)) so in this case the *changes* in stresses from one side of the element to the opposite side such as $\partial\sigma_x/\partial x\,dx$ are the important quantities. The area on which $\partial\sigma_x/\partial x\,dx$ acts is $dy\,dz$, so the net force acting is $(\partial\sigma_x/\partial x\,dx)\,dy\,dz$. Taking similar account of the changes in the two other stresses which act in the x direction, σ_{xy} and σ_{xz} (which is shown in the three-dimensional view, Fig. 3.1), and considering that there may also be a 'body force' B_x per unit volume in this direction (that is, a force which acts throughout the volume of the material, such as a gravitational, magnetic or 'inertia' force), the equilibrium equation for the x direction is:

$$\left(\frac{\partial \sigma_x}{\partial x} dx\right) dy\, dz + \left(\frac{\partial \sigma_{xy}}{\partial y} dy\right) dx\, dz + \left(\frac{\partial \sigma_{xz}}{\partial z} dz\right) dx\, dy + (B_x)\, dx\, dy\, dz = 0 \quad (3.3)$$

Cancelling $dx\, dy\, dz$ and treating the other directions similarly, the three equations of equilibrium of forces for the classical elasticity theory are therefore:

$$\begin{aligned} \frac{\partial \sigma_x}{\partial x} + \frac{\partial \sigma_{xy}}{\partial y} + \frac{\partial \sigma_{xz}}{\partial z} + B_x = 0 \\ \frac{\partial \sigma_y}{\partial y} + \frac{\partial \sigma_{yz}}{\partial z} + \frac{\partial \sigma_{xy}}{\partial x} + B_y = 0 \\ \frac{\partial \sigma_z}{\partial z} + \frac{\partial \sigma_{xz}}{\partial x} + \frac{\partial \sigma_{yz}}{\partial y} + B_z = 0 \end{aligned} \quad (3.4)$$

where B_y, B_z are the body forces per unit volume in the y, z directions.

Hooke's Law

The above equilibrium equations and the strain–displacement relations which are developed later in Eqs. (3.6) are not confined to the elastic case and apply to plastic and other material behavior as well. For elasticity theory we of course use the elastic stress–strain relations, or Hooke's Law. For isotropic materials this has been found from experiments to be:

$$\begin{aligned} &\epsilon_x = \frac{1}{E}(\sigma_x - \nu\sigma_y - \nu\sigma_z), \qquad \epsilon_y = \frac{1}{E}(\sigma_y - \nu\sigma_z - \nu\sigma_x), \\ &\epsilon_z = \frac{1}{E}(\sigma_z - \nu\sigma_x - \nu\sigma_y) \\ &\epsilon_{xy} = \frac{1}{G}\sigma_{xy} = \frac{2(1+\nu)}{E}\sigma_{xy}, \qquad \epsilon_{yz} = \frac{1}{G}\sigma_{yz} = \frac{2(1+\nu)}{E}\sigma_{yz}, \\ &\epsilon_{xz} = \frac{1}{G}\sigma_{xz} = \frac{2(1+\nu)}{E}\sigma_{xz} \end{aligned} \quad (3.5)$$

where the 'normal strains' $\epsilon_x, \epsilon_y, \epsilon_z$, are the changes in length per unit original length in the x, y, z directions, the 'shear strain' ϵ_{xy} is the change in the angle between the sides whose normals were originally in the x and y directions, and similarly for ϵ_{yz} and ϵ_{xz}; as in the case of the shear stresses, the order of the subscripts of the shear strains is immaterial. The positive sense of normal and shear strain is taken the same as that of the corresponding normal and shear stress. E, ν, G are the modulus of elasticity (Young's modulus), Poisson's ratio, and shear modulus respectively.

Poisson's ratio ν, defined as the absolute value of the ratio between the lateral and the longitudinal strains due to a longitudinal normal stress, has values ranging from almost zero for some porous materials to almost one half (the value required for zero volume change) for rubber or for the equivalent concept in plastically flowing materials. For most structural materials used in engineering it has a value close to 0·3; that value is used in this book when formulas involving ν are reduced to approximate numerical values, unless otherwise noted.

Relation between *G*, *E*, and ν

The relation implied in the last three of Eqs. (3.5) can be derived from the geometric relation between the change in the angles of a square element under a shear stress σ_s, Fig. 3.3(a), and the changes in length of its diagonals, which will be

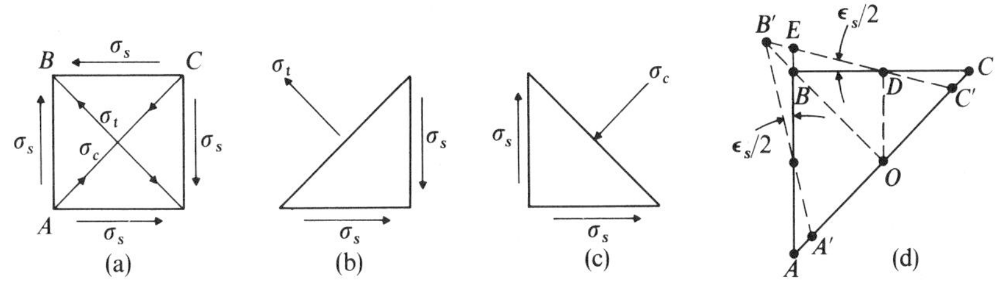

FIGURE 3.3

under tensile and compressive stress σ_t and σ_c respectively. From the equilibrium in the direction of σ_t of the triangular element shown at (b) (remembering to multiply the stresses shown by the areas on which they act, and taking the diagonal dimension and the dimension perpendicular to the paper as unity) it can be seen that $\sigma_t = (\sigma_s \cos 45°) \sin 45° + (\sigma_s \sin 45°) \cos 45° = \sigma_s$. Similarly, using the element shown at (c), $\sigma_c = \sigma_s$.

The part $\overline{ABC}$ of the element shown at (a) will deform to $A'B'C'$, Fig. 3.3(d). The length $\overline{OB}$ will be subject to a unit lengthening σ_t/E due to σ_t as well as a unit lengthening $\nu\sigma_c/E$ due to σ_c, so that its total increase in length will be $\overline{BB'} = \overline{OB}(\sigma_t + \nu\sigma_c)/E = \overline{OB}(1+\nu)\sigma_s/E$. $\overline{AO}$ and $\overline{OC}$ will suffer equal shortenings $\overline{AA'}$ and $\overline{CC'}$. If the strains are small we can take the point *D* to be in the middle of *BC*, and $\overline{BE} = \overline{BB'} \sin 45°$, $\overline{BD} = \overline{OB} \sin 45°$. Using these relations we then find $\epsilon_s/2 = \sigma_s/2G = \overline{BE}/\overline{BD} = \overline{BB'}/\overline{OB} = (1+\nu)\sigma_s/E$, from which:

$$G = \frac{E}{2(1+\nu)} \tag{3.5a}$$

as assumed in Eq. (3.5). For $\nu = 0{\cdot}3$, $G = 0{\cdot}385E$.

Continuity Condition

All the elements of a body must not only be in equilibrium but also the changes in their shape due to the strains in them must be such that the elements still fit together after deformation; otherwise there would be cracks opening up between elements, or overlaps (parts of elements occupying the same space at the same time). This continuity or 'compatibility' condition is imposed by satisfying the geometric relation between the strains and a set of displacements u_x, u_y, u_z in the x, y, z directions which are continuous functions of x, y, z. Solutions of the equilibrium conditions (3.4) alone do not necessarily represent possible stress distributions; to be possible they must also satisfy the continuity conditions represented by Eqs. (3.5) (or whatever stress–strain conditions apply to the material considered) taken together with the following strain–displacement relations.

Strain–Displacement Relations

Figure 3.4(a) shows the sides AB and BC of an element viewed in the direction of the z axis, as in Fig. 3.2. A typical point in the unstressed body has an initial position B, with coordinates x, y, z. When the body is strained the point B is displaced to its final position B'. We designate the components of its displacement in the x, y, z directions by u_x, u_y, u_z respectively (the displacement u_z is at right angles to the paper and can not be shown). Since the displacements are continuous functions of the coordinates, the components of the displacement of point C, which was a distance dx further on in the x direction, will be $u_x + \partial u_x/\partial x\, dx$, $u_y + \partial u_y/\partial x\, dx$, $u_z + \partial u_z/\partial x\, dx$, and similarly for point A, as shown. The effect of these various displacements is clarified if we separate them into two parts, the displacements u_x, u_y, u_z which are common to all the points and are shown at Fig. 3.4(b) (this is a 'rigid body' displacement which leaves the sides with their original lengths dx, dy, dz and with the original right angles between them) and the *changes* in the displacements shown at Fig. 3.4(c) which produce all the strains.

From Fig. 3.4(c) the strain ϵ_x in the x direction is, remembering that C' is also displaced a distance $\partial u_z/\partial x\, dx$ perpendicular to the figure, and using the

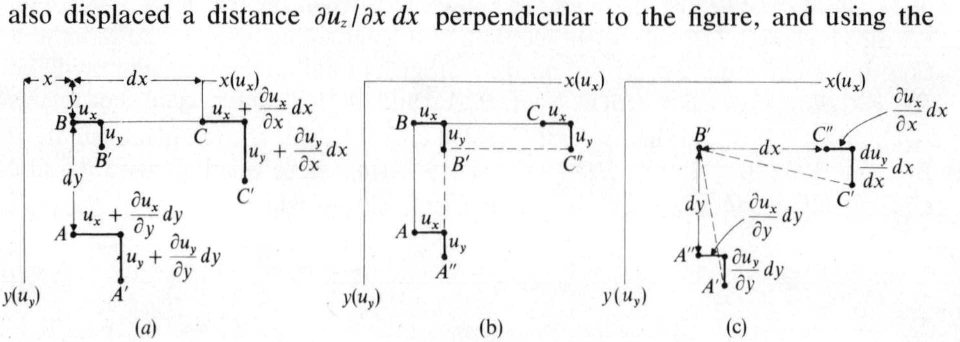

FIGURE 3.4

three-dimensional extension of the theorem of Pythagoras (see Fig. 4.6 and the accompanying text, Sec. 4.2), and the binomial theorem:

$$\epsilon_x = \frac{B'C' - BC}{BC} = \frac{\left[\left(dx + \frac{\partial u_x}{\partial x}\,dx\right)^2 + \left(\frac{\partial u_y}{\partial x}\,dx\right)^2 + \left(\frac{\partial u_z}{\partial x}\,dx\right)^2\right]^{1/2} - dx}{dx}$$

$$= \left[1 + 2\frac{\partial u_x}{\partial x} + \left(\frac{\partial u_x}{\partial x}\right)^2 + \left(\frac{\partial u_y}{\partial x}\right)^2 + \left(\frac{\partial u_z}{\partial x}\right)^2\right]^{1/2} - 1 \qquad (3.5b)$$

$$= \frac{\partial u_x}{\partial x} + \frac{1}{2}\left(\frac{\partial u_x}{\partial x}\right)^2 + \frac{1}{2}\left(\frac{\partial u_y}{\partial x}\right)^2 + \frac{1}{2}\left(\frac{\partial u_z}{\partial x}\right)^2 - \frac{1}{2}\left(\frac{\partial u_x}{\partial x}\right)^2 + \cdots \approx \frac{\partial u_x}{\partial x}$$

where the dots represent still higher powers of the derivatives. The error in this approximation tends to zero for the theoretical infinitesimal strains which are postulated in classical elasticity theory, and is small when we apply this theory to hard engineering materials in the elastic range, since the derivatives of the displacements represent strains or parts of strains, which have maximum values of around 0·001 for mild steel and 0·01 for the strongest steels or aluminum alloys, hard plastics, etc. Making similar approximations, we see that the total change in the angle $ABC = \epsilon_{xy} \approx (\partial u_x/\partial y)\,dy/dy + (\partial u_y/\partial x)\,dx/dx = \partial u_x/\partial y + \partial u_y/\partial x$, and similarly for the other strains. We therefore take the strain–displacement relations as:

$$\epsilon_x = \frac{\partial u_x}{\partial x}, \qquad \epsilon_y = \frac{\partial u_y}{\partial y}, \qquad \epsilon_z = \frac{\partial u_z}{\partial z}$$
$$\epsilon_{xy} = \frac{\partial u_x}{\partial y} + \frac{\partial u_y}{\partial x}, \qquad \epsilon_{yz} = \frac{\partial u_y}{\partial z} + \frac{\partial u_z}{\partial y}, \qquad \epsilon_{xz} = \frac{\partial u_z}{\partial x} + \frac{\partial u_x}{\partial z} \qquad (3.6)$$

Rigid Body Displacements

It can be seen from Eqs. (3.6) that the displacements:

$$u_x = u_{x0} - \theta_z y + \theta_y z, \qquad u_y = u_{y0} - \theta_x z + \theta_z x, \qquad u_z = u_{z0} - \theta_y x + \theta_x y \qquad (3.6a)$$

where u_{x0}, u_{y0}, u_{z0}, θ_x, θ_y, θ_z are constants, produce no strains. These are the most general 'rigid body' displacements, which move the body as a whole through distances u_{x0}, u_{y0}, u_{z0} in the x, y, z directions and rotate it as a whole through angles θ_x, θ_y, θ_z about the x, y, z axes. Such displacements can be added to the strain-producing displacements as required for the satisfaction of boundary conditions, without affecting the strains (and therefore the stresses) in the body. In applying energy methods, rigid body displacements can be ignored in calculating the work done by the loads, since the work done on such displacements by a balanced system of forces cancels.

Simplification of Elasticity Equations

Equations (3.4), (3.5), (3.6) are the basic equations of classical elasticity theory, as summarized in Table 1.2, Sec. 1.2; these, together with the boundary conditions, must be satisfied by the stresses, strains, and displacements in an elastic body. We are usually not interested in the strains as such, whereas both the stresses and the displacements are commonly required in practical problems, for determining the possibility of failure and the deformation of the bodies studied as well as for satisfying the boundary conditions. We can eliminate the strains by equating expressions (3.5) for the strains in terms of the stresses to expressions (3.6) for the strains in terms of the displacements, which gives:

$$\frac{\partial u_x}{\partial x}=\frac{1}{E}(\sigma_x-\nu\sigma_y-\nu\sigma_z),\ \frac{\partial u_y}{\partial y}=\frac{1}{E}(\sigma_y-\nu\sigma_z-\nu\sigma_x),\ \frac{\partial u_z}{\partial z}=\frac{1}{E}(\sigma_z-\nu\sigma_x-\nu\sigma_y)$$
$$\frac{\partial u_x}{\partial y}+\frac{\partial u_y}{\partial x}=\frac{2(1+\nu)}{E}\sigma_{xy},\ \frac{\partial u_y}{\partial z}+\frac{\partial u_z}{\partial y}=\frac{2(1+\nu)}{E}\sigma_{yz},\ \frac{\partial u_z}{\partial x}+\frac{\partial u_x}{\partial z}=\frac{2(1+\nu)}{E}\sigma_{xz} \tag{3.7a}$$

or, solving these equations simultaneously for the stresses:

$$\sigma_x=\frac{E}{1+\nu}\left(\frac{\partial u_x}{\partial x}+\frac{\nu e}{1-2\nu}\right),\ \sigma_y=\frac{E}{1+\nu}\left(\frac{\partial u_y}{\partial y}+\frac{\nu e}{1-2\nu}\right),\ \sigma_z=\frac{E}{1+\nu}\left(\frac{\partial u_z}{\partial z}+\frac{\nu e}{1-2\nu}\right)$$
$$\sigma_{xy}=\frac{E}{2(1+\nu)}\left(\frac{\partial u_x}{\partial y}+\frac{\partial u_y}{\partial x}\right),\ \sigma_{yz}=\frac{E}{2(1+\nu)}\left(\frac{\partial u_y}{\partial z}+\frac{\partial u_z}{\partial y}\right),\ \sigma_{xz}=\frac{E}{2(1+\nu)}\left(\frac{\partial u_z}{\partial x}+\frac{\partial u_x}{\partial z}\right) \tag{3.7b}$$

where $e=\partial u_x/\partial x+\partial u_y/\partial y+\partial u_z/\partial z$; e is thus the sum of the normal strains, which (for small strains) equals the unit increase in volume and is called the 'dilation' (or 'dilatation'). Using expressions (3.7a) for $\partial u_x/\partial x, \partial u_y/\partial y, \partial u_z/\partial z$, we can relate e to the sum of the normal stresses Θ:

$$e=\frac{\partial u_x}{\partial x}+\frac{\partial u_y}{\partial y}+\frac{\partial u_z}{\partial z}=\frac{1-2\nu}{E}(\sigma_x+\sigma_y+\sigma_z)=\frac{1-2\nu}{E}\Theta \tag{3.7c}$$

The basic equations have now been reduced to the equilibrium equations (3.4) and either (3.7a) or (3.7b), which involve only the stresses and the displacements. They can be further simplified by eliminating either the stresses or the displacements. Thus, substituting expressions (3.7b) into Eqs. (3.4), considering for the present the case of no body forces, we eliminate the stresses and obtain the three basic equations of elasticity theory in terms of the displacements u_x, u_y, u_z. These can be written compactly in the form:

$$\nabla^2 u_\alpha+\frac{1}{1-2\nu}\frac{\partial e}{\partial \alpha}=0, \qquad \text{where } \alpha=x, y, z \tag{3.8}$$

In this expression $\nabla^2=\partial^2/\partial x^2+\partial^2/\partial y^2+\partial^2/\partial z^2$ is an 'operator', that is an instruction to perform certain specified operations, in this case to add together the second derivatives with respect to x, y, and z of the function to which the operator is applied, which here is respectively u_x in the first, u_y in the second, and u_z in the third equation. The simpler and more familiar operators $\partial/\partial x, \partial/\partial y, \partial/\partial z$ are respectively applied to e in the first, second, and third equations.

The operator ∇^2 is pronounced 'del square' and is called the 'Laplacian operator' after the great French mathematician. If we apply the operators $\partial/\partial x$, $\partial/\partial y$, $\partial/\partial z$ to the first, second, and third of Eqs. (3.8) respectively, as described below, and add the equations together, we obtain $\nabla^2 e = 0$, and hence from Eq. (3.7c) $\nabla^2\Theta = 0$. Such an equation, which states that the Laplacian of a function is zero, is called 'Laplace's equation', and functions such as e and Θ which satisfy it are called 'harmonic' functions; many examples will be given later.

The three differential equations (3.8) involve three functions u_x, u_y, u_z. If we have three algebraic equations involving three quantities we can in general eliminate two of the quantities and 'solve for' each of the three quantities by well known processes. We can do something similar with differential equations and the functions which they involve. For example the first two of equations (3.8) have only one term each involving u_z, $\partial^2 u_z/\partial z\,\partial x$ in the first equation and $\partial^2 u_z/\partial y\,\partial z$ in the second. We can then apply the operator $\partial/\partial y$ to the first equation, that is take the derivative with respect to y of both the left-hand side and the right-hand side (which of course is zero in this case) and they will still be equal to each other. Similarly applying the operator $\partial/\partial x$ to the second equation, the terms involving u_z both become $\partial^3 u_z/\partial x\,\partial y\,\partial z$, and by subtracting one equation from the other we obtain an equation involving only u_x and u_y. Applying suitable operators to the first or second equation and to the third equation we can similarly eliminate one after the other of the terms involving u_z from the third equation, and thus obtain two equations involving u_x, u_y only. Eliminating the terms involving u_y from these in somewhat similar manner we finally obtain the equation $\nabla^4 u_x = 0$ which involves u_x only, where ∇^4 is defined below. Similarly we can show that $\nabla^4 u_y = 0$, $\nabla^4 u_z = 0$.

We use the symbol ∇^{2m} to represent the Laplacian operator ∇^2 applied m times in succession. In the cases just considered $m = 2$ and $\nabla^{2m} = \nabla^4$ (pronounced 'del fourth') $= (\partial^2/\partial x^2 + \partial^2/\partial y^2 + \partial^2/\partial z^2)(\partial^2/\partial x^2 + \partial^2/\partial y^2 + \partial^2/\partial z^2)$. Functions such as u_x, u_y, u_z which satisfy the equation $\nabla^4 u_\alpha = 0$ are called 'biharmonic functions'. In the present case they must also be related to each other as specified by Eqs. (3.8), but biharmonic functions in general represent a very broad class of functions, including obviously all harmonic functions; examples will also be given later. All this symbolism may seem rather unnecessary but it is worth while because there is much use for it here and in physics generally.

The sum of several harmonic functions (or of several biharmonic functions) is evidently also an harmonic (or biharmonic) function. Similarly the result of applying to a harmonic (or biharmonic) function any operator which gives zero when applied to zero and whose order of application with ∇^2 or ∇^4 is immaterial, gives another harmonic (or biharmonic) function. Thus, since the displacements are biharmonic functions, the stresses σ_x, σ_y, etc., must, from Eqs. (3.7b), also be biharmonic functions.

Biharmonic functions can be related to harmonic functions in the following way. Assume that $\phi = f\psi$, where $\phi(x, y, z)$ is a biharmonic function, $\psi(x, y, z)$ is a harmonic function, and $f(x, y, z)$ is a function to be determined. Substituting $f\psi$

into the equation $\nabla^4\phi = 0$ and making use of $\nabla^2\psi = 0$, we find:

$$\begin{aligned}\nabla^4\phi = \psi\nabla^4 f &+ 4[(\partial\psi/\partial x)\nabla^2(\partial f/\partial x) + (\partial\psi/\partial y)\nabla^2(\partial f/\partial y) + (\partial\psi/\partial z)\nabla^2(\partial f/\partial z)] \\ &+ 4[(\partial^2 f/\partial x^2)(\partial^2\psi/\partial x^2) + (\partial^2 f/\partial y^2)(\partial^2\psi/\partial y^2) + (\partial^2 f/\partial z^2)(\partial^2\psi/\partial z^2)] \\ &+ 8[(\partial^2 f/\partial x\,\partial y)(\partial^2\psi/\partial x\,\partial y) + (\partial^2 f/\partial y\,\partial z)(\partial^2\psi/\partial y\,\partial z) + (\partial^2 f/\partial z\,\partial x) \\ &\times (\partial^2\psi/\partial z\,\partial x)] = 0\end{aligned}$$

Since ψ and its derivatives are not in general zero or related to each other except by the relation $\nabla^2\psi = 0$, the above expression can in general only be zero if $\partial^2 f/\partial x^2 = \partial^2 f/\partial y^2 = \partial^2 f/\partial z^2 =$ a constant (which means that $\nabla^2 f$ is also a constant) and $\partial^2 f/\partial x\,\partial y$, $\partial^2 f/\partial y\,\partial z$, $\partial^2 f/\partial z\,\partial x = 0$. These conditions limit f to a constant or the functions x, y, z or $x^2 + y^2 + z^2$ (or x, y, $x^2 + y^2$ for functions of x, y only). Thus the product of any harmonic function by one of these functions is a biharmonic function. It has been shown that the most general biharmonic function can be represented by the sum of such a product with another independent harmonic function, for example $\phi = x\psi_1 + \psi_2$ where ψ_1 and ψ_2 are independent harmonic functions.

More generally, it can be shown that if ψ satisfies the condition $\nabla^{2m+2}\psi = 0$ then $f\psi$ satisfies the condition $\nabla^{2m+4}(f\psi) = 0$, where f has the values cited above. The case considered above is for $m = 0$.

When the Laplacian operator ∇^2 is applied to a function of only two variables such as x and y, then derivatives with respect to z are of course zero, and the above operators reduce to $\nabla^2 = \partial^2/\partial x^2 + \partial^2/\partial y^2$ and $\nabla^4 = (\partial^2/\partial x^2 + \partial^2/\partial y^2)(\partial^2/\partial x^2 + \partial^2/\partial y^2) = \partial^4/\partial x^4 + 2\partial^4/\partial x^2\,\partial y^2 + \partial^4/\partial y^4$. We will have many applications of such operators.

It should be pointed out here that the artificial 'escalation'* of the order of a differential equation, by application to it of a differential operator, produces an equation which is not identical to the original one in that it has more solutions than the original equation had. The additional solutions have no connection with the physical relation which the original equation may have described. For example, a general solution of Eq. (2.2), $M_x = -EI\,d^2w/dx^2$, contains two arbitrary constants of integration which can be used in satisfying end conditions. If we should apply the operator d/dx to Eq. (2.2) it would become a third order equation whose general solution contains three constants. However the third constant is related only to the mathematical operation and cannot be meaningfully used to satisfy the physical end conditions. Other effects of escalation are discussed in connection with Eqs. (6.36) and (7.3e).

Equations (3.8) apply to the case of zero body forces. If body forces $B_\alpha = B_x, B_y, B_z$ are retained in the equilibrium equations (3.4) then, using Eqs. (3.7b) for the stresses, we find instead of Eqs. (3.8):

$$\nabla^2 u_\alpha + \frac{1}{1-2\nu}\frac{\partial e}{\partial \alpha} = -\frac{2(1+\nu)}{E}B_\alpha, \qquad \alpha = x, y, z \tag{3.8a}$$

*This term and some of its implications were suggested to the author in correspondence with S. B. Batdorf.

For the dynamics case, using d'Alembert's principle, $B_\alpha = -\rho\, \partial^2 u_\alpha / \partial t^2$, where ρ is the mass per unit volume of the material and t is time.

If we wish to simplify the basic equations by eliminating the displacements instead of the stresses, we can apply the operator ∇^2 to Eqs. (3.7a) and operators $\partial/\partial x$, $\partial/\partial y$, $\partial/\partial z$ to Eqs. (3.8), using Eq. (3.7c) to replace e by Θ, after which the resulting relations can easily be combined to eliminate u_x, u_y, u_z, giving the six conditions:

$$\begin{aligned} &(1+\nu)\nabla^2\sigma_\alpha + \partial^2\Theta/\partial\alpha^2 = 0, \qquad \text{where } \alpha = x, y, z \\ &(1+\nu)\nabla^2\sigma_{\alpha\beta} + \partial^2\Theta/\partial\alpha\,\partial\beta = 0, \qquad \text{where } \alpha\beta = xy, yz, zx \end{aligned} \tag{3.8b}$$

However, in the general case this procedure has the disadvantage of resulting in nine conditions, the six continuity equations (3.8b) plus the three equilibrium equations (3.4), relating the six stresses (these nine conditions are presumably not independent, but there seems to be no accepted way to reduce them to six all-sufficient conditions), instead of only three basic equations (3.8). When the displacements are required they can be obtained from the stresses by using Eqs. (3.7a), but this necessitates the relatively onerous task of integrating the first three of these equations and then determining the functions of integration so as to satisfy the last three in the most general way.

It is much simpler when we start the problem by finding the displacements from Eqs. (3.8), because the stresses can then be obtained very easily from the displacements by using Eqs. (3.7b). In the special case of two-dimensional elasticity, a relatively simple single equation (approximate in most applications) which defines the stresses can be developed, and this is traditionally used even when it is necessary to solve for the displacements. However, we will find that when the displacements are needed, as is usually the case, it is simpler to start by finding the displacements even in this case.

Simple Solutions of the Elasticity Equations

Many solutions have been found to Eqs. (3.8). Some simple solutions are rather obvious, such as the rigid body displacements of Eqs. (3.6a).

Another set of simple but important solutions are those which involve only linear variation of the stresses. For instance, it can easily be checked that Eqs. (3.8) and (3.7b) are satisfied by the solutions:

$$u_x = \frac{s_x}{E}x, \qquad u_y = -\frac{\nu s_x}{E}y, \qquad u_z = -\frac{\nu s_x}{E}z, \qquad \sigma_x = s_x \tag{3.8c}$$

$$u_x = \frac{(1+\nu)s_{xy}}{E}y, \qquad u_y = \frac{(1+\nu)s_{xy}}{E}x, \qquad u_z = 0, \qquad \sigma_{xy} = s_{xy} \tag{3.8d}$$

$$u_x = \frac{s_x}{E}zx, \qquad u_y = -\frac{\nu s_x}{E}zy, \qquad u_z = \frac{s_x}{2E}(-x^2 + \nu y^2 - \nu z^2), \qquad \sigma_x = s_x z \tag{3.8e}$$

$$u_x = \frac{(1+\nu)s_{xy}}{E}yz, \qquad u_y = \frac{(1+\nu)s_{xy}}{E}xz, \qquad u_z = -\frac{(1+\nu)s_{xy}}{E}xy, \qquad \sigma_{xy} = s_{xy}z \tag{3.8f}$$

where s_x, s_{xy} are constants and all the stresses not given are zero. Similar solutions can readily be derived for σ_x proportional to y, σ_y constant or proportional to z or x, σ_z constant or proportional to x or y, σ_{yz} constant or proportional to x, and σ_{xz} constant or proportional to y. The solutions given above and the similar solutions obtainable by interchanging x and y are particularly useful in two-dimensional plate problems discussed later (for example in satisfying boundary conditions) as well as in rectangular beam problems as discussed in Sec. 3.3 (see Fig. 3.8).

More General Elasticity Solutions

All the solutions described above are merely special cases of the following solutions of a more general type. The presence of the operator ∇^2 in all of Eqs. (3.8) immediately suggests seeking solutions in terms of harmonic functions, or biharmonic functions, or even higher multiharmonic functions similarly defined.

Considering harmonic functions first, simply taking the displacements as such functions obviously does not provide a solution, due to the presence of the second terms in Eqs. (3.8); hence we are forced to try expressions for the displacements involving sums of products of powers of the coordinates by harmonic functions or derivatives of these functions with respect to the coordinates. Since each term in such a summation must represent a length, dimensional consistency requires that the powers of the coordinates in such terms increase step by step with the order of the derivatives, as for example in the hypothetical expression $\psi + x(\partial\psi/\partial y) - y^2(\partial^2\psi/\partial y\,\partial z)$.

The two simplest expressions of this type involve zero powers of coordinates (that is constants) times first derivatives of a harmonic function, and first powers of the coordinates times the function. The first of these can be expressed in the most general way by:

$$u_\alpha = a_\alpha \frac{\partial\psi}{\partial x} + b_\alpha \frac{\partial\psi}{\partial y} + c_\alpha \frac{\partial\psi}{\partial z}, \qquad \text{where } \alpha = x, y, z, \qquad \nabla^2\psi = 0 \qquad (3.8g)$$

and a_α, b_α, c_α are numerical coefficients to be determined. The dilation is then:

$$e = a_x \frac{\partial^2\psi}{\partial x^2} + b_y \frac{\partial^2\psi}{\partial y^2} + c_z \frac{\partial^2\psi}{\partial z^2} + (b_x + a_y)\frac{\partial^2\psi}{\partial x\,\partial y} + (c_y + b_z)\frac{\partial^2\psi}{\partial y\,\partial z} + (a_z + c_x)\frac{\partial^2\psi}{\partial z\,\partial x} \qquad (3.8h)$$

Substituting these expressions into Eqs. (3.8) we see that, since $\nabla^2(\partial\psi/\partial\alpha) = (\partial/\partial\alpha)\nabla^2\psi = 0$, the elasticity equations are satisfied only if e is zero or constant. Considering that the second derivatives of ψ are in general not zero or constant, we see that there are only four solutions (beside the trivial one in which all coefficients are zero): (1) $b_x = -a_y$, (2) $c_y = -b_z$, (3) $a_z = -c_x$, (4) $a_x = b_y = c_z$ (so that $e = a_x\nabla^2\psi = 0$), the remaining coefficients being zero in each solution. These are shown as solutions 1–4 in Table 3.1 (a_y, b_z, c_x, c_z being taken as unity).

Table 3.1 SOME SOLUTIONS OF EQS. (3.8)

No.	1	2	3	4	5	6	7	8	9	10	11
$u_x =$	$\frac{\partial\psi}{\partial y}$	0	$-\frac{\partial\psi}{\partial z}$	$\frac{\partial\psi}{\partial x}$	$x\frac{\partial\psi}{\partial x} - \gamma\psi$	$y\frac{\partial\psi}{\partial x}$	$z\frac{\partial\psi}{\partial x}$	$y\frac{\partial\psi}{\partial y} + z\frac{\partial\psi}{\partial z} + 2\lambda\psi$	$-x\frac{\partial\psi}{\partial y}$	$-x\frac{\partial\psi}{\partial z}$	$y\frac{\partial\psi}{\partial z} - z\frac{\partial\psi}{\partial y}$
$u_y =$	$-\frac{\partial\psi}{\partial x}$	$\frac{\partial\psi}{\partial z}$	0	$\frac{\partial\psi}{\partial y}$	$x\frac{\partial\psi}{\partial y}$	$y\frac{\partial\psi}{\partial y} - \gamma\psi$	$z\frac{\partial\psi}{\partial y}$	$-y\frac{\partial\psi}{\partial x}$	$z\frac{\partial\psi}{\partial z} + x\frac{\partial\psi}{\partial x} + 2\lambda\psi$	$-y\frac{\partial\psi}{\partial z}$	$z\frac{\partial\psi}{\partial x} - x\frac{\partial\psi}{\partial z}$
$u_z =$	0	$-\frac{\partial\psi}{\partial y}$	$\frac{\partial\psi}{\partial x}$	$\frac{\partial\psi}{\partial z}$	$x\frac{\partial\psi}{\partial z}$	$y\frac{\partial\psi}{\partial z}$	$z\frac{\partial\psi}{\partial z} - \gamma\psi$	$-z\frac{\partial\psi}{\partial x}$	$-z\frac{\partial\psi}{\partial y}$	$x\frac{\partial\psi}{\partial x} + y\frac{\partial\psi}{\partial y} + 2\lambda\psi$	$x\frac{\partial\psi}{\partial y} - y\frac{\partial\psi}{\partial x}$

where $\nabla^2\psi(x, y, z) = 0$, $\gamma = (3 - 4\nu)$, $\lambda = 2(1 - \nu)$

No.	12	13	14	15	16	17	18
$u_x =$	$\frac{\partial^2\phi}{\partial x^2} - \lambda\nabla^2\phi$	$\frac{\partial^2\phi}{\partial x\partial y}$	$\frac{\partial^2\phi}{\partial z\partial x}$	$x\frac{\partial^2\phi}{\partial x\partial y} - y\frac{\partial^2\phi}{\partial x^2} + \lambda y\nabla^2\phi - \gamma\frac{\partial\phi}{\partial y}$	$y\frac{\partial^2\phi}{\partial z\partial x} - z\frac{\partial^2\phi}{\partial x\partial y}$	$z\frac{\partial^2\phi}{\partial x^2} - x\frac{\partial^2\phi}{\partial z\partial x} - \lambda z\nabla^2\phi + \gamma\frac{\partial\phi}{\partial z}$	$\frac{\partial\Phi}{\partial x} + \gamma\frac{\partial\phi}{\partial x} - \lambda x\nabla^2\phi$
$u_y =$	$\frac{\partial^2\phi}{\partial x\partial y}$	$\frac{\partial^2\phi}{\partial y^2} - \lambda\nabla^2\phi$	$\frac{\partial^2\phi}{\partial y\partial z}$	$x\frac{\partial^2\phi}{\partial y^2} - y\frac{\partial^2\phi}{\partial x\partial y} - \lambda x\nabla^2\phi + \gamma\frac{\partial\phi}{\partial x}$	$y\frac{\partial^2\phi}{\partial y\partial z} - z\frac{\partial^2\phi}{\partial y^2} + \lambda z\nabla^2\phi - \gamma\frac{\partial\phi}{\partial z}$	$z\frac{\partial^2\phi}{\partial x\partial y} - x\frac{\partial^2\phi}{\partial y\partial z}$	$\frac{\partial\Phi}{\partial y} + \gamma\frac{\partial\phi}{\partial y} - \lambda y\nabla^2\phi$
$u_z =$	$\frac{\partial^2\phi}{\partial z\partial x}$	$\frac{\partial^2\phi}{\partial y\partial z}$	$\frac{\partial^2\phi}{\partial z^2} - \lambda\nabla^2\phi$	$x\frac{\partial^2\phi}{\partial y\partial z} - y\frac{\partial^2\phi}{\partial z\partial x}$	$y\frac{\partial^2\phi}{\partial z^2} - z\frac{\partial^2\phi}{\partial y\partial z} - \lambda y\nabla^2\phi + \gamma\frac{\partial\phi}{\partial y}$	$z\frac{\partial^2\phi}{\partial z\partial x} - x\frac{\partial^2\phi}{\partial z^2} + \lambda x\nabla^2\phi - \gamma\frac{\partial\phi}{\partial x}$	$\frac{\partial\Phi}{\partial z} + \gamma\frac{\partial\phi}{\partial z} - \lambda z\nabla^2\phi$

where $\nabla^4\phi(x, y, z) = 0$, $\Phi = x\frac{\partial\phi}{\partial x} + y\frac{\partial\phi}{\partial y} + z\frac{\partial\phi}{\partial z}$

The second of the above-mentioned types can be expressed by:

$$u_\alpha = (a_\alpha x + b_\alpha y + c_\alpha z)\psi, \qquad \text{where } \alpha = x, y, z, \qquad \nabla^2\psi = 0 \tag{3.8i}$$

and a_α, b_α, c_α are numerical coefficients to be determined. Substituting these expressions into Eqs. (3.8) and setting up the conditions for satisfying these equations, by separation of variables and use of $\nabla^2\psi = 0$ as was done in the case just considered, we find that no solution of this type is possible.

The most general remaining expressions for the displacements which involve no higher than the first powers of the coordinates and first derivatives of the harmonic function can be written in general form as:

$$u_\alpha = a_\alpha\psi + (b_\alpha x + c_\alpha y + d_\alpha z)\frac{\partial\psi}{\partial x} + (e_\alpha x + f_\alpha y + g_\alpha z)\frac{\partial\psi}{\partial y} + (h_\alpha x + i_\alpha y + j_\alpha z)\frac{\partial\psi}{\partial z},$$
$$\text{where } \alpha = x, y, z, \qquad \nabla^2\psi = 0 \tag{3.8j}$$

and a_α, b_α, etc., are numerical coefficients to be determined. Substituting these expressions into Eqs. (3.8) and again setting up and solving simultaneously the relations which must be satisfied between the coefficients, we find seven more solutions which are shown as solutions 5–11 in Table 3.1.

It is interesting to note that solutions are of two types. One type occurs as a triplet of three related solutions which can be obtained one from another by 'cyclic interchange' of x, y, z (that is by changing x to y, y to z, z to x) in both the subscripts of the displacements and in the expressions for the displacements; solutions 1–3, 5–7, and 8–10 are of this type. The other type is a single solution which remains unchanged by cyclic interchange; solutions 4 and 11 are of this type.

Any solution of Eqs. (3.8) can obviously be multiplied by any constant (that is the expressions for u_x, u_y, u_z can be multiplied by the same constant) and remain a solution. Similarly, solutions can be superposed by adding the expression for u_x of one solution to that of another solution and similarly for u_y, u_z. New forms of solutions can also be obtained by applying the same differential operator to the expression for u_x, u_y, u_z, by changes of variable, by substituting for ψ a function $\Psi(\psi, x, y, z)$ for which $\nabla^2\Psi = 0$ when $\nabla^2\psi = 0$ (for example $\Psi = x\,\partial\psi/\partial x + y\,\partial\psi/\partial y + z\,\partial\psi/\partial z$), etc.

Solutions 1–4 were published by J. Dougall* in 1914, while solutions 4–7 were published by P. F. Papkovich** in 1932 (and a few years later independently by H. Neuber), but some of these had been used by others earlier. These first seven solutions are the most important, the first three because of their simplicity, and the next four, 4–7, because it was proved that together (using independent harmonic functions in each) they represent a complete general solution of Eqs. (3.8), that is they theoretically contain all possible solutions. However, although such a complete solution should theoretically suffice for all purposes, other solutions, particularly such simple ones as solutions 1–3, may be much more convenient for specific purposes.

**Trans. Roy. Soc. of Edinburgh*, vol. 49, p. 895.

***Izv. Akad. Nauk SSSR, Phys.-Math. Ser.*, vol. 10, p. 1425.

A similar survey of all possible solutions involving squares or products of the coordinates and no higher than first derivatives of the harmonic function yields no solutions. A similar survey for no higher than first powers of the coordinates and second derivatives of the harmonic function yields many solutions, including those given by Dougall in addition to his solutions 1–4. However the usefulness of such solutions is questionable, since all of them proved to be obtainable as superpositions of solutions 1–4 with derivatives of solutions 5–11.

Because of the fact that Eqs. (3.8) are second-order differential equations, any solutions in terms of biharmonic functions ϕ (that is solutions which use and depend upon the fourth-order differential equation $\nabla^4\phi = 0$) must involve at least second derivatives of ϕ in the expressions for displacements. The most general expression for such solutions with constant coefficients and second derivatives of ϕ is:

$$u_\alpha = a_\alpha \frac{\partial^2\phi}{\partial x^2} + b_\alpha \frac{\partial^2\phi}{\partial y^2} + c_\alpha \frac{\partial^2\phi}{\partial z^2} + d_\alpha \frac{\partial^2\phi}{\partial x\,\partial y} + e_\alpha \frac{\partial^2\phi}{\partial y\,\partial z} + f_\alpha \frac{\partial^2\phi}{\partial z\,\partial x}, \quad \alpha = x, y, z,$$

$$\nabla^4\phi = 0 \qquad (3.8\text{k})$$

where a_α, b_α, etc., are to be determined. Substituting these expressions into Eqs. (3.8), setting up the required conditions for a_α, b_α, etc., by separating variables and using $\nabla^4\phi = 0$, we obtain the three solutions 12–14, Table 3.1. These were published by the Italian mathematician C. Somigliana in 1898, who proved that together they provide a complete general solution of Eqs. (3.8).

The three biharmonic functions involved in the Somigliana general solution presumably are not completely independent, because, as was mentioned above, each biharmonic function can be expressed in terms of two independent harmonic functions, and the Papkovich solution shows that a general solution need involve only four rather than six independent harmonic functions. This relationship between the Somigliana and the Papkovich general solutions (the only general solutions which have been universally accepted as satisfactorily established) can perhaps be clarified by the following observations.

While a biharmonic function ϕ can be expressed in *terms* of two harmonic functions ψ_1, ψ_2 as, say, $\phi = x\psi_1 + \psi_2$, the entire function $x\psi_1$ is actually also a biharmonic function, that is the original biharmonic function is really expressed as the sum of another biharmonic function $x\psi_1$ and a harmonic function ψ_2. If this is done for each of the three solutions 12, 13, 14, the three biharmonic solutions of the type of $x\psi_1$ (involving three independent harmonic functions) are apparently independent of each other, whereas the harmonic solutions of the type of ψ_2 are probably not independent and involve only one independent harmonic function. Thus the complete solution involves only four independent harmonic functions, the same as for the Papkovich solution.

Questions of independence may be extremely important in searching for a solution to a specific problem, because if independence is disregarded it is common experience that much time can be lost by investigating numerous apparently different solutions which can not be used together to satisfy boundary conditions because they are actually related.

The importance of the four relatively simple solutions 1–4 is indicated by the fact that they can readily be derived from the general solution represented by 12–14. If we use the symbol $\partial(13)/\partial x$ for a solution derived by applying the operator $\partial/\partial x$ to the expressions for u_x, u_y, u_z in solution 13, and so on, it can readily be checked that, replacing $\nabla^2\phi$ by ψ, solution 1 = $[\partial(13)/\partial x - \partial(12)/\partial y]/\lambda$, solution 2 = $[\partial(14)/\partial y - \partial(13)/\partial z]/\lambda$, solution 3 = $[\partial(12)/\partial z - \partial(14)/\partial x]/\lambda$, solution 4 = $[\partial(12)/\partial x + \partial(13)/\partial y + \partial(14)/\partial z]/(1-\lambda)$. This of course does not mean that the combination of solutions 1–4 necessarily represents a general solution (a new solution obtained by applying a derivative such as $\partial/\partial x$ to an old solution is less general than the old solution, because, for example, this eliminates functions of y, z only) but it does suggest that together they represent a very broad class of solutions.

It is of interest to note that, besides this relation between the systems 1–4 and 12–14, solutions in each of the systems can be similarly related to each other. Thus $\partial(1)/\partial z + \partial(2)/\partial x + \partial(3)/\partial y = 0$; $\partial(1)/\partial x - \partial(2)/\partial z = \partial(4)/\partial y$, with two more relations obtainable by cyclical interchange of x, y, z and solutions 1, 2, 3; and $\partial^2(12)/\partial x^2 + \partial^2(13)/\partial x\partial y + \partial^2(14)/\partial z\partial x + (1-2\nu)\nabla^2(12) = 0$, with two more relations obtainable by cyclical interchange of x, y, z and solutions 12, 13, 14.

A further survey, similar to those which have been outlined above, of all solutions involving no higher than first powers of the coordinates and second derivatives of a biharmonic function yields solutions 15–18, Table 3.1. Solutions 12–14 and 15–17 are of the triplet type mentioned above, while 18 is of the single type unaffected by cyclical interchange of x, y, z. Probably an indefinite number of more complex solutions exist but their usefulness is also questionable.

Since elasticity solutions based on polyharmonic functions become more general as the latter's order increases, a similar but partial survey was made of solutions of Eqs. (3.8) in terms of 'triharmonic functions' Ψ satisfying the relation $\nabla^6\Psi = 0$. Because of the increasing complexity of complete surveys of such solutions this study was limited to solutions symmetrical with respect to x and y, that is solutions such as 4 and 14, or a combination of 12 and 13 or of 1, 2, and 3, which remain unchanged when x and y are interchanged in the subscripts and in the derivatives. Such solutions are the most useful in the study of plates having the thickness in the z direction. This study produced the negative but perhaps not unimportant result that all such solutions involving derivatives of Ψ no higher than the fourth or no higher than the fifth (as discussed in the preceding case, they must involve at least fourth derivatives) were found to be resolvable into combinations of simpler known solutions such as 12, 13, and 14 in which ϕ is replaced by $\nabla^2\Psi$.

Solutions of other types than those given in Table 3.1 are also possible and may be convenient for special purposes. For example solutions can be obtained in terms of what might be called 'plu-harmonic' or 'plu-biharmonic' functions ψ, ϕ defined by the relations $\nabla^2\psi = f(x, y, z)$ or $\nabla^4\phi = f(x, y, z)$ respectively. Such functions include harmonic or biharmonic functions when f has the value of zero, plus other functions when f is not zero. For example it can be checked that if $\psi(x, y)$ is a function of x and y satisfying the condition:

$$\nabla^2\psi = (A + B)(x^2 + y^2) + (C + D)xy + E(x + y) + F + G \tag{3.8 l}$$

then the elasticity conditions (3.8) are satisfied by the displacements:

$$u_x = \frac{\partial\psi}{\partial x} - \frac{Ax^3}{3} - \left(\frac{D}{12} + \frac{C}{3(1-\nu)}\right)y^3 - \frac{D}{4}x^2y - Bxy^2 + \frac{\nu}{2(1-\nu)}Cyz^2 - \frac{E}{2}y^2 - Exy - \frac{F}{2}x$$

$$u_z = -\frac{\nu}{1-\nu}(Cxy + G)z \tag{3.8m}$$

the expression for u_y being the same as for u_x with x, y interchanged. Such solutions can be derived by assuming general expressions for the function f and for the displacement with unknown coefficients, and using conditions (3.8) to solve for the coefficients.

More practical examples of this type of solution will be given later, such as Eqs. (3.12a) in the next article. A practical application is given in Eqs. (5.79e) in Chap. 5, where this type of solution permits a simpler derivation in an important case than more conventional solutions.

Application of Solutions to Physical Problems

The distribution of displacements and stresses must satisfy Eqs. (3.8) and (3.7b) and the boundary conditions that the stresses or displacements at the surfaces corresponding to the boundaries of the physical body have the values prescribed for the distributed forces or displacements at these boundaries in the physical problem. Such a distribution can be taken as an 'exact' solution of the physical problem. However, if the distribution involves a 'singular point' within the boundaries of the body (for instance stresses becoming infinite at the center of a disc), or if the body is a 'multiply-connected' body such as a ring, then some solutions obtained for the stresses will not represent the stresses due to the external loads only but will include 'initial stresses' such as could be produced by expanding or contracting the material locally in some way, say at the center of the disc or at some section of the ring. Initial stresses can (and generally do) exist in all bodies, usually due to forming operations as was discussed in Sec. 1.7, but they do not enter into solutions of elasticity theory except in the cases mentioned. It is interesting to note here that if we start the solution by finding single valued expressions for the displacements satisfying Eqs. (3.8) then no difficulty with multiply-connected bodies can arise.

Elasticity Solutions Including Body Forces

Solutions of Eqs. (3.8a) for problems involving body forces can be obtained as the general solution of the homogeneous equations obtained by setting the right-hand sides equal to zero, that is solutions of Eqs. (3.8) which have been discussed above and shown in Table 3.1, plus a particular solution of Eqs. (3.8a).

As a simple example of such a particular solution of Eqs. (3.8a), consider the case of gravitational body force or weight on a body of homogeneous material. For this case B_x, $B_y = 0$, $B_z = \gamma$, where γ is the constant density (weight per unit volume) of the material. A simple solution for this case is:

$$u_x = Azx, \qquad u_y = Byz, \qquad u_z = Cx^2 + Dy^2 + Fz^2, \qquad \text{where:}$$
$$A + B + 2(1-2\nu)(C+D) + 4(1-\nu)F = -2(1+\nu)(1-2\nu)\gamma/E \tag{3.8n}$$

All but one of the coefficients $A, B, \ldots, F$ can be taken as zero, or chosen to help satisfy boundary conditions together with solutions of Eqs. (3.8).

A more complex solution of Eqs. (3.8a):

$$u_x = \frac{(1+\nu)\gamma}{4Ec^2}[-6c^2 + (2-\nu)z^2 - (1-\nu)x^2]zx$$
$$u_y = \frac{(1+\nu)\gamma}{4Ec^2}[-6c^2 + (2-\nu)z^2 - (1-\nu)y^2]yz \tag{3.8o}$$
$$u_z = \frac{(1+\nu)\gamma}{16Ec^2}\left[\frac{4(1+4\nu)c^2}{1-\nu}z^2 - 2(1+\nu)z^4 + (1-\nu)(x^4+y^4) + 6\nu(x^2+y^2)z^2\right]$$

satisfies both B_x, $B_y = 0$, $B_z = \gamma$ and the conditions $z = \pm c$: $\sigma_z, \sigma_{yz}, \sigma_{xz} = 0$, and would be useful for considering the effect of the weight (or uniform vertical acceleration) of a horizontal plate of thickness $h = 2c$. An example of a solution for variable body force will be given later in deriving the first terms of the series of Eqs. (5.32).

Cylindrical Coordinates

Elasticity theory can be derived in a similar manner using other coordinate systems than rectangular. Alternatively, the above relations can be converted to other systems by using the relations between the coordinates and the displacements and the stresses corresponding to the two systems. As an example, Fig. 3.5(a) shows the relation between the 'cylindrical' coordinate system and the rectangular. The z coordinate is the same for both systems, and the x and y coordinates are replaced by radial and angular coordinates r and θ, as in two-dimensional polar coordinates. The small squares in some of the corners indicate the angles which are right angles, since this is not evident in such a perspective view.

From the figure, the coordinates of a typical point P in the two systems are related by:

$$x = r\cos\theta, \qquad y = r\sin\theta, \qquad r = \sqrt{x^2+y^2}, \qquad \theta = \tan^{-1}\frac{y}{x} \tag{3.9a}$$

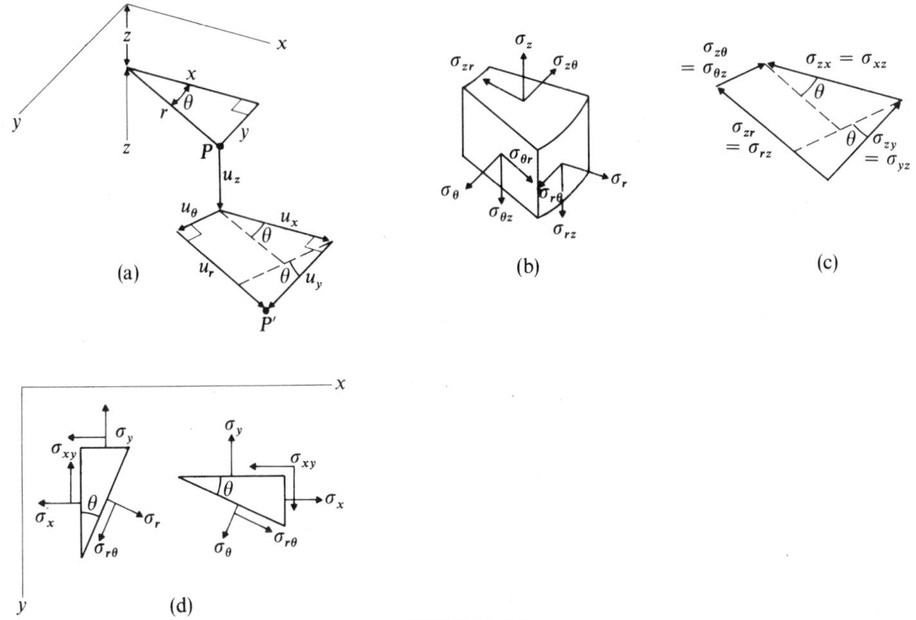

FIGURE 3.5

The displacement of P to its final position P' can be described by either the components u_x, u_y, u_z or u_r, u_θ, u_z. Drawing the dotted lines shown parallel to u_r and u_θ we find, besides the obvious $u_z = u_z$:

$$u_r = u_x \cos\theta + u_y \sin\theta, \qquad u_\theta = u_y \cos\theta - u_x \sin\theta$$

or solving for u_x, u_y: (3.9b)

$$u_x = u_r \cos\theta - u_\theta \sin\theta, \qquad u_y = u_r \sin\theta + u_\theta \cos\theta$$

Figure 3.5(b) shows an infinitesimal element bounded by surfaces normal to the r, θ, z directions, with the normal and shear stresses acting upon them. Comparing this with the element of Fig. 3.1, we see that the σ_z stresses are the same in both coordinate systems. The shear stresses acting on planes normal to the z direction can be described either by σ_{rz}, $\sigma_{\theta z}$ or by σ_{xz}, σ_{yz}; since these act on the same area the resultant of σ_{rz}, $\sigma_{\theta z}$ must equal the resultant of σ_{xz}, σ_{yz} as shown vectorially in Fig. 3.5(c). Hence $\sigma_z = \sigma_z$ and:

$$\sigma_{rz} = \sigma_{xz} \cos\theta + \sigma_{yz} \sin\theta, \qquad \sigma_{\theta z} = \sigma_{yz} \cos\theta - \sigma_{xz} \sin\theta$$

or solving for σ_{xz}, σ_{yz}: (3.9c)

$$\sigma_{xz} = \sigma_{rz} \cos\theta - \sigma_{\theta z} \sin\theta, \qquad \sigma_{yz} = \sigma_{rz} \sin\theta + \sigma_{\theta z} \cos\theta$$

Assuming for convenience that the infinitesimal elements shown in Fig. 3.5(d) have unit thickness in the z direction and an hypotenuse of unit length, their equilibrium in the r and θ directions evidently requires that:

$$\begin{aligned}
\sigma_r &= \sigma_x \cos^2\theta + \sigma_y \sin^2\theta + 2\sigma_{xy} \sin\theta \cos\theta \\
\sigma_\theta &= \sigma_y \cos^2\theta + \sigma_x \sin^2\theta - 2\sigma_{xy} \sin\theta \cos\theta \qquad (3.9\text{d}) \\
\sigma_{r\theta} &= (\sigma_y - \sigma_x) \sin\theta \cos\theta + \sigma_{xy} (\cos^2\theta - \sin^2\theta)
\end{aligned}$$

Solving these equations simultaneously for σ_x, σ_y, σ_{xy} we obtain the same equations with σ_θ and σ_x, σ_r and σ_y, $\sigma_{r\theta}$ and σ_{xy} interchanged. The differences between the shear stresses on the top and bottom surfaces of these elements also enters their equilibrium but are small quantities of higher order compared to the terms shown.

Equations (3.9b, c, d) permit a specific solution for the displacements and stresses in rectangular coordinates to be changed to cylindrical coordinates. To obtain a complete theory in cylindrical coordinates we need the relations between derivatives with respect to the two types of coordinates. Using Eqs. (3.9a) we find:

$$\frac{\partial r}{\partial x} = \frac{x}{r} = \cos\theta, \qquad \frac{\partial r}{\partial y} = \frac{y}{r} = \sin\theta, \qquad \frac{\partial \theta}{\partial x} = \frac{-y}{x^2+y^2} = \frac{-y}{r^2} = \frac{-\sin\theta}{r},$$

$$\frac{\partial \theta}{\partial y} = \frac{x}{x^2+y^2} = \frac{x}{r^2} = \frac{\cos\theta}{r}$$

$$\frac{\partial}{\partial x} = \frac{\partial}{\partial r}\frac{\partial r}{\partial x} + \frac{\partial}{\partial \theta}\frac{\partial \theta}{\partial x} = \cos\theta\frac{\partial}{\partial r} - \frac{\sin\theta}{r}\frac{\partial}{\partial \theta}, \qquad \frac{\partial}{\partial y} = \frac{\partial}{\partial r}\frac{\partial r}{\partial y} + \frac{\partial}{\partial \theta}\frac{\partial \theta}{\partial y}$$

$$= \sin\theta\frac{\partial}{\partial r} + \frac{\cos\theta}{r}\frac{\partial}{\partial \theta}$$

$$\begin{aligned}
\frac{\partial^2}{\partial x^2} &= \frac{\partial}{\partial x}\frac{\partial}{\partial x} = \left(\cos\theta\frac{\partial}{\partial r} - \frac{\sin\theta}{r}\frac{\partial}{\partial \theta}\right)\left(\cos\theta\frac{\partial}{\partial r} - \frac{\sin\theta}{r}\frac{\partial}{\partial \theta}\right) \qquad (3.9\text{e}) \\
&= \cos^2\theta\frac{\partial^2}{\partial r^2} + \sin^2\theta\left(\frac{1}{r}\frac{\partial}{\partial r} + \frac{1}{r^2}\frac{\partial^2}{\partial \theta^2}\right) - 2\sin\theta\cos\theta\left(\frac{1}{r}\frac{\partial^2}{\partial r\partial\theta} - \frac{1}{r^2}\frac{\partial}{\partial\theta}\right) \\
\frac{\partial^2}{\partial y^2} &= \frac{\partial}{\partial y}\frac{\partial}{\partial y} = \sin^2\theta\frac{\partial^2}{\partial r^2} + \cos^2\theta\left(\frac{1}{r}\frac{\partial}{\partial r} + \frac{1}{r^2}\frac{\partial^2}{\partial \theta^2}\right) + 2\sin\theta\cos\theta\left(\frac{1}{r}\frac{\partial^2}{\partial r\partial\theta} - \frac{1}{r^2}\frac{\partial}{\partial\theta}\right) \\
\frac{\partial^2}{\partial x\partial y} &= \frac{\partial}{\partial x}\frac{\partial}{\partial y} = \sin\theta\cos\theta\left(\frac{\partial^2}{\partial r^2} - \frac{1}{r}\frac{\partial}{\partial r} - \frac{1}{r^2}\frac{\partial^2}{\partial \theta^2}\right) + \left(\cos^2\theta - \sin^2\theta\right)\left(\frac{1}{r}\frac{\partial^2}{\partial r\partial\theta} - \frac{1}{r^2}\frac{\partial}{\partial\theta}\right) \\
\nabla^2 &= \frac{\partial^2}{\partial x^2} + \frac{\partial^2}{\partial y^2} + \frac{\partial^2}{\partial z^2} = \frac{\partial^2}{\partial r^2} + \frac{1}{r}\frac{\partial}{\partial r} + \frac{1}{r^2}\frac{\partial^2}{\partial \theta^2} + \frac{\partial^2}{\partial z^2}
\end{aligned}$$

The order of applying several differentiations is of course immaterial, and hence the order of application of successively applied operators which involve only derivatives is also immaterial. In general the order of application of operators some of which contain operations involving the variables other than derivatives (for instance ∇^2 in cylindrical coordinates contains the factors $1/r$, $1/r^2$) does make a difference, and the correct order of application must be strictly observed. Thus $(\partial/\partial r)[(1/r)\partial/\partial r]$ is obviously not the same thing as $[(1/r)\partial/\partial r](\partial/\partial r)$. An exception to this rule is the case when all the operators involved have equivalents in another coordinate system which contain only derivatives; then the order of application must be immaterial even when the operators contain other things. For instance, since $(\partial/\partial x)(\partial/\partial y)$ is the same as $(\partial/\partial y)(\partial/\partial x)$, then the equivalent in cylindrical coordinates $[\cos\theta\,\partial/\partial r-(\sin\theta/r)\partial/\partial\theta]\,[\sin\theta\,\partial/\partial r+(\cos\theta/r)\partial/\partial\theta]$ is the same when the order is reversed (as can easily be checked) although this would not be obvious otherwise.

Using Eqs. (3.9b) and (3.9e) and $\sin^2\theta+\cos^2\theta=1$:

$$
\begin{aligned}
e=\frac{\partial u_x}{\partial x}+\frac{\partial u_y}{\partial y}+\frac{\partial u_z}{\partial z}&=\left(\cos\theta\frac{\partial}{\partial r}-\frac{\sin\theta}{r}\frac{\partial}{\partial\theta}\right)(u_r\cos\theta-u_\theta\sin\theta)\\
&\quad+\left(\sin\theta\frac{\partial}{\partial r}+\frac{\cos\theta}{r}\frac{\partial}{\partial\theta}\right)(u_r\sin\theta+u_\theta\cos\theta)+\frac{\partial u_z}{\partial z}\\
&=\frac{\partial u_r}{\partial r}+\frac{u_r}{r}+\frac{1}{r}\frac{\partial u_\theta}{\partial\theta}+\frac{\partial u_z}{\partial z}
\end{aligned}
\tag{3.9f}
$$

In similar manner the first of Eqs. (3.8) becomes:

$$
\begin{aligned}
&\left(\frac{\partial^2}{\partial r^2}+\frac{1}{r}\frac{\partial}{\partial r}+\frac{\partial^2}{\partial z^2}\right)(u_r\cos\theta-u_\theta\sin\theta)+\frac{1}{r^2}\left[\left(u_\theta-2\frac{\partial u_r}{\partial\theta}-\frac{\partial^2u_\theta}{\partial\theta^2}\right)\sin\theta\right.\\
&\left.-\left(u_r+2\frac{\partial u_\theta}{\partial\theta}-\frac{\partial^2u_r}{\partial\theta^2}\right)\cos\theta\right]+\frac{1}{1-2\nu}\left(\cos\theta\frac{\partial e}{\partial r}-\frac{\sin\theta}{r}\frac{\partial e}{\partial\theta}\right)=0
\end{aligned}
$$

Transforming the second of Eqs. (3.8) in the same manner, multiplying the first of these equations by $\cos\theta$ and the second by $\sin\theta$ and adding the two equations, and then multiplying the first equation by $\sin\theta$ and the second by $\cos\theta$ and subtracting them, we obtain two equations which, with the third of Eqs. (3.8), form the three basic equations of elasticity theory in cylindrical coordinates:

$$
\begin{aligned}
&\nabla^2u_r+\frac{1}{1-2\nu}\frac{\partial e}{\partial r}-\frac{u_r}{r^2}-\frac{2}{r^2}\frac{\partial u_\theta}{\partial\theta}=0\\
&\nabla^2u_\theta+\frac{1}{1-2\nu}\frac{1}{r}\frac{\partial e}{\partial\theta}-\frac{u_\theta}{r^2}+\frac{2}{r^2}\frac{\partial u_r}{\partial\theta}=0,\qquad \nabla^2u_z+\frac{1}{1-2\nu}\frac{\partial e}{\partial z}=0
\end{aligned}
\tag{3.9g}
$$

The relations between the stresses and the displacements, Eqs. (3.7b) are converted in a similar way to:

$$
\begin{aligned}
\sigma_r&=\frac{E}{1+\nu}\left(\frac{\partial u_r}{\partial r}+\frac{\nu e}{1-2\nu}\right), &\sigma_\theta&=\frac{E}{1+\nu}\left(\frac{1}{r}\frac{\partial u_\theta}{\partial\theta}+\frac{u_r}{r}+\frac{\nu e}{1-2\nu}\right)\\
\sigma_z&=\frac{E}{1+\nu}\left(\frac{\partial u_z}{\partial z}+\frac{\nu e}{1-2\nu}\right); &\sigma_{r\theta}&=\frac{E}{2(1+\nu)}\left(\frac{1}{r}\frac{\partial u_r}{\partial\theta}+\frac{\partial u_\theta}{\partial r}-\frac{u_\theta}{r}\right),\\
\sigma_{\theta z}&=\frac{E}{2(1+\nu)}\left(\frac{1}{r}\frac{\partial u_z}{\partial\theta}+\frac{\partial u_\theta}{\partial z}\right), &\sigma_{rz}&=\frac{E}{2(1+\nu)}\left(\frac{\partial u_z}{\partial r}+\frac{\partial u_r}{\partial z}\right)
\end{aligned}
\tag{3.9h}
$$

It can easily be checked that Eq. (3.7c) applies also to cylindrical coordinates.

Axisymmetric Case

If the body and the loading are symmetrical about the z axis, then it is safe to assume that the stresses and strains are also symmetrical about this axis, except possibly in stability and vibration problems which will not be considered here. In this case of axisymmetry the circumferential displacement u_θ can be taken as zero and $\partial u_r/\partial\theta$, $\partial u_z/\partial\theta = 0$. The basic Eqs. (3.9g) and stress–displacement relations Eqs. (3.9h) then reduce to:

$$\nabla^2 u_r + \frac{1}{1-2\nu}\frac{\partial e}{\partial r} - \frac{u_r}{r^2} = 0, \qquad \nabla^2 u_z + \frac{1}{1-2\nu}\frac{\partial e}{\partial z} = 0 \tag{3.10a}$$

$$\text{where } \nabla^2 = \frac{\partial^2}{\partial r^2} + \frac{1}{r}\frac{\partial}{\partial r} + \frac{\partial^2}{\partial z^2}, \qquad e = \frac{\partial u_r}{\partial r} + \frac{u_r}{r} + \frac{\partial u_z}{\partial z}$$

$$\sigma_r = \frac{E}{1+\nu}\left(\frac{\partial u_r}{\partial r} + \frac{\nu e}{1-2\nu}\right), \quad \sigma_\theta = \frac{E}{1+\nu}\left(\frac{u_r}{r} + \frac{\nu e}{1-2\nu}\right), \quad \sigma_z = \frac{E}{1+\nu}\left(\frac{\partial u_z}{\partial z} + \frac{\nu e}{1-2\nu}\right)$$

$$\sigma_{rz} = \frac{E}{2(1+\nu)}\left(\frac{\partial u_z}{\partial r} + \frac{\partial u_r}{\partial z}\right), \qquad \sigma_{r\theta},\ \sigma_{\theta z} = 0 \tag{3.10b}$$

The last equations check the previous conclusion that shear stresses on planes of symmetry (in this case any plane through the z axis) are zero. The physical meaning of the important term u_r/r in the expression for σ_θ is evident; if the radius is increased by u_r the circumference is increased from $2\pi r$ to $2\pi(r+u_r)$, giving a circumferential strain of $[2\pi(r+u_r) - 2\pi r]/2\pi r = u_r/r$.

It should be mentioned that *all* derivatives with respect to θ are not necessarily zero in the axisymmetric case. For instance $\partial u_x/\partial\theta$ obviously need not be zero; actually $u_x = u_r \cos\theta - u_\theta \sin\theta$ in the general case and hence $\partial u_x/\partial\theta = -u_r \sin\theta$ in the present case. This should be remembered in making transformations directly from the rectangular to the axisymmetric case.

As with Eqs. (3.8c, etc.) in rectangular coordinates, simple solutions involving only linear variation of the stresses can be developed in cylindrical coordinates, and are important, for example the axisymmetric solutions:

$$u_r = (1-\nu)\frac{s}{E}r, \qquad u_z = -2\nu\frac{s}{E}z; \qquad \sigma_r = \sigma_\theta = s \tag{3.10c}$$

$$u_r = (1-\nu)\frac{s}{E}rz, \qquad u_z = -\frac{s}{E}\left(\frac{1-\nu}{2}r^2 + \nu z^2\right); \qquad \sigma_r = \sigma_\theta = sz \tag{3.10d}$$

where s is a constant, and all the stresses not shown are zero.

General Elasticity Solutions in Cylindrical Coordinates

Table 3.1(a) shows the correspondingly numbered solutions of Table 3.1 converted to cylindrical coordinates by the methods which have been outlined above. The remainder of the solutions given in Table 3.1 contain $\sin\theta$ or $\cos\theta$ in

Table 3.1a SOME SOLUTIONS OF EQS. (3.9g)

No.	1	4	7	10	11	14	15	18
$u_r =$	$\frac{1}{r}\frac{\partial\psi}{\partial\theta}$	$\frac{\partial\psi}{\partial r}$	$z\frac{\partial\psi}{\partial r}$	$-r\frac{\partial\psi}{\partial z}$	$-\frac{z}{r}\frac{\partial\psi}{\partial\theta}$	$\frac{\partial^2\phi}{\partial z\,\partial r}$	$\frac{\partial^2\phi}{\partial r\,\partial\theta}-\frac{2\lambda}{r}\frac{\partial\phi}{\partial\theta}$	$r\frac{\partial^2\phi}{\partial r^2}+z\frac{\partial^2\phi}{\partial z\,\partial r}$ $+2\lambda\frac{\partial\phi}{\partial r}-\lambda r\nabla^2\phi$
$u_\theta =$	$-\frac{\partial\psi}{\partial r}$	$\frac{1}{r}\frac{\partial\psi}{\partial\theta}$	$\frac{z}{r}\frac{\partial\psi}{\partial\theta}$	0	$z\frac{\partial\psi}{\partial r}$ $-r\frac{\partial\psi}{\partial z}$	$\frac{1}{r}\frac{\partial^2\phi}{\partial\theta\,\partial z}$	$2\lambda\frac{\partial\phi}{\partial r}+\frac{1}{r}\frac{\partial^2\phi}{\partial\theta^2}$ $-\lambda r\nabla^2\phi$	$\frac{z}{r}\frac{\partial^2\phi}{\partial\theta\,\partial z}+\frac{\partial^2\phi}{\partial r\,\partial\theta}$ $+\frac{\gamma}{r}\frac{\partial\phi}{\partial\theta}$
$u_z =$	0	$\frac{\partial\psi}{\partial z}$	$z\frac{\partial\psi}{\partial z}$ $-\gamma\psi$	$r\frac{\partial\psi}{\partial r}$ $+2\lambda\psi$	$\frac{\partial\psi}{\partial\theta}$	$\frac{\partial^2\phi}{\partial z^2}$ $-\lambda\nabla^2\phi$	$\frac{\partial^2\phi}{\partial\theta\,\partial z}$	$r\frac{\partial^2\phi}{\partial z\,\partial r}+z\frac{\partial^2\phi}{\partial z^2}$ $+2\lambda\frac{\partial\phi}{\partial z}-\lambda z\nabla^2\phi$

where $\nabla^2\psi(r,\theta,z)=0$, $\nabla^4\phi(r,\theta,z)=0$, $\lambda=2(1-\nu)$

the expressions for the displacements and stresses when they are converted to cylindrical coordinates, which makes them less convenient and general than those shown.

General Axisymmetric Elasticity Solutions

For the axisymmetric case the only displacements are u_r and u_z, and $\psi = \psi(r, z)$, $\phi = \phi(r, z)$ are not functions of θ so all the terms in Table 3.1(a) involving $\partial\psi/\partial\theta$ or $\partial\phi/\partial\theta$ become zero. This means that in solutions 1, 11, and 15 u_r, $u_z = 0$ and these solutions can provide no axisymmetric solutions. However, solutions 4, 7, 10, 14, and 18 satisfy conditions (3.10a) for an axisymmetric elasticity solution if the expressions for u_r and u_z are taken as in Table 3.1(a); u_θ is of course zero for this case, while $\nabla^2\psi(r, z) = 0$ and $\nabla^4\phi(r, z) = 0$.

Saint-Venant's Principle

Before considering the next case, a very important generalization regarding the effect of certain types of loads on elastic stress distribution should be mentioned, which is very helpful in judging the degree of error which certain common types of simplification introduce. Saint-Venant's principle, enunciated by the famous nineteenth century engineer and scientist, states that if a system of loads applied to a small region of an elastic body is replaced by a statically equivalent system, similarly applied, the effect on the stresses in the body is local, being negligible at distances from the region large compared to the region's dimensions; it follows that if the load system is self balanced it can be replaced by zero and so can be removed or applied, with only a local effect upon the stress distribution.

A statement which is more rigorous in that it avoids or warns against certain rather exotic classes of exceptions, is:* 'A self balanced system of forces applied to a small region of a homogeneous elastic body is ordinarily resisted principally by the material in the neighborhood of the region, the resulting stresses falling off rapidly away from the region and becoming negligible at distances large compared to the dimensions of the region; it is safe to assume this condition if the shape of the body is not such as to obviously favor the stressing of more distant material. With the same reservations, if a system of forces with unbalanced *force* components (it is not always true for unbalanced moments, as was shown by von Mises**) balanced by forces relatively far away, is replaced by a statically equivalent system, the forces of both systems being of the same order of magnitude and intensity and applied to the same small region of the body,† the

*L. H. Donnell, 'About Saint-Venant's principle', *J. Appl. Mech., Trans. ASME*, vol. 84, 1962, p. 752.

***Bull. Amer. Math. Soc.*, vol. 51, 1945, p. 555.

†If an unbalanced system f is applied to a region of maximum dimension s and balanced by forces at distances large compared to s, the addition of a balanced system F applied to the same region ordinarily does not appreciably change the stresses at distances from the region

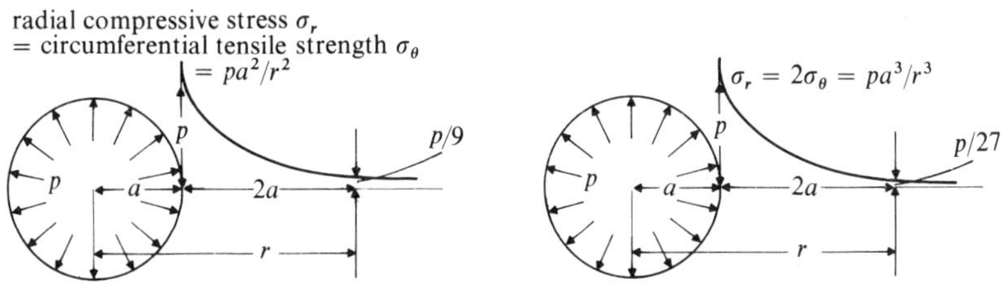

hole in plate
(a)

spherical cavity in solid
(b)

FIGURE 3.6

stresses at distances large compared to the dimensions of the region will not be significantly changed.'

Figure 3.6 shows two illustrations of Saint-Venant's principle, taken from well known elasticity solutions, which exemplify what the principle means quantitatively. Figure 3.6(a), an application of the well known Lamé solution for thick rings under internal pressure (see Eqs. (5.79e), Sec. 5.4), shows a self-balanced system of forces consisting of a uniform pressure p acting on a region consisting of the circumference of a circular hole in an infinite plate. It will be seen that the circumferential and radial stresses (the only stresses produced) vary from p at the edge of the region inversely as the square of the distance from the center, and have already dropped off to one-ninth of the value p at a distance from the region equal to its principal dimension, the diameter. Figure 3.6(b) shows the corresponding three-dimensional case of uniform pressure on the inside of a spherical cavity in an infinite body; in this case the stresses vary inversely as the cube of the distance. These results are typical for two- and for three-dimensional cases. Other illustrations of the principle will be given in Secs. 3.4 and 5.3.

3.2 TWO-DIMENSIONAL ELASTICITY THEORY

With the limitations which have been mentioned, the displacements and stresses which satisfy Eqs. (3.8) and (3.7b) together with the boundary conditions are 'exact' solutions insofar as a linear continuum theory can be exact. Such three-dimensional elasticity solutions are not easy to obtain. In many applications

---

large compared to s. However, even a balanced system produces some stress throughout a finite body, and if no limit is placed on the magnitude of the F forces compared to the f forces, it will always be possible to specify F forces which produce stresses of the same or greater magnitude than those produced by f at any finite distance from the region. The phenomenon studied by Sternberg and Eubanks (*J. Rat. Mech. Anal.*, vol. 4, 1955, p. 135) that different kinds of singular points, equivalent in that they represent the same concentrated force, may produce different stress systems in a finite region, seems to be an example of this class in which the F system becomes very large as the dimensions of the region tend to zero.

the stresses are largely or wholly two-dimensional and a simplified theory can be used in which we satisfy the conditions of equilibrium and continuity in two dimensions only. There are two types of practical problems in which such a two-dimensional condition can be assumed. One is the case of a thin flat sheet with unloaded faces, under edge loads which are in the direction of the plane xy of the middle surface and uniformly distributed over the thickness. The other is the case of a long cylindrical body with loads on its sides which act in the planes of cross sections and uniformly along the length; a thin slice between adjacent cross sections is in a similar condition to the thin sheet of the first case except that there will generally be normal stresses on its faces.

Plane Stress

In the first case of the thin sheet, we assume that the σ_x, σ_y, σ_{xy} stresses (which are of course uniform over the thickness at the edges) are everywhere uniform over the thickness, and that the σ_z, σ_{xz}, σ_{yz} stresses (which are of course zero on the faces) are everywhere zero. This is called the case of 'plane stress'.

Some understanding of the physical conditions which determine how accurate this assumption will be in any particular case can be gained from the following discussion. In general there will be ϵ_z strains in the thickness direction, due principally to Poisson's ratio effects of σ_x and σ_y. If the ϵ_z strains are zero or uniform over the sheet, so that the outer surfaces and other surfaces parallel to the middle surface remain plane, then it is not hard to see that if we satisfy equilibrium and continuity in the x, y directions, then equilibrium and continuity can also be satisfied in the z direction with σ_z, σ_{xz}, σ_{yz} zero and σ_x, σ_y, σ_{xy} uniform across the thickness as we have assumed; we will show later that in such a case this assumption represents an exact three-dimensional solution.

However, in general the ϵ_z strains and hence changes in thickness will not be constant and will vary in some nonlinear fashion along the sheet as suggested in exaggerated manner by the dotted lines in Fig. 3.7(a). It can be seen that an originally plane section such as AB cannot remain plane as shown, since it would then intersect the outer surfaces at angles different from a right angle, and there would be ϵ_{xz}, ϵ_{yz} strains at the faces. Since σ_{xz}, σ_{yz} and hence ϵ_{xz}, ϵ_{yz} are zero at the unloaded faces, cross sections must actually become curved like CD, so that they intersect the faces at right angles. This curvature of the cross sections must vary

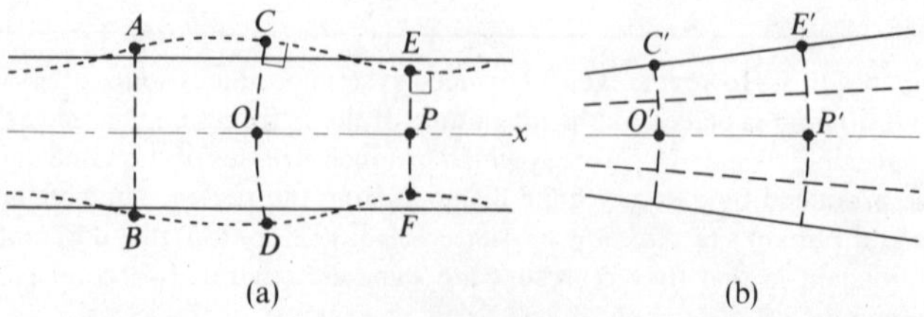

FIGURE 3.7

along the sheet with the rate of change of the thickness, and consequently distances like *OP* and *CE*, which were originally equal, will not in general remain equal, and ϵ_x strains will not be the same at the middle and outer surfaces. There will be variations of ϵ_x, ϵ_y, and ϵ_{xy}, and so of $\sigma_x, \sigma_y, \sigma_{xy}$ over the thickness with possible accompanying $\sigma_z, \sigma_{xz}, \sigma_{yz}$ stresses in the interior of the sheet, all of which are neglected in plane stress theory.

The error in making this neglection will be small if Poisson's ratio of the material is very small (as is the case for some porous materials) so that changes of thickness are small. The errors will be small for other materials if the distances in the plane of the sheet over which σ_x and σ_y undergo changes of the order of 100 percent are large compared to the thickness, since in this case thickness changes will be gradual. In cases of stress concentration around a hole or notch or other discontinuity having dimensions comparable to the thickness the error will be important.

If the change in thickness varies *linearly* with x and y, so that surfaces originally parallel to the middle surface remain plane as indicated in Fig. 3.7(b), then cross sections can become arcs which intersect all the surfaces at right angles, and distances such as $O'P'$ and $C'E'$ can remain equal. Thus it can be shown that a solution is also an exact three-dimensional solution, no matter how large the thickness is, provided that ϵ_z is zero or constant or varies linearly along the sheet, so that outer surfaces remain plane. However, it does not follow that this is the only type of exact solution for plates with unloaded faces—we will see in Sec. 5.4 that other types of exact solutions are also possible.

By Saint-Venant's principle the required uniform distribution of edge forces can be replaced by any distribution which is statically equivalent over each elementary length of the edge, without great effect on the stresses except in an edge zone of width equal to the thickness; in this zone local corrections such as those discussed in Sec. 3.4 can be applied if required.

Plane Strain

By the same reasoning used above, solutions will be exact if there are σ_z stresses on the faces which are such as to make ϵ_z zero or uniform or linear with x and y. In such a case, since the faces of many similarly loaded sheets remain plane, they could be stacked together to form a long cylindrical body with the loads on its sides acting in the planes of cross sections and uniform in the z direction; the sheets would fit together and the σ_z stresses would form action-and-reaction pairs on matching faces, so that only the σ_z forces on the end cross sections would have to be supplied by external loads. If, due to the particular end conditions prevailing, the loads on the end cross sections have a different but statically equivalent distribution, then by Saint-Venant's principle this will not be significant except in end zones having lengths of the order of magnitude of cross-sectional dimensions.

Furthermore, if ϵ_z is not equal to zero it can be made equal to zero by superposing the loading condition given by Eqs. (3.18) in Sec. 3.3, consisting only of linearly varying σ_z stresses applied to each end cross section; this loading produces only the same linear distribution of σ_z stresses on all cross sections, which can afterwards be subtracted from the other σ_z stresses found. Hence there will be no loss in generality if we restrict ourselves to the case when $\epsilon_z = 0$, which is called the case of 'plane strain'.

Stress–Strain Relations in Plane Stress and Plane Strain

For the case of plane stress, taking σ_z, σ_{xz}, σ_{yz} equal to zero in Eqs. (3.5), we find:

$$\epsilon_x = \frac{1}{E}(\sigma_x - \nu\sigma_y), \qquad \epsilon_y = \frac{1}{E}(\sigma_y - \nu\sigma_x), \qquad \epsilon_z = -\frac{\nu}{E}(\sigma_x + \sigma_y),$$
$$\epsilon_{xy} = \frac{2(1+\nu)}{E}\sigma_{xy} \tag{3.11a}$$

or, solving the first two equations simultaneously for σ_x, σ_y:

$$\sigma_x = \frac{E}{1-\nu^2}(\epsilon_x + \nu\epsilon_y), \qquad \sigma_y = \frac{E}{1-\nu^2}(\epsilon_y + \nu\epsilon_x), \qquad \sigma_{xy} = \frac{E}{2(1+\nu)}\epsilon_{xy} \tag{3.11b}$$

The transverse strain ϵ_z is not needed for solving two-dimensional elasticity problems, but it is useful for checking whether a two-dimensional solution happens to be exact; from the discussion above it follows that a plane stress solution is exact if $\sigma_x + \sigma_y$ is a linear function of x and y. The transverse strain has also been used for determining the sum of the two principal stresses experimentally by measuring changes in the thickness, for the purpose of calculating principal stresses in combination with photoelasticity results which give the difference between the principal stresses.

For the case of plane strain, taking ϵ_z equal to zero in Eq. (3.5) and eliminating σ_z from the first two equations by using the third equation, the equations for ϵ_x, ϵ_y, ϵ_{xy} can be put in the following form; ϵ_{xz}, ϵ_{yz}, σ_{xz}, σ_{yz} will be zero for this case while:

$$\sigma_z = \nu(\sigma_x + \sigma_y) \tag{3.11c}$$

$$\epsilon_x = \frac{1-\nu^2}{E}\left(\sigma_x - \frac{\nu}{1-\nu}\sigma_y\right), \qquad \epsilon_y = \frac{1-\nu^2}{E}\left(\sigma_y - \frac{\nu}{1-\nu}\sigma_x\right)$$
$$\epsilon_{xy} = \frac{2[1+\nu/(1-\nu)]}{E/(1-\nu^2)}\sigma_{xy} \tag{3.11d}$$

It will be seen that relations (3.11d) are the same as those in Eqs. (3.11a) for the case of plane stress except for the substitution of the modified elastic constants $E/(1-\nu^2)$ and $\nu/(1-\nu)$ for the usual elastic constants E and ν; for the value of ν of most engineering materials, around 0·3, the modified constants are about 10 and 40 percent higher, respectively.

Exact Two-Dimensional Solutions

It has been intuitively argued that two-dimensional solutions are also exact three-dimensional solutions when ϵ_z varies linearly with x and y, say if $\epsilon_z = \mathbf{a}x + \mathbf{b}y + \mathbf{c}$ where $\mathbf{a}, \mathbf{b}, \mathbf{c}$ are constants. For a solution to be two-dimensional we should have $\sigma_{xz}, \sigma_{zy} = 0$ and $\partial/\partial z(\sigma_x, \sigma_y, \sigma_z, \sigma_{xy}) = 0$, that is the stresses should be functions of x, y only or zero.

These conditions and the exact elasticity conditions (3.8) and (3.7b) can be satisfied in a number of ways. For instance it can be checked by substitution that they are all satisfied if we take:

$$\frac{E}{1+\nu} u_x = -\frac{\partial\psi}{\partial x} + 2\mathbf{b}xy + \frac{\nu}{2}\mathbf{a}z^2, \qquad \frac{E}{1+\nu} u_y = -\frac{\partial\psi}{\partial y} + 2\mathbf{a}xy + \frac{\nu}{2}\mathbf{b}z^2,$$

$\frac{E}{1+\nu} u_z = -\nu(\mathbf{a}x + \mathbf{b}y + \mathbf{c})z$, from which, using Eqs. (3.7b):

$$\sigma_x = \frac{\partial^2\psi}{\partial y^2} - \mathbf{a}x + \mathbf{b}y + \mathbf{c}, \qquad \sigma_y = \frac{\partial^2\psi}{\partial x^2} + \mathbf{a}x - \mathbf{b}y + \mathbf{c},$$

$$\sigma_{xy} = -\frac{\partial^2\psi}{\partial x\,\partial y} + \mathbf{b}x + \mathbf{a}y, \qquad \sigma_z, \sigma_{xz}, \sigma_{yz} = 0 \tag{3.12a}$$

where $$\nabla^2\psi = (1+\nu)(\mathbf{a}x + \mathbf{b}y) - (1-\nu)\mathbf{c}$$

Since $\sigma_z, \sigma_{xz}, \sigma_{yz} = 0$ the solutions described by Eqs. (3.12a) are plane stress solutions. A function such as $\psi = \psi(x, y)$ is analogous to what is traditionally called a 'stress function', although it might as logically be called a 'displacement function'. In any case it is a function which determines the stresses and displacements through relations such Eqs. (3.12a).

From the last of Eqs. (3.12a) ψ can represent a rather wide range of functions, which includes all harmonic functions, plus quite a few more since $\mathbf{a}, \mathbf{b}, \mathbf{c}$ can have any values including zero; however, this range is much more restricted than that of biharmonic functions. Equations (3.12a) thus apply to a considerable range of physical problems, which probably includes all those plane stress problems for which such a simple exact solution is possible. We will later develop approximate solutions for the more numerous plane stress problems for which no such simple exact solution has been found.

For the particular case when $\mathbf{a}, \mathbf{b}, \mathbf{c}$ are zero, solution (3.12c) reduces to:

$$\frac{E}{(1+\nu)} u_x = -\frac{\partial\psi}{\partial x}, \qquad \frac{E}{(1+\nu)} u_y = -\frac{\partial\psi}{\partial y}, \qquad u_z = 0; \qquad \sigma_x = \frac{\partial^2\psi}{\partial y^2}, \qquad \sigma_y = \frac{\partial^2\psi}{\partial x^2} \tag{3.12b}$$

$$\sigma_{xy} = -\frac{\partial^2\psi}{\partial x\,\partial y}, \qquad \sigma_z, \sigma_{xz}, \sigma_{yz}, \epsilon_z = 0, \qquad \text{where} \qquad \nabla^2\psi(x, y) = 0 \tag{3.12c}$$

Since both $\sigma_z, \epsilon_z = 0$, this slightly more restricted solution satisfies the conditions of both plane stress and plane strain, and for this case the stress–strain relations of Eqs. (3.11a) and (3.11d) coincide.

Solutions such as the above are found by trial and error. After promising forms for the displacement expressions are found, the proper numerical coefficients can of course be found by representing them by symbols, substituting the expressions in the required conditions and solving for the coefficients. The quest can also be simplified and systematized in a number of ways. Thus, applying $\partial/\partial z$ to the first four of Eqs. (3.7a) and using the condition $\partial/\partial z(\sigma_x, \sigma_y, \sigma_{xy}, \sigma_z) = 0$, we find that $\partial^2 u_x/\partial x\,\partial z$, $\partial^2 u_y/\partial y\,\partial z$, $\partial^2 u_z/\partial z^2 = 0$, $\partial^2 u_x/\partial y\,\partial z = -\partial^2 u_y/\partial z\,\partial x$. Using the condition $\sigma_{xz}, \sigma_{yz} = 0$ in the last two of Eqs. (3.7a) we find that $\partial u_y/\partial z = -\partial u_z/\partial y$, $\partial u_x/\partial z = -\partial u_z/\partial x$. Applying $\partial/\partial x$ and $\partial/\partial y$ to these last equations and combining them with the previous relations, it can readily be shown that $\partial^2 u_z/\partial x^2$, $\partial^2 u_z/\partial y^2$, $\partial^2 u_z/\partial x\,\partial y$, $\partial^2 u_x/\partial y\,\partial z$, $\partial^2 u_y/\partial z\,\partial x = 0$. Using these findings it is an easy step to show that the first of the basic relations (3.8) can be written in the form:

$$2(1-\nu)\frac{\partial^2 u_x}{\partial x^2} + (1-2\nu)\frac{\partial^2 u_x}{\partial y^2} + \frac{\partial^2 u_y}{\partial x\,\partial y} = -2\nu\mathbf{a} \tag{3.12d}$$

while the second is the same except that x, y and $\mathbf{a}$, $\mathbf{b}$ are interchanged and the third is satisfied identically. The basic conditions are thus simplified to two equations in u_x, u_y.

If we relax the condition that $\sigma_z = 0$, we can obtain a more general solution in the form:

$$\frac{E}{(1+\nu)}u_x = \frac{\partial^3\phi}{\partial y^3} - \nu\nabla^2\frac{\partial\phi}{\partial y} - \frac{\mathbf{a}z^2}{2(1+\nu)}, \qquad \frac{E}{(1+\nu)}u_y = \frac{\partial^3\phi}{\partial x^3} - \nu\nabla^2\frac{\partial\phi}{\partial x} - \frac{\mathbf{b}z^2}{2(1+\nu)},$$

$$Eu_z = (\mathbf{a}x + \mathbf{b}y + \mathbf{c})z, \qquad \text{from which, using Eqs. (3.7b):}$$

$$\sigma_x = \frac{\partial^4\phi}{\partial x\,\partial y^3} + \frac{\nu(\mathbf{a}x + \mathbf{b}y + \mathbf{c})}{(1+\nu)(1-2\nu)}, \qquad \sigma_y = \frac{\partial^4\phi}{\partial x^3\,\partial y} + \frac{\nu(\mathbf{a}x + \mathbf{b}y + \mathbf{c})}{(1+\nu)(1-2\nu)},$$

$$\sigma_{xy} = -\frac{\partial^4\phi}{\partial x^2\,\partial y^2} + \frac{1-\nu}{2}\nabla^4\phi, \qquad \sigma_z = \nu\nabla^2\frac{\partial^2\phi}{\partial x\,\partial y} + \frac{(1-\nu)(\mathbf{a}x + \mathbf{b}y + \mathbf{c})}{(1+\nu)(1-2\nu)}$$

$$\sigma_{xz}, \sigma_{yz} = 0 \tag{3.13a}$$

where: $\nabla^4\phi(x, y) = -\dfrac{2\nu(\mathbf{b}x + \mathbf{a}y)}{(1-\nu^2)(1-2\nu)} + \mathbf{c}'$,

and where $\mathbf{c}'$ is an independent constant quantity. From this we can derive the plane strain case $\epsilon_z = \partial u_z/\partial z = 0$ by taking $\mathbf{a}, \mathbf{b}, \mathbf{c} = 0$. For the plain strain case therefore:

$$\frac{E}{(1+\nu)}u_x = \frac{\partial^3\phi}{\partial y^3} - \nu\nabla^2\frac{\partial\phi}{\partial y}, \qquad \frac{E}{(1+\nu)}u_y = \frac{\partial^3\phi}{\partial x^3} - \nu\nabla^2\frac{\partial\phi}{\partial x}; \qquad u_z, \epsilon_z = 0$$

$$\sigma_x = \frac{\partial^4\phi}{\partial x\,\partial y^3}, \qquad \sigma_y = \frac{\partial^4\phi}{\partial x^3\,\partial y}, \qquad \sigma_{xy} = -\frac{\partial^4\phi}{\partial x^2\,\partial y^2} + \frac{1-\nu}{2}\mathbf{c}', \tag{3.13b}$$

$$\sigma_z = \nu\nabla^2\frac{\partial^2\phi}{\partial x\,\partial y}, \qquad \sigma_{xz}, \sigma_{yz} = 0$$

where: $\nabla^4\phi(x, y) = \mathbf{c}'$.

The function $\phi(x, y)$ in this solution can represent any biharmonic function solution covers all plane strain problems. Furthermore, as has been discussed earlier, it can readily be shown that the more general solution (3.13a) is actually only a superposition of the plane strain solution (3.13b) with simple solutions which we will cover later, principally the case of a cylinder with zero lateral loading and with the same linearly varying normal stress $\sigma_z = \mathbf{a}x + \mathbf{b}y + \mathbf{c}$ acting on all cross sections including the ends, given by Eqs. (3.18). Hence there is no loss in generality in confining ourselves to the simpler plane strain solution (3.13b).

Approximate General Plane Stress Solution

For studying the stresses in rectangular beams which are loaded on the top, bottom or end surfaces but not on the sides or 'faces', we need a plane stress solution, and most cases of interest are not covered by the exact solutions (3.12a, b, c). To obtain an approximate (but more exact than classical beam theory) general plane stress solution, we start by assuming that $\sigma_z, \sigma_{xz}, \sigma_{yz} = 0$. Then for the case of zero body force in the z direction the third of the equilibrium equations (3.4) is satisfied identically, and the first two become:

$$\frac{\partial \sigma_x}{\partial x} + \frac{\partial \sigma_{xy}}{\partial y} + B_x = 0, \qquad \frac{\partial \sigma_y}{\partial y} + \frac{\partial \sigma_{xy}}{\partial x} + B_y = 0 \tag{3.14a}$$

The equilibrium equations (3.14a), the stress–strain relations (3.11b) and the strain–displacement relations (3.6) are the basic equations of plane stress theory. Substituting expressions (3.6) for the strains in terms of the displacements into (3.11b) we obtain:

$$\sigma_x = \frac{E}{1-\nu^2}\left(\frac{\partial u_x}{\partial x} + \nu \frac{\partial u_y}{\partial y}\right), \qquad \sigma_y = \frac{E}{1-\nu^2}\left(\frac{\partial u_y}{\partial y} + \nu \frac{\partial u_x}{\partial x}\right),$$
$$\sigma_{xy} = \frac{E}{2(1+\nu)}\left(\frac{\partial u_x}{\partial y} + \frac{\partial u_y}{\partial x}\right) \tag{3.14b}$$

These relations can also be obtained from Eqs. (3.7b) by setting $\sigma_z = 0$ in the third equation, eliminating $\partial u_z/\partial z$ between this equation and that defining e, which gives $e = [(1-2\nu)/(1-\nu)](\partial u_x/\partial x + \partial u_y/\partial y)$, and using this value of e in the first two equations.

Substituting expressions (3.14b) for the stresses into Eqs. (3.14a), we obtain two basic equations for the two displacements u_x, u_y:

$$2\frac{\partial^2 u_x}{\partial x^2} + (1-\nu)\frac{\partial^2 u_x}{\partial y^2} + (1+\nu)\frac{\partial^2 u_y}{\partial x\, \partial y} = -2(1-\nu^2)B_x/E$$
$$2\frac{\partial^2 u_y}{\partial y^2} + (1-\nu)\frac{\partial^2 u_y}{\partial x^2} + (1+\nu)\frac{\partial^2 u_x}{\partial x\, \partial y} = -2(1-\nu^2)B_y/E \tag{3.14c}$$

These equations can be satisfied in several ways. The following appears to be as simple as any general solution. It can be checked by substitution that for

$B_x, B_y = 0$ Eqs. (3.14c) are satisfied if:

$$Eu_x = \frac{\partial^3\phi}{\partial y^3} - \nu\frac{\partial^3\phi}{\partial x^2\,\partial y}, \qquad Eu_y = \frac{\partial^3\phi}{\partial x^3} - \nu\frac{\partial^3\phi}{\partial x\,\partial y^2} \tag{3.15a}$$

where $\nabla^4\phi(x, y) = \mathbf{c}$ and $\mathbf{c}$ is an arbitrary quantity. Then from Eqs. (3.14b):

$$\sigma_x = \frac{\partial^4\phi}{\partial x\,\partial y^3}, \qquad \sigma_y = \frac{\partial^4\phi}{\partial x^3\,\partial y}, \qquad \sigma_{xy} = -\frac{\partial^4\phi}{\partial x^2\,\partial y^2} + \frac{\mathbf{c}}{2(1+\nu)}, \qquad \sigma_z = 0 \tag{3.15b}$$

If constant body forces B_x, B_y are present we must add $\nu B_y xy - B_x(x^2 + \nu y^2)/2$ and $\nu B_x xy - B_y(y^2 + \nu x^2)$ to expressions (3.15a) for Eu_x, Eu_y, and we must add $-B_x x$ and $-B_y y$ to expressions (3.15b) for σ_x, σ_y respectively. For variable B_x, B_y we can obtain particular solutions of Eqs. (3.14c) as described later in the development of Eqs. (5.32) and add these to solutions (3.15a, b) of the homogeneous equations obtained by setting the right-hand sides of Eqs. (3.14c) equal to zero.

The displacement u_z is generally not important in plane stress problems but we need it to fully understand the approximations which we are making. From Eqs. (3.11a), (3.15b) $\epsilon_z = \partial u_z/\partial z = -(\nu/E)(\sigma_x + \sigma_y) = -(\nu/E)\nabla^2\partial^2\phi/\partial x\,\partial y$, from which $Eu_z = -\nu z\nabla^2\,\partial^2\phi/\partial x\,\partial y + f(x, y)$. If we assume symmetry about the middle surface so that u_z is zero when $z = 0$, then the function of integration $f(x, y)$ will be zero and $Eu_z = -\nu z\nabla^2\,\partial^2\phi/\partial x\,\partial y$. The dilation then becomes $e = [(1-2\nu)/E]\nabla^2\,\partial^2\phi/\partial x\,\partial y$. Using the above values of u_x, u_y, u_z, e we find that the third of the exact three-dimensional equations (3.8) is satisfied identically, but the left-hand sides of the first two become $\nabla^4\,\partial\phi/\partial y - \nu\nabla^2\,\partial^3\phi/\partial x^2\,\partial y$ and $\nabla^4\,\partial\phi/\partial x - \nu\nabla^2\,\partial^3\phi/\partial x\,\partial y^2$, which in general would be zero if and only if $\nu = 0$; similarly, the first four of Eqs. (3.7b) are satisfied (including the important $\sigma_z = 0$) but the last two become $\sigma_{xz} = [-\nu z/2(1+\nu)]\nabla^2\,\partial^3\phi/\partial x^2\,\partial y$, $\sigma_{yz} = [-\nu z/2(1+\nu)]\nabla^2\,\partial^3\phi/\partial x\,\partial y^2$, and σ_{xz} and σ_{yz} would also in general be zero only if $\nu = 0$.

Thus we see that the plane stress solution (3.15a, b) would be exact for a material with zero Poisson's ratio. For other materials equilibrium is satisfied, but the continuity condition and the condition that the faces are free of shear stresses are only approximately satisfied. The errors are proportional to $\nu\nabla^2\,\partial^3\phi/\partial x^2\,\partial y$ and $\nu\nabla^2\,\partial^3\phi/\partial x\,\partial y^2$, which in turn are proportional to ν and to the rates of change in the x and y directions of σ_x, σ_y, σ_{xy}; all this is in complete agreement with the intuitive arguments advanced at the beginning of this section.

If we substitute the modified elastic constants $E/(1-\nu^2)$, $\nu/(1-\nu)$ for E and ν in the plane stress equations (3.15a, b), the expressions for $\sigma_x, \sigma_y, \sigma_{xy}, u_x, u_y$ become the same as for the plane strain case, Eqs. (3.13b). But outside the distribution of stresses and displacements in the x and y directions the two cases are quite different.

An alternative solution of Eqs. (3.14c) with zero body forces is:

$$Eu_x = -(1-\nu)\frac{\partial^2\phi}{\partial x^2} - 2\frac{\partial^2\phi}{\partial y^2}, \qquad Eu_y = (1+\nu)\frac{\partial^2\phi}{\partial x\,\partial y} \tag{3.15c}$$

where $\nabla^4\phi = 0$, from which, using Eqs. (3.14b):

$$(1+\nu)\sigma_x = -\frac{\partial^3\phi}{\partial x^3} - (2+\nu)\frac{\partial^3\phi}{\partial x\,\partial y^2}, \qquad (1+\nu)\sigma_y = -\nu\frac{\partial^3\phi}{\partial x^3} + \frac{\partial^3\phi}{\partial x\,\partial y^2},$$
$$(1+\nu)\sigma_{xy} = \nu\frac{\partial^3\phi}{\partial x^2\,\partial y} - \frac{\partial^3\phi}{\partial y^3} \tag{3.15d}$$

This solution has considerably more complex expressions for the stresses but has the advantage of involving lower derivatives. Since it satisfies the same basic relations, it gives the same displacements and stresses as the solution of Eqs. (3.15a, b). Another solution of the same type can be obtained by interchanging x and y in both the derivatives and the subscripts of expressions (3.15c) and (3.15d).

Traditional Plane Stress Elasticity Theory

It can be checked by substitution that the equilibrium equations (3.14a) without body forces are satisfied if we take:

$$\sigma_x = \frac{\partial^2\phi}{\partial y^2}, \qquad \sigma_y = \frac{\partial^2\phi}{\partial x^2}, \qquad \sigma_{xy} = -\frac{\partial^2\phi}{\partial x\,\partial y} \tag{3.16a}$$

where $\phi(x, y)$ is called the 'Airy stress function' after its originator. The condition of continuity is complied with by satisfying relation (3.6) between the strains and continuous displacements. For the plane stress case we eliminate the strains by equating expressions (3.6) for the strains in terms of the displacements to (3.11a) for the strains in terms of the stresses. Using expressions (3.16a) for the stresses this gives:

$$E\frac{\partial u_x}{\partial x} = \frac{\partial^2\phi}{\partial y^2} - \nu\frac{\partial^2\phi}{\partial x^2}, \quad E\frac{\partial u_y}{\partial y} = \frac{\partial^2\phi}{\partial x^2} - \nu\frac{\partial^2\phi}{\partial y^2}, \quad E\left(\frac{\partial u_x}{\partial y} + \frac{\partial u_y}{\partial x}\right) = -2(1+\nu)\frac{\partial^2\phi}{\partial x\,\partial y} \tag{3.16b}$$

Next we eliminate the displacements by applying the operators $\partial^2/\partial y^2$, $\partial^2/\partial x^2$, and $-\partial^2/\partial x\,\partial y$ to the first, second, and third of these equations respectively and adding them together, which gives:

$$\nabla^4\phi = 0 \tag{3.16c}$$

If the displacements are needed they must be obtained by integrating the first two equations (3.16b) with respect to x and y respectively, substituting the resulting expressions for u_x, u_y into the third equation, and determining the most general form of the functions of integration which will satisfy the latter. Since we are interested in displacements for their practical importance as well as for satisfying boundary conditions, we will generally find it more convenient to use Eqs. (3.15a, b), which are essentially the same as Eqs. (3.16a, b) to which the operator $\partial^2/\partial x\,\partial y$ has been applied, which enables us to have expressions for the displacements without integration; of course they involve the same approximations and yield the same results.

If ϕ satisfies $\nabla^2\phi = 0$, that is if ϕ is a harmonic as well as a biharmonic function, then solution (3.16a, b) coincides with (3.12b, c) and is exact. In this case also $\sigma_x + \sigma_y = 0$ in the other plane stress solutions (3.15a, b) as well as (3.15c, d) and its alternate, and so these all are exact three-dimensional solutions.

General Exact and Antisymmetric Plate Solutions

Besides the above for most applications approximate plane stress solutions, general exact three-dimensional elasticity solutions for the plate with unloaded faces are possible; these involve nonlinear distributions of the displacements and stresses in the z direction. Also, in addition to the cases which we have so far considered in which the loads are symmetrical about the middle surface (the 'membrane' case) analogous approximate as well as exact solutions can be derived for cases in which the loads are antisymmetric with respect to the middle surface (the 'flexure' case). Since all these types of solutions are particularly useful for the satisfaction of the edge boundary conditions for plates under lateral loading they will be discussed later in Chap. 5, Sec. 5.4.

The greater accuracy of exact solutions is usually of little significance unless exact boundary conditions on the stresses or displacements at every point of the boundaries are also satisfied, rather than the gross boundary conditions involving only the resultants of the stresses or the displacements of the middle surface to which we usually restrict ourselves. This question is discussed in Sec. 5.5.

3.3 PLANE STRESS ELASTICITY THEORY APPLIED TO BEAM PROBLEMS

In the following we will interchange y and z in order to match the notation used in previous beam studies. Equations (3.15a, b) then become:

$$Eu_x = \frac{\partial^3 \phi}{\partial z^3} - \nu \frac{\partial^3 \phi}{\partial x^2\, \partial z}, \quad Eu_z = \frac{\partial^3 \phi}{\partial x^3} - \nu \frac{\partial^3 \phi}{\partial x\, \partial z^2}, \quad \nabla^4 \phi = \frac{\partial^4 \phi}{\partial x^4} + \frac{2\partial^4 \phi}{\partial x^2\, \partial z^2} + \frac{\partial^4 \phi}{\partial z^4} = \mathbf{c}$$

$$\sigma_x = \frac{\partial^4 \phi}{\partial x\, \partial z^3}, \quad \sigma_z = \frac{\partial^4 \phi}{\partial x^3\, \partial z}, \quad \sigma_{xz} = -\frac{\partial^4 \phi}{\partial x^2\, \partial z^2} + \frac{\mathbf{c}}{2(1+\nu)}, \quad \sigma_y = 0 \qquad (3.17)$$

Any function $\phi(x, z)$ which satisfies the condition $\nabla^4 \phi = \mathbf{c}$ represents, when interpreted by relations (3.17), a possible plane stress condition. To represent the solution of a particular physical problem it is only necessary (with the reservations previously noted regarding singular points and multiply connected bodies) that the boundary conditions be satisfied; that is, the stresses or displacements at the surfaces corresponding to the boundaries of the physical body must have the values prescribed for the distributed forces or displacements at these boundaries in the physical problem. There is no universal, direct way of finding a function ϕ which will do this in a particular case. The solutions which have been found are obtained by trying suitable functions and using ingenuity to combine them so as to satisfy the required boundary conditions.

The general solution of $\nabla^4 \phi = \mathbf{c}$ can be taken as the general solution to

$\nabla^4\phi = 0$, that is all biharmonic functions, plus particular solutions which make $\nabla^4\phi$ constant. There are only three of the latter: x^4, x^2z^2, and z^4 with arbitrary coefficients. All three give zero σ_x and σ_z and a constant σ_{xz}, and differ only by rigid body displacements; hence we can choose one of them, say x^2z^2, and ignore the other two. Thus we see that $\nabla^4\phi = \mathbf{c}$ can be replaced by $\nabla^4\phi = 0$ except where a constant component of σ_{xz} is involved, when we can add x^2z^2 with an arbitrary coefficient to the expression for ϕ, and $4/(1+\nu)$ times this coefficient to σ_{xz} (as $\mathbf{c} = \nabla^4 x^2z^2 = 8$).

Power Function Solutions

The simplest solutions of $\nabla^4\phi = 0$ are combinations of power functions of the form:

$$\phi = \sum_p \sum_q c_{pq} x^p z^q \tag{3.17a}$$

where the c_{pq}'s are coefficients to be determined. Table 3.2 shows various combinations of stresses and displacements which satisfy Eqs. (3.17), obtained by using such power functions for $\phi = \phi_m$. Each of these, represented by one line of the table, can of course be multiplied through by an arbitrary constant, and the resulting solutions can be combined in any way required.

Those for which $p + q < 4$ give no stresses, and at most only rigid body displacements, so they are omitted. The first three solutions given ($m = 1, 2, 3$) are for $p + q = 4$, the next four are for $p + q = 5$, the next four for $p + q = 6$, and the last four for $p + q = 7$; the table could be continued indefinitely, but the solutions given are sufficient for studying rectangular beam problems involving zero or uniform lateral loading. The first two solutions satisfy $\nabla^4\phi = 0$ identically while the third satisfies the condition that $\nabla^4\phi$ equals a constant, as discussed above. For all the rest several power terms must be combined in order to satisfy $\nabla^4\phi = 0$ and this requirement reduces the number of independent solutions to four for each value of $p + q$; other forms of solutions could be found, but they would merely be combinations of the four given for each $p + q$.

The forms of the functions ϕ_m from which these solutions were obtained are not needed for applications. However, for purposes of checking they are as follows:

$$\phi_1 = xz^3/6, \qquad \phi_3 = -(1+\nu)x^2z^2/4\nu,$$

$$\phi_4 = (5x^4z - z^5)/120, \qquad \phi_6 = (5x^2z^3 - z^5)/60, \qquad \phi_8 = (x^5z - xz^5)/60,$$

$$\phi_9 = (3x^5z - 10x^3z^3 + 3xz^5)/180, \qquad \phi_{10} = (15x^4z^2 - 2x^6 - z^6)/360,$$

$$\phi_{12} = (2x^7 - 21x^5z^2 + 7xz^6)/840, \qquad \phi_{14} = (x^7 - 35x^3z^4 + 14xz^6)/840$$

while ϕ_2, ϕ_5, ϕ_7, ϕ_{11}, ϕ_{13}, ϕ_{15} are the same as ϕ_1, ϕ_4, ϕ_6, ϕ_{10}, ϕ_{12}, ϕ_{14} respectively with x and z interchanged.

Many other types of functions have been found which satisfy Laplace's equation $\nabla^2\psi = 0$, that is harmonic functions, and which therefore also satisfy $\nabla^4\psi = 0$, such as $\tan^{-1}(x/z)$, $\log(x^2 + z^2)$, $x/(x^2 + z^2)$, $(x^2 - z^2)/(x^2 + z^2)^2$, $(x^4 -$

Table 3.2 PLANE STRESS SOLUTIONS OBTAINED FROM POWER FUNCTIONS

m	σ_x	σ_z	σ_{xz}	Eu_x	Eu_z
1*	1			x	$-\nu z$
2*		1		$-\nu x$	z
3*			1	$(1+\nu)z$	$(1+\nu)x$
4*		x		$-\dfrac{z^2+\nu x^2}{2}$	xz
5*	z			xz	$-\dfrac{x^2+\nu z^2}{2}$
6*	x		$-z$	$\dfrac{x^2-(2+\nu)z^2}{2}$	$-\nu xz$
7*		z	$-x$	$-\nu xz$	$\dfrac{z^2-(2+\nu)x^2}{2}$
8	$-z^2$	x^2		$-xz^2-\dfrac{\nu}{3}x^3$	$x^2z+\dfrac{\nu}{3}z^3$
9*	z^2-x^2	x^2-z^2	$2xz$	$(1+\nu)\left(xz^2-\dfrac{x^3}{3}\right)$	$(1+\nu)\left(x^2z-\dfrac{z^3}{3}\right)$
10		$2xz$	$-x^2$	$-\dfrac{z^3}{3}-\nu x^2z$	$xz^2-\dfrac{2+\nu}{3}x^3$
11	$2xz$		$-z^2$	$x^2z-\dfrac{2+\nu}{3}z^3$	$-\dfrac{x^3}{3}-\nu xz^2$
12	z^3	$-3x^2z$	x^3	$xz^3+\nu x^3z$	$\dfrac{(2+\nu)}{4}x^4-\dfrac{3}{2}x^2z^2-\dfrac{\nu z^4}{4}$
13	$-3xz^2$	x^3	z^3	$\dfrac{(2+\nu)}{4}z^4-\dfrac{3}{2}x^2z^2-\dfrac{\nu}{4}x^4$	$x^3z+\nu xz^3$
14	$2z^3-3x^2z$	$-z^3$	$3xz^2$	$-x^3z+(2+\nu)xz^3$	$\dfrac{x^4}{4}+\dfrac{3\nu}{2}x^2z^2-\dfrac{1+2\nu}{4}z^4$
15	$-x^3$	$2x^3-3xz^2$	$3x^2z$	$\dfrac{z^4}{4}+\dfrac{3\nu}{2}x^2z^2-\dfrac{1+2\nu}{4}x^4$	$-xz^3+(2+\nu)x^3z$

* These are exact three-dimensional solutions, since $\sigma_x+\sigma_z$ is linear in x and z.

$6x^2z^2+z^4)/(x^2+z^2)^4$. As was discussed previously, these can be multiplied by $f=x$, z, or x^2+z^2 to give biharmonic functions.

Exponential Function Solutions

We can also obtain solutions by the separation of variables, as was discussed in Sec. 2.3 under mathematical solutions. Thus we can take ϕ as the product of an unknown function of z by an exponential function of x or a function which can be expressed in terms of exponential functions such as trigonometric or hyperbolic functions, even derivatives of all of which have the same general form as the original function. The unknown function of z can then be determined by solving the ordinary differential equation which results when the function of x is cancelled from the equation.

In this way, taking the function of x as $\cos c_m x$, we find that $\nabla^4\phi=0$ is satisfied by taking

$$\phi=\sum_m (\cos c_m x)/c_m^3 \quad [(A_m-2D_m)\sinh c_m z+(B_m-2C_m)\cosh c_m z + C_m c_m z\sinh c_m z + D_m c_m z\cosh c_m z] \tag{3.17b}$$

where c_m, A_m, B_m, C_m, D_m are arbitrary constants (the combination of these constants shown is used to simplify the expressions for stresses and displacements). All derivatives of ϕ have a similar 'trigonometric–hyperbolic' form but with different coefficients. Table 3.3 shows these coefficients in the expressions for the displacements and stresses, which are found by using Eqs. (3.17). If $\sin c_m x$ is used instead of $\cos c_m x$, the expressions for the stresses and displacements are the same except that $\sin c_m x$ and $\cos c_m x$ are interchanged and the signs of the stresses are reversed.

If C_m, $D_m=0$ then the solution is an exact three-dimensional solution, since $\sigma_x+\sigma_z=0$. That is, if we use only the first two terms of ϕ in the brackets then ϕ is also a solution of $\nabla^2\phi=0$. Two more terms $E_m(c_m z)^2\sinh c_m z$ and $F_m(c_m z)^2\cosh c_m z$ can be added to ϕ in the brackets to obtain a solution for $\nabla^6\phi=0$, two more with third powers of $c_m z$ for a solution to $\nabla^8\phi=0$, etc. (as was

Table 3.3 TRIGONOMETRIC–HYPERBOLIC PLANE STRESS SOLUTION

( ) =	$\sum_m$ ( )	[( ) $\sinh c_m z$ +	( ) $\cosh c_m z$ +	( ) $c_m z \sinh c_m z$ +	( ) $c_m z\cosh c_m z$]
Eu_x	$\cos c_m x$	$(1+\nu)B_m+(1-\nu)C_m$	$(1+\nu)A_m+(1-\nu)D_m$	$(1+\nu)D_m$	$(1+\nu)C_m$
Eu_z	$\sin c_m x$	$(1+\nu)A_m-2D_m$	$(1+\nu)B_m-2C_m$	$(1+\nu)C_m$	$(1+\nu)D_m$
σ_x	$c_m\sin c_m x$	$-B_m-C_m$	$-A_m-D_m$	$-D_m$	$-C_m$
σ_z	$c_m\sin c_m x$	B_m-C_m	A_m-D_m	D_m	C_m
σ_{xz}	$c_m\cos c_m x$	A_m	B_m	C_m	D_m

discussed earlier, powers of x or of x^2+z^2 could be used for the same purpose, but this would not usually be convenient).

Similar solutions can be obtained by substituting $e^{-c_m z}$ for $\sinh c_m z$ and $e^{c_m z}$ for $\cosh c_m z$, or by interchanging x and y, or by interchanging the trigonometric and the hyperbolic or exponential functions.

Application of Power Function Solutions

Figure 3.8 shows the physical application of some of the solutions given in Table 3.2. The first solution ($m = 1$) can evidently represent a plate under uniform horizontal tension, as shown by Fig. 3.8(a). In this case we could deduce that this is an exact solution without using any mathematics at all, because if the plate is divided into equal rectangular elements, as suggested in the figure, and if they are all subjected to the same tensile stress in the x direction, they will all be in equilibrium; and since they will all stretch the same amount in the direction of the stress and contract the same amount perpendicular to the stress they will also fit together. It is not difficult to see that the same conclusion would hold for any bar of cylindrical shape with horizontal generators.

Similarly, the fifth solution ($m = 5$) corresponds to a plate under a horizontal stress which is constant in the horizontal direction but varies linearly in the vertical direction. If the x axis is taken in the middle of a rectangular plate, this case is one of pure bending, as shown in Fig. 3.8(b). If the x axis is not in the middle, the plate can be considered to be in a combination of uniform axial loading and pure bending, as shown in Fig. 3.8(c). Again, as the figure suggests, it is not difficult to see that if the plate is divided into equal rectangular elements, the assumption that linearly varying σ_x stresses on the ends results in the same σ_x stresses on all cross sections satisfies equilibrium (except for the vertical components of the σ_x stresses due to the curvature, which classical elasticity theory assumes to be infinitesimal) and the condition that the deformed elements fit together, and that this conclusion applies to any bar of cylindrical shape with horizontal generators.

Stated more generally, if the same linear distribution of normal stresses acts upon both end cross sections of a cylindrical bar, then there will be the same distribution of normal stresses on all intermediate cross sections, and these will be

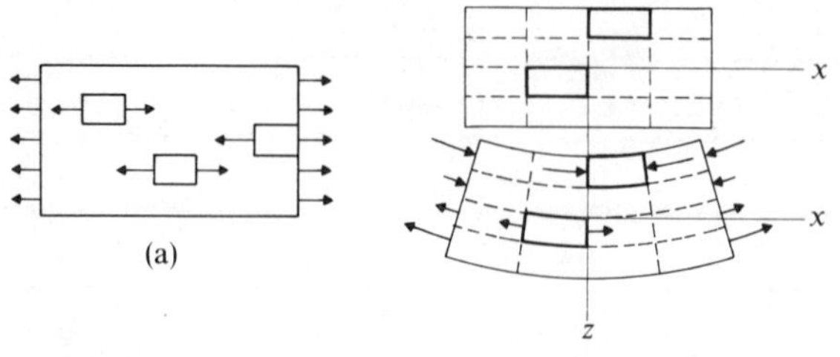

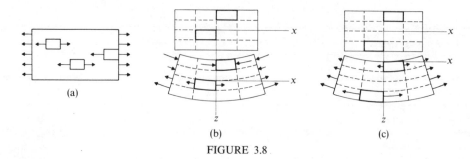

FIGURE 3.8

the only stresses acting. More formally, as in Eqs. (3.8c, e) it can be checked by substitution that the following solution for linearly distributed σ_z stresses:

$$Eu_x = -\nu\left[\frac{\mathbf{a}}{2}\left(x^2 - y^2 + \frac{z^2}{\nu}\right) + \mathbf{b}xy + \mathbf{c}x\right], \qquad Eu_y = -\nu\left[\frac{\mathbf{b}}{2}\left(y^2 - x^2 + \frac{z^2}{\nu}\right) + \mathbf{a}xy + \mathbf{c}y\right]$$

$$Eu_z = (\mathbf{a}x + \mathbf{b}y + \mathbf{c})z, \qquad \sigma_z = \mathbf{a}x + \mathbf{b}y + \mathbf{c}, \qquad \sigma_x, \sigma_y, \sigma_{xy}, \sigma_{yz}, \sigma_{xz} = 0 \tag{3.18}$$

satisfies the exact three-dimensional Eqs. (3.8), (3.7b); if the bar is cylindrical with generators parallel to the z direction there will be no loading on its sides and a linearly varying load on its end cross sections, and similarly for the x, y directions.

If bodies such as those shown in Fig. 3.8(a, b, c) are long and slender and they are loaded on their ends not by uniformly or linearly distributed forces but by statically equivalent forces, then by Saint-Venant's principle the stresses will be practically the same as discussed above except in end regions of lengths equal to the width.

Simply Supported Beam of Rectangular Cross Section Under Uniform Load

While there are other cases where the simple physical reasoning used above can suffice to determine the correct solution, we generally need mathematics to guide us. Although none of the solutions given in Table 3.2 describes by itself a practical problem such as the case shown in Fig. 3.9(a), we can find combinations of them which do. To go about this systematically we can use our knowledge of the distributed forces on the boundaries and of the elementary classical beam theory solution to conclude that σ_x should have a term which varies linearly in the z

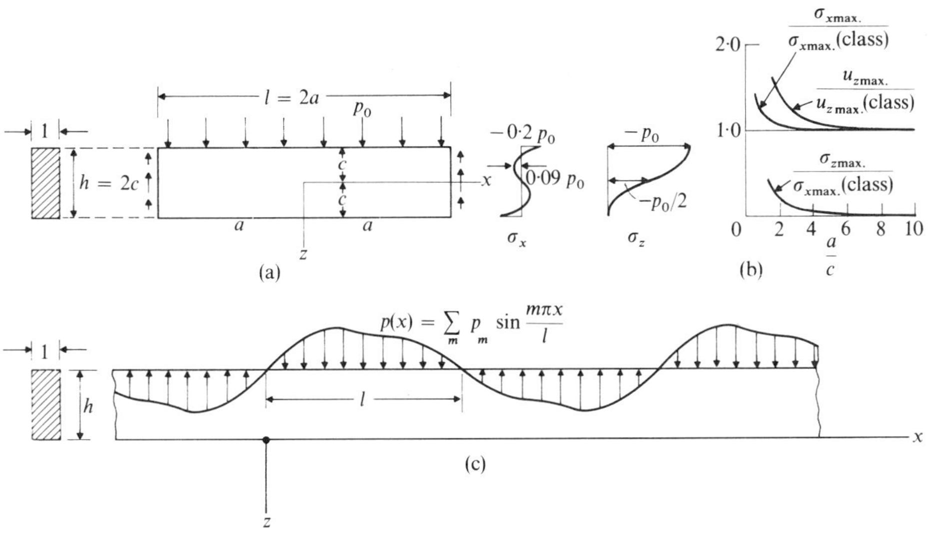

FIGURE 3.9

direction and nonlinearly in the x direction, σ_z is probably constant in the x direction and varying in the z direction, while σ_{xz} probably varies linearly in the x direction and has components which are constant and varying with z^2.

Looking over Table 3.2 we see that solutions $m = 2, 5, 7, 14$ involve such stresses, so we tentatively assume that $\phi = C_2\phi_2 + C_5\phi_5 + C_7\phi_7 + C_{14}\phi_{14}$. Then from Table 3.2 the stresses are:

$$\sigma_x = C_5 z + C_{14}(2z^3 - 3x^2 z), \qquad \sigma_z = C_2 + C_7 z - C_{14} z^3,$$
$$\sigma_{xz} = -C_7 x + 3C_{14} x z^2 \tag{3.19}$$

Substituting these expressions into the boundary conditions $x = \pm a$: $\int_{-c}^{c} \sigma_x \, dz = 0$, $\int_{-c}^{c} \sigma_x z \, dz = 0$, $\int_{-c}^{c} \sigma_{xz} \, dz = \mp p_0 a$, and $z = c$: $\sigma_z, \sigma_{xz} = 0$, $z = -c$: $\sigma_z = -p_0$, $\sigma_{zx} = 0$, we find that the following five equations must be satisfied:

$$C_5 - 3C_{14}\left(a^2 - \frac{2c^2}{5}\right) = 0, \qquad 2C_7 c - 2C_{14} c^3 = p_0,$$
$$C_2 + C_7 c - C_{14} c^3 = 0, \qquad C_7 - 3C_{14} c^2 = 0, \qquad C_2 - C_7 c + C_{14} c^3 = -p_0 \tag{3.20}$$

We have only four unknowns to satisfy five equations, but we would have had five unknowns if we had taken the vertical reaction at the ends as another unknown, and we would have obtained the correct value for it since the solutions of Table 3.2 satisfy equilibrium. In any case we find that Eqs. (3.20) can all be satisfied by the values:

$$C_2 = -\frac{p_0}{2}, \qquad C_5 = \frac{3p_0}{4c^3}\left(a^2 - \frac{2c^2}{5}\right), \qquad C_7 = \frac{3p_0}{4c}, \qquad C_{14} = \frac{p_0}{4c^3} \tag{3.21}$$

Multiplying these coefficients by the corresponding expressions for stresses and displacements in Table 3.2, and combining (and adding a constant rigid body displacement to u_z sufficient to make u_z zero at the middle of the ends $x = \pm a$, $z = 0$) we obtain:

$$Eu_x = \frac{3p_0 c}{4}\left[\left(*\mathbf{a}^2 - \frac{\mathbf{x}^2}{3}* - \frac{2+5\nu}{5} + \frac{2+\nu}{3}\mathbf{z}^2\right)\mathbf{z} + \frac{2\nu}{3}\right]\mathbf{x}$$
$$Eu_z = \frac{3p_0 c}{8}\left[\left(*\frac{5\mathbf{a}^2 - \mathbf{x}^2}{6}* + \frac{8+5\nu}{5} - \nu\mathbf{z}^2\right)(\mathbf{a}^2 - \mathbf{x}^2) + \left(\frac{5+2\nu}{5} - \frac{1+2\nu}{6}\mathbf{z}^2\right)\mathbf{z}^2 - \frac{4\mathbf{z}}{3}\right]$$
$$\sigma_x = \frac{3p_0}{4}\left(*\mathbf{a}^2 - \mathbf{x}^2* - \frac{2}{5} + \frac{2\mathbf{z}^2}{3}\right)\mathbf{z} \tag{3.22}$$
$$\sigma_z = \frac{p_0}{4}(-2 + 3\mathbf{z} - \mathbf{z}^3), \qquad \sigma_{xz} = \frac{3p_0}{4}(\mathbf{z}^2 - 1)\mathbf{x}$$

where $\mathbf{z}$, $\mathbf{x}$, $\mathbf{a}$ represent the dimensionless ratios z/c, x/c, a/c.

The classical beam theory solution for this case gives the same σ_{xz} as above, no σ_z stresses at all, and the parts between the asterisks in the above expressions for σ_x, Eu_x, and Eu_z (the elementary value of u_x is taken as that due to rotating the cross sections about their centers, or $u_x = -z\, du_z/dx$). Considering the fact that $\mathbf{z}$ can have values from zero to unity and $\mathbf{x}$ from zero to $\mathbf{a}$, it can be seen that the additional terms in the elasticity solution (3.22) which are neglected in the classical beam theory solution have values of the order of $1/\mathbf{a}^2$ times the values of the terms which are considered in the classical solution. The ratio $\mathbf{a} = a/c = 2a/2c$ is the ratio of the length of the beam to its height and is a measure of its 'slenderness'. For slenderness $\mathbf{a} \geq 6$ the terms neglected in the classical solution have magnitudes only a few percent of those which are considered, and can be ignored in most practical problems. For $\mathbf{a} < 6$, that is for short, stocky beams, the classical solution may not be good enough, and for values of $\mathbf{a}$ close to unity, that is for what would usually be called a 'block' rather than a beam, the classical beam theory gives completely misleading values for stresses and displacements. Figure 3.9(b) shows how the maximum values of σ_x, σ_z, and u_z compare with classical beam theory values for various values of the slenderness ratio $\mathbf{a}$.

This elasticity solution itself has limitations for describing practical physical problems, principally because, while we have satisfied the required conditions for the *resultant* forces and moments on the ends, the actual distribution of forces on the ends is not what is likely to be found in a practical problem. From Eqs. (3.22), at the ends $x = \pm a$:

$$\sigma_x = \frac{p_0}{10}(5\mathbf{z}^3 - 3\mathbf{z}) \tag{3.22a}$$

and the vertical reactions are supplied by σ_{xz} shear stresses distributed parabolically over the end cross sections in the same manner that they are over interior cross sections, as indicated in Fig. 3.9(a). By Saint-Venant's principle we could eliminate this balanced distribution of σ_x forces and replace the σ_{xz} forces by more practical reactions, such as concentrated or nearly concentrated vertical forces at the bottom corners, without appreciably changing the conditions except in end zones of length equal to the height $2c$. In Sec. 3.4 we will develop 'local stress fields' which correct the stress distribution in these end zones, and Eqs. (3.22) with such corrections should give a good approximation for values of the slenderness ratio $\mathbf{a}$ as small as one or two. These results illustrate and are typical of the way in which stresses and displacements predicted by the elementary classical theories differ from more exact values.

Beams of Rectangular Cross Section Under End Moments and Shears

This case of a beam with zero loading on the sides, and end moments and transverse shear forces $x = 0, l$: $M_x = M_1, M_2$; $F_{xz} = (M_2 - M_1)/l$ was shown in Fig. 2.12, p. 62. As in the above case, taking into consideration our knowledge of the boundary conditions and the classical solution, we surmise that σ_x should vary linearly in the z direction, with components constant with x and varying linearly with x, σ_z is probably zero, while σ_{xz} is probably constant with x, with components constant with z and varying with z^2. These requirements are filled by the solutions $m = 3, 5, 11$ of Table 3.2, so we tentatively take $\phi = C_3\phi_3 + C_5\phi_5 + C_{11}\phi_{11}$.

The boundary conditions are $x = 0, l$: $\int_{-c}^{c} \sigma_x\, dz = 0$, $\int_{-c}^{c} \sigma_x z\, dz = M_1, M_2$,

$\int_{-c}^{c} \sigma_{zx}\, dz = (M_2 - M_1)/l$; $z = \pm c$: $\sigma_z, \sigma_{xz} = 0$. These give us four equations:

$$2C_5\frac{c^3}{3} = M_1, \qquad 2(C_5 + 2C_{11}l)\frac{c^3}{3} = M_2, \qquad 2\left(C_3c - C_{11}\frac{c^3}{3}\right) = (M_2 - M_1)/l, \qquad C_3 - C_{11}c^2 = 0 \tag{3.23}$$

which are satisfied by the values:

$$C_3 = \frac{3(M_2 - M_1)}{4cl}, \qquad C_5 = \frac{3M_1}{2c^3}, \qquad C_{11} = \frac{3(M_2 - M_1)}{4c^3l} \tag{3.24}$$

Using Table 3.2 and adding a rigid body rotation θ_y, Eq. (3.6a), (that is adding $\theta_y z$ to u_x and $-\theta_y x$ to u_z) calculated to make u_z zero when $z = 0, x = 0, l$, we obtain the stresses and displacements:

$$\begin{aligned}
&\sigma_x = \frac{3}{2c^2}\left[M_1 + (M_2 - M_1)\frac{\mathbf{x}}{\mathbf{l}}\right]\mathbf{z}, \qquad \sigma_z = 0, \qquad \sigma_{xz} = \frac{3(M_2 - M_1)}{4c^2\mathbf{l}}(1 - \mathbf{z}^2)\\
&u_x = \frac{1}{4El}\{(M_2 - M_1)[6(1 + \nu) + 3\mathbf{x}^2 - (2 + \nu)\mathbf{z}^2 - \mathbf{l}^2] + 3M_1\mathbf{l}(2\mathbf{x} - \mathbf{l})\}\mathbf{z}\\
&u_z = \frac{1}{4El}\{(M_2 - M_1)\mathbf{x}(\mathbf{l}^2 - \mathbf{x}^2 - 3\nu\mathbf{z}^2) + 3M_1\mathbf{l}(\mathbf{l}\mathbf{x} - \mathbf{x}^2 - \nu\mathbf{z}^2)\} \qquad (\mathbf{l} = l/c)
\end{aligned} \tag{3.25}$$

It can easily be checked, using Eqs. (2.5), (2.5a), that the values for the stresses are the same as those given by the classical beam theory, while the value of Eu_z when $z = 0$ is the same as the value given by classical beam theory, designated as w in Eq. (2.37), although the vertical displacement of points not on the middle surface contains other terms which are due to transverse strains and are not considered in classical beam theory.

For the above to be true the moments and transverse forces on the ends of the beam must be applied by distributed normal and shear forces varying in the manner prescribed by the classical theory for the stresses on internal cross sections.

In general *all* the classical solutions, for beams and plates and shells presume that end or edge reactions are applied by parabolically distributed transverse shear forces and that end or edge moments are applied by linearly distributed normal forces, like those on internal sections. These end distributions will be statically equivalent to the actual distributions of the end forces and moments, whatever these may be, and hence by Saint-Venant's principle the stresses and relative displacements will be little affected except near the ends or edges, where the distribution can be corrected by the superposition of local stress fields such as will be discussed in the next section.

The solutions defined by Eqs. (3.22) and (3.25) can be combined to obtain solutions for uniformly loaded beams under other end conditions, as discussed in Sec. 2.7, but these will satisfy only gross end conditions, as has been noted before. In particular, for fixed ends consideration will have to be given to the horizontal displacement u_x at the ends, part of which is nonlinear and so distorts the end cross section, in order to get more accurate values for the deflections and end stresses than were obtained with classical beam theory; this question will also be discussed in Sec. 3.4.

Trigonometric–Hyperbolic Elasticity Solutions

Consider the case shown in Fig. 3.9(c) of an infinitely long beam of height $h = 2c$ loaded by a cyclically distributed pressure $p(x)$ on the top surface, with half cycle length l. Taking the origin of coordinates on the bottom surface as shown in the figure, the boundary conditions are:

$$z = 0:\ \sigma_z,\ \sigma_{xz} = 0; \qquad z = -h:\ \sigma_z = -\sum_m p_m \sin\frac{m\pi x}{l},\ \sigma_{xz} = 0 \qquad (m = 1, 2, 3, \ldots) \qquad (3.26)$$

Substituting the expressions for σ_z, σ_{xz} given in Table 3.3 into these conditions, we find that they are satisfied if:

$$c_m = \frac{m\pi}{l}, \qquad B_m = 0, \qquad A_m = D_m = \alpha_m p_m / c_m, \qquad C_m = \gamma_m p_m / c_m, \qquad \text{where:}$$

$$\alpha_m = \frac{\lambda_m \sinh \lambda_m}{\lambda_m^2 - \sinh^2 \lambda_m}, \qquad \gamma_m = \frac{\sinh \lambda_m + \lambda_m \cosh \lambda_m}{\lambda_m^2 - \sinh^2 \lambda_m}, \qquad \lambda_m = hc_m = \frac{m\pi h}{l} \qquad (3.27)$$

The stresses and displacements can then be obtained by substituting these values of A_m, B_m, C_m, D_m into the expressions given in Table 3.3.

As was discussed in Sec. 2.4, the loading function $-\Sigma_m p_m \sin m\pi x/l$ can represent any distribution of load on the top surface over the interval $x = 0$ to $x = l$ (the distribution from $x = l$ to $x = 2l$ being antisymmetrical to this about $x = l$; if a cosine series were added, making a complete Fourier series, any distribution between $x = 0$ and $x = 2l$ could be represented). The corresponding coefficients p_m can be found by harmonic analysis, and will have the values given in Eq. (2.11) for a beam of length l. It can easily be checked that the stresses acting upon the ends of the portion of the beam from $x = 0$ to $x = l$ will be statically equivalent to simple supporting forces at the ends of a beam with hinged ends at these points. If l is large compared to h then by Saint-Venant's principle the exact kind of hinged support will be important only near the ends. If l is not large compared to h the exact distribution of end supporting forces may be important, and local stress fields such as those discussed in Sec. 3.4 may have to be superposed at the ends to give the end conditions required.

For the infinite beam under a cyclically varying load with cycle length $2l$, as the cycle length becomes smaller compared to h the stresses become more and more concentrated toward the loaded surface of the beam, and in the limit become important only in a narrow layer having a depth of the order of l, with values which can be found more easily from Eqs. (3.32), (3.33) given later. This case is discussed in more detail in Sec. 3.5, Fig. 3.21.

Series Solution in the Loading Function

The above methods of determining displacements and stresses not considered in classical beam theory are rather laborious. Another type of solution, which is especially convenient for finding the most important corrections to the classical theory, is given by series expressions for the deflections and stresses in continuously loaded beams of rectangular cross section in terms of the top and bottom loading.* In such series the first terms represent the values given by the classical beam theory, the next terms are the most important corrections on these and involve higher derivatives of the loading (that is, finer detail in the way the loading varies), the next terms involve still higher derivatives of the loading, and so on. All the terms are calculated to approach a plane-stress elasticity solution in the limit. This is evidently an application of the general 'method of successive approximation'.

Let $t_z(x)$ and $b_z(x)$ represent distributed forces per unit area in the z direction acting uniformly across the width of the top $z = -c$ and the bottom $z = c$ surfaces respectively of a rectangular beam of height $h = 2c$, Fig. 3.10. For

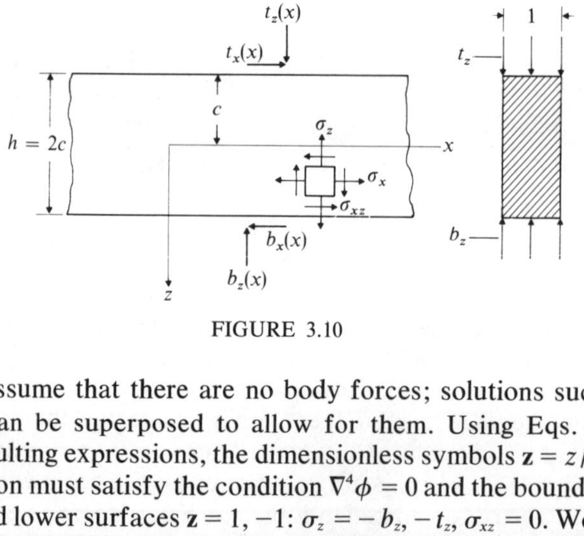

FIGURE 3.10

simplicity we assume that there are no body forces; solutions such as those of Eqs. (3.8n, o) can be superposed to allow for them. Using Eqs. (3.17) and, to simplify the resulting expressions, the dimensionless symbols $\mathbf{z} = z/c$ and $\mathbf{x} = x/c$, the series solution must satisfy the condition $\nabla^4\phi = 0$ and the boundary conditions on the upper and lower surfaces $\mathbf{z} = 1, -1$: $\sigma_z = -b_z, -t_z$, $\sigma_{xz} = 0$. We also have the relations from Eq. (2.3c): $-dF_{xz}/dx = -d^2M_x/dx^2 = t_z - b_z$, where $F_{xz} = \int_{-c}^{c} \sigma_{xz}\,dz$ and $M_x = \int_{-c}^{c} \sigma_x z\,dz$ are the transverse shear force and bending moment per unit width of the beam; these equilibrium conditions are redundant as far as determining the solution is concerned, because any solution of Eqs. (3.17) satisfies equilibrium, but they are useful for replacing integrals of the loading functions in the solution by the more familiar concepts of the transverse shear and bending moment.

Trial shows that the above conditions can be satisfied if we take ϕ in the form:

*L. H. Donnell, 'Bending of rectangular beams', *J. Appl. Mech., Trans. ASME*, vol. 74, 1952, p. 123.

$$\phi = a_1 d^{-7}(t_z - b_z)/d\mathbf{x}^{-7} + (b_1\mathbf{z}^2 + c_1)\, d^{-5}(t_z - b_z)/d\mathbf{x}^{-5}$$
$$+ (d_1\mathbf{z}^4 + e_1\mathbf{z}^2 + f_1)d^{-3}(t_z - b_z)/d\mathbf{x}^{-3} + \cdots$$
$$+ a_2\mathbf{z}\, d^{-3}(t_z + b_z)/d\mathbf{x}^{-3} + (b_2\mathbf{z}^3 + c_2\mathbf{z})\, d^{-1}(t_z + b_z)/d\mathbf{x}^{-1}$$
$$+ (d_2\mathbf{z}^5 + e_2\mathbf{z}^3 + f_2\mathbf{z})\, d(t_z + b_z)/d\mathbf{x} + \cdots$$

The negatively ordered derivatives of course really represent integrations, and are given in this form for convenience; that is, the conventional definition $d^m(d^n\alpha/d\beta^n)/d\beta^m = d^{m+n}\alpha/d\beta^{m+n}$, which is generally used only with positive values of m, n is applied to both positive and negative values. The question of constants of integration which naturally arises in adopting such a convention will be discussed later.

Substituting this expression for ϕ into the above conditions, using the formulas of Eqs. (3.17) for $\sigma_x, \sigma_z, \sigma_{xz}$, we obtain relations which are easily solved step by step for the coefficients $a_1, b_1, \ldots,$ and we find that there are just enough relations to determine all the coefficients. We thus find successively $b_1 = 0$, $d_1 = -1/16$, $e_1 = 3/8$, $a_1 = 3/2$, etc., which with Eqs. (3.17) and the above definitions of F_{xz} and M_x yield the following series expressions for the displacements and stresses:

$$u_x = -\mathbf{z}\frac{dw_{c1}}{d\mathbf{x}}* - \left(\frac{2+\nu}{4}\mathbf{z}^3 - \frac{3(2+5\nu)}{20}\mathbf{z}\right)\frac{F_{xz}}{E}$$
$$+\frac{c}{E}\left\{\frac{\nu}{2}\int (t_z + b_z)\, d\mathbf{x} + \left[\left(-\frac{3+2\nu}{80}\mathbf{z}^5 - \frac{1-2\nu}{40}\mathbf{z}^3 + \frac{87-70\nu}{2800}\mathbf{z}\right)\frac{d(t_z - b_z)}{d\mathbf{x}}\right.\right.$$
$$\left.\left.+\frac{3\mathbf{z}^2-1}{12}\frac{d(t_z + b_z)}{d\mathbf{x}} + \cdots\right]\right\}$$
$$u_z = w_{c1}* - \left(\frac{3\nu}{4}\mathbf{z}^2 - \frac{3(8+5\nu)}{20}\right)\frac{M_x}{Ec}$$
$$+\frac{c}{E}\left\{\left(-\frac{1+2\nu}{16}\mathbf{z}^4 + \frac{3(5+2\nu)}{40}\mathbf{z}^2 - \frac{227+70\nu}{2800}\right)(t_z - b_z) - \frac{\mathbf{z}}{2}(t_z + b_z)\right.$$
$$+\left[\left(\frac{2+3\nu}{480}\mathbf{z}^6 - \frac{2-\nu}{160}\mathbf{z}^4 + \frac{70-87\nu}{5600}\mathbf{z}^2 - \frac{26-765\nu}{252\,000}\right)\right.$$
$$\left.\left.\times\frac{d^2(t_z - b_z)}{d\mathbf{x}^2} - \frac{\nu(\mathbf{z}^3 - \mathbf{z})}{12}\frac{d^2(t_z + b_z)}{d\mathbf{x}^2} + \cdots\right]\right\} \tag{3.28}$$
$$\sigma_x = \frac{3\mathbf{z}}{2}\frac{M_x}{c^2}* + \frac{5\mathbf{z}^3 - 3\mathbf{z}}{10}(t_z - b_z) + \left[\left(-\frac{3\mathbf{z}^5}{80} - \frac{\mathbf{z}^3}{40} + \frac{87\mathbf{z}}{2800}\right)\frac{d^2(t_z - b_z)}{d\mathbf{x}^2} + \frac{3\mathbf{z}^2 - 1}{12}\frac{d^2(t_z + b_z)}{d\mathbf{x}^2}\cdots\right]$$
$$\sigma_z = * - \frac{\mathbf{z}^3 - 3\mathbf{z}}{4}(t_z - b_z) - \frac{1}{2}(t_z + b_z) + \left[\frac{\mathbf{z}^5 - 2\mathbf{z}^3 + \mathbf{z}}{40}\frac{d^2(t_z - b_z)}{d\mathbf{x}^2} + \cdots\right]$$
$$\sigma_{xz} = -\frac{3(\mathbf{z}^2 - 1)}{4}\frac{F_{xz}}{c}* + \left[\left(-\frac{\mathbf{z}^4}{8} + \frac{3\mathbf{z}^2}{20} - \frac{1}{40}\right)\frac{d(t_z - b_z)}{d\mathbf{x}} + \cdots\right]$$

In the above expressions, using Eqs. (2.2) and (2.4a),

$$w_{cl} = (3/2c^3E)\, d^{-4}(t_z - b_z)/dx^{-4} = -(3/2c^3E)\, d^{-2}M_x/dx^{-2}$$
$$= -(3/2c^3E)\int dx \int M_x\, dx$$

is the vertical deflection of the middle surface given by the classical beam theory. Other ways for determining w_{cl} are discussed in Chapter 2 and in books and courses on elementary strength of materials. The terms in the series before the asterisks are those given by the classical beam theory; the next terms in the series for the stresses were given by F. Seewald in 1927.* Using the above: $\int_{-c}^{c} \sigma_x \, dz = F_x = 0$, $\int_{-c}^{c} \sigma_x z \, dz = M_x$, $\int_{-c}^{c} \sigma_{xz} \, dz = F_{xz}$.

For distributed shear forces per unit area $t_x(x)$ and $b_x(x)$ acting uniformly across the width of the top and bottom surfaces of the beam in the x direction, Fig. 3.10,** the conditions to be satisfied are the same as above except that the boundary conditions are changed to $\mathbf{z} = 1, -1$: $\sigma_z = 0$, $\sigma_{xz} = -b_x, -t_x$. Trial shows that these and the remaining conditions can be satisfied if ϕ has the form:

$$\phi = (a_1\mathbf{z}^3 + b_1\mathbf{z})d^{-2}(t_x - b_x)/d\mathbf{x}^{-2} + (c_1\mathbf{z}^5 + d_1\mathbf{z}^3 + e_1\mathbf{z})(t_x - b_x) + \cdots$$
$$+ a_2 d^{-6}(t_x + b_x)/d\mathbf{x}^{-6} + (b_2\mathbf{z}^2 + c_2)\, d^{-4}(t_x + b_x)/d\mathbf{x}^{-4} + \cdots$$

Substituting this expression into all the conditions and solving for the coefficients $a_1, b_1, \ldots$ as before, we obtain for this case:

$$Eu_x = -\frac{1}{2}\int d\mathbf{x}\int \{(t_x - b_x) - 3\mathbf{z}(t_x + b_x)\}\, d\mathbf{x} + \left(\frac{2+\nu}{4}\mathbf{z}^2 - \frac{2+3\nu}{12}\right)(t_x - b_x)$$
$$- \left(\frac{2+\nu}{4}\mathbf{z}^3 + \frac{6+5\nu}{20}\mathbf{z}\right)(t_x + b_x)$$
$$+ \left[\left(-\frac{3+2\nu}{48}\mathbf{z}^4 + \frac{1+2\nu}{24}\mathbf{z}^2 + \frac{1+3\nu}{720}\right)\frac{d^2(t_x - b_x)}{d\mathbf{x}^2}\right.$$
$$\left. + \left(\frac{3+2\nu}{80}\mathbf{z}^5 - \frac{7+6\nu}{120}\mathbf{z}^3 + \frac{53+70\nu}{2800}\mathbf{z}\right)\frac{d^2(t_x + b_x)}{d\mathbf{x}^2}\cdots\right] \tag{3.29}$$

$$Eu_z = -\frac{3}{2}\int d\mathbf{x}\int d\mathbf{x}\int (t_x + b_x)\, d\mathbf{x} + \frac{1}{2}\int\left\{\nu\mathbf{z}(t_x - b_x) - \left(\frac{3\nu}{2}\mathbf{z}^2 - \frac{4+5\nu}{10}\right)(t_x + b_x)\right\} d\mathbf{x}$$
$$+ \left[\frac{-(1+2\nu)\mathbf{z}^3 + (3+2\nu)\mathbf{z}}{12}\frac{d(t_x - b_x)}{d\mathbf{x}}\right.$$
$$\left. + \left(\frac{1+2\nu}{16}\mathbf{z}^4 - \frac{5+6\nu}{40}\mathbf{z}^2 + \frac{87+70\nu}{2800}\right)\frac{d(t_x + b_x)}{d\mathbf{x}}\cdots\right]$$

$$\sigma_x = -\frac{1}{2}\int [(t_x - b_x) - 3\mathbf{z}(t_x + b_x)]\, d\mathbf{x} + \left[\frac{3\mathbf{z}^2 - 1}{6}\frac{d(t_x - b_x)}{d\mathbf{x}} - \frac{5\mathbf{z}^3 - 3\mathbf{z}}{10}\frac{d(t_x + b_x)}{d\mathbf{x}}\cdots\right]$$

$$\sigma_z = \left[-\frac{\mathbf{z}^2 - 1}{4}\frac{d(t_x - b_x)}{d\mathbf{x}} + \frac{\mathbf{z}^3 - \mathbf{z}}{4}\frac{d(t_x + b_x)}{d\mathbf{x}}\right.$$
$$\left. + \frac{\mathbf{z}^4 - 2\mathbf{z}^2 + 1}{24}\frac{d^3(t_x - b_x)}{d\mathbf{x}^3} - \frac{\mathbf{z}^5 - 2\mathbf{z}^3 + \mathbf{z}}{40}\frac{d^3(t_x + b_x)}{d\mathbf{x}^3}\cdots\right]$$

$$\sigma_{xz} = \frac{\mathbf{z}}{2}(t_x - b_x) - \frac{3\mathbf{z}^2 - 1}{4}(t_x + b_x) + \left[\frac{-\mathbf{z}^3 + \mathbf{z}}{6}\frac{d^2(t_x - b_x)}{d\mathbf{x}^2} + \left(\frac{\mathbf{z}^4}{8} - \frac{3\mathbf{z}^2}{20} + \frac{1}{40}\right)\frac{d^2(t_x + b_x)}{d\mathbf{x}^2}\cdots\right]$$

**Abh. Aerodyn. Inst., Tech. Hochschule*, Aachen, vol. 7, p. 11. See also Timoshenko, *Theory of Elasticity*, 1st edn, p. 44, New York.

**B. A. Boley and I. S. Tolins, 'On the stresses and deflections of rectangular beams', *J. Appl. Mech., Trans. ASME*, 1956, p. 339.

Convergence and Application of Series Solution in Loading Functions

A question immediately arises regarding the convergence of such series expressions. We are interested in what might be called 'practical convergence', that is whether a good approximation can be obtained with a comparatively few terms, rather than 'mathematical convergence' which sometimes means little in terms of practical applications. A simple and practical test for such convergence can be made by comparing the results obtained by using Eqs. (3.28) and (3.29) with exact values obtained from Eqs. (3.27) for harmonically varying normal or tangential loadings such as $t_z = t_1 \cos \pi x/l$, for various ratios of the half wave length l of the loading to the height $h = 2c$ of the beam. This comparison reveals that the parts of the series involving derivatives of t_z, b_z or t_x, b_x with respect to x diverge for values of l/h less than about one, that is for alternating loads which reverse in direction at intervals of about the height or less. All the terms involving such derivatives have been enclosed in square brackets in the above series expressions.

For harmonically varying loading of longer wave length the series converge; they converge more rapidly as the wave length increases, so that for $l/h \gg 1$ the last terms in the series given above are not needed to obtain an excellent approximation. Furthermore, these series give exact closed (non-series) solutions for any loading which is a power function of x, because eventually the derivatives of such a function become zero. If we make a harmonic analysis of such a power function, as in Eq. (2.11), we of course will find that it contains an infinite number of harmonic components having decreasing amplitudes as their wave lengths decrease. This shows that the mere presence of harmonic components with l/h less than unity does not necessarily mean that the series diverge for the loading as a whole; whether they diverge or not depends upon the relative amplitudes of the various harmonic components of which the loading is composed.

General terms which have been calculated for such series* show that the series of Eqs. (3.28) and (3.29) are not of the type of certain rather exotic mathematical series whose successive terms at first decrease in magnitude but after a while start to increase. Hence, it is safe to use the whole of Eqs. (3.28), (3.29) for any loading for which the successive terms of the series are found to decrease steadily in importance until the last terms are, or promise to be, negligible. Moreover, even if this is not true the terms of the series up to the square brackets can be used, and give a considerably better approximation than the classical beam theory, even for discontinuous loading (for which the derivatives in the brackets would not exist anyway).

A better approximation for discontinuous loading can be obtained by replacing it by a harmonic series, and applying the whole of expressions (3.28), (3.29) to the harmonic components for which l/h is considerably greater than one (say for $l/h \geqslant 1{\cdot}5$), and the parts of (3.28), (3.29) up to the brackets to the harmonic components of shorter wave length. In this way a good approximation can be obtained for any loading condition, except in the immediate neighborhood

*C. W. Lee and L. H. Donnell, 'A study of thick plates under tangential loads applied on the faces', *Proc. 3rd U.S. Nat. Cong. of Appl. Mech., ASME*, 1958, p. 401.

of extreme discontinuities such as concentrated loads, corrections for which are given later in Sec. 3.4 (Figs. 3.14, 3.15).

Summarizing, such series solutions in terms of the loading function provide: (1) exact closed solutions for power loading functions; (2) series solutions converging rapidly to exact solutions for reasonably 'smooth' loading; (3) simple improvements on classical beam theory, or still better but more complex solutions, for general types of loading.

End Conditions and Examples

In the preceding we have only discussed satisfaction of the boundary conditions on the sides of a beam. For satisfying end conditions we can use the constants of integration corresponding to the negative order terms (integrations) in the above expressions for ϕ, retaining only those which do not violate the other conditions ($\nabla^4\phi = 0$ and the boundary conditions on the top and bottom surfaces). What is left turns out to be the same as what we have found for the other beam solutions we have studied, namely equivalent to the superposition of arbitrary axial forces or end moments with corresponding transverse shear given by Eqs. (3.25), plus arbitrary rigid body displacements. This is sufficient to satisfy the gross end conditions, that is required integrals of the stresses on the end surfaces or required displacements and slopes of the middle surface. Satisfaction of more detailed requirements for stresses or displacements over the entire end surfaces can be achieved, as previously discussed, by the superposition of local stress fields such as are given in Sec. 3.4.

As an example of the application of Eqs. (3.28), for a uniform loading p_0 on the top surface of a hinged-end beam, Fig. 3.9(a), we set $t_z = p_0$, $b_z = 0$. Using $F_{xz} = -c\int(t_z - b_z)\,d\mathbf{x}$, $M_x = c\int F_{xz}\,d\mathbf{x}$, $w_{cl} = -(3/2cE)\int d\mathbf{x}\,M_x\,d\mathbf{x}$, we have $F_{xz} = -p_0c(\mathbf{x}+C_1)$, $M_x = -p_0c^2(\mathbf{x}^2/2 + C_1\mathbf{x} + C_2)$, $w_{cl} = (3p_0c/2E)(\mathbf{x}^4/24 + C_1\mathbf{x}^3/6 + C_2\mathbf{x}^2/2 + C_3\mathbf{x} + C_4)$. Choosing $C_1, C_2, \ldots$ to satisfy the boundary conditions for hinged ends $\mathbf{x} = \pm\mathbf{a}$: $M_x = 0$, $w_{cl} = u_{z(z=0)} = 0$ (the latter requires also a vertical rigid body displacement), we obtain expressions identical to those of Eqs. (3.22), which were found by a different and somewhat more laborious method.

We can obtain exact closed solutions for linearly and quadratically varying loads in the same manner from Eqs. (3.28), but higher powered load functions require more terms of the series than are shown. Confining ourselves to the stresses to save space, for a linearly varying pressure on the top surface $t_z = p_1\mathbf{x}$, $b_z = 0$, $F_{xz} = -cp_1(\mathbf{x}^2/2 + C_1)$, $M_x = -c^2p_1(\mathbf{x}^3/6 + C_1\mathbf{x} + C_2)$. Choosing C_1, C_2 to satisfy the conditions for hinged ends at $x = 0, l$, we find:

$$\sigma_x = \frac{p_1}{4}(2\mathbf{z}^2 + \mathbf{l}^2 - \mathbf{x}^2 - 6/5)\mathbf{zx}, \qquad \sigma_z = \frac{p_1}{4}(3\mathbf{z} - \mathbf{z}^3 - 2)\mathbf{x}$$

$$\sigma_{xz} = \frac{p_1}{8}[(\mathbf{z}^2 - 1)(3\mathbf{x}^2 - \mathbf{l}^2) - \mathbf{z}^4 + 6\mathbf{z}^2/5 - 1/5] \tag{3.30a}$$

where $\mathbf{l} = l/c$. In similar manner for a load $t_z = (1 - \mathbf{x}^2/\mathbf{a}^2)p_2$ on the top surface, which varies parabolically from p_2 at the center to zero at the ends, on a beam with hinged ends at $x = \pm a$, we find

$$F_{xz} = -cp_2(\mathbf{x} - \mathbf{x}^3/3\mathbf{a}^2 + C_1), \qquad M_x = -c^2p_2(\mathbf{x}^2/2 - \mathbf{x}^4/12\mathbf{a}^2 + C_1\mathbf{x} + C_2),$$

from which:

$$\begin{aligned}
\sigma_x &= p_2\left[\frac{3\mathbf{z}}{2}\left(\frac{\mathbf{x}^4}{12\mathbf{a}^2} - \frac{\mathbf{x}^2}{2} + \frac{5\mathbf{a}^2}{12}\right) + \frac{5\mathbf{z}^3 - 3\mathbf{z}}{10}\left(1 - \frac{\mathbf{x}^2}{\mathbf{a}^2}\right) + \frac{105\mathbf{z}^5 + 70\mathbf{z}^3 - 87\mathbf{z}}{1400\mathbf{a}^2} - \frac{3\mathbf{z}^2 - 1}{6\mathbf{a}^2}\right] \\
\sigma_z &= p_2\left[\frac{\mathbf{z}^3 - 3\mathbf{z} + 2}{4}\left(\frac{\mathbf{x}^2}{\mathbf{a}^2} - 1\right) - \frac{\mathbf{z}^5 - 2\mathbf{z}^3 + \mathbf{z}}{20\mathbf{a}^2}\right] \qquad (3.30b) \\
\sigma_{xz} &= p_2\left[\frac{3(\mathbf{z}^2 - 1)}{4}\left(1 - \frac{\mathbf{x}^2}{3\mathbf{a}^2}\right) + \frac{5\mathbf{z}^4 - 6\mathbf{z}^2 + 1}{20\mathbf{a}^2}\right]\mathbf{x}
\end{aligned}$$

where $\mathbf{a} = a/c$.

Similar exact closed solutions for tangential loads which are a power function of x are contained in Eqs. (3.29). For example if we take all constants of integration as zero we find the following stresses in beams which are fixed at the right end and free except for balanced stress systems on the left end $x = 0$, and loaded by a uniform tangential force per unit area $t_x = q_0$ on the top surface:

$$\sigma_x = \frac{q_0}{2}(3\mathbf{z} - 1)\mathbf{x}, \qquad \sigma_z = 0, \qquad \sigma_{xz} = -\frac{q_0}{4}(3\mathbf{z}^2 - 2\mathbf{z} - 1) \qquad (3.31a)$$

or loaded by a linearly varying tangential force per unit area $t_x = q_1\mathbf{x}$ on the top surface:

$$\begin{aligned}
\sigma_x &= \frac{q_1}{20}\left[5(3\mathbf{z} - 1)\mathbf{x}^2 - 10\mathbf{z}^3 + 10\mathbf{z}^2 + 6\mathbf{z} - \frac{10}{3}\right] \\
\sigma_z &= \frac{q_1}{4}(\mathbf{z}^3 - \mathbf{z}^2 - \mathbf{z} + 1), \qquad \sigma_{xz} = -\frac{q_1}{4}(3\mathbf{z}^2 - 2\mathbf{z} - 1)\mathbf{x}
\end{aligned} \qquad (3.31b)$$

and so on. Other end conditions can be obtained by superposing known solutions for axial loads, end moments, local stress fields on the ends such as those discussed in the next section, and rigid body displacements.

3.4 SOME USEFUL LOCAL STRESS FIELDS FOR BEAMS

Plane stress solutions of the type discussed in the last section permit the solving of many practical beam problems almost exactly, except that the solutions usually involve loads which are different from but statically equivalent to the actual loading over a small part of the surface of the beam. By Saint-Venant's principle the difference between the stresses caused by actual loading and by the statically equivalent loading for which we have a solution is a 'local stress field', that is a local stress distribution.

Harmonic Loading on Semi-Infinite Plate

The following solutions of Eq. (3.17) represent a very simple illustration of such local stress fields. They can be derived in the same way as (or derived from) the solutions of Eqs. (3.26) and (3.27), and when they apply represent special simpler cases of these solutions. They are:

$$\phi = \frac{l^4 p}{\pi^4} \sin\frac{\pi x}{l} e^{-\pi z/l}\left(2+\frac{\pi z}{l}\right), \qquad Eu_x = \frac{lp}{\pi} \sin\frac{\pi x}{l} e^{-\pi z/l}\left[(1-\nu)-(1+\nu)\frac{\pi z}{l}\right],$$
$$Eu_z = \frac{lp}{\pi}\cos\frac{\pi x}{l} e^{-\pi z/l}\left[-2-(1+\nu)\frac{\pi z}{l}\right], \qquad \sigma_x = p\cos\frac{\pi x}{l} e^{-\pi z/l}\left(1-\frac{\pi z}{l}\right), \tag{3.32}$$
$$\sigma_z = p\cos\frac{\pi x}{l} e^{-\pi z/l}\left(1+\frac{\pi z}{l}\right), \qquad \sigma_{xz} = p\sin\frac{\pi x}{l} e^{-\pi z/l}\frac{\pi z}{l}$$

which describes the displacements and stresses under a harmonically distributed normal loading on the edge of a 'semi-infinite' plate, Fig. 3.11(a), and:

$$\phi = -\frac{l^4 q}{\pi^4}\sin\frac{\pi x}{l} e^{-\pi z/l}\left(1+\frac{\pi z}{l}\right), \qquad Eu_x = \frac{lq}{\pi}\sin\frac{\pi x}{l} e^{-\pi z/l}\left[-2+(1+\nu)\frac{\pi z}{l}\right],$$
$$Eu_z = \frac{lq}{\pi}\cos\frac{\pi x}{l} e^{-\pi z/l}\left[(1-\nu)+(1+\nu)\frac{\pi z}{l}\right], \qquad \sigma_x = -q\cos\frac{\pi x}{l} e^{-\pi z/l}\left(2-\frac{\pi z}{l}\right), \tag{3.33}$$
$$\sigma_z = -q\cos\frac{\pi x}{l} e^{-\pi z/l}\frac{\pi z}{l}, \qquad \sigma_{xz} = q\sin\frac{\pi x}{l} e^{-\pi z/l}\left(1-\frac{\pi z}{l}\right)$$

which describes the displacements and stresses under a similar tangential loading, illustrated in Fig. 3.11(b).

It will be seen that the effects of such self-balancing load systems die out rapidly with the depth, and would be negligible at depths of around the wave length $2l$. These solutions could therefore be used with good approximation for loads acting on one side of a beam of depth of the order $2l$ or more, since the stresses required to be acting upon the opposite side could be ignored in most practical cases; they can be combined to give other load distributions. Such solutions are also useful for satisfying edge boundary conditions of plates under lateral loading, as will be discussed in Sec. 5.3; in such an application it is the usually large width of the plate which must be equal to or greater than $2l$.

It should be remembered that, as discussed in Sec. 3.2, these are approximate solutions (since ϕ is a biharmonic but not a harmonic function of x and z) which will not be very good approximations unless the *width* of the beam (or thickness of the plate) is *small* compared to $2l$. Solutions not subject to this limitation will be developed later in Sec. 5.4.

Concentrated Loads

As can be checked from Eqs. (3.9g, h) a concentrated force (concentrated in the plane of the plate but uniformly distributed across the thickness) acting upon the

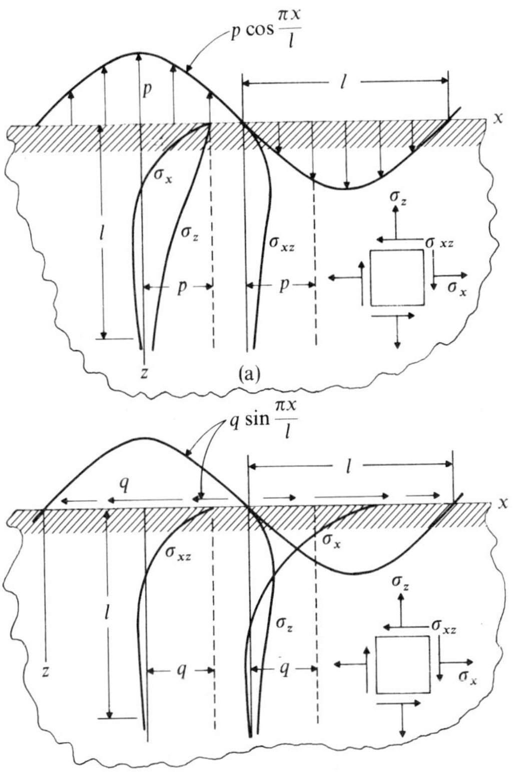

FIGURE 3.11

apex of a wedge-shaped plate which extends indefinitely, Fig. 3.12(a), or upon the edge of a semi-infinite plate (which can be considered to be a wedge with an angle of 180°), produces the simple 'radial' stress system shown, consisting only of radially directed normal stresses whose magnitude varies as the cosine of the angle with the line of action of the force, and inversely as the distance from the apex. If the apex or point of application of the force is taken as the origin of rectangular x, z coordinates, this stress distribution and corresponding displacements can be described by the solution of Eqs. (3.17):

$$\phi = [(3Ax^2z + Az^3 + 3Bxz^2 + Bx^3)\tan^{-1} x/z + (Ax^3 - Bz^3)\log(x^2 + z^2) + Bx^2z - Axz^2]/6$$

from which:

$$Eu_x = A(1-\nu)\tan^{-1}\frac{x}{z} - B\log(x^2+z^2) - \frac{3(1+\nu)Azx + 8Bx^2 + (11+3\nu)Bz^2}{3(x^2+z^2)}$$

$$\sigma_x = \frac{2x^2(Az - Bx)}{(x^2+z^2)^2}, \qquad \sigma_z = \frac{2z^2(Az - Bx)}{(x^2+z^2)^2}, \qquad \sigma_{xz} = \frac{2zx(Az - Bx)}{(x^2+z^2)^2} \qquad (3.34)$$

where ϕ satisfies $\nabla^4\phi = 0$, and A and B depend upon the magnitude and direction

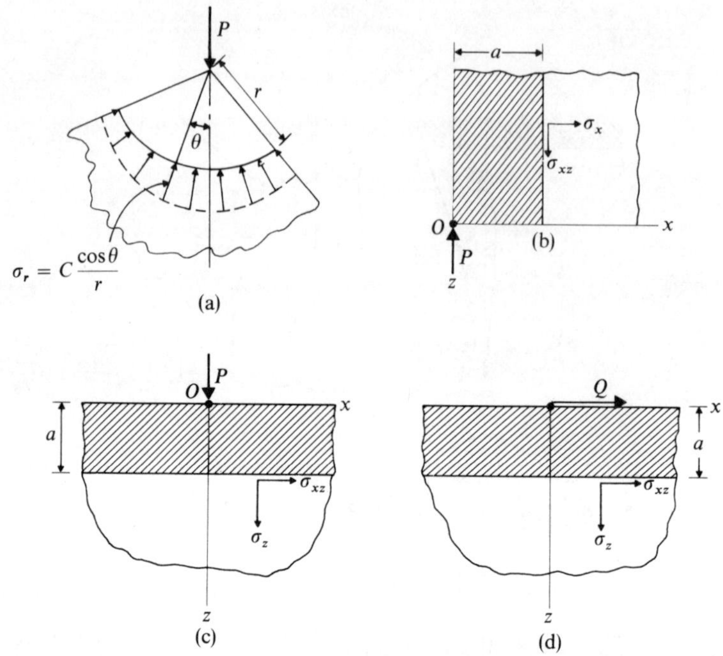

FIGURE 3.12

of the force and upon the directions of the straight boundaries of the body. Eu_z is the same (except for a rigid body displacement) as $-Eu_x$ with x, z and A, B interchanged.

Concentrated Load on Corner

Consider for example the case of a vertical reaction P (force per unit thickness of the plate) which acts upon a lower corner of a beam, which we will have to assume for the present to be indefinitely high and long, as indicated in Fig. 3.12(b). We can determine A and B from the equilibrium condition of a free body on which P acts, for example the portion of the body shown shaded in the figure to the left of the line $x = a$. The distance a must be finite because the stresses approach infinity at the origin, so we would have indeterminate quantities to deal with if a is infinitesimal.

The only forces acting on this body are P and the stresses σ_x and σ_{xz} acting upon the side $x = a$. Our solution gives zero forces on radial lines through the apex such as $0x$ and $0z$, and the forces acting upon a horizontal top surface an infinite distance above the $0x$ line can be taken as zero because the area of this surface is finite and the stresses tend to zero at infinity. Hence, equilibrium requires that, for each unit thickness of the plate:

$$0=\int_{-\infty}^{0}\sigma_{x(x-a)}\,dz=\int_{-\infty}^{0}\frac{2a^2(Az-Ba)}{(a^2+z^2)^2}\,dz=\left|_{-\infty}^{0}\frac{-Aa^2-Baz}{a^2+z^2}-B\tan^{-1}\frac{z}{a}=-A-\frac{\pi}{2}B\right.$$

$$P=\int_{-\infty}^{0}\sigma_{xz(x-a)}\,dz=\int_{-\infty}^{0}\frac{2a(Az^2-Baz)}{(a^2+z^2)^2}\,dz=\left|_{-\infty}^{0}\frac{-Aaz+Ba^2}{a^2+z^2}+A\tan^{-1}\frac{z}{a}=\frac{\pi}{2}A+B\right. \tag{3.35}$$

from which:

$$A=\frac{2\pi P}{\pi^2-4}\approx 1{\cdot}0705\,P,\qquad B=\frac{-4P}{\pi^2-4}\approx -0{\cdot}6815\,P \tag{3.36}$$

These results will be used later in this section.

Concentrated Load on Semi-Infinite Plate

In the case shown in Fig. 3.12(c), B is zero from symmetry about the z axis, because σ_x and σ_z must be symmetrical and σ_{xz} must be antisymmetrical about this axis. From vertical equilibrium of the element shown shaded:

$$P=-\int_{-\infty}^{\infty}\sigma_{z(z=a)}\,dx=\int_{-\infty}^{\infty}\frac{-2Aa^3\,dx}{(x^2+a^2)^2}=\left|_{-\infty}^{\infty}\frac{-Aax}{x^2+a^2}-A\tan^{-1}\frac{x}{a}=-\pi A,\right.$$

$$A=-\frac{P}{\pi} \tag{3.37}$$

Similarly, for the case shown in Fig. 3.12(d), $A=0$ from antisymmetry, and horizontal equilibrium of the shaded element requires that $Q=-\int_{-\infty}^{\infty}\sigma_{xz(z=a)}\,dx$, from which $B=Q/\pi$. These problems were solved by Boussinesq and others around 1890.

Although the stresses from such concentrated loads decrease inversely as the distance from the point of application, such a stress distribution due to an unbalanced load is not a 'local' stress distribution like that from a balanced system of forces, which produces stresses decreasing with a higher power of the distance. However we can construct useful balanced systems using these results.

Uniformly Distributed Load on Edge of Semi-Infinite Plate

Consider a semi-infinite plate on the edge of which a uniformly distributed compression p acts over the distance $-a<x'<a$, Fig. 3.13(a). Then the function ϕ or the displacements or stresses corresponding to the part of the load $p\,dx'$ can be found from Eqs. (3.34), (3.37) by substituting $(x-x')$ for x, $p\,dx'$ for P and $A=-P/\pi$, $B=0$. The same quantities due to the entire load can then be found by integrating these expressions as x' goes from $-a$ to a. For example, the σ_x stress at point O with coordinates x, z, due to the whole load, is:

$$\sigma_x=-\frac{2pz}{\pi}\int_{-a}^{a}\frac{(x-x')^2\,dx'}{[(x-x')^2+z^2]^2}=-\frac{2pz}{\pi}\int_{-a}^{a}\frac{[(x^2)-(2x)x'+x'^2]\,dx'}{[(x^2+z^2)-(2x)x'+x'^2]^2} \tag{3.38}$$

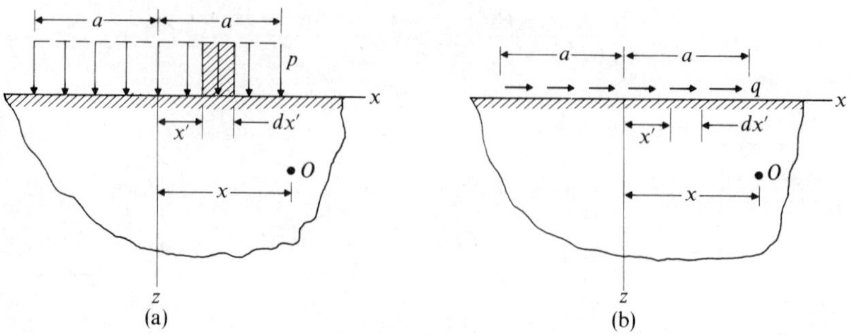

FIGURE 3.13

where x, z are considered as constants during the integration. In this way we find, using the symbols $\alpha = (x - a)/z$, $\beta = (x + a)/z$:

$$\begin{aligned} \sigma_x &= \frac{p}{\pi}\left(\tan^{-1}\alpha - \tan^{-1}\beta - \frac{\alpha}{1+\alpha^2} + \frac{\beta}{1+\beta^2}\right) \\ \sigma_z &= \frac{p}{\pi}\left(\tan^{-1}\alpha - \tan^{-1}\beta + \frac{\alpha}{1+\alpha^2} - \frac{\beta}{1+\beta^2}\right) \\ \sigma_{xz} &= \frac{p}{\pi}\left(\frac{-1}{1+\alpha^2} + \frac{1}{1+\beta^2}\right) \end{aligned} \tag{3.38a}$$

Superposing solution (3.34) with $B = 0$ and $A = -P/\pi = -2000a/\pi$ from Eq. (3.37) for a concentrated normal compressive force of $2000a$ units, upon solution (3.38a) for $p = -1000$, we obtain the local stress field due to the balanced load system shown in Fig. 3.14(a). The net stresses for points at intervals of $2a$ are shown in Fig. 3.14(b), calculated to the nearest whole number (that is to a tenth of one percent of the maximum stress). Slide rule calculations are sufficient for this order of accuracy. Figures 3.14(c), (d) show a similar field obtained by superposing two distributed loads, one for $p = 2000$ distributed over the width a, and the other for $p = -1000$ distributed over $2a$. It will be seen that these results can be applied to loads on one surface of a rectangular beam whose height is of the order of $8a$ or more without serious error, since the stresses on the opposite side will be negligible; they can be combined in various ways, and the values at other points than those shown can be found by cross plotting.

Similarly, for a distributed tangential load, Fig. 3.13(b):

$$\begin{aligned} \sigma_x &= \frac{q}{\pi}\left(\frac{1}{1+\alpha^2} - \frac{1}{1+\beta^2} - \log\frac{1+\beta^2}{1+\alpha^2}\right) \\ \sigma_z &= \frac{q}{\pi}\left(\frac{-1}{1+\alpha^2} + \frac{1}{1+\beta^2}\right) \\ \sigma_{xz} &= \frac{q}{\pi}\left(\tan^{-1}\alpha - \tan^{-1}\beta - \frac{\alpha}{1+\alpha^2} + \frac{\beta}{1+\beta^2}\right) \end{aligned} \tag{3.38b}$$

Figures 3.14(e), (f) show the local stress field due to this kind of load balanced by a concentrated tangential force at its middle. In this case we have infinite stresses

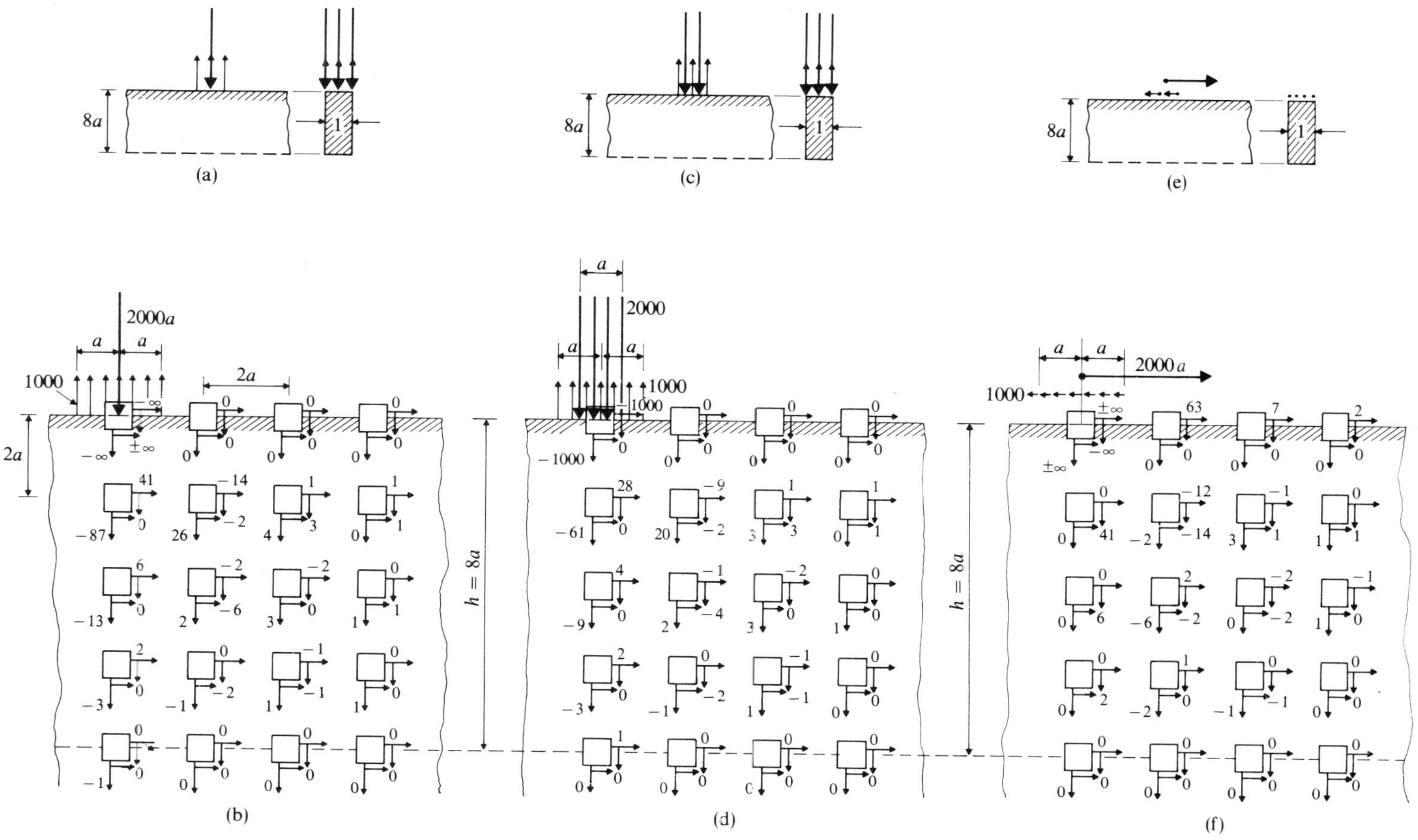

FIGURE 3.14

not only at the point of application of the concentrated load but also, due to the last term in the expression for σ_x, Eqs. (3.38b), at the ends of the distributed load.

Corrections to Classical Beam Stresses for Concentrated Loads

Classical beam theory satisfies equilibrium conditions, and the difference between the actual stresses produced by localized loads acting on the side of a beam and the stresses predicted by the classical theory for the same case forms a local stress field. Such local stress fields added to classical solutions give the correct stresses around the loads.

Such a correction field for a concentrated load could be found by taking a beam with a length of about four times the height, outside of which the local stresses will be negligible. The Boussinesq solution (3.34) and (3.37) can then be used for a concentrated load on the middle of one side, cancelling the stresses which this solution requires on the other side of the beam by using Eqs. (3.28), (3.29), then subtracting the classical solution and removing axial forces and moments on the ends by superposing the simple elasticity solutions which have been discussed for uniform axial stress and pure bending. A calculation of this type for the analogous plate case will be presented in Sec. 5.3.

This problem has been solved by von Kármán and Seewald* in another manner by means of trigonometric–hyperbolic series, Eq. (3.27), using Fourier integrals to sum the series, which converge rather slowly in this case. Figure 3.15 shows their results, from which the solution of Fig. 3.14(b) has been subtracted, so as to convert the concentrated load to a short, uniformly distributed load. This eliminates the physically unrealistic stress situation around a concentrated load, where the stresses tend to infinity. The solutions of Fig. 3.14(b) or (d) can of course be added to Fig. 3.15 to convert it back to the concentrated load case or to a different distributed load case. Equations (3.34), (3.37), and (3.38a) then permit the detailed evaluation of the stresses near the load.

End Corrections

The problem of satisfaction of realistic end conditions is somewhat more difficult than that of dealing with the middle part of a beam, because any local stress field which is introduced to modify end conditions must involve no normal or shearing stresses on the adjacent top and bottom surfaces of the beam. By Saint-Venant's principle, end corrections must be balanced, that is have zero resultants, in order to be represented by a local stress field.

The ideal solution to the problem would be to have a series of local stress fields which would satisfy the above conditions on the top and bottom surfaces, and would produce balanced normal or tangential stress distributions on the end surface, which could be combined to give any desired balanced force distributions

*See Timoshenko, *Theory of Elasticity*, 1st edn., p. 99, McGraw-Hill, New York.

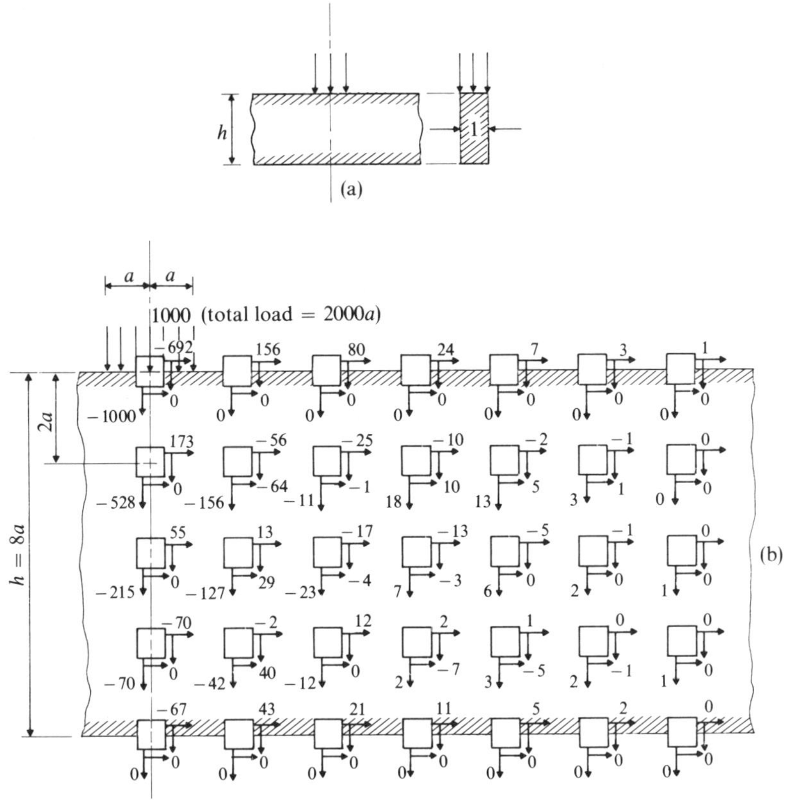

FIGURE 3.15
Actual stresses minus classical beam theory stresses

or corresponding displacements on the ends. Preferably the terms of the series should be orthogonal over this interval so that the required combination of terms could easily be determined.

Horvay* has evolved a series involving Airy stress functions ϕ_n, Eqs. (3.16a), each of which is in the form of a product of a function of x by a function of z, and which satisfy all of the above conditions except the condition of continuity Eq. (3.16c), which is satisfied approximately by using energy principles to determine suitable forms and coefficients for the functions of x. Comparison with more exact solutions indicate that the approximations are close. Tables of values of the required functions have also been provided to facilitate their use.

Referring to Fig. 3.16(a), this series can be written as: $\phi = \Sigma_n \phi_n = \Sigma_n (a_n g_n + b_n h_n) f_n$, where $n = 2, 3, 4, 5, \ldots$, the even values applying to symmetrical (membrane) loading and the odd values to antisymmetrical (flexural) loading.

*G. Horvay, 'Tables of self-equilibrating functions', *J. Math. and Physics*, vol. 33, no. 4, Jan. 1955, p. 360.

The parts $a_n g_n f_n$ give normal loads on the end $x = 0$ and the parts $b_n h_n f_n$ give tangential loads on the end. Each f_n is a function of z only, g_n and h_n are functions of x only, and a_n and b_n are numerical coefficients which can be determined by a method analogous to harmonic analysis to make the series as a whole yield any desired balanced stress distribution on the end cross section. In some of the applications which we will make this requirement can be satisfied with one or two terms only, so that we will not need to make use of this orthogonality of the terms to determine the a_n, b_n coefficients.

The first eight f_n's are, using $\mathbf{z} = z/c$:

$$f_2 = \frac{3}{16}\sqrt{5 \cdot 7}(1-\mathbf{z}^2)^2, \qquad f_9 = \frac{1}{16}\sqrt{\frac{11 \cdot 13 \cdot 23}{5}}(-5\mathbf{z} + 85\mathbf{z}^3 - 323\mathbf{z}^5 + 323\mathbf{z}^7)f_2$$

$$f_4 = \frac{1}{2}\sqrt{\frac{13}{5}}(-1 + 11\mathbf{z}^2)f_2, \qquad f_7 = \frac{1}{8}\sqrt{11 \cdot 13 \cdot 19}(\mathbf{z} - 10\mathbf{z}^3 + 17\mathbf{z}^5)f_2$$

$$f_6 = \frac{1}{8}\sqrt{\frac{11 \cdot 17}{5}}(1 - 26\mathbf{z}^2 + 65\mathbf{z}^4)f_2, \qquad f_5 = \frac{1}{2}\sqrt{11}(-3\mathbf{z} + 13\mathbf{z}^3)f_2 \tag{3.39}$$

$$f_8 = \frac{1}{16}\sqrt{11 \cdot 13}(-1 + 45\mathbf{z}^2 - 255\mathbf{z}^4 + 323\mathbf{z}^6)f_2, \qquad f_3 = \sqrt{11}\mathbf{z}f_2$$

The formulas for g_n and h_n are, using $\mathbf{x} = x/c$:

$$g_n = e^{-\alpha_n \mathbf{x}}\left(\cos \beta_n \mathbf{x} + \frac{\alpha_n}{\beta_n}\sin \beta_n \mathbf{x}\right), \qquad h_n = e^{-\alpha_n \mathbf{x}}\left(\frac{1}{\beta_n}\sin \beta_n \mathbf{x}\right) \tag{3.40}$$

where $\alpha_n = \alpha_2, \alpha_3, \ldots = 2{\cdot}075\,149$, $3{\cdot}655\,963$, $5{\cdot}258\,543$, $6{\cdot}942\,082$, $8{\cdot}728\,814$, $10{\cdot}627\,227$, $12{\cdot}640\,404$, $14{\cdot}769\,022$, and $\beta_n = \beta_2, \beta_3, \ldots = 1{\cdot}142\,910$, $1{\cdot}538\,202$ $2{\cdot}062\,104$, $2{\cdot}681\,885$, $3{\cdot}404\,732$, $4{\cdot}235\,321$, $5{\cdot}174\,922$, $6{\cdot}222\,861$.

The stresses are given by Eqs. (3.16a) with y replaced by $\mathbf{z}$:

$$\sigma_x = \partial^2\phi/\partial\mathbf{z}^2 = \sum_n (a_n g_n + b_n h_n)\, \partial^2 f_n/\partial\mathbf{z}^2,$$

$$\sigma_z = \partial^2\phi/\partial\mathbf{x}^2 = \sum_n (a_n\, \partial^2 g_n/\partial\mathbf{x}^2 + b_n\, \partial^2 h_n/\partial\mathbf{x}^2)f_n,$$

$$\sigma_{xz} = -\partial^2\phi/\partial\mathbf{z}\,\partial\mathbf{x} = -\sum_n (a_n\, \partial g_n/\partial\mathbf{x} + b_n\, \partial h_n/\partial\mathbf{x})\, \partial f_n/\partial\mathbf{z}.$$

At the end $x = 0$, h_n and $\partial g_n/\partial\mathbf{x}$ are zero while g_n and $\partial h_n/\partial\mathbf{x}$ equal unity, so that these expressions for σ_x and σ_{xz} reduce to:

$$\sigma_{x(x=0)} = \sum_n a_n\, \partial^2 f_n/\partial\mathbf{z}^2, \qquad \sigma_{xz(x=0)} = -\sum_n b_n\, \partial f_n/\partial\mathbf{z}.$$

The shapes of these normal and tangential end loads for the first four terms are shown in Figs. 3.16(b), (c) respectively. It can easily be checked that the boundary conditions $z = \pm c$, $\mathbf{z} = \pm 1$: $\sigma_z, \sigma_{xz} = 0$ are satisfied.

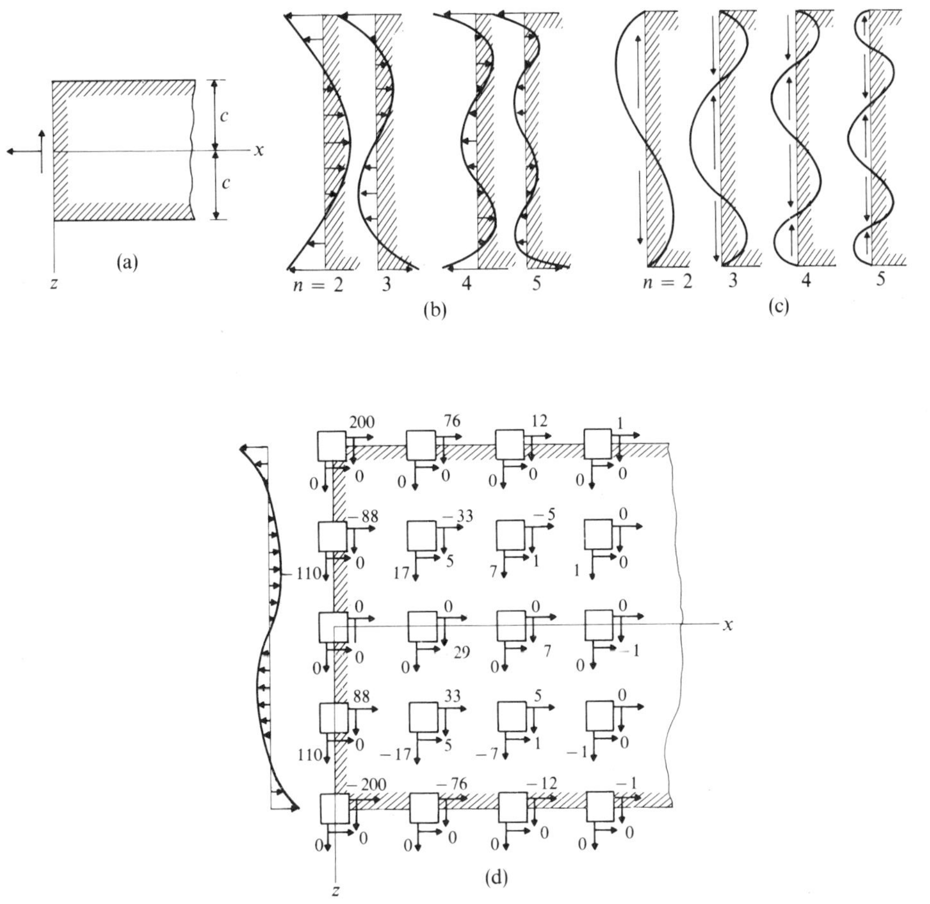

FIGURE 3.16

It can also be checked that:

$$\int_{-1}^{1} d\mathbf{z}\, f_n \cdot f_{n'} = 1 \qquad \text{when} \quad n = n' \tag{3.40a}$$

$$\int_{-1}^{1} d\mathbf{z}\, f_n \cdot f_{n'} = 0 \qquad \text{when} \quad n \neq n' \tag{3.40b}$$

In general we wish to use this orthogonality of the f_n functions in the interval -1 to 1 represented by Eq. (3.40b) to determine the coefficients a_n and b_n so as to make the series of functions as a whole yield the desired distribution of $\sigma_{x(x=0)}$ and $\sigma_{xz(x=0)}$ on the end surface.

To do this it is necessary first to integrate the expressions for these stresses in the z direction to reduce them to a form involving the f_n's themselves rather than their derivatives, since the derivatives of the f_n's are not orthogonal. For this purpose we must therefore integrate the expression for σ_x twice and the expression for σ_{xz} once with respect to z. It is determined in the reference cited above that for the membrane case of symmetry about the middle surface (n even) the limits of this integration should be as follows:

$$\int_{-1}^{z} d\mathbf{z}' \int_{0}^{z'} d\mathbf{z}''\sigma_x(\mathbf{z}'')_{(x=0)} = \sum_n a_n f_n, \qquad \int_{-1}^{z} d\mathbf{z}'\sigma_{xz}(\mathbf{z}')_{(x=0)} = -\sum_n b_n f_n \tag{3.40c}$$

In a practical problem the left sides of these equations will be known functions of z, call them $s_x(z)$ and $s_{xz}(z)$ respectively. Then, in the same way as in harmonic analysis, we determine each particular coefficient a_n and b_n by multiplying both sides of these equations by dz and by the corresponding f_n, and integrating both sides over the interval -1 to 1. By Eqs. (3.40a, b) we will thus obtain:

$$a_n = \int_{-1}^{1} d\mathbf{z} f_n s_x, \qquad b_n = \int_{-1}^{1} d\mathbf{z} f_n s_{xz} \tag{3.40d}$$

For the flexure case of antisymmetry (n odd) the limits of integration should be:

$$\begin{gathered}\int_{0}^{z} d\mathbf{z}' \int_{-1}^{z'} d\mathbf{z}''\sigma_x(\mathbf{z}'')_{(x=0)} = \sum_n a_n f_n,\\ \int_{0}^{z} d\mathbf{z}'\sigma_{xz}(\mathbf{z}')_{(x=0)} = \sum_n b_n f_n\end{gathered} \tag{3.40e}$$

As a simple application of these functions, a comparison of Eq. (3.22a) with Eq. (3.39) shows that the addition of the local stress field defined by:

$$\phi_3 = (p_0/40) f_3 g_3 \tag{3.41}$$

to each end of the uniformly loaded, hinged-end beam whose solution was given by Eq. (3.22) eliminates the undesired σ_x stresses at the ends. Figure 3.16(d) shows this local stress field for $p_0 = 1000$.

Concentrated Reaction on the Corner

To make elasticity solutions for beams complete, it is necessary to replace the end reactions produced by parabolically distributed shear forces on the ends by a somewhat more realistic reaction such as a concentrated reaction at the corner. To demonstrate how such results can be achieved we will now work out the local stress field due to the balanced load system shown in Fig. 3.17(a), which, when superposed upon upward shear produced reactions of total magnitude P, converts the reactions to concentrated corner reactions.*

Such a load system can not be produced by superposing the shear distributions shown in Fig. 3.16(c), since these all must have zero magnitudes at the corners. We therefore start out with the case shown in Fig. 3.12(b), which

*L. H. Donnell, 'End reactions in beams', pp. 161–5, *Problems of Mechanics*, volume commemorating the sixtieth anniversary of V. V. Novoshilov, Shipbuilding Pub. House, Leningrad, 1970 (in Russian).

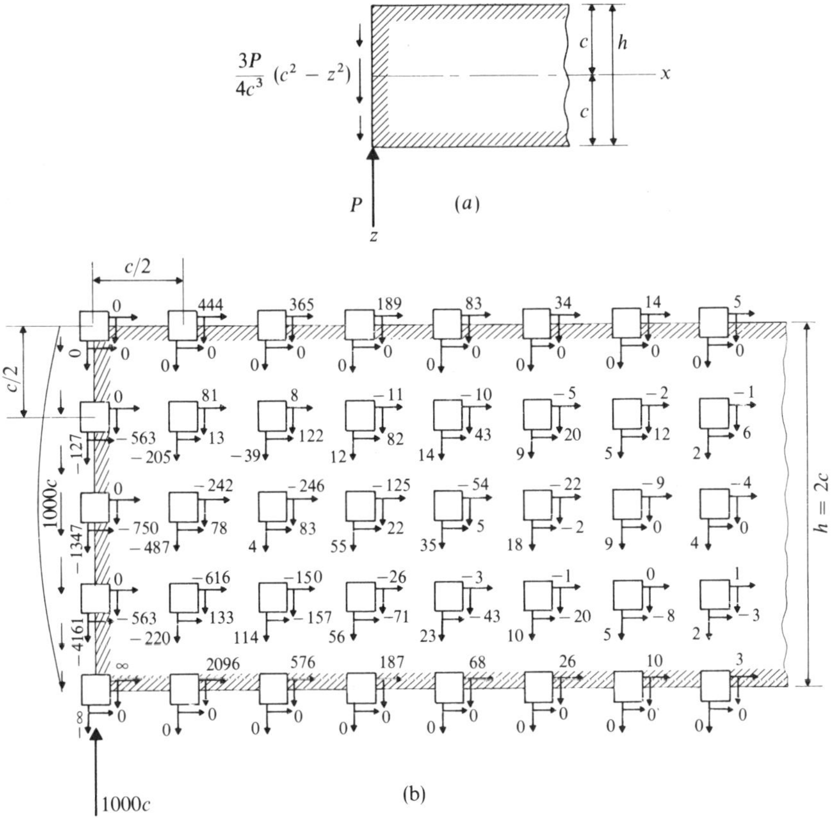

FIGURE 3.17 (from *Problems of Mechanics*, V. V. Novozhilov Commemorative Volume, p. 161, 1970, courtesy of Shipbuilding Pub. House, Leningrad)

produces stresses defined by Eqs. (3.34) and (3.36), plus the case of Fig. 2.12, Eq. (3.25), using $M_1 = 0$ and $M_2 = -Pl$, $F = -P$. This gives us the desired end loads, but the first case also produces undesired σ_z and σ_{xz} stresses on the top surface of the beam; these are obtained by substituting $z = -h$ and expressions (3.36) into (3.34):

$$\sigma_{z\,(\text{top})} = \frac{-4Ph^2}{(\pi^2 - 4)} \frac{(\pi h - 2x)}{(x^2 + h^2)^2},$$
$$\sigma_{xz\,(\text{top})} = \frac{4Phx}{(\pi^2 - 4)} \frac{(\pi h - 2x)}{(x^2 + h^2)^2} \tag{3.42}$$

To eliminate these distributed forces on the top surface, we add the terms before the square brackets of Eqs. (3.28), (3.29), Fig. 3.10, taking b_z, $b_x = 0$, $t_z = \sigma_{z\,(\text{top})}$, $t_x = \sigma_{xz\,(\text{top})}$, Eq. (3.42). The values of t_z and its integrals required in Eqs. (3.28) are then, using the dimensionless $\mathbf{x} = x/c$, $\mathbf{z} = z/c$ and $P' =$

$16P/[(\pi^2-4)c]\approx 2{\cdot}725\,P/c$, and Eqs. (2.3), (2.4):

$$\frac{M_x}{c^2}=\int\frac{F_{xz}}{c}\,d\mathbf{x}=\frac{P'}{8}(4+\pi\mathbf{x})\tan^{-1}\frac{\mathbf{x}}{2}+C_1\mathbf{x}+C_2$$
$$\frac{F_{xz}}{c}=-\int t_z\,d\mathbf{x}=\frac{P'}{4}\left[\frac{4+\pi\mathbf{x}}{(\mathbf{x}^2+4)}+\frac{\pi}{2}\tan^{-1}\frac{\mathbf{x}}{2}\right]+C_1 \tag{3.43}$$
$$t_z=-2P'\frac{\pi-\mathbf{x}}{(\mathbf{x}^2+4)^2}$$

Similarly t_x and its integral and derivative required in Eqs. (3.29) are:

$$\int t_x\,d\mathbf{x}=-\frac{P'}{4}\left[\frac{2(\pi-\mathbf{x})}{\mathbf{x}^2+4}+\tan^{-1}\frac{\mathbf{x}}{2}\right]+C_3 \tag{3.44}$$
$$t_x=P'\frac{\mathbf{x}(\pi-\mathbf{x})}{(\mathbf{x}^2+4)^2},\qquad \frac{dt_x}{d\mathbf{x}}=P'\frac{4\pi-8\mathbf{x}-3\pi\mathbf{x}^2+2\mathbf{x}^3}{(\mathbf{x}^2+4)^3}$$

We have now eliminated the forces from the top surface of the beam and there are still no forces on the bottom surface, but we have added unwanted σ_x and σ_{xz} stresses on the end surface. These σ_x and σ_{xz} stresses on the end, obtained by setting $x=0$ in Eqs. (3.43) and (3.44) and substituting these values into the first and third of Eqs. (3.28) and (3.29) respectively, are:

$$\sigma_x=\frac{\pi P'}{480}(25-63\mathbf{z}+15\mathbf{z}^2-45\mathbf{z}^3)+(3C_2\mathbf{z}-C_3+3C_3\mathbf{z})/2+\cdots$$
$$\sigma_{xz}=3(P'+4C_1)(1-\mathbf{z}^2)/16 \tag{3.45}$$

We will use the constants of integration C_1, C_2, C_3 to make the resultant forces and moments of these stresses over the end equal to zero, so that what is left will be a self-balanced system of forces, which can be annulled by the self-balanced systems of Eqs. (3.39), (3.40). Setting $\int_{-1}^{1}\sigma_x\,d\mathbf{z}$, $\int_{-1}^{1}\sigma_x\mathbf{z}\,d\mathbf{z}$, $\int_{-1}^{1}\sigma_{xz}\,d\mathbf{z}=0$ and solving, we find:

$$C_1=-P'/4,\qquad C_2=0,\qquad C_3=\pi P'/8 \tag{3.46}$$

Putting these values in Eq. (3.45) we obtain:

$$\sigma_{xz}=0,\qquad \sigma_x=\frac{\pi P}{30(\pi^2-4)c}(-5+27\mathbf{z}+15\mathbf{z}^2-45\mathbf{z}^3)$$
$$=\left(\frac{\pi P}{24(\pi^2-4)c}\right)4(-1+3\mathbf{z}^2)+\left(\frac{3\pi P}{40(\pi^2-4)c}\right)4(3\mathbf{z}-5\mathbf{z}^3) \tag{3.47}$$

When the even and odd powers of $\mathbf{z}$ in the expression for σ_x are separated as shown in Eq. (3.47), and compared with Eq. (3.39), it becomes evident that the stresses on the end $x=0$ can be eliminated by adding stress fields described by the stress functions $\phi_2+\phi_3=a_2f_2g_2+a_3f_3g_3$, Eqs. (3.39), (3.40), where:

$$a_2=-\frac{\pi P}{24(\pi^2-4)c},\qquad a_3=-\frac{3\pi P}{40(\pi^2-4)c} \tag{3.48}$$

We now have the complete solution of the case shown in Fig. 3.17(a), in the form of the sum of the stresses corresponding to: (1) Eqs. (3.34) and (3.36); (2) Eqs. (3.25), with $M_1 = 0$, $M_2 = -Pl$; (3) Eqs. (3.28), (3.29), (3.43), (3.44), (3.46), with b, $b' = 0$; (4) stress functions $\phi_2 + \phi_3$ given by Eqs. (3.39), (3.40), (3.48). It must be kept in mind that the origin of coordinates is at the loaded corner for part (1), and at the center of the end for the other parts. Carrying out these calculations for a point on the middle surface a distance c from the end, for example, the stress components for $P/c = 1000$ are, in the above order:

$$\sigma_x = (-194) + (0) + (0 + 0 - 85 + 10) + (23 + 0) = -246$$

$$\sigma_z = (-194) + (0) + (233 - 16) + (-19 + 0) = 4$$

$$\sigma_{xz} = (194) + (-750) + (591 + 59) + (0 - 11) = 83$$

Figure 3.17(b) shows the stresses for this case calculated at intervals of $c/2$. It will be seen that they become very small at a distance from the end of the order of twice the beam height.

As an illustration of the application of this local stress field, Fig. 3.18 shows the actual stress distribution for the case of a short rectangular beam of length

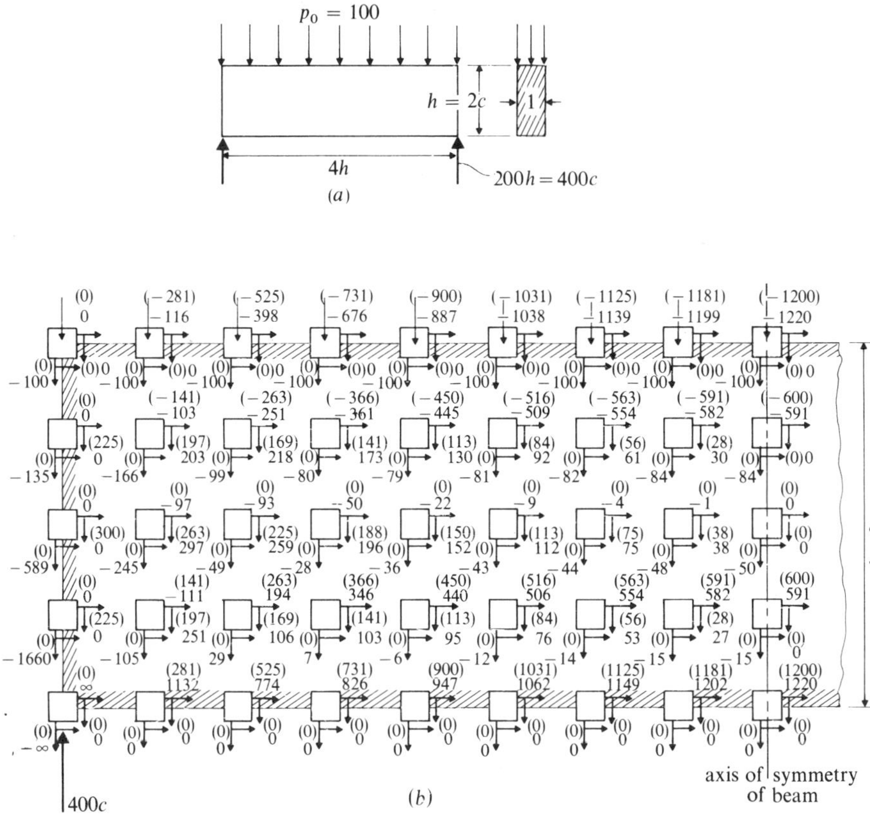

FIGURE 3.18

four times the height under a uniformly distributed pressure on the top surface of $p_0 = 100$ and supported at the bottom corners. This was obtained by superposing the solution of Eq. (3.25), or Eq. (3.28) with $t_z = p_0$ and $b_z = 0$, with local stress fields of Fig. 3.16(d) multiplied by 0·1 and of Fig. 3.17 multiplied by 0·4 applied at the ends. For comparison, the stresses predicted by the classical theory are also shown in parentheses. The differences between the actual and classical values would be somewhat less important for longer beams.

Application of Beam End Corrections to Plate and Shell Edge Correction

Local stress fields such as Eqs. (3.39) and (3.40) and the case just considered are designed to correct beam end conditions, by superposing them on solutions which satisfy only gross boundary conditions, to at least approximately satisfy the actual boundary conditions at every point of the ends. In plate and shell theory we have the same problem that our solutions usually satisfy only gross boundary conditions, and the plane strain analog of the above plane stress beam end corrections can be superposed on such plate or shell solutions at each section of the edges to similarly satisfy the actual boundary conditions at each point of the section. Such applications to plates and shells will be less accurate than for beams, especially if the edges are curved or conditions vary rapidly along the edge, but they are better than nothing when exact edge conditions are important and, as is usually the case, no better correction method is available; they should give a reasonably good approximation if the radii of curvature and the distances over which edge conditions vary appreciably are large compared to the thickness, as is usually the case even for quite 'thick' plates and shells.

Using Eqs. (3.5) or (3.11c), the plane strain analog consists of the same $\sigma_x, \sigma_z, \sigma_{xz}$ stresses as in the plane stress solution, plus a σ_y stress equal to $\nu(\sigma_x + \sigma_z)$, to make $\epsilon_y = (\sigma_y - \nu\sigma_x - \nu\sigma_z)/E = 0$. Since the σ_x and σ_z stresses and hence the added σ_y stresses in such local stress fields will be balanced and highly localized near the edge it is safe to assume that they produce negligible general ϵ_y strains; to be sure some local ϵ_y strains will be produced if the conditions to be corrected vary in the y direction, but the effect of these should not be important if the variation is reasonably gradual, as it will be in most practical applications. These questions will be further discussed and illustrated in Secs. 5.2 and 5.3.

Deflection of Cantilever Beam

It was stated in the discussion of Eq. (3.25), which gives the displacements for the case shown in Fig. 2.12, that the distortion of end cross sections complicates the satisfaction of realistic fixed end conditions. We are now in a position to study such a case. As an example consider the case of a cantilever beam loaded by a force F per unit width at the right end $x = l$, Fig. 2.12, and made of a relatively flexible material cemented to a rigid wall so that we can assume that the displacements u_x, u_z are zero at the wall $x = 0$. Setting $M_2 = 0$, $M_1 = -Fl$ and $x = 0$ in Eqs. (3.25) we obtain:

$$u_{x(x=0)} = \frac{F}{4E}[6(1+\nu) - (2+\nu)\mathbf{z}^2 + 2\mathbf{l}^2]\mathbf{z}, \qquad u_{z(x=0)} = \frac{3\nu F\mathbf{l}}{4E}\mathbf{z}^2 \tag{3.49}$$

To make these displacements at the wall zero we can use the local stress fields defined by Eqs. (3.39), (3.40). We cannot find the displacements corresponding to these stress fields from Eqs. (3.16b) because they do not satisfy elasticity theory exactly. But a good approximation to u_x and u_z can be obtained by integrating the normal strains over the horizontal and vertical directions respectively. Thus, taking $x = \infty$ and $z = 0$, respectively, as fixed:

$$\begin{aligned} u_x &= -\frac{1}{E}\int_\infty^x (\sigma_x - \nu\sigma_z)\, dx' = -\frac{c}{E}\sum_n \left[\frac{d^2 f_n}{d\mathbf{z}^2}\int_\infty^{\mathbf{x}} (a_n g_n + b_n h_n)\, d\mathbf{x}' \right. \\ &\quad \left. - \nu f_n \Big|_\infty^{\mathbf{x}} \frac{d}{d\mathbf{x}}(a_n g_n + b_n h_n)\right] \\ u_z &= \frac{1}{E}\int_0^z (\sigma_z - \nu\sigma_x)\, dz' = \frac{c}{E}\sum_n \left[\frac{d^2}{d\mathbf{x}^2}(a_n g_n + b_n h_n)\int_0^{\mathbf{z}} f_n d\mathbf{z}' \right. \\ &\quad \left. - \nu(a_n g_n + b_n h_n)\Big|_0^{\mathbf{z}} \frac{df_n}{d\mathbf{z}}\right] \end{aligned} \tag{3.50}$$

To annul both u_x and u_z at the wall would require use of the whole series of local stress fields. However, only the u_x displacement of the beam cross section at the wall appreciably affects its rotation at the wall and hence its deflection (from Eq. (3.49) $u_z = 0$ at $x = 0$, $z = 0$). Comparison of the first of Eqs. (3.49) and (3.50) indicates that the u_x displacement at the wall can be annulled by using only the term having the coefficient a_3. With this term, using Eqs. (3.39), (3.40) in the first equation of (3.50), setting $x = 0$, and adding the first of Eqs. (3.49) and a rigid body displacement u_{x0} and rotation θ_y, Eq. (3.6a), the total u_x displacement at the wall becomes:

$$u_{x(x=0)} = \frac{F}{4E}[6(1+\nu) - (2+\nu)\mathbf{z}^2 + 2\mathbf{l}^2]\mathbf{z} + \frac{c}{E}4(3\mathbf{z} - 5\mathbf{z}^3)a_3\frac{2\alpha_3}{\alpha_3^2 + \beta_3^2} + u_{x0} + \theta_y \mathbf{z} \tag{3.51}$$

Equating the sum of the terms involving $\mathbf{z}^3$, $\mathbf{z}$ and the constant term separately to zero and solving for a_3, u_{x0}, and θ_y, we find that the u_x displacement at the wall will be zero if:

$$u_{x0} = 0, \qquad a_3 = -0{\cdot}0269(2+\nu)\frac{F}{c}, \qquad \theta_y = -\frac{F}{20Ec}[3(8+9\nu) + 10\,\mathbf{l}^2] \tag{3.52}$$

We can now determine the total deflection of the middle surface, that is the value of u_z for $z = 0$. Using Eqs. (3.39), (3.40) in the second of Eqs. (3.50) and setting $z = 0$ we find that this is zero for the local stress field. Adding $-\theta_y x$, Eq. (3.6a), to the third of Eqs. (3.25) and setting $z = 0$ we obtain the deflection w of the middle surface of the cantilever with a load F on its end:

$$\begin{aligned} w = u_{z(z=0)} &= \frac{4Fx}{Eh^3}\left(\frac{3lx}{2} - \frac{x^2}{2} + \frac{3(8+9\nu)h^2}{40}\right) \\ w_{max} = w_{(x=l)} &= \frac{4Fl^3}{Eh^3}\left(1 + \frac{3(8+9\nu)}{40}\frac{h^2}{l^2}\right) \end{aligned} \tag{3.53}$$

The parenthesis in the expression for w_{max} represents a correction factor for the classical value $4Fl^3/Eh^3$. As can be seen, it is unimportant for slender beams, adding only about 3 percent to the classical value of the deflection for $l/h = 5$. This correction factor is about half way between the values which would be obtained by adding to the deflection the length of the beam times the average and the maximum transverse shear strain.

Actual 'Concentrated' Loads

In previous discussions the theoretical concept of a 'concentrated' load has been taken for granted; this is very convenient, and is sufficiently accurate for calculating the conditions a small distance away (say a distance greater than $2\sqrt{A}$, where A is one of the areas defined below) from the theoretical point of application. In the immediate neighborhood of this point, however, it should be recognized that the load must actually be distributed over a small but finite area. The magnitude of this area may be determined by yielding of the material of one or both bodies in the neighborhood of the contact point, in which case the area can be taken as $A_y = P/\sigma_y$, where P is the normal load and σ_y is the lower of the yield points of the two materials.

If the material remains elastic, the area is determined by Hertz' well known theory for elastic contact stresses,* which is based on the assumption that near a point of contact of curved surfaces the Boussinesq solution for a load on a semi-infinite solid can be used without serious error to satisfy the required displacement conditions at the surface of contact. In Hertz' solution the area of contact depends somewhat on the angle between the planes of the principal curvatures of the two bodies in contact but has an approximate value of $A_{el} = 2{\cdot}6(P/E'c)^{2/3}$. Here E' is the arithmetical average of the values of $E/(1-\nu^2)$ for the two materials, and c is the average curvature (the reciprocal of the radius of curvature) of the bodies at the point of contact, that is the arithmetical average of four curvatures consisting of the two principal curvatures of each body (taking a concave curvature as negative). In this elastic case the maximum stress is at the center of the elliptical contact area and is 1·5 times the average stress P/A_{el}; however, it is probably safe to use the average stress for considering possibilities of failure, since there is a favorable triaxial compression condition in the middle part of the area (due to the compressed material tending to expand but being restrained by surrounding material). If $P/A_{el} < \sigma_y$, that is if $P/A_{el} < P/A_y$ or $A_{el} > A_y$, the materials should remain elastic. If $A_{el} < A_y$ it may be assumed that local yielding will occur.

*See Timoshenko, *Theory of Elasticity*, 1st edn, McGraw-Hill, New York, p. 345.

3.5 CORRECTIONS TO CLASSICAL BEAM DEFLECTIONS

The Love–Kirchhoff approximation which is the basis of all the classical theories produces two potentially important types of error. The first is the error in the stress predicted, which has been discussed for beams in the last two sections; such errors can generally be ignored in the static loading of structures made of ductile materials, as was discussed in Sec. 1.7, but consideration should be given to them if brittle materials or fatigue conditions are involved.

The second error is in the strains predicted; this usually is important only for its effect upon the deflections. In general the neglected effects *increase* the deflections predicted by classical theories, which consider only the deflection resulting from flexure. That is, beams as well as plates and shells are actually more flexible, deflecting more under lateral loading and having smaller buckling resistance and lower natural frequencies of vibration than we predict if we use the classical theories alone.

For beams of compact cross section and made of homogeneous, isotropic material this error is negligible if the beams are slender and under the usual types of loading; however, the error can be serious if the beams are very short or are under closely spaced loads which alternate in direction. The error is also likely to be important for any length or loading of beams made of 'sandwich' material, or of flanged or latticed construction, in which most of the material is concentrated in the outer faces where it is most effective in resisting bending moment. When the inner part of a beam has been lightened until it is just strong enough to resist the transverse shear forces, the transverse shear stresses and strains will be of the same order of magnitude as the bending stresses and strains; hence it must be expected that the additional deflections due to transverse stresses and strains will also be of the same order of magnitude as the deflections due to bending which are given by the classical theory.

The methods used in Sec. 3.4, based on elasticity theory, of course give correct deflections as well as correct stresses, but such methods are usually unnecessarily complex for studying deflections and are rather impractical for such cases as sandwich and latticed construction. A good approximation to correct deflections can be obtained by a relatively simple correction to classical beam theory, and this is the subject of the present section.

Rough Analysis of Beam with Center Load

The effect of the strains neglected in the classical theory can be studied in various ways. The following analysis of a simply supported beam of rectangular cross section having a concentrated load P at the center, Fig. 3.19(a), is quite crude, but it is valuable because it clearly shows the relative magnitudes of the various effects, which tend to be masked in more sophisticated analyses. The total center deflection can be considered to be the sum of three parts, δ_f, δ_s, δ_n, due to three causes which will be considered to act independently of each other, as indicated in Fig. 3.19(b), (c), (d). δ_f is the flexural or bending deflection, due to

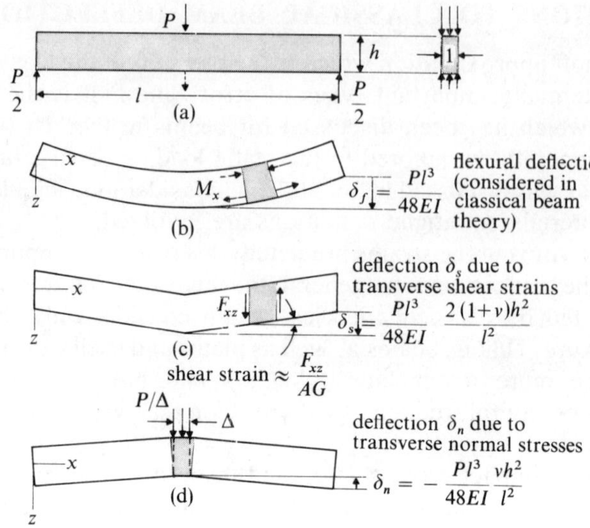

FIGURE 3.19

curving of the beam by the lengthening of the bottom fibers and the shortening of the top fibers accompanying the longitudinal bending stresses, as shown at (b). This is the deflection which is predicted by the classical beam theory studied in elementary strength of materials, and it is given in handbooks as $\delta_f = Pl^3/48EI = Pl^3/4Eh^3$ and was also derived in Eq. (2.16).

δ_s is the deflection caused by transverse shear strains, as shown at (c). If we make the rough approximation that transverse shear stresses are distributed uniformly over the cross section (instead of coming to zero at the top and bottom, as they actually must) the cross sections would remain plane and vertical. However, the angles between the cross sections and the top and bottom surfaces would be changed by the shear strain $\epsilon_s = P/(2hG)$, where $P/2$ is the total shear force, $P/(2h)$ is the assumed uniform shear stress, and G is the modulus of elasticity in shear. Then it can be seen from the figure, using Eq. (3.5a), that $\delta_s = \epsilon_s l/2 = Pl/(4hG) = (1+\nu)Pl/(2hE)$.

δ_n is the deflection caused by transverse normal stresses, as shown at (d). The deflections caused by transverse normal *strains*, which are produced by both the transverse and the longitudinal stresses and result only in slight changes in vertical distances from the middle surface, are very small. However, an appreciable effect is produced by the longitudinal expansion, due to Poisson's ratio effect, of the material directly under the load P (there is a similar expansion under the end reactions $P/2$ but it has negligible effect on deflection). To make the stresses and strains finite we consider P as uniformly distributed over a small width Δ. The material directly under the load will then be subjected to a vertical compressive stress of P/Δ, while the vertical stress on the bottom surface will of course be zero. The stress distribution between the top and the bottom surfaces is complex (see Fig. 3.15) but it is sufficient for the present purposes to make the rough assumption that there is a simple vertical stress distribution on the small rectangular portion of the beam of width Δ and height h, Fig. 3.19(d), which varies linearly from P/Δ at the top to zero at the bottom. Under this assumption due to Poisson's ratio effect the top will expand horizontally a distance $(P/\Delta)(\nu/E)(\Delta/2) = \nu P/2E$ on each side of the centerline, with expansions decreasing linearly to zero as we go from the top to the bottom. The vertical side will rotate through the angle $(\nu P/2E)/h = \nu P/2hE$, and the right half of the beam will rotate through the same angle, producing the deflection due to the transverse normal stresses $\delta_n = -(\nu P/2hE)(l/2) = -\nu Pl/4hE$, the minus sign indicating that the deflection is in the *upward* direction. If the same load had been applied to the bottom surface of the beam, as indicated by the dotted arrow in Fig. 3.19(a), the material next to it would have been shortened horizontally and the resulting deflection would have had the same magnitude and direction.

The above expressions for the deflections δ_f, δ_s, δ_n and the total deflection δ_t can be written in the form:

$$\delta_f = \frac{Pl^3}{4Eh^3}, \qquad \delta_s = \frac{Pl^3}{4Eh^3}\left(\frac{2(1+\nu)h^2}{l^2}\right), \qquad \delta_n = \frac{Pl^3}{4Eh^3}\left(-\frac{\nu h^2}{l^2}\right)$$
$$\delta_t = \delta_f + \delta_s + \delta_n = \frac{Pl^3}{4Eh^3}\left(1 + \frac{(2+\nu)h^2}{l^2}\right) \tag{3.54}$$

The second term in the parenthesis of δ_t represents the correction to the classical elementary formula. It will be noted that the deflection due to transverse normal stress is proportional to and much smaller than that due to transverse shear strain. The former can therefore be taken into account by multiplying the deflection due to transverse shear by a numerical factor of the order of unity; experience shows that this conclusion is not limited to this case but is quite general.

General Analysis, Method of Timoshenko*

Hence we will take the total deflection of the middle surface of a general beam of uniform cross section as $w_t = w_f + w_s$. We define w_f as the flexural deflection considered in the classical theory, due to the lengthening and shortening of longitudinal fibers by the longitudinal bending stresses. We define w_s as the deflection due to transverse shear strains alone, calculated on the approximate assumption that the shear stresses are uniformly distributed over the cross section; however, we will later introduce a numerical factor to allow both for the deflection due to transverse normal stresses and for the error made in taking the shear stress and strain as uniformly instead of parabolically distributed over the cross section.

Since the w_s displacements produce no relative rotation of cross sections, such rotation is due entirely to w_f, which is the same thing as what was called simply w in Chap. 2. Hence Eq. (2.2) becomes $M_x = -EI d^2 w_f/dx^2$ (which is then

**Strength of Materials*, Part 1, 3rd edn., Van Nostrand, New York.

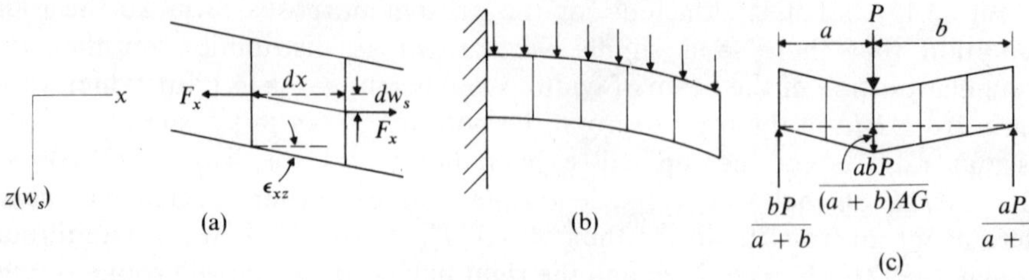

FIGURE 3.20

no longer an approximation) and, similarly, the transverse component of F_x on an element of length dx, Fig. 2.1(c), becomes $F_x(d^2w_f/dx^2)\,dx$. There is also one more term in the equilibrium equations than was considered before, namely the moment $F_x\,dw_s$ due to the shear displacement of cross sections, as indicated in Fig. 3.20(a).

The equation of moments of an elementary length dx of the beam is then changed from Eq. (2.3) to:

$$F_{xz}\,dx - dM_x - F_x\,dw_s = 0, \qquad \text{or } F_{xz} = \frac{dM_x}{dx} + F_x\frac{dw_s}{dx} = -EI\frac{d^3w_f}{dx^3} + F_x\frac{dw_s}{dx} \tag{3.55a}$$

while the equation of equilibrium of transverse forces becomes:

$$dF_{xz} + p\,dx + F_x\frac{d^2w_f}{dx^2}\,dx = 0, \qquad \text{or } -\frac{dF_{xz}}{dx} = p + F_x\frac{d^2w_f}{dx^2} \tag{3.55b}$$

Eliminating F_{xz} between these equations and remembering that $w_t = w_f + w_s$, we obtain:

$$EI\frac{d^4w_f}{dx^4} - F_x\frac{d^2w_s}{dx^2} = p + F_x\frac{d^2w_f}{dx^2}, \qquad \text{or } EI\frac{d^4w_f}{dx^4} = p + F_x\frac{d^2w_t}{dx^2} \tag{3.56}$$

This result can be compared to Eq. (2.4) in the more approximate classical theory.

Turning now from equilibrium conditions to displacements, since the shear displacement w_s leaves the cross sections parallel to each other after deformation, the slope dw_s/dx at any point would be equal to the shear strain ϵ_{xz} at that point if the cross sections also remain vertical. If the cross sections do not remain vertical the slope will equal ϵ_{xz} plus some constant C_1. Taking $\epsilon_{xz} = \sigma_{xz}/G = F_{xz}/GA_s$, where G and A_s are the shear modulus and cross-sectional area of the part of the beam which is assumed to be subjected to uniform shear stress, and using Eq. (3.55a) for F_{xz}, we find:

$$\frac{dw_s}{dx} = \epsilon_{xz} + C_1 = -\frac{EI}{GA_s}\frac{d^3w_f}{dx^3} + \frac{F_x}{GA_s}\frac{dw_s}{dx} + C_1, \qquad \text{or}$$

$$\frac{dw_s}{dx}\left(1 - \frac{F_x}{GA_s}\right) \approx \frac{dw_s}{dx} = -\frac{EI}{GA_s}\frac{d^3w_f}{dx^3} + C_1 \tag{3.57}$$

In deriving Eq. (3.57) F_x/GA_s is discarded because it is of the order of magnitude of a unit strain, which is very small compared to unity for any ordinary engineering material in the elastic range. Integrating Eq. (3.57) over x, we find:

$$w_s = -\frac{EI}{GA_s}\frac{d^2w_f}{dx^2} + C_1x + C_0 \tag{3.58}$$

and the total deflection is:

$$w_t = w_f + w_s = \left(1 - \frac{EI}{GA_s}\frac{d^2}{dx^2}\right) w_f + C_1x + C_0 \tag{3.59}$$

The terms C_1x, C_0 represent a rigid body rotation and translation respectively, and they are generally zero because such rigid body motions are prevented in practical beams by the boundary constraints. While they could conceivably be different from zero in highly unusual cases, they are zero in ordinary problems which we will consider, such as the cases illustrated in Figs. 3.20(b), (c), so we will assume them to be zero in the following.

General Loading and Shape of Cross Section

The second term in the parenthesis, Eq. (3.59), represents a correction to the classical w_f based on flexure alone. As discussed at the beginning of this section, we have found the effects of transverse normal stresses to be proportional to those of transverse shear, and we can allow for them, as well as for the error which we have made by taking the transverse shear stress to be uniformly distributed over the cross section, by multiplying the above correction by some number α, which should have a magnitude close to unity. Hence we write:

$$w_t = \left(1 - \frac{\alpha EI}{GA_s}\frac{d^2}{dx^2}\right) w_f \tag{3.60}$$

General Loading, Rectangular Cross Section

For the case of a homogeneous beam of rectangular cross section we obtained a similar result in Eqs. (3.28), which were derived from elasticity theory considering all the stresses and strains. If in Eqs. (3.28) we take $z = 0$: $u_z = w_t$, and $w_{cl} = w_f$, $M_x = -EI\,d^2w_f/dx^2 = -(Eh^3/12)\,d^2w_f/dx^2$, $h = 2c$, we obtain, using only the first two terms of the series:

$$w_t = w_f - \frac{8+5\nu}{40}h^2\frac{d^2w_f}{dx^2} = \left(1 - \frac{8+5\nu}{40}h^2\frac{d^2}{dx^2}\right) w_f \tag{3.60a}$$

Comparing Eqs. (3.60) and (3.60a) for the case of rectangular cross sections we find α for this case, and so at least approximately for other cases, using Eq. (3.5a):

$$\frac{\alpha EI}{GA_s} = \frac{\alpha\, 2(1+\nu)h^3}{h\,12} = \frac{8+5\nu}{40}h^2, \qquad \alpha = \frac{3(8+5\nu)}{20(1+\nu)} \approx 1{\cdot}1 \tag{3.60b}$$

For any given problem Eq. (3.60) with this value of α can be solved simultaneously with the equilibrium condition (3.56) to relate the actual total deflection w_t to the loadings p and F_x.

Harmonic Loading, Rectangular Cross Section

Equations (3.60) and (3.60b) are of course based upon some approximations. They tell nothing about how the deflection varies in the z direction, and they do not apply accurately to extremely small l/h ratios. To determine their limitations we need to compare them with an 'exact' theory. We have such a theory in the trigonometric–hyperbolic elasticity solution given by Table 3.3, which was applied to the case of a cyclically distributed load on the upper surface of a long beam of rectangular cross section in Eq. (3.27). For the sinusoidal distribution of load $p = p_1 \sin \pi x/l$, Fig. 3.21(a), we have $m = 1$, $c_m = \pi/l$, $\lambda = \pi h/l$ and the vertical displacement becomes:

$$u_z = \frac{p_1 l}{\pi E} \sin \frac{\pi x}{l} \left\{ \left[2 \cosh \frac{z}{h} \lambda - (1+\nu) \frac{z}{h} \lambda \sinh \frac{z}{h} \lambda \right] (\sinh \lambda + \lambda \cosh \lambda) \right.$$

$$\left. + \left[(1-\nu) \sinh \frac{z}{h} \lambda - (1+\nu) \frac{z}{h} \lambda \cosh \frac{z}{h} \lambda \right] \lambda \sinh \lambda \right\} \Big/ (\sinh^2 \lambda - \lambda^2) \tag{3.61a}$$

The deflection predicted for this case by Eqs. (3.60) and (3.56) can be found by taking $w_f = W_f \sin \pi x/l$ and $w_t = W_t \sin \pi x/l$. Substituting these into the two equations, using $F_x = 0$, $p = p_1 \sin \pi x/l$, $I = h^3/12$, $A_s = h$, $G = E/[2(1+\nu)]$ and expression (3.60b) for α and eliminating W_f, we find:

$$w_t = u_{z(z=0)} = \frac{12 p_1 l}{\pi E \lambda^3} \left(1 + \frac{8 + 5\nu}{40} \lambda^2 \right) \sin \pi x/l \tag{3.61b}$$

The deflection predicted by the classical theory is given by Eq. (2.10) for $m = 1$:

$$w_{cl} = w = \frac{12 p_1 l}{\pi E \lambda^3} \sin \pi x/l \tag{3.61c}$$

Figure 3.21(b) shows values of u_z/w_{c1} for different values of $h/l = \lambda/\pi$. The deflections given by the exact Eq. (3.61a) are shown for three surfaces, for the top surface by a dash-dot line, for the middle surface by a dashed line and for the bottom surface by a dash-double dot line. The deflections given by Eq. (3.61b) derived from Eq. (3.60) are shown by a dashed line. Within the range of h/l shown, from 0 to 0·3, that is from very long waves to a half wave length of about three times the height of the beam, the deflection given by Eq. (3.60) is indistinguishable from the exact deflection of the middle surface, and at $h/l = 0{\cdot}3$ is 21 percent greater than the deflection given by the classical theory. At this value of h/l the exact deflections of the top and bottom surfaces are beginning to diverge from the middle surface deflection but are still only a few percent different from it.

Figure 3.21(c) shows these results for the entire range of h/l from zero to infinity (this type of chart* is very convenient, permitting such a complete range of values to be plotted in one continuous curve). It can be seen that when h/l is greater than 0·3 the deflections of the top, middle, and bottom surfaces begin to

*L. H. Donnell, 'A chart for plotting relations between variables over their entire real range'. *Quart. Appl. Math.*, vol. I, 1943, p. 276.

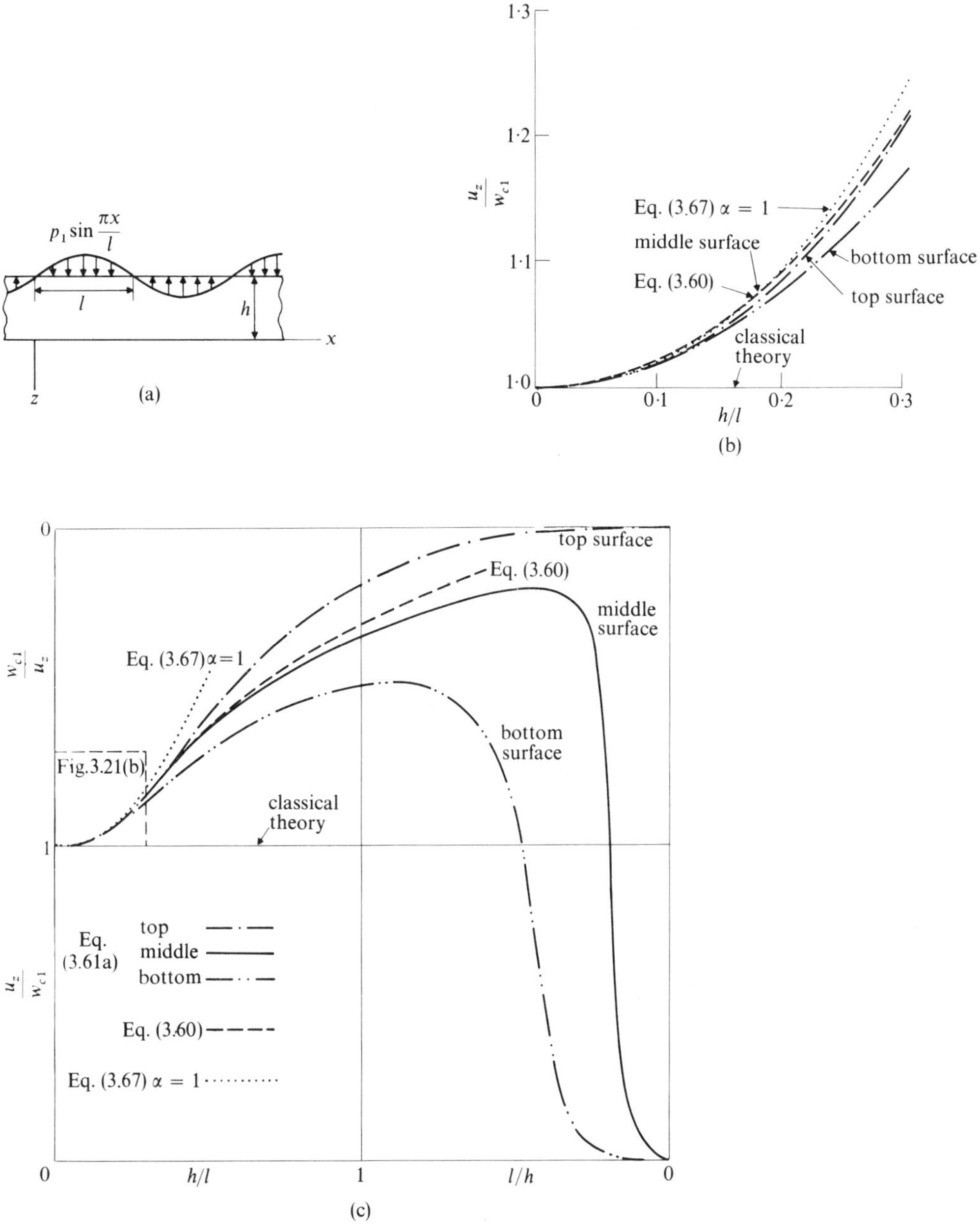

$p_1 \sin \frac{\pi x}{l}$
l
h
x
z
(a)
1·3
1·2
1·1
1·0
$\frac{u_z}{w_{c1}}$
Eq. (3.67) α = 1
middle surface
Eq. (3.60)
bottom surface
top surface
classical theory
0
0·1
0·2
0·3
h/l
(b)
top surface
Eq. (3.60)
middle surface
Eq. (3.67) α=1
bottom surface
Fig.3.21(b)
classical theory
$\frac{w_{c1}}{u_z}$
$\frac{u_z}{w_{c1}}$
Eq. (3.61a)
top
middle
bottom
Eq. (3.60)
Eq. (3.67) α = 1
0
1
h/l
l/h
(c)

FIGURE 3.21

diverge greatly, and as l/h approaches zero, that is for very short load-cycle lengths, the deflections of the middle and bottom surfaces approach zero, and the displacements, like the stresses, are concentrated in a thin layer near the loaded surface. While Eq. (3.60) with (3.60b) gives the deflection of the middle surface quite closely for h/l up to unity, it does not describe the whole picture beyond $h/l = 0{\cdot}3$ since it does not differentiate between the widely divergent deflections of the top, middle and bottom surfaces.

Applications of Eqs. (3.60), (3.60a)

As was pointed out in Sec. 3.3, a section of the beam, Fig. 3.21(a), from $x = 0$ to $x = l$ is approximately in the condition of a beam of length l with hinged ends. The above results indicate then that Eq. (3.60a), as well as the more general Eq. (3.60) with α taken as 1·1, should give the deflections with reasonable accuracy for beams as short as about three times the height under a loading which does not alternate in direction.

For flanged or I-beams or sandwich beams, the shear stress is actually nearly uniformly distributed over the area of the web or core, which take almost all of the transverse shear load, and the effects of transverse normal stresses will be even less important in comparison to the greatly increased effects of transverse shear. For a flanged beam such as an I-beam, then, A_s should be taken as the cross-sectional area of the web and $E/G = 2(1+\nu)$. For a sandwich beam, A_s should be taken as the cross-sectional area of the core and G as the shear modulus of the core material in the direction of the shear, while EI is the bending stiffness of the beam as a whole, which can usually be calculated from the faces alone.

For a latticed beam, such as shown in Fig. 3.22, EI is again the bending stiffness of the beam as a whole, calculated from the flanges. GA_s is calculated as its value for an equivalent beam which would have the same stiffness against transverse shear. Take A_1, A_2 as the areas (the total area of the diagonals on *both* sides in the case shown in Fig. 3.22(a)) and θ_1, θ_2 as the angles of the diagonal members as shown. Assume that the diagonals resist all of the transverse shear force F_{xz} and that they are hinged at their ends (which is on the conservative side). The axial stresses in the diagonals, from the equilibrium in the transverse direction of a portion of the beam on one side of a section which cuts the diagonals, are then $s_1 = F_{xz}/(A_1 \cos\theta_1)$, $s_2 = F_{xz}/(A_2 \cos\theta_2)$ respectively. The changes in lengths of the diagonals are $\delta_1 = s_1 h'/(\cos\theta_1 E) = F_{xz}h'/(A_1 E \cos^2\theta_1)$, $\delta_2 = F_{xz}h'/(A_2 E \cos^2\theta_2)$. From Fig. 3.22(d), ignoring any change in the length of the relatively heavy flange due to transverse shear, it can be seen that $h'(\tan\theta_1 + \tan\theta_2)\epsilon_{xz} = \delta_1/\cos\theta_1 + \delta_2/\cos\theta_2$. Solving for ϵ_{xz} from this relation, and setting it equal to F_{xz}/GA_s as before, we obtain:

$$GA_s = \frac{E(\tan\theta_1 + \tan\theta_2)}{1/(A_1 \cos^3\theta_1) + 1/(A_2 \cos^3\theta_2)} \tag{3.62}$$

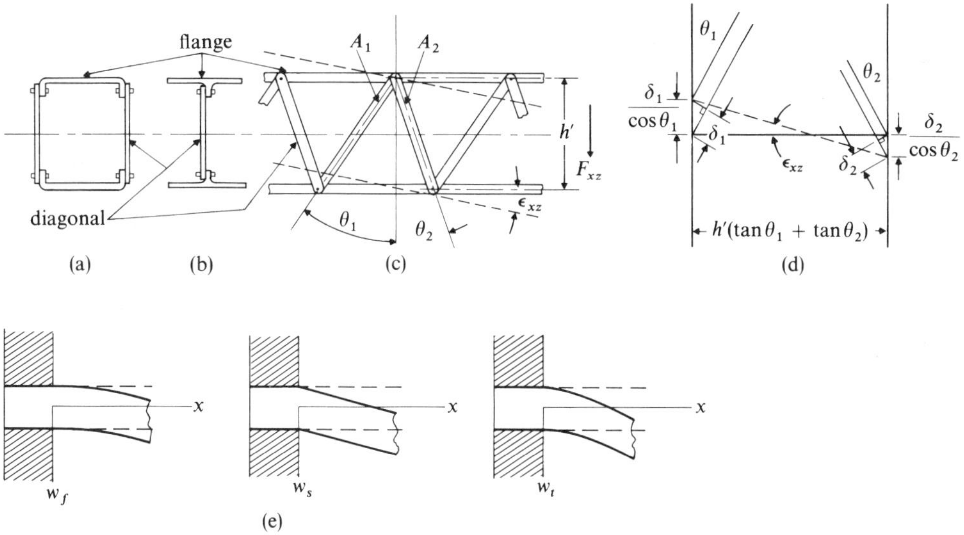

FIGURE 3.22

As an example of the application of Eq. (3.62), consider a hinged-end beam of length l which has a deflection due to flexure of $w_f = W_f \sin \pi x/l$ produced by a loading $p = p_1 \sin \pi x/l$. Substituting these expressions into Eq. (3.60a) we find that the total deflection of such a beam with a rectangular cross-section and made of a homogeneous isotropic material would be $w_t = [1 + \pi^2(8+5\nu)h^2/40l^2]w_f \approx (1 + 2{\cdot}34h^2/l^2)w_f$. Consider now a uniform latticed beam which has been designed so that the maximum stress in the diagonals is equal to the maximum bending stress in the flanges. For simplicity take $A_1 = A_2 = A_d$, $\theta_1 = \theta_2 = \theta$. The condition that the stresses are equal is, using Eqs. (3.55a) and (2.5):

$$-\left(\frac{F_{xz}}{A_d \cos\theta}\right)_{max} = \left(\frac{EI}{A_d \cos\theta}\frac{d^3 w_f}{dx^3}\right)_{max} = -\frac{\pi^3 EIW_f}{l^3 A_d \cos\theta}$$
$$= \left(\frac{Eh'}{2}\frac{d^2 w_f}{dx^2}\right)_{max} = -\frac{\pi^2 Eh' W_f}{2l^2} \tag{3.63}$$

Solving, we find $A_d = 2\pi I/(lh' \cos\theta)$. Using this in Eqs. (3.62), (3.60), (3.60b), we obtain $w_{tot} = [1 + 1{\cdot}1\pi h'/(2l \sin\theta \cos\theta)]w_f \approx (1 + 4{\cdot}0h'/l)w_f$ for $\theta = 30°$. Using these results we find that the error we make in calculating deflections by elementary beam theory for the solid bar is 3 percent for a slenderness ratio $l/h = 9$, and 9 percent for $l/h = 5$. On the other hand, for the latticed bar described there would be a 3 percent error for a slenderness ratio of $l/h' = 133$ and a 9 percent error for $l/h' = 45$; for $l/h' = 5$ the total deflection is 1·85 times the flexural deflection predicted by the classical theory. The buckling strength of a hinged-end strut is reduced by the same factors, since it deflects in the same half sine wave shape assumed above.

Lateral Vibration of Beams

This case is a little more complex, because when wave lengths of vibration modes are short enough for the effects of transverse stresses and strains to be important then the so-called 'rotary inertia' or 'rotatory inertia', that is the inertia due to the rotation of the beam cross sections, which involves longitudinal accelerations, will also be important and must be considered along with the usual transverse accelerations. An element of a homogeneous beam of length dx has a moment of inertia of mass about the neutral surface of $I\,dx\rho$, where I is the moment of inertia of the area of the cross section and ρ is the mass of the beam material per unit volume. The angle of rotation of the element is $\partial w_f/\partial x$ in the clockwise direction (Fig. 2.1); hence there is a clockwise angular acceleration of $\partial^3 w_f/\partial x\,\partial t^2$, resulting in a counterclockwise 'inertia moment' of $I\rho\,\partial^3 w_f/\partial x\,\partial t^2\,dx$, which must be included in the moment equation.

The moment and transverse equilibrium equations (3.55a, b) become, using A, k for the total area and radius of gyration of the cross section:

$$F_{xz} = -EI\frac{\partial^3 w_f}{\partial x^3} + I\rho\frac{\partial^3 w_f}{\partial x\,\partial t^2}$$

$$-\frac{\partial F_{xz}}{\partial x} = EI\frac{\partial^4 w_f}{\partial x^4} - I\rho\frac{\partial^4 w_f}{\partial x^2\,\partial t^2} = -A\rho\frac{\partial^2 w_t}{\partial t^2}$$

or

$$\frac{Ek^2}{\rho}\frac{\partial^4 w_f}{\partial x^4} = k^2\frac{\partial^4 w_f}{\partial x^2\,\partial t^2} - \frac{\partial^2 w_t}{\partial t^2} \tag{3.64}$$

Letting $w_f = w_m \sin m\pi x/l \sin 2\pi N_m t$ and substituting this into Eq. (3.60) we find $w_t = [1 + (\alpha EI/GA_s)(m^2\pi^2/l^2)]w_f$. Putting these values into Eq. (3.64) and solving for the frequency N_m, we find:

$$N_m = \frac{m^2\pi k}{2l^2}\sqrt{\frac{E}{\rho}}\left[1 + \frac{m^2\pi^2k^2}{l^2}\left(1 + \frac{\alpha EA}{GA_s}\right)\right]^{-1/2} = \frac{m^2\pi k}{2l^2}\sqrt{\frac{E}{\rho}}\,\beta \tag{3.65}$$

Comparing this result with Eq. (2.20) we see that the frequency is reduced by transverse strains and rotary inertia by the correction factor β. For a solid rectangular bar, using $k^2 = I/A = h^2/12$ and Eq. (3.60b), $\alpha EA/GA_s = 3(8+5\nu)/10 \approx 2{\cdot}85$, which gives the correction factor $(1 + 3{\cdot}17m^2h^2/l^2)^{-1/2}$. For $l/m = 10h$, that is if the half wave length of the vibration mode is ten times the beam height h, the correction factor β is about $0{\cdot}985$; for $l/m = 3h$ it is $0{\cdot}860$. Beyond the latter value vibration as a beam may not be the most important mode, other kinds of vibratory motion occurring which must be studied by elasticity theory; such vibrations are beyond the scope of this book.

Strictly speaking, in making the above derivation, the expression for F_{xz} in Eq. (3.55a), and hence in Eq. (3.60) which was derived from it, should also have been modified by adding the 'rotary inertia' term $I\rho\,\partial^3 w_f/\partial x\,\partial t^2$ to it, as was done in Eq. (3.64). This would have slightly changed the second term in the parenthesis in Eq. (3.60). However, this term in itself is a minor correction term, and to make minor corrections to a minor correction term would be an unjustified refinement, since it could only be important beyond the limits of application of the theory, which were discussed in the last paragraph.

General Loading Solutions

Equations (3.60) and (3.56) can be used for any kind of loading. For hinged-end beams, for example, we can assume that

$$w_f = \sum_m w_{fm} \sin m\pi x/l, \qquad w_t = \sum_m w_{tm} \sin m\pi x/l.$$

Substituting these in Eq. (3.60) we find

$$w_{fm} = w_{tm}/(1 + \alpha\pi^2 m^2 EI/l^2 GA_s).$$

Substituting these values and

$$p = \sum_m p_m \sin m\pi x/l$$

into Eq. (3.56) and using Eq. (2.11) for p_m, we obtain:

$$\frac{\pi^2}{2l}\left(\frac{\pi^2 EI}{l^2}\frac{m^4}{1+\dfrac{\alpha\pi^2 m^2 EI}{l^2 GA_s}} + m^2 F_x\right) w_{tm} = \int_0^l dx\, p(x) \sin\frac{m\pi x}{l} \tag{3.66}$$

For the case $p = 0$, this is satisfied by an axial buckling compressive force $P = -F_x = (\pi^2 EI/l^2)/(1 + \alpha\pi^2 EI/l^2 GA_s)$, as was suggested above.

If $F_x = 0$ and $p(x)$ represents a concentrated transverse load P at the center of the beam, we find, as in the derivation of Eq. (2.16):

$$w_{max} = \frac{2Pl^3}{\pi^4 EI}\sum_m \frac{1+\dfrac{m^2\pi^2\alpha EI}{l^2 GA_s}}{m^4}\sin^2\frac{m\pi}{2} \approx \frac{Pl^3}{48EI}\left(1 + 12\frac{\alpha EI}{l^2 GA_s}\right) \tag{3.66a}$$

Taking $\alpha EI/GA_s = (8 + 5\nu)h^2/40$ for a solid rectangular beam from Eq. (3.60b), the parenthesis in Eq. (3.66a) is approximately $(1 + 2{\cdot}85h^2/l^2)$, which can be compared with the rough value obtained in Eq. (3.54).

Boundary Conditions

In the above examples the ends were all hinged and the boundary conditions were easily satisfied by taking w_f, w_s, w_t all as sine functions. More generally and more precisely than in Eqs. (2.6), at a hinged support w_t, $M_x = 0$, whence w_t, $d^2w_f/dx^2 = 0$; at an end which is clamped horizontally as illustrated in Fig. 3.22(e) w_t, $dw_f/dx = 0$; at a free end M_x, F_{xz}, $F_x = 0$, whence d^2w_f/dx^2, $d^3w_f/dx^3 = 0$, etc.

For example, for a cantilever beam horizontally clamped at $x = 0$ and loaded at $x = l$ by an end load $P = F_{xz}$, the boundary conditions are $x = 0$: w_t, $dw_f/dx = 0$; $x = l$: $d^2w_f/dx^2 = 0$, $d^3w_f/dx^3 = -P/EI$. Equations (3.56) and (3.60) become for this case $d^4w_f/dx^4 = 0$ and $w_t = w_f - \gamma d^2w_f/dx^2$, where $\gamma = \alpha EI/GA_s$. The general solution of the first of these equations can be written as $w_f = C_0 + C_1x + C_2x^2 + C_3x^3$; then from the second equation: $w_t = (C_0 - 2\gamma C_2) + (C_1 - 6\gamma C_3)x + C_2x^2 + C_3x^3$. The boundary conditions then become $C_0 - 2\gamma C_2 = 0$, $C_1 = 0$, $2C_2 + 6C_3l = 0$, $C_3 = -P/6EI$. Solving, we find $C_0 = \gamma lP/EI$, $C_2 = lP/2EI$. The deflection is then $w_t = (P/2EI)(lx^2 - x^3/3 + 2\gamma x)$. Taking α as in Eq. (3.60b) we find that the third term in the last parenthesis is $(8+5\nu)/(8+9\nu)$ smaller than the value given by Eq. (3.53) for a beam whose rectangular cross-section is constrained to remain flat and vertical at the wall. If we replace the boundary condition $dw_f/dx = 0$ by $dw_f/dx = -(\nu/10)d^3w_f/dx^3$ then we get the same deflection as Eq. (3.53); the latter may therefore be taken as the boundary condition for the end fixity considered in deriving Eq. (3.53).

Elastic supports or statically indeterminate problems present no difficulties. Consider a beam with uniform loading p per unit length with the boundary conditions $x = 0$: $w_t = 0$, $\beta' dw_f/dx = EI d^2w_f/dx^2$ (representing elastic clamping, β' being the moment per radian rotation); $x = l$: $\beta w_t = EI d^3w_f/dx^3$, $d^2w_f/dx^2 = 0$ (elastic support, β being the force per unit deflection of the support). Equations (3.56), (3.60) are satisfied by $w_f = C_0 + C_1x + C_2x^2 + C_3x^3 + px^4/24EI$ and $w_t = (C_0 - 2\gamma C_2) + (C_1 - 6\gamma C_3)x + (C_2 - \gamma p/2EI)x^2 + C_3x^3 + px^4/24EI$. Substituting these expressions in the four boundary conditions, they can readily be solved for C_0, C_1, C_2, C_3. It should be noted that statically indeterminate structures such as this may be initially stressed without any external loading such as p, say by raising the right hand support a distance δ, which would change the third boundary condition to $x = l$: $\beta(w_t + \delta) = EI\, d^3w_f/dx^3$.

Direct Relation Between Loading and w_t

Equation (3.60) gives a relation between w_t and w_f, and we have seen how this relation can be used with Eq. (3.56) to obtain w_t for any loading. But it is usually more convenient to have a single equation which relates the total deflection w_t directly to the loading, without having to bother with w_f. The obvious way to obtain this is to solve Eq. (3.60) for w_f and substitute the resulting expression into Eq. (3.56), but we cannot do this directly because of the form of Eq. (3.60). However if w_t and w_f are related as in Eq. (3.60) it should be possible to express w_f in the series form $w_f = w_t + a\, d^2w_t/dx^2 + b\, d^4w_t/dx^4 + \cdots$. Substituting expression (3.60) into this, we obtain $w_f = w_f + (a - \beta)\, d^2w_f/dx^2 + (b - a\beta)\, d^4w_f/dx^4 + \cdots$, where $\beta \equiv \alpha EI/GA_s$. This is satisfied in general only if $a = \beta$, $b = a\beta = \beta^2$, etc., that is if $w_f = [1 + (\alpha EI/GA_s)\, d^2/dx^2 + (\alpha EI/GA_s)^2\, d^4/dx^4 + \cdots]w_t$. Substituting this in Eq. (3.56) we obtain:

$$EI\left(1 + \frac{\alpha EI}{GA_s}\frac{d^2}{dx^2} + \cdots\right)\frac{d^4w_t}{dx^4} = p + F_x\frac{d^2w_t}{dx^2} \tag{3.67}$$

When Eq. (3.67), with only the one correction term shown, is applied to the sinusoidal loading case and compared in the same way that Eq. (3.56) was to the exact solution for this case, it is found that the divergence is much more rapid than it is for Eq. (3.56) as the beams become shorter. This result is not surprising since Eq. (3.56) is equivalent to Eq. (3.67) with the entire series of correction terms, and the series converges rapidly only for slender beams. However, if the correction factor is made about 10 percent smaller than it was for Eqs. (3.60), (3.60b) by taking $\alpha = 3(8+5\nu)/22(1+\nu)$ for a homogeneous rectangular beams and $\alpha = 1$ for other beams, the deflections obtained with Eq. (3.67) check the exact theory within a few percent for height-length ratios from zero to 0·3, beyond which even Eqs. (3.60) and (3.56) cannot describe the complete picture adequately. Equation (3.67) with the above modifications is shown by a dotted line on Figs. 3.21(b), (c).

Thus Eq. (3.67), with one correction term as shown, can be used within this range, by taking $\alpha EI/GA_s = (8+5\nu)h^2/44$ for homogeneous rectangular beams, $\alpha = 1$ for beams with lightened centers, A_s as the area of the web for flanged beams such as I-beams, GA_s as the core shear modulus times area of the core for sandwich beams, and as given by Eq. (3.62) for latticed beams, etc. As might be expected, the correction for the effect of transverse strains and stresses needs to be applied only to the flexural term, that is the term involving the flexural stiffness EI, no correction being required for the $F_x\, d^2w_t/dx^2$ term. Equation (3.67) can be used for solving beam problems in the same way that Eq. (2.4) has been used in the preceding section, the solution involving somewhat more arithmetic but no greater mathematical difficulties, at least for hinged-end beams using a series solution; for more complex boundary conditions involving both w_t and w_f it may be just as easy to use Eqs. (3.60) and (3.56).

Solutions of the fourth order Eq. (3.56) give four constants of integration which can be used for satisfying end boundary conditions, the same as for Eq. (2.4). As we have seen, the constants of integration involved in developing Eq. (3.60) involve only rigid body motion, and since Eq. (3.67), even with an infinite number of terms, is physically equivalent to Eq. (3.60) we cannot expect any more help from it for satisfying end conditions. Thus the new Eqs. (3.56) and (3.60) or (3.67), like their less accurate classical counterparts, contain in themselves only the means of satisfying gross end conditions involving resultant forces and moments on the ends or deflections of the middle surface. Satisfaction of more detailed end conditions can be achieved by using supplementary solutions derived from elasticity theory, as discussed in the first part of this chapter.

An alternative form of direct relation between lateral loading p and w_t can be obtained more simply from Eqs. (3.28). If, instead of writing the first two terms of the expression for $u_z = w_t$ in terms of $w_{cl} = w_f$ as we did in developing Eq. (3.60a), we use Eqs. (2.2) and (2.4a) to write them in terms of p, we obtain:

$$\frac{2Ec^3}{3}\frac{d^4w_t}{dx^4} = p - \frac{8+5\nu}{10}c^2\frac{d^2p}{dx^2} \tag{3.67a}$$

which may be convenient to use in some problems.

Range of Application of Various Approximations

The range of application of elementary classical beam theory, and of the various adaptations and improvements of it which have been discussed, can be summarized in the relation between three quantities: the thickness h; the maximum deflection w_{max}; and the 'primary half wave length of the deflection' l, which can be taken as the distance between points of inflection, that is points where the curvature changes sign. For hinged-end beams under transverse loads which all act in the same direction or under a buckling load, l can be taken as the length of the beam; for similarly loaded fixed end beams it can be taken as half the length of the beam; for loads cyclically alternating in direction it can be taken as half the cycle length.

If the ends of the beam are prevented from moving axially, the nonlinear term $\sigma_{xm} h\, d^2w/dx^2$ must be considered as it was in Sec. 2.6, unless w_{max} is small compared to h ($w_{max} \ll h$), the dividing line being about $w_{max} = 0{\cdot}1h$ for 3 percent error, and $w_{max} = 0{\cdot}2h$ for 10 percent error. For $w_{max} \gg h$ the flexural term may be ignored and the beam treated as a filament, the dividing line being about $w_{max} = 2h$ for 10 percent error and $w_{max} = 3h$ for 3 percent error. The error due to using the tangent of the angle of slope for the angle, etc., can be ignored as long as $w_{max} \ll l$, say $w_{max} \leq 0{\cdot}1l$.

The classical beam theory is sufficiently accurate if $l \gg h$, the dividing line being about $l = 10h$ for 3 percent error in deflection, buckling load and vibration frequency and 0·3 percent error in bending stress, or $l = 5h$ for 10 percent error in deflection, buckling load and vibration frequency and 1 percent error in bending stress.

4

CLASSICAL PLATE THEORY

4.1 INTRODUCTION TO PLATE THEORY

Forces and Moments on Elements of Plates

As has been noted, the use of the Love–Kirchhoff approximation makes classical beam theory essentially a one-dimensional phenomenon, in which the normal and transverse forces F_x, F_{xz} and bending moment M_x on cross sections can be found in terms of the displacement of the middle surface, which is assumed to be a function only of the axial coordinate x. Taking the plane of the middle surface of a plate of uniform thickness $h = 2c$ as the x, y plane, as shown in Fig. 4.1, with z the transverse coordinate as in beams, we have normal and transverse forces F_x, F_{xz} and bending moments M_x on sections normal to the x direction the same as in beams, and in addition normal and transverse forces F_y, F_{yz} and bending moments M_y on sections normal to the y direction. Besides these forces and moments

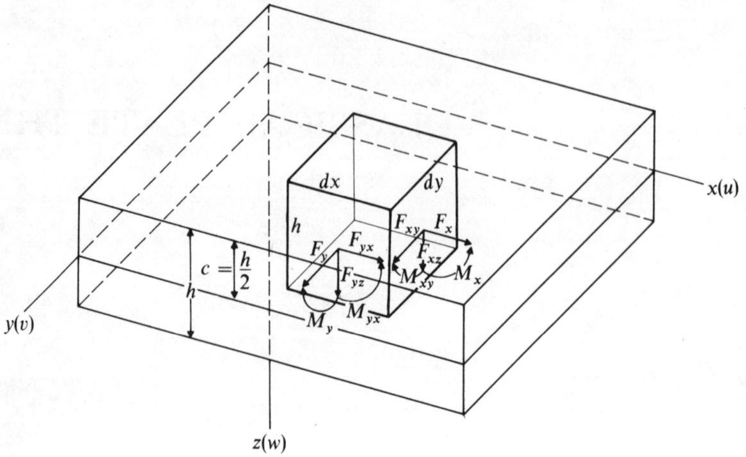

FIGURE 4.1

which are analogous to those usually present in beams, plates in general have shearing forces F_{xy} and F_{yx}, and twisting moments M_{xy} and M_{yx}, as shown in the figure. By using the Love–Kirchhoff approximation all these forces and moments can be found in terms of the displacement of the middle surface, which is a function only of two coordinates, x and y; thus classical plate theory is essentially a two-dimensional phenomenon. As with beams, the symbols F_x, M_x, etc., are defined as the forces or moments *per unit length* of the section on which they act; hence if the elements have dimensions dx, dy in the x, y directions, as shown in Fig. 4.1 the total forces acting on the element's sides are $F_x\,dy$, $M_x\,dy$, $F_y\,dx$, $M_{xy}\,dy$, $F_{yz}\,dx$, etc.

The forces F_x, F_y, F_{xy}, F_{yx} are 'membrane' forces, which are due to stretching and shearing of the plate *in* the x, y plane; they are the same as the forces obtained by integrating over the thickness the stresses of plane stress elasticity theory. The shearing forces F_{xy}, F_{yx} are in general of the same order of magnitude as the normal forces F_x, F_y. The moments M_x, M_y, M_{xy}, M_{yx}, together with the transverse shear forces F_{xz}, F_{yz} (which balance changes in these moments) are 'flexural' moments and forces, which are due to flexing of the plates *out* of the x, y plane. The twisting moments M_{xy}, M_{yx} are in general of the same order of magnitude as the bending moments M_x, M_y.

Like the σ_x stresses in beam theory, the stresses σ_x, σ_y, σ_{xy} which act on the cross sections of plates can be separated into uniform components like those in plane stress elasticity theory, which compose the membrane forces F_x, F_{xy}, etc., and components which vary linearly from zero at the middle surface, which compose the flexural moments M_x, M_{xy}, etc. The transverse shear forces F_{xz} and F_{yz} are composed of parabolically distributed shear stresses, as in beams.

Although the forces on sections of plates are somewhat similar to those on sections of beams except for the added dimension, it is not sufficient to think of and to treat a plate as if it were a complex of beams running in two directions; plates differ basically from such a complex of independent beams in many ways besides the obvious one that the bending in two directions and the twisting of plates are inherently related to one another. The material in narrow beams is free to expand or contract in the width direction, corresponding to the Poisson's ratio effect of the longitudinal stresses, but the elements of plates cannot expand or contract freely in this direction; because of this constraint the modulus E which describes the stiffness of the material of beams in general has to be replaced by the somewhat larger 'plate modulus' $E/(1-\nu^2)$ in describing the corresponding behavior of plates. The case of a wide beam, or a plate with free edges loaded as a beam, requires something between these moduli, as will be discussed in Sec. 4.5, Fig. 4.22.

There is an even more important way in which plates differ basically from beams. We can have membrane forces in plates produced by edge loads in the plane of the middle surface as in plane elasticity problems, the same as for axial loads acting on beams. But in beams, membrane forces can be produced by the transverse displacements only if the beam supports are such that they constrain axial movement, as in the case discussed in Sec. 2.6. On the other hand, membrane forces are in general produced by transverse displacement of plates with or without such constraints. This is because movements in the plane of a plate cannot in general take place freely like the axial movement of a beam on rollers, due to the fact that different parts of the plate tend to move in different ways so that such displacements interfere with each other. For example, consider a circular plate under lateral load; diametral elements of the plate such as are shown in Fig. 4.2(a) become curved transversely and consequently tend to pull in at the ends, as shown in Fig. 4.2(b). Even if this radial movement is not resisted by the edge supports it is resisted by the outer part of the plate acting as a ring, as suggested in Fig. 4.2(c); as a result, transverse loads put the inner part of the plate in radial tension and the outer part in circumferential compression.

Another way of describing these effects of transverse displacements of plates is to say that when a plate is bent into a nondevelopable shape, some parts of it will have to be stretched and others compressed or sheared in the plane of the plate to deform the originally flat plate into a non-developable deflected shape. (It follows that membrane stresses will *not* be produced by a deformation which *is* developable into the original surface, unless, like the beam of Sec. 2.6, the edges

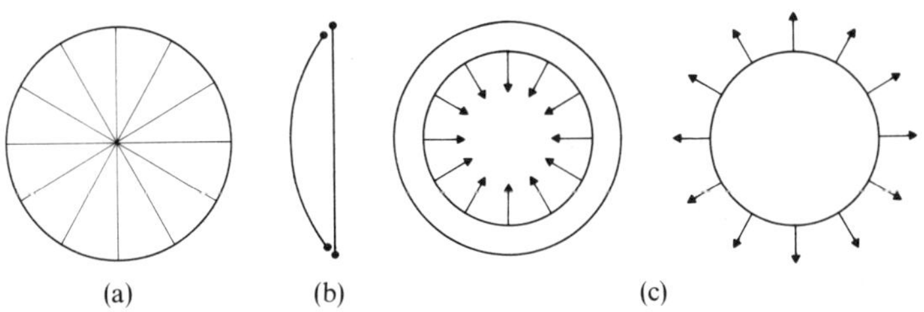

(a) (b) (c)

FIGURE 4.2

are constrained against movement in the plane of the plate.) We will find that the membrane forces produced by transverse displacements are nonlinear effects, increasing with a higher power than one of the deflection–thickness ratio and hence (as in the beam considered in Sec. 2.6) unimportant when this ratio is small compared to unity, and becoming the dominant factor when this ratio is large compared to unity.

4.2 DISPLACEMENT–STRAIN RELATIONS

Approximate Analysis

We designate the components in the x, y, z direction of the displacements of points in the middle surface of the plate by u, v, w, as indicated in the parentheses after the axis designations in Fig. 4.1. These displacements u, v, w are of course in general functions of both x and y.

In Fig. 4.3 p is a point in the middle surface, and P is a general point in the plate a distance z from p along a normal to the middle surface drawn at p; p' and P' represent the same points after undergoing displacements in accordance with the Love–Kirchhoff assumption. If there were no w displacements, P would undergo the same displacement u (or v, not shown) as p does, as illustrated in Fig. 4.3(a). If there are w displacements also and they vary with x and y, the normal to the middle surface will be rotated approximately through the angle $\partial w/\partial x$ in the plane shown in Fig. 4.3(b), and P will suffer an additional displacement $z\,\partial w/\partial x$ in the x direction in the negative sense. The total displacement of P in the x direction, U, and similarly its total displacement in the y direction, V, are therefore:

$$U = u - z\,\partial w/\partial x, \qquad V = v - z\,\partial w/\partial y \tag{4.1}$$

The strains in a layer of the plate parallel to and a distance z from the middle surface will therefore be, replacing u_x, u_y by U, V in Eq. (3.6):

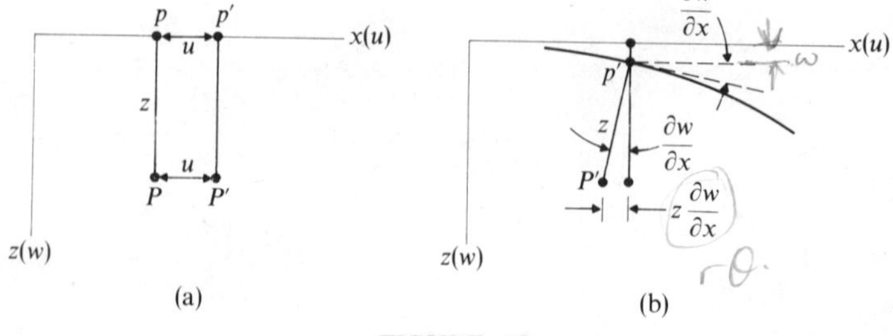

FIGURE 4.3

$$\epsilon_x = \partial U/\partial x = \partial u/\partial x - z\,\partial^2 w/\partial x^2,$$
$$\epsilon_y = \partial V/\partial y = \partial v/\partial y - z\,\partial^2 w/\partial y^2,$$
$$\epsilon_{xy} = \partial U/\partial y + \partial V/\partial x = \partial u/\partial y + \partial v/\partial x - 2z\,\partial^2 w/\partial x\,\partial y$$

However, there is another obvious component of strain due to w which the above analysis has not considered. This is strain of the type which was considered in the case of the beam discussed in Sec. 2.6 and illustrated in Fig. 2.11(b), which is caused by the tipping of elements. Figure 4.4(a) shows adjacent general points O, P, initially a distance dx apart, and their positions O', P' after varying w displacements have occurred. It will be seen that the element OP has undergone a unit tensile strain in the x direction, additional to that considered above, which is equal to, using the binomial theorem: $(\overline{O'P'} - \overline{OP})/\overline{OP} = [1 + (\partial w/\partial x)^2]^{1/2} - 1 \approx (\partial w/\partial x)^2/2$. There of course may be a similar component of strain in the y direction of $(\partial^2 w/\partial y^2)/2$.

There will also in general be a similarly caused shear strain, which is a little harder to evaluate, largely because of the difficulty of drawing a three-dimensional diagram on paper. Figure 4.4(b) shows two sides OP and OQ of a general element with dimensions dx, dy in a plane parallel to the middle surface; the angle POQ is of course a right angle before displacement. It can be seen that after unequal w displacements of the points P, O, Q (for simplicity, the constant part of the displacement is not shown) the angle $P'OQ'$ will no longer be a right angle. The change in this angle, which is the shear strain due to the variation in w, can be evaluated as follows. The angle QOP' will still be a right angle, as symbolized by the small square in the corner. Now if we rotate QOP' about OP' as an axis until Q reaches Q'', the angle $Q''QQ'$ will equal $POP' = \partial w/\partial x$ (assuming as usual that the slopes are small enough so that angles and their tangents and sines are interchangeable). Then, to the usual approximation, $QQ' = (\partial w/\partial y)\,dy$, $Q''Q' = (\partial w/\partial x)QQ' = (\partial w/\partial x)(\partial w/\partial y)\,dy$, and the angle $Q''OQ' = \epsilon_{xy} = Q''Q'/dy = (\partial w/\partial x)(\partial w/\partial y)$.

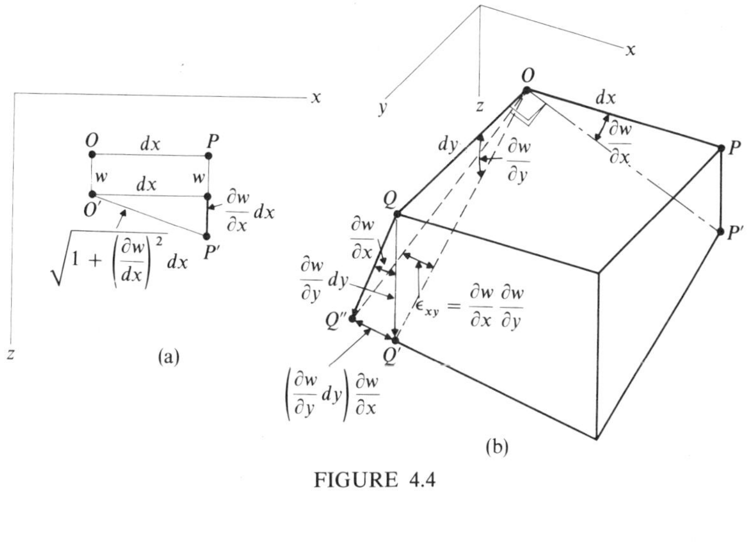

FIGURE 4.4

The total strains in the layer of the plate parallel to and a distance z from the middle surface can then be written as:

$$\begin{aligned}
\epsilon_x &= \frac{\partial u}{\partial x} + \frac{1}{2}\left(\frac{\partial w}{\partial x}\right)^2 - \frac{\partial^2 w}{\partial x^2} z \\
\epsilon_y &= \frac{\partial v}{\partial y} + \frac{1}{2}\left(\frac{\partial w}{\partial y}\right)^2 - \frac{\partial^2 w}{\partial y^2} z \qquad (4.2) \\
\epsilon_{xy} &= \frac{\partial u}{\partial y} + \frac{\partial v}{\partial x} + \frac{\partial w}{\partial x} \cdot \frac{\partial w}{\partial y} - 2\frac{\partial^2 w}{\partial x\, \partial y} z
\end{aligned}$$

The last terms in these expressions contain the factor z, and represent *flexural* strains, zero at the middle surface and varying linearly with the distance z from it; we will designate these parts of the strains as ϵ_{xf}, ϵ_{yf}, ϵ_{xyf}. The other terms, constant with respect to z, are the *membrane* strains, which we will designate by ϵ_{xm}, ϵ_{ym}, ϵ_{xym}. Thus $\epsilon_x = \epsilon_{xm} + \epsilon_{xf}$, etc.

Exact Analysis of Displacement–Strain Relations

The simple derivation above shows clearly the most important terms which are present in the expressions for the strains, but it involves numerous approximations in addition to the basic Love–Kirchhoff assumption, including the assumption that the different components of the strain are independent of each other. The following exact (within the Love–Kirchhoff assumption) analysis of the strains gives very complex expressions, which include many terms of no importance in most practical problems, but it is of value to enable us to appreciate better what is really happening and what we are leaving out.

Figure 4.5(a) shows a point o and adjacent points p, q at distances dx, dy from o in the x, y directions, all in the undisplaced middle surface which is taken as the x, y plane, with the origin taken at o for convenience. It also shows corresponding general points O, P, Q, which are on normals to the middle surface erected at o, p, q, and in a plane a distance z from the middle surface. Figure 4.5(b) shows the displaced positions o', p', q', O' of the points o, p, q, O. If the

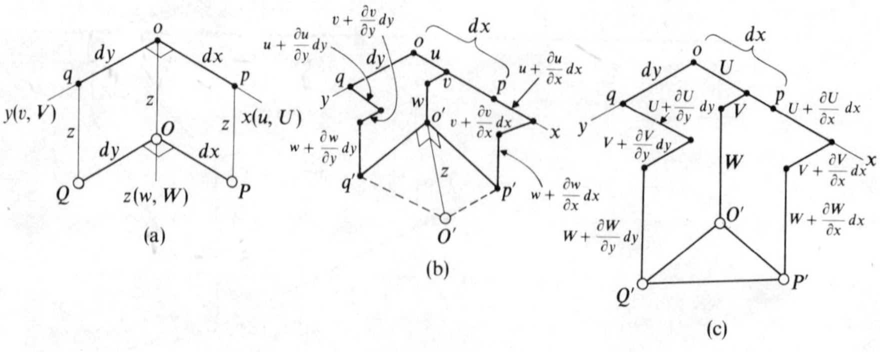

FIGURE 4.5

displacement of o in the x direction is u, the displacement of p, which is the same as o except that its x coordinate is increased by dx, must be u plus the rate at which u varies with x times the amount that x has increased, that is $u + (\partial u/\partial x)\,dx$, and similarly for the other displacements and for the point q. The x, y, z, coordinates of o', p', q' are tabulated in the upper half of Table 4.1.

Table 4.1

	x	y	z
o'	u	v	w
p'	$u+\left(1+\frac{\partial u}{\partial x}\right)dx$	$v+\frac{\partial v}{\partial x}dx$	$w+\frac{\partial w}{\partial x}dx$
q'	$u+\frac{\partial u}{\partial y}dy$	$v+\left(1+\frac{\partial v}{\partial y}\right)dy$	$w+\frac{\partial w}{\partial y}dy$
O'	U	V	W
P'	$U+\left(1+\frac{\partial U}{\partial x}\right)dx$	$V+\frac{\partial V}{\partial x}dx$	$W+\frac{\partial W}{\partial x}dx$
Q'	$U+\frac{\partial U}{\partial y}dy$	$V+\left(1+\frac{\partial V}{\partial y}\right)dy$	$W+\frac{\partial W}{\partial y}dy$

We can now calculate the exact coordinates of O' by using the Love–Kirchhoff assumption that the length $\overline{o'O'}$ remains equal to z and that the angles $\overline{p'o'O'}$ and $\overline{q'o'O'}$ remain right angles. By the theorem of Pythagoras, $\overline{O'p'}^2 = \overline{o'p'}^2 + z^2$, $\overline{O'q'}^2 = \overline{o'q'}^2 + z^2$. By the same theorem $\overline{O'p'}^2$, $\overline{o'p'}^2$, $\overline{O'q'}^2$, $\overline{o'q'}^2$, $\overline{O'o'}^2$ can be obtained as the sum of the squares of the differences between the x, y and z coordinates of the end points; thus in Fig. 4.6: $\overline{AB}^2, = \overline{AC}^2 + \overline{BC}^2 = (\overline{AD}^2 + \overline{CD}^2) + \overline{BC}^2 = (x_B - x_A)^2 + (y_B - y_A)^2 + (z_B - z_A)^2$. Designating the unknown coordinates of O' by U, V, W, as shown in Fig. 4.5(c), and using Table 4.1 for the coordinates of o', p', q', the equations $\overline{O'p'}^2 = \overline{o'p'}^2 + z^2$, $\overline{O'q'}^2 = \overline{o'q'}^2 + z^2$, $\overline{O'o'}^2 = z^2$ can be written as:

$$\left[\left(1+\frac{\partial u}{\partial x}\right)dx + (u-U)\right]^2 + \left[\frac{\partial v}{\partial x}dx + (v-V)\right]^2 + \left[\frac{\partial w}{\partial x}dx + (w-W)\right]^2$$
$$= \left[\left(1+\frac{\partial u}{\partial x}\right)^2 + \left(\frac{\partial v}{\partial x}\right)^2 + \left(\frac{\partial w}{\partial x}\right)^2\right]dx^2 + z^2$$
$$\left[\frac{\partial u}{\partial y}dy + (u-U)\right]^2 + \left[\left(1+\frac{\partial v}{\partial y}\right)dy + (v-V)\right]^2 + \left[\frac{\partial w}{\partial y}dy + (w-W)\right]^2$$
$$= \left[\left(\frac{\partial u}{\partial y}\right)^2 + \left(1+\frac{\partial v}{\partial y}\right)^2 + \left(\frac{\partial w}{\partial y}\right)^2\right]dy^2 + z^2 \tag{4.3}$$
$$(u-U)^2 + (v-V)^2 + (w-W)^2 = z^2$$

These three equations must now be solved for the three unknowns U, V, W. Expanding the left sides of the first two equations and subtracting from them the

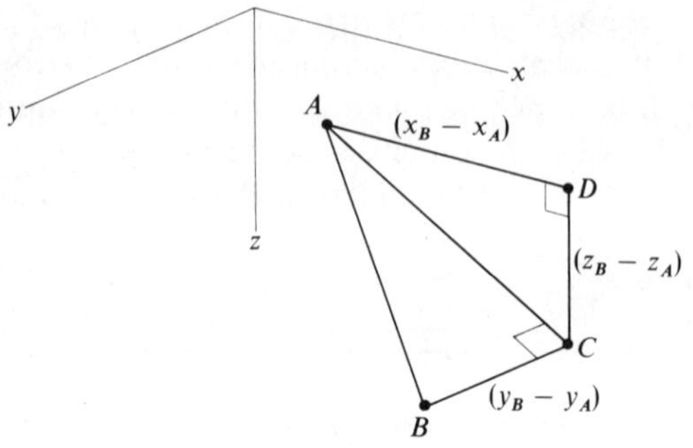

FIGURE 4.6

last equation, we obtain

$$
\begin{aligned}
&(u-U)\left(1+\frac{\partial u}{\partial x}\right)+(v-V)\frac{\partial v}{\partial x}+(w-W)\frac{\partial w}{\partial x}=0\\
&(u-U)\frac{\partial u}{\partial y}+(v-V)\left(1+\frac{\partial v}{\partial y}\right)+(w-W)\frac{\partial w}{\partial y}=0
\end{aligned}
\tag{4.4}
$$

Solving Eqs. (4.4) for $(u - U)$ and $(v - V)$ in terms of $(w - W)$, substituting these into the last of Eq. (4.3) and solving for $(w - W)$, etc., we obtain

$$
\begin{aligned}
&U = u + smz, \qquad V = v + snz, \qquad W = w + s(1+l)z, \qquad \text{where}\\
&m = -\frac{\partial w}{\partial x}\left(1+\frac{\partial v}{\partial y}\right)+\frac{\partial w}{\partial y}\frac{\partial v}{\partial x}, \qquad n = -\frac{\partial w}{\partial y}\left(1+\frac{\partial u}{\partial x}\right)+\frac{\partial w}{\partial x}\frac{\partial u}{\partial y}\\
&l = \frac{\partial u}{\partial x}+\frac{\partial v}{\partial y}+\frac{\partial u}{\partial x}\frac{\partial v}{\partial y}-\frac{\partial u}{\partial y}\frac{\partial v}{\partial x}, \qquad s = [m^2+n^2+(1+l)^2]^{-1/2}
\end{aligned}
\tag{4.5}
$$

Since the point P differs from the point O only in the small increase dx in its x coordinate, the displacement in the x direction of P must be $U + (\partial U/\partial x)\,dx$, etc., as shown in Fig. 4.5(c) and in Table 4.1. Finally, having obtained exact expressions for the x, y, z coordinates of O', P', Q' we can use the theorem of Pythagoras to calculate expressions for the lengths $\overline{O'P'}$, $\overline{O'Q'}$, $\overline{P'Q'}$, and hence with the aid of the law of cosines the angle $P'O'Q'$, after displacement. Then we can find the strains $\epsilon_x = (\overline{O'P'} - dx)/dx$, $\epsilon_y = (\overline{O'Q'} - dy)/dy$, $\epsilon_{xy} = \pi/2 - \angle P'O'Q' = \sin^{-1}[(\overline{O'P'}^2 + \overline{O'Q'}^2 - \overline{P'Q'}^2)/(2\overline{O'P'}\cdot\overline{O'Q'})]$.

However, when the values of U, V, W from Eq. (4.5) are substituted into such expressions, the results are extremely complex and it is hard to see what they mean. To express the strains in the form of sums of simple terms as was done in Eq. (4.2), we have to expand the function which is raised to a minus one half power in a power series by using the binomial theorem, and cut the series off by making the assumption that the derivatives of the displacements are small compared to unity. Thus in Eq. (4.5) we obtain $s = [1 + (2l + l^2 + m^2 + n^2)]^{-1/2} \approx 1 - (2l + l^2 + m^2 + n^2)/2 + 3(2l + l^2 + m^2 + n^2)^2/8 \ldots.$ Retaining only terms up to the second degree (that is squares or products) of displacement derivatives, we find $s = 1-(\partial u/\partial x)-(\partial v/\partial y)-(\partial w/\partial x)^2/2-(\partial w/\partial y)^2/2+(\partial u/\partial x)(\partial v/\partial y)+(\partial u/\partial y)(\partial v/\partial x)+(\partial u/\partial x)^2+(\partial v/\partial y)^2$; there are fourteen third degree terms and larger and larger numbers of higher degree terms. Continuing this process we find

$$U = u + [-(\partial w/\partial x) + (\partial u/\partial x)(\partial w/\partial x) + (\partial v/\partial x)(\partial w/\partial y)]z,$$
$$V = v + [-(\partial w/\partial y) + (\partial u/\partial y)(\partial w/\partial x) + (\partial v/\partial y)(\partial w/\partial y)]z,$$
$$W = w + [1 - (\partial w/\partial x)^2/2 - (\partial w/\partial y)^2/2]z,$$

and the strains are

$$\epsilon_x = \frac{\partial u}{\partial x} + \frac{1}{2}\left(\frac{\partial w}{\partial x}\right)^2 + \frac{1}{2}\left(\frac{\partial v}{\partial x}\right)^2 + \left[-\frac{\partial^2 w}{\partial x^2} + \frac{\partial^2 u}{\partial x^2}\frac{\partial w}{\partial x} + \frac{\partial^2 v}{\partial x^2}\frac{\partial w}{\partial y} + \frac{\partial u}{\partial x}\frac{\partial^2 w}{\partial x^2}\right]z$$

$$\epsilon_y = \frac{\partial v}{\partial y} + \frac{1}{2}\left(\frac{\partial w}{\partial y}\right)^2 + \frac{1}{2}\left(\frac{\partial u}{\partial y}\right)^2 + \left[-\frac{\partial^2 w}{\partial y^2} + \frac{\partial^2 u}{\partial y^2}\frac{\partial w}{\partial x} + \frac{\partial^2 v}{\partial y^2}\frac{\partial w}{\partial y} + \frac{\partial v}{\partial y}\frac{\partial^2 w}{\partial y^2}\right]z$$

$$\begin{aligned}\epsilon_{xy} = {} & \frac{\partial u}{\partial y} + \frac{\partial v}{\partial x} + \frac{\partial w}{\partial x}\frac{\partial w}{\partial y} - \frac{\partial u}{\partial x}\frac{\partial v}{\partial x} - \frac{\partial u}{\partial y}\frac{\partial v}{\partial y} \\ & + \left[-2\frac{\partial^2 w}{\partial x\,\partial y} + 2\frac{\partial^2 u}{\partial x\,\partial y}\frac{\partial w}{\partial x} + 2\frac{\partial^2 v}{\partial x\,\partial y}\frac{\partial w}{\partial y} + 2\left(\frac{\partial u}{\partial x} + \frac{\partial v}{\partial y}\right)\frac{\partial^2 w}{\partial x\,\partial y}\right. \\ & \left. + \left(\frac{\partial u}{\partial y} + \frac{\partial v}{\partial x}\right)\nabla^2 w\right]z\end{aligned} \tag{4.6}$$

Besides the higher powers and products of the displacement derivatives which have been omitted in Eq. (4.6), terms involving z^2 have also been omitted, for instance $(\partial^2 w/\partial x\,\partial y)^2 z^2/2$ in the expression for ϵ_x.

Comparing Eqs. (4.6) with Eqs. (4.2), we see that the correct expressions for strains contain, as expected, a number of additional terms of the second degree in addition to those of still higher degree and those involving z^2 which have been omitted. It is important to have an understanding of the relative importance of these different terms in practical problems. We will confine the argument to cases where both the strains and the deflection slopes $\partial w/\partial x$ and $\partial w/\partial y$ are small compared to unity. Strains must be small in the elastic straining of hard materials, and are usually small in thin plates and shells even if plastic flow occurs or the material is rubber-like, because strains involving compression in any direction are limited by buckling; only in cases such as the inflation of rubber membranes, where the principal membrane stresses are tensile, can large strains occur. And, as was discussed in Sec. 2.2, large slopes could only occur in very thin plates which are bent into a developable shape, or in plates made of rubber-like materials or deformed plastically as in drawing operations; in the plates used in machines and structures the allowable strains and deflection slopes are generally very small compared to unity even in what we will call the 'large deflection' range.

Consider first the membrane part of the strains. If the loads are only applied on the edge of and in the plane of the plate, and there are no w displacements (that is, the plane stress elasticity case) then the first terms in the strain expressions, $\partial u/\partial x$, $\partial v/\partial y$, $\partial u/\partial y$, $\partial v/\partial x$, will be the important terms and in the cases we are considering will each be very small compared to unity. The squares or products of these terms will then be negligible compared to the terms themselves, and represent refinements which were neglected in the classical linear elasticity theory.

On the other hand for the case when transverse displacements occur due to transverse loading or during the process of buckling (with or without edge loads due to edge constraints) the u and v displacements will be somewhere between zero and the amounts which are necessary to cancel the second order strains due to slope. In ϵ_x, for example, $\partial u/\partial x$ will not exceed $(\partial w/\partial x)^2/2$ and must therefore be very small since $\partial w/\partial x$ is small; this means that such u and v displacements are much smaller than the w displacements. Among the terms in the flexural strains, therefore, terms such as $(\partial u/\partial x)(\partial^2 w/\partial x^2)$ must be very small compared to the principal flexural terms such as $\partial^2 w/\partial x^2$; similar arguments can be presented for the relative insignificance of terms such as $(\partial^2 u/\partial x^2)(\partial w/\partial x)$ and terms containing z^2.

For a combination of plane stress and transverse loading it is not possible to cite such a clear-cut argument, but experience indicates that for such cases also, if the strains and displacement slopes are small, the terms given in Eqs. (4.2) are the only important ones. For extreme cases such as inflation of a rubber membrane or drawing operations, when strains and slopes are of the order of magnitude of unity, relations such as Eqs. (4.6) which retain only second degree terms will probably also be insufficient, and it may be necessary to use the exact relations given in Eqs. (4.5), without attempting to put the strain expressions in the form of a sum of simple terms. In Chap. 6 it will be shown that it is possible to obtain numerical solutions with the corresponding exact expressions for shells of which Eqs. (4.5) represent a special case. In the present chapter we will confine ourselves to Eqs. (4.2), which are very adequate for most engineering applications.

4.3 EQUILIBRIUM EQUATIONS

Expressions for Forces and Moments on Sections

The second step in plate analysis is to substitute expressions (4.2) for the strains into the stress–strain relations for the materials being considered, to obtain the stresses σ_x, σ_y, σ_{xy} in a layer parallel to the middle surface and a distance z from it, Fig. 4.7. For the elastic case the stress–strain relations are of course Hooke's Law. Since expansion or contraction in the transverse direction of layers of the plate parallel to the middle surface is relatively free, such layers can be considered to be in a state of plane stress; the stress–strain relations are therefore given by Eqs. (3.11b), Sec. 3.2.*

* This question is considered more completely in the discussion of Eqs. (6.21), Sec. 6.4.

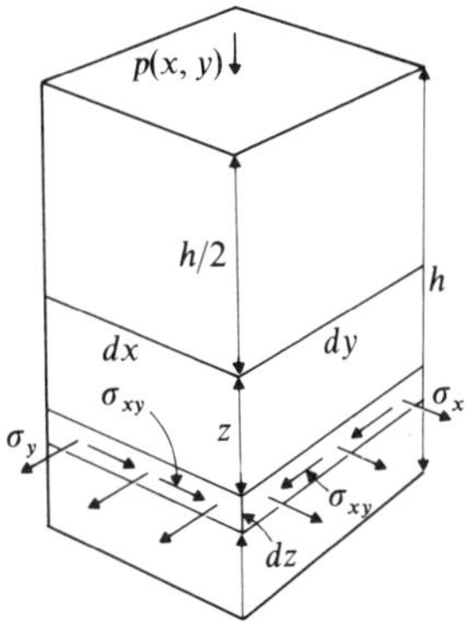

FIGURE 4.7

The third step is to obtain the forces F_x, F_y, F_{xy}, F_{yx} and the moments about the middle surface M_x, M_y, M_{xy}, M_{yx} by integrating over the height of the cross section the forces on an element of the cross section of width dz, and the moments of these forces about the middle surface. Remembering that the symbol F_x represents forces per unit width of section, the total force in the x direction on the side of the element on which it acts, Fig. 4.7, is: $F_x\,dy = \int \sigma_x\,dy\,dz = dy \int \sigma_x\,dz$, from which $F_x = \int \sigma_x\,dz$, the integrations being from $-c$ to c, and so on. We thus find:

$$F_x = \int_{-c}^{c} \sigma_x\,dz = h\sigma_{xm}, \qquad F_y = \int_{-c}^{c} \sigma_y\,dz = h\sigma_{ym}, \qquad F_{xy} = F_{yx} = \int_{-c}^{c} \sigma_{xy}\,dz = h\sigma_{xym}$$

$$M_x = \int_{-c}^{c} \sigma_x z\,dz, \qquad M_y = \int_{-c}^{c} \sigma_y z\,dz, \qquad M_{xy} = M_{yx} = \int_{-c}^{c} \sigma_{xy} z\,dz \tag{4.7}$$

It will be seen that as we have defined the symbols with double subscripts F_{xy}, F_{yx}, M_{xy}, M_{yx}, the order in which the subscripts are written is immaterial in the case of plates, as it is for the stress σ_{xy}; hence we will hereafter use only the symbols F_{xy}, M_{xy} for plates. We have no equations of the above sort for the transverse shear forces F_{xz} and F_{yz} because we are neglecting the shear strains corresponding to these forces and have no way as yet to determine them. Later we will determine F_{xz} and F_{yz} and the distribution of the shear stresses composing them from equilibrium conditions, as is done in the case of beams.

Equilibrium Equations

The last basic relations which we need are found by considering the equilibrium of an element of the plate of the dimensions dx, dy, h as illustrated in Fig. 4.7. For simplicity we show only the middle surface of the element in Fig. 4.8, with all the forces acting on the element at (c) and the moments about the middle surface of the forces on the edge cross sections at (d). We need equations of equilibrium of forces in the x, y, z directions Σf_x, Σf_y, $\Sigma f_z = 0$, and of moments of the forces

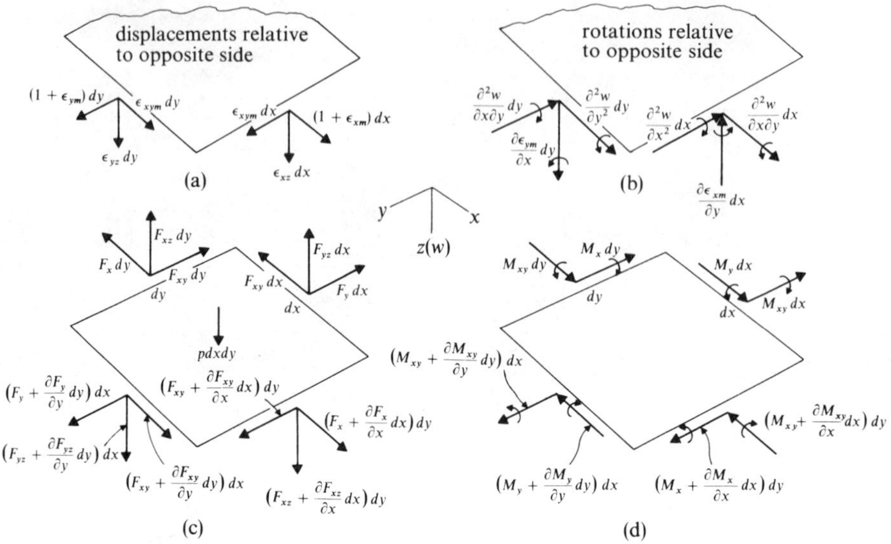

FIGURE 4.8

about axes in the x, y, z directions, say through the center of the element, $\Sigma\, m_x$, $\Sigma\, m_y$, $\Sigma\, m_z = 0$.

We show the *total* forces and moments acting on the element, obtained by multiplying the forces and moments per unit length of section by the original length of section on which they act. As was discussed in Sec. 3.1 for the stresses, the effects of the variation of the forces and moments along the sides of the element can be ignored as small quantities of higher order, and the forces and moments and their changes from one side to the opposite side can be considered to act at the centers of the sides, as suggested by the figures.

Although the element actually has a new deformed shape when these forces and moments are acting upon it, Fig. 4.8(c) shows the correct magnitude of the forces, because we have defined a stress on a surface as the force acting divided by its original area and we used the original dimensions in calculating the forces. The expressions for moments in Fig. 4.8(d) are not quite exact because although the forces used are correct, the thickness of the plate h and the moment arms of the forces are changed by the ϵ_z strains. The error is very slight because the effect of ϵ_z strains due to positive and negative flexural stresses (which are usually the biggest stresses acting) largely cancel each other; in any case changes in the moment arms are limited to a small fraction of one percent for hard materials in the elastic range.

The deformation of the element also slightly displaces and rotates the lines of action of the forces and changes the arms of the moments, and this of course affects equilibrium. Most of these displacements and rotations are unimportant in practical problems but it is instructive to consider them before discarding any of them. Since rigid body movements do not affect equilibrium, only the displacements and rotations of the center of one side relative to the center of the opposite side, shown in Figs. 4.8(a) and (b), need to be considered.

It may help to clarify consideration of this problem to note that if we wished we could take the directions in which forces are summed and the axes about which moments are taken as the x', y', z' directions tangent to and normal to the middle surface of the displaced element at its center, as shown in Fig. 4.9, rather than the x, y, z directions, so that the resultant of the external pressure p is in the z' direction and the vectors representing the forces and moments on the edges are at nearly equal angles to the z' direction. Another important point is that all the displacements and rotations contain a factor dx or dy and only those effects which involve their first power need to be considered; thus the differences between the cosine of an angle of rotation and unity, or between the angle and the sine or tangent of the angle are small quantities of higher order since they are nearly proportional to the squares of the angles.

Figure 4.8(a) shows the components of the displacements and Fig. 4.8(b) the components of the rotations of the sides on which they are shown relative to the opposite sides. To describe moments and rotations we use vectors with small curved arrows which show the direction of the moment or rotation; the vectors are drawn in the direction of the axis of the moment or rotation, the sense of the arrow heads being consistent with the right hand rule that the arrow points in the direction of the thumb of the right hand when the moment or rotation is in the direction of the fingers. Moment vectors, and vectors representing small rotations such as are considered here can be separated into components the same as other vector quantities, although this would not be true for large rotations.

In Fig. 4.8(a) ϵ_{xm}, ϵ_{ym}, ϵ_{xym} represent the membrane parts of ϵ_x, ϵ_y, ϵ_{xy} respectively, that is the strains at the middle surface. The symbols ϵ_{xz} and ϵ_{yz} represent average transverse strains, that is the strains which we would have if the transverse shear stresses were uniformly distributed over the cross section. Although we have no values for ϵ_{xz} and ϵ_{yz}, we could determine them from the transverse shear forces F_{xz}, F_{yz} obtained from a first approximation neglecting ϵ_{xz} and ϵ_{yz}, and then use these values for a second approximation (see Eqs. (6.23b)).

As has been discussed previously, the most important of the displacements and rotations due to the deformation shown in Figs. 4.8(a), (b) are those due to

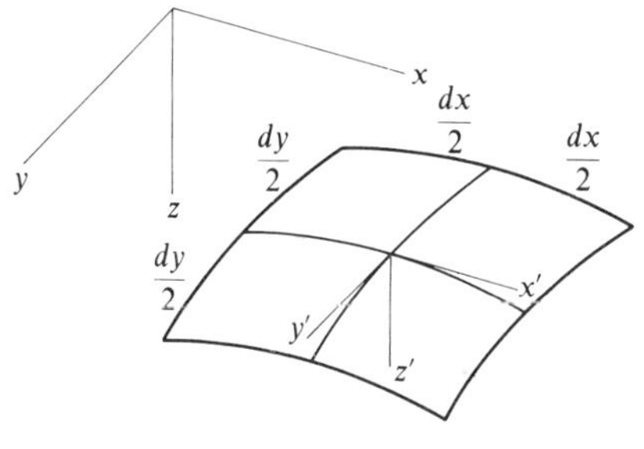

FIGURE 4.9

curvature and twist, containing $\partial^2 w/\partial x^2$, $\partial^2 w/\partial y^2$, $\partial^2 w/\partial x\,\partial y$, because even a comparatively thick plate is far more flexible in flexure than it is for strains in its own plane. The most important effect of these rotations furthermore is that they result in transverse components (like the $F_x\,\partial^2 w/\partial x^2$ term in beam theory) of the membrane forces F_x, F_y, F_{xy}, which in buckling problems are very large compared to the other forces and moments, as was discussed in Sec. 2.5. We will set up the equilibrium equations considering only these important effects of the deformation of the element, which obviously must be considered in buckling problems, but we will also discuss the importance of the terms omitted.

The terms which are involved in the equilibrium equations are of the following types:

(1) External forces which act on the element, for example the force $p\,dx\,dy$ which enters the equation of equilibrium of forces in the transverse direction, $\Sigma f_z = 0$, where $p(x, y)$ is a distributed force per unit area acting normal to the surface in the positive direction of the z axis, as shown in Fig. 4.8(c). To the approximation which we obtain by using the Love–Kirchhoff assumption, we can not distinguish between such a force acting on the upper or lower surface or on an intermediate surface, or a body force per unit volume p/h acting over the volume $h\,dx\,dy$. The effects of changes in area or volume due to deformations are negligible. Similar distributed external forces may act tangentially in the x or y directions, or exert moments because of the distance h between such tangential forces acting on the top and bottom surfaces; for example if distributed tangential forces in the x direction t_x and b_x act on the top and bottom surfaces, there would be terms $(t_x + b_x)\,dx\,dy$ in $\Sigma f_x = 0$, and $(t_x - b_x)h\,dx\,dy$ in $\Sigma m_y = 0$. Such forces can easily be included, but for simplicity we will include only the transverse loading p, which is the most common loading in practical problems, and also the most important because it acts in the weak transverse direction.

For problems in which the effect of displacements on the action of the loads has to be considered, we will consider that p remains normal to the surface after displacement, as would be the case if p is due to fluid pressure. Loads which act at an angle to the surface initially or after deformation can be studied by combining p with tangential loads such as t_x, whose directions are also assumed to remain tangential after deformation.

(2) Changes in the forces or moments on cross sections from one side to the opposite side, the main parts of the forces or moments cancelling out; for example we would have $(\partial F_{xz}/\partial x)\,dx\,dy$ and $(\partial F_{yz}/\partial y)\,dy\,dx$ in $\Sigma f_z = 0$. We can ignore the fact that the force or moment vectors are actually at small angles to the directions in which equilibrium is being measured because of the rotations of the sections on which they act, since the differences between the cosines of these angles and unity are small quantities of higher order.

(3) Components of the force or moment vectors which are nominally at right angles to the direction in which equilibrium is measured but which have small components in that direction due to the rotations of the sections on which they

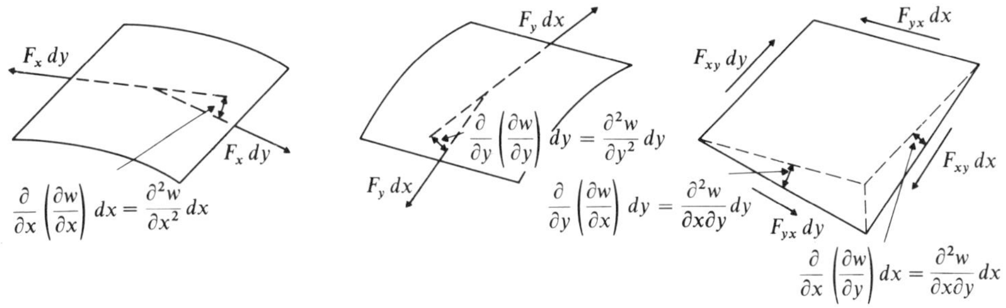

FIGURE 4.10

act. In such terms the changes in the vectors from one side to the opposite side can be ignored because the main parts of the vectors are involved and the changes are small quantities of higher order compared to them. Thus in the most important equation of equilibrium $\Sigma f_z = 0$ in the transverse direction, we have $F_x\,dy(\partial^2 w/\partial x^2)\,dx$ (like the $F_x(\partial^2 w/\partial x^2)\,dx$ term in Eq. (2.4) for beams) and $F_y\,dx(\partial^2 w/\partial y^2)\,dy$, as well as $F_{xy}\,dx(\partial^2 w/\partial x\,\partial y)\,dy + F_{xy}\,dy(\partial^2 w/\partial x\,\partial y)\,dx = 2F_{xy}(\partial^2 w/\partial x\,\partial y)\,dx\,dy$, as is demonstrated in Fig. 4.10. These are important in stability problems even when the displacements are small, because F_x, F_y, F_{xy} have finite values before the deflection starts. In other problems F_x, F_y, F_{xy} are caused by the displacements, and these terms are then nonlinear terms which are important only when the deflections are of the order of magnitude of the thickness or more. The terms $F_{xz}\,dy(\partial^2 w/\partial x^2)\,dx$ in $\Sigma f_x = 0$ and $F_{yz}\,dx(\partial^2 w/\partial y^2)\,dy$ in $\Sigma f_y = 0$ will be ignored because F_{xz} and F_{yz} are zero before displacement starts and are much smaller than F_x, F_y, F_{xy} in ordinary problems of the type which we will consider, as was discussed in connection with Eq. (2.3b) for beams.

(4) Finally, in the moment equations, we have the moments of opposite forces on opposite sides of the element, which form couples with moment arms equal to the distances between the lines of action of the forces. The most important of such couples are those which have the sides of the element as moment arms, for example $(F_{xz}\,dy)\,dx$ in $\Sigma m_y = 0$. Again we can ignore the change $(\partial F_{xz}/\partial x)\,dx$, the differences between the cosines of the small angles of rotation and unity, and the small change in the length of the moment arm which is due to its curvature, Fig. 4.11(a), as small quantities of higher order. The effect on the moment arm of the strain $\epsilon_{xm}\,dx$ is of the same order of magnitude as the small moment $F_x\,dy\,\epsilon_{xz}\,dx$ in

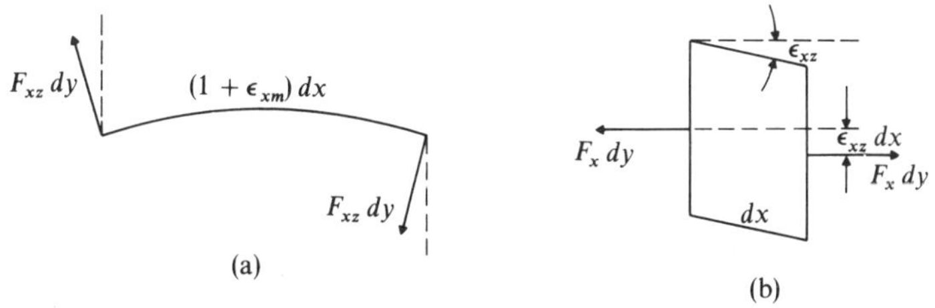

FIGURE 4.11

the same equation, illustrated in Fig. 4.11(b), and both are unimportant for hard materials in the elastic range since strains are limited to a small fraction of one percent of unity; they will also be ignored.

It will be noticed that all the terms in the equations of equilibrium contain the factor $dx\,dy$. Dividing through by this factor, the six equations of equilibrium are then:

$$\sum f_x = 0: \quad \frac{\partial F_x}{\partial x} + \frac{\partial F_{xy}}{\partial y} = 0, \qquad \sum f_y = 0: \quad \frac{\partial F_y}{\partial y} + \frac{\partial F_{xy}}{\partial x} = 0$$

$$\sum f_z = 0: \quad p + \frac{\partial F_{xz}}{\partial x} + \frac{\partial F_{yz}}{\partial y} + F_x \frac{\partial^2 w}{\partial x^2} + F_y \frac{\partial^2 w}{\partial y^2} + 2F_{xy} \frac{\partial^2 w}{\partial x\,\partial y} = 0$$

$$\sum m_x = 0: \quad \frac{\partial M_y}{\partial y} + \frac{\partial M_{xy}}{\partial x} = F_{yz} \qquad \sum m_z = 0: \quad F_{xy} - F_{xy} \equiv 0 \tag{4.8}$$

$$\sum m_y = 0: \quad \frac{\partial M_x}{\partial x} + \frac{\partial M_{xy}}{\partial y} = F_{xz}$$

Among terms which have been neglected other than those which have been mentioned are the terms of type (3)

$$M_{xy}(\partial^2 w/\partial y^2 - \partial^2 w/\partial x^2) + (M_x - M_y)\,\partial^2 w/\partial x\,\partial y \qquad \text{in} \qquad \sum m_z = 0,$$

which would then cease to be an identity. Terms such as these are very small when strains and deflection slopes are small, and experience indicates that they are all unimportant compared to the terms which we have considered in ordinary problems such as we will study.

Consider the first two equations of Eq. (4.8), $\Sigma f_x = 0$, $\Sigma f_y = 0$. Using Eqs. (4.7) and dividing through by h, these become:

$$\frac{\partial \sigma_{xm}}{\partial x} + \frac{\partial \sigma_{xym}}{\partial y} = 0, \qquad \frac{\partial \sigma_{ym}}{\partial y} + \frac{\partial \sigma_{xym}}{\partial x} = 0 \tag{4.9}$$

These are the same except for the subscript m as Eqs. (3.14a), and they can be satisfied if we take, as in Eq. (3.16a):

$$\sigma_{xm} = \frac{\partial^2 \phi}{\partial y^2}, \qquad \sigma_{ym} = \frac{\partial^2 \phi}{\partial x^2}, \qquad \sigma_{xym} = -\frac{\partial^2 \phi}{\partial x\,\partial y} \tag{4.10}$$

For the elastic case, using Eqs. (3.11a), (4.10) and (4.2) to equate the membrane strains in terms of the membrane stresses to the same strains in terms of the displacements, we have:

$$\left[\epsilon_{xm} = \frac{1}{E}(\sigma_{xm} - \nu\sigma_{ym}) = \right] \frac{1}{E}\left(\frac{\partial^2 \phi}{\partial y^2} - \nu \frac{\partial^2 \phi}{\partial x^2}\right) = \frac{\partial u}{\partial x} + \frac{1}{2}\left(\frac{\partial w}{\partial x}\right)^2$$

$$\left[\epsilon_{ym} = \frac{1}{E}(\sigma_{ym} - \nu\sigma_{xm}) = \right] \frac{1}{E}\left(\frac{\partial^2 \phi}{\partial x^2} - \nu \frac{\partial^2 \phi}{\partial y^2}\right) = \frac{\partial v}{\partial y} + \frac{1}{2}\left(\frac{\partial w}{\partial y}\right)^2 \tag{4.11}$$

$$\left[\epsilon_{xym} = \frac{2(1+\nu)}{E}\sigma_{xym} = \right] -\frac{2(1+\nu)}{E}\frac{\partial^2 \phi}{\partial x\,\partial y} = \frac{\partial u}{\partial y} + \frac{\partial v}{\partial x} + \frac{\partial w}{\partial x}\frac{\partial w}{\partial y}$$

We next apply the operators $\partial^2/\partial y^2$, $\partial^2/\partial x^2$ and $-\partial^2/\partial x\,\partial y$ to the first, second and third of these equations respectively and add them together, in order to eliminate u and v as we did in the development of Eq. (3.16c); we thus obtain:

$$\begin{aligned}
\frac{1}{E}\left(\frac{\partial^4\phi}{\partial y^4}-\nu\frac{\partial^4\phi}{\partial x^2\,\partial y^2}\right) &= \frac{\partial^3 u}{\partial x\,\partial y^2}+\left(\frac{\partial^2 w}{\partial x\,\partial y}\right)^2+\frac{\partial w}{\partial x}\frac{\partial^3 w}{\partial x\,\partial y^2}\\
\frac{1}{E}\left(\frac{\partial^4\phi}{\partial x^4}-\nu\frac{\partial^4\phi}{\partial x^2\,\partial y^2}\right) &= \frac{\partial^3 v}{\partial x^2\,\partial y}+\left(\frac{\partial^2 w}{\partial x\,\partial y}\right)^2+\frac{\partial w}{\partial y}\frac{\partial^3 w}{\partial x^2\,\partial y}\\
\frac{2(1+\nu)}{E}\frac{\partial^4\phi}{\partial x^2\,\partial y^2} &= -\frac{\partial^3 u}{\partial x\,\partial y^2}-\frac{\partial^3 v}{\partial x^2\,\partial y}-\frac{\partial w}{\partial y}\frac{\partial^3 w}{\partial x^2\,\partial y}\\
&\quad -\frac{\partial^2 w}{\partial x^2}\frac{\partial^2 w}{\partial y^2}-\left(\frac{\partial^2 w}{\partial x\,\partial y}\right)^2-\frac{\partial w}{\partial x}\frac{\partial^3 w}{\partial x\,\partial y^2}
\end{aligned} \tag{4.12}$$

Adding and multiplying by E we find:

$$\frac{\partial^4\phi}{\partial x^4}+2\frac{\partial^4\phi}{\partial x^2\,\partial y^2}+\frac{\partial^4\phi}{\partial y^4}=\nabla^4\phi=E\left[\left(\frac{\partial^2 w}{\partial x\,\partial y}\right)^2-\frac{\partial^2 w}{\partial x^2}\frac{\partial^2 w}{\partial y^2}\right] \tag{4.13}$$

This important equation (4.13) was derived by von Kármán.* Comparing it with Eq. (3.16c), we see that the membrane stresses in plates can be taken as a superposition of any stresses satisfying plane elasticity (that is $\nabla^4\phi=0$ and the boundary conditions) and stresses satisfying Eq. (4.13), because such a superposition would evidently also satisfy Eq. (4.13). This can be taken to mean that we take $\phi=\phi_p+\phi_h$, where ϕ_p is a particular solution, that is any function which satisfies Eq. (4.13) irrespective of boundary conditions, and ϕ_h is the solution of the homogeneous equation obtained from Eq. (4.13) by retaining only the terms in ϕ, that is setting the right-hand side equal to zero. The latter, by Eq. (3.16c), includes all plane stress solutions, which theoretically should involve enough 'variables of integration' to satisfy the gross boundary conditions on the forces or displacements in the plane of the plate along all four sides of a rectangular plate, or along corresponding boundaries if it is of some other shape. The most commonly used and useful plane stress solutions of this homogeneous equation, involving power functions and trigonometric–hyperbolic functions, were discussed previously in Sec. 3.3.

A further very important conclusion which we can draw from Eq. (4.13) is that since the particular solution part of the membrane stresses represented by ϕ_p is connected to the transverse deflection w by squares or products of derivatives of w, this part of the membrane stresses is a 'large deflection' or 'finite displacement' effect, which is negligible when w is small and only becomes important when w is large; how large w must be for it to be important is difficult to determine in a simple manner, but experience indicates that, as in the corresponding beam case considered in Sec. 2.6, this particular solution part of the membrane stresses becomes important only when w reaches the order of magnitude of the thickness. That is if $w \leqslant 0{\cdot}2h$, say, we can ignore such membrane stresses, and

* *Enzyklopädie der Math. Wiss.*, vol. 4, art. 27, 1910, p. 349.

take the right side of Eq. (4.13) as zero. When this is the case there can still be membrane stresses as in plane elasticity, produced by edge loads in the plane of the plate, but this plane stress condition *will be independent of the transverse loading* and the deflections caused by it. That is, when w is small the two stress conditions in the plate, membrane stresses and the flexural stresses due to transverse loading, can be studied separately and merely superimposed.

We could have used Eqs. (3.15b) instead of the Airy stress function of Eqs. (3.16a) to satisfy the equilibrium conditions (4.9) in the above analysis, but there would be no advantage to this because, as can be seen from Eqs. (4.11), the displacements in the plane of the plate depend upon w as well as upon σ_x, σ_y, σ_{xy} in the present case and so could not be obtained from Eqs. (3.15a).

Consider now the remainder of the equilibrium equations (4.8). For the elastic case, using the stress–strain relations (3.11b) and strain–displacement relations (4.2) in expressions (4.7) for M_x, M_y, M_{xy}, we find:

$$\begin{aligned} M_x &= \int_{-c}^{c} \sigma_x z\, dz = [E/(1-\nu^2)] \int_{-c}^{c} (\epsilon_x - \nu\epsilon_y) z\, dz \\ &= -[Eh^3/12(1-\nu^2)](\partial^2 w/\partial x^2 + \nu \partial^2 w/\partial y^2), \end{aligned}$$

etc. We can write these expressions in the form:

$$\begin{aligned} M_x &= -D\left(\frac{\partial^2 w}{\partial x^2} + \nu \frac{\partial^2 w}{\partial y^2}\right), \qquad M_y = -D\left(\frac{\partial^2 w}{\partial y^2} + \nu \frac{\partial^2 w}{\partial x^2}\right) \\ M_{xy} &= -D(1-\nu)\frac{\partial^2 w}{\partial x\, \partial y}, \qquad \text{where } D = \frac{Eh^3}{12(1-\nu^2)} = \frac{2Ec^3}{3(1-\nu^2)} \end{aligned} \tag{4.14}$$

The quantity D is the modified modulus of elasticity $E/(1-\nu^2)$ times the moment of inertia per unit length of section about the middle surface of cross sections of the plate; it corresponds to the EI of beam theory per unit width of the beam. D is a measure of and is called the 'flexural stiffness' of the plate. Substituting expressions (4.14) into $\Sigma\, m_x = 0$, $\Sigma\, m_y = 0$ of Eqs. (4.8), we obtain the following expressions for the transverse shear forces F_{xz}, F_{yz}:

$$\begin{aligned} F_{xz} &= -D\left(\frac{\partial^3 w}{\partial x^3} + \frac{\partial^3 w}{\partial x\, \partial y^2}\right) = -D\frac{\partial}{\partial x}\nabla^2 w, \\ F_{yz} &= -D\left(\frac{\partial^3 w}{\partial x^2\, \partial y} + \frac{\partial^3 w}{\partial y^3}\right) = -D\frac{\partial}{\partial y}\nabla^2 w \end{aligned} \tag{4.15}$$

Expressions for Stresses

We can now assemble expressions for the various stresses in plates, including the transverse shear stresses. Using Eqs. (3.11b), (4.2), (4.10), (4.14), the normal and shear stresses in the x, y directions are:

$$\sigma_x = \sigma_{xm} + \sigma_{xf},$$

$$\sigma_{xm} = \frac{\partial^2 \phi}{\partial y^2} = \frac{E}{1-\nu^2}\left[\frac{\partial u}{\partial x} + \nu\frac{\partial v}{\partial y} + \frac{1}{2}\left(\frac{\partial w}{\partial x}\right)^2 + \frac{\nu}{2}\left(\frac{\partial w}{\partial y}\right)^2\right] = \frac{F_x}{h}$$

$$\sigma_{xf} = -\frac{Ez}{1-\nu^2}\left(\frac{\partial^2 w}{\partial x^2} + \nu\frac{\partial^2 w}{\partial y^2}\right) = \frac{12M_x z}{h^3}$$

$$\sigma_{xy} = \sigma_{xym} + \sigma_{xyf},$$

$$\sigma_{xym} = -\frac{\partial^2 \phi}{\partial x\,\partial y} = \frac{E}{2(1+\nu)}\left(\frac{\partial u}{\partial y} + \frac{\partial v}{\partial x} + \frac{\partial w}{\partial x}\frac{\partial w}{\partial y}\right) = \frac{F_{xy}}{h} \tag{4.16}$$

$$\sigma_{xyf} = -\frac{Ez}{1+\nu}\frac{\partial^2 w}{\partial x\,\partial y} = \frac{12M_{xy}z}{h^3}$$

The expressions for σ_y are the same as those for σ_x with u and v, x and y interchanged.

The transverse shear stresses can be found, as in beam theory, from the equilibrium of a portion of an element of a plate on one side of a plane parallel to the middle surface, as shown in Fig. 4.12. For simplicity only the forces in the x direction are shown in the figure. From equilibrium in this direction: $\sigma_{xz}\,dx\,dy = \int_z^c (\partial\sigma_x/\partial x + \partial\sigma_{xy}/\partial y)\,dx\,dy\,dz'$. From Eq. (4.9) the membrane part of the expression in the parentheses is zero, and using Eq. (4.16)

$$\sigma_{xz} = \int_z^c (\partial\sigma_{xf}/\partial x + \partial\sigma_{xyf}/\partial y)\,dz'$$

$$= -E/(1-\nu^2)[\partial^3 w/\partial x^3 + \nu\,\partial^3 w/\partial x\,\partial y^2 + (1-\nu)\,\partial^3 w/\partial x\,\partial y^2]\int_z^c z'\,dz',$$

and using Eq. (4.15):

$$\sigma_{xz} = -\frac{E}{2(1-\nu^2)}(c^2 - z^2)\frac{\partial}{\partial x}\nabla^2 w = \frac{6F_{xz}}{h^3}(c^2 - z^2) \tag{4.17}$$

The expression for σ_{yz} is the same as for σ_{xz} with x, y interchanged.

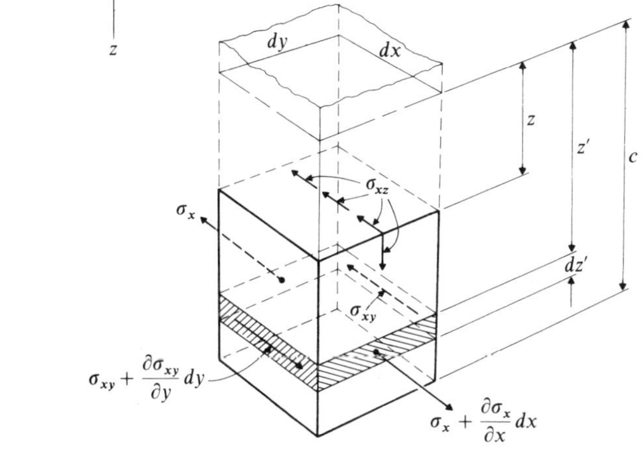

FIGURE 4.12

Transverse Equilibrium Equation

Finally, substituting expressions (4.15) into $\Sigma f_z = 0$ of Eqs. (4.8) and using Eqs. (4.7) and (4.10) we can write the important transverse equilibrium equation as:

$$D\left(\frac{\partial^4 w}{\partial x^4} + 2\frac{\partial^4 w}{\partial x^2\,\partial y^2} + \frac{\partial^4 w}{\partial y^4}\right) = p + F_x\frac{\partial^2 w}{\partial x^2} + F_y\frac{\partial^2 w}{\partial y^2} + 2F_{xy}\frac{\partial^2 w}{\partial x\,\partial y}$$

or:

$$D\nabla^4 w = p + h\left(\sigma_{xm}\frac{\partial^2 w}{\partial x^2} + \sigma_{ym}\frac{\partial^2 w}{\partial y^2} + 2\sigma_{xym}\frac{\partial^2 w}{\partial x\,\partial y}\right) \tag{4.18}$$

$$= p + h\left(\frac{\partial^2 \phi}{\partial y^2}\frac{\partial^2 w}{\partial x^2} + \frac{\partial^2 \phi}{\partial x^2}\frac{\partial^2 w}{\partial y^2} - 2\frac{\partial^2 \phi}{\partial x\,\partial y}\frac{\partial^2 w}{\partial x\,\partial y}\right)$$

It will be seen that Eq. (2.4) for beams is a special case of Eq. (4.18) for the case when $\partial/\partial y = 0$ (that is, there is no variation in the y direction), except that for the beam E is substituted for $E/(1-\nu^2)$ because of the beam's freedom to expand or contract laterally. Other equations for beams can similarly be seen to be special cases of the corresponding plate equations.

General Solutions of Plate Problems

The stress function $\phi(x, y)$ defines the membrane stresses, while the deflection $w(x, y)$ defines the flexural stresses. These two functions must satisfy the two Eqs. (4.13) and (4.18), together with all the boundary conditions on forces or displacements in the plane of the plate and out of this plane. In the general case the boundary conditions along each of the four sides of a rectangular plate define two conditions on the membrane forces or the u and v displacements in the plane of the plate; these are the 'membrane' conditions associated with ϕ, which must be satisfied by the solution of Eq. (4.13). In addition, for the case of hinged or fixed edges (other edge conditions involve complications which will be considered in Sec. 4.5), the boundary conditions along each side define two conditions on the flexural moments or forces or on the transverse displacement or rotation *out* of the plane of the plate; these are the 'flexural' conditions associated with w, which must be satisfied by the solution of Eq. (4.18). As an example, for a completely fixed edge parallel to the x axis we should satisfy the conditions $u, v, w, \partial w/\partial y = 0$ along the edge, the first two being membrane and the last two flexural conditions.

Equations (4.13) and (4.18) are each of the fourth order in x and y, that is they contain fourth derivatives with respect to x and y. They theoretically determine the two functions ϕ and w with enough functions of integration to satisfy the four conditions outlined along each of the four sides of a rectangular plate. These conditions can be satisfied by solutions of the homogeneous equations obtained by retaining only the terms in ϕ and w in Eqs. (4.13), (4.18) respectively; these solutions are added to particular solutions of these equations to obtain complete solutions.

For the present we will confine ourselves to plates of rectangular shape, using rectangular coordinates. For other shaped plates it is usually more convenient to use a coordinate system such that one of the coordinates is constant along the boundary, for instance polar coordinates for circular or pie-shaped plates. The basic equations for plates in any coordinate system can readily be derived from the general shell theory presented in Chap. 6, and some discussion of these equation will be found there and for circular plates at the end of this chapter.

Direct solution of nonlinear simultaneous differential equations such as Eqs. (4.13) and (4.18) is not generally practical, but a number of indirect methods of obtaining solutions have been developed, such as series solutions and applications of energy methods. Such nonlinear 'large deflection' cases, as well as small deflection improvements on classical plate theory (which also permit satisfaction of more complete boundary conditions than those on resultant forces and moments and deflections of the middle surface discussed above) will be considered in Chap. 5. In the remainder of this chapter we will consider the many practical problems for which linear, small deflection solutions of classical plate theory can be used.

4.4 SMALL DEFLECTIONS OF HINGED-EDGE PLATES

If thc transverse deflection w is small (which means, as has been discussed, small compared to the thickness, h) then the right side of Eq. (4.13) can be taken as zero, and the membrane stresses, if any, can be calculated independently of the deflection w. If there are edge loads in the plane of the middle surface, the membrane stresses can be calculated as a plane stress problem independent of the transverse deflection, for example to establish the distribution of membrane stresses tending to produce buckling, prior to a classical stability study. Equation (4.18) then becomes linear in w, since σ_{xm}, σ_{ym}, σ_{xym}, if present, can be considered to be constant with respect to w when w is small. Solutions of Eq. (4.18) then are not very different in principle from solutions of Eq. (2.4) for beams, which were studied in Chap. 2; the following examples will illustrate such differences as there are.

Transverse Load Case, Hinged Edges

For the case when the membrane stresses σ_{xm}, σ_{ym}, σ_{xym} are zero, Eq. (4.18) becomes:

$$D\nabla^4 w = p \tag{4.19}$$

For a rectangular plate such as shown in Fig. 4.13, with hinged edges at $x = 0, a$ and $y = 0, b$ the gross boundary conditions are:

$$\begin{aligned} x = 0, a: &\quad w, M_x = 0, \quad \text{or} \quad w, \frac{\partial^2 w}{\partial x^2} + \nu \frac{\partial^2 w}{\partial y^2} = 0 \\ y = 0, b: &\quad w, M_y = 0, \quad \text{or} \quad w, \frac{\partial^2 w}{\partial y^2} + \nu \frac{\partial^2 w}{\partial x^2} = 0 \end{aligned} \tag{4.20}$$

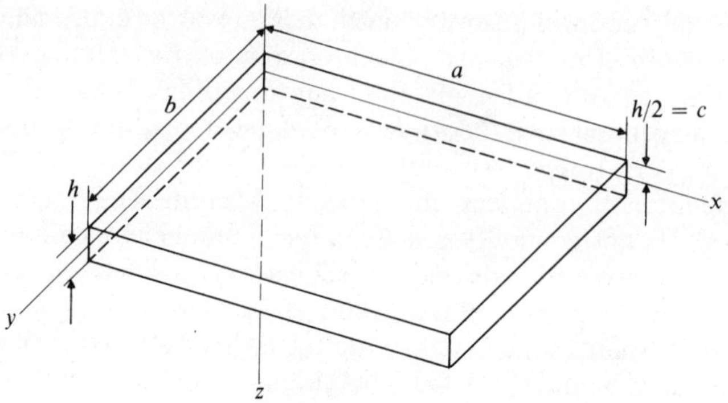

FIGURE 4.13

These conditions are satisfied if:

$$w = \sum_m \sum_n w_{mn} \sin \frac{m\pi x}{a} \sin \frac{n\pi y}{b} \quad (m, n = 1, 2, 3, \ldots) \tag{4.21}$$

Equation (4.19) can also be satisfied if:

$$p = \sum_m \sum_n p_{mn} \sin \frac{m\pi x}{a} \sin \frac{n\pi y}{b} \tag{4.22}$$

where the coefficients p_{mn} can be determined to make expression (4.22) conform to any desired distribution of $p(x, y)$ by an harmonic analysis analogous to that used with beams, using the fact that the terms of such a series are orthogonal in the interval $0<x<a$, $0<y<b$. To determine any particular coefficient p_{mn} we multiply both sides of Eq. (4.22) by $\sin(m\pi x/a)\sin(n\pi y/b)$ and integrate over the area of the plate:

$$\int_0^a dx \int_0^b dy\, p(x, y) \sin \frac{m\pi x}{a} \sin \frac{n\pi y}{b} = \int_0^a dx \int_0^b dy$$

$$\left(p_{11} \sin \frac{\pi x}{a} \sin \frac{\pi y}{b} \cdots + p_{mn} \sin \frac{m\pi x}{a} \sin \frac{n\pi y}{b} \cdots\right) \sin \frac{m\pi x}{a} \sin \frac{n\pi y}{b} = \frac{a}{2} \cdot \frac{b}{2} p_{mn}$$

whence

$$p_{mn} = \frac{4}{ab} \int_0^a dx \int_0^b dy\, p(x, y) \sin \frac{m\pi x}{a} \sin \frac{n\pi y}{b} \tag{4.23}$$

Substituting expressions (4.21), (4.22), (4.23) in Eq. (4.19), equating the coefficients of corresponding terms, solving for w_{mn} and substituting back in Eq. (4.21), as in the derivation of Eq. (2.12) for beams, we obtain:

$$w = \frac{4a^3}{\pi^4 bD} \sum_m \sum_n \frac{\sin \frac{m\pi x}{a} \sin \frac{n\pi y}{b}}{\left(m^2 + \frac{a^2}{b^2} n^2\right)^2} \int_0^a dx \int_0^b dy\, p(x, y) \sin \frac{m\pi x}{a} \sin \frac{n\pi y}{b} \tag{4.24}$$

which can be used for any lateral loading in the same general way as Eq. (2.12) was used for beams.

Uniform Loading on Hinged-Edge Plate

For a plate under a uniform loading $p = p_0$ and Eq. (4.24) reduces to:

$$w = \frac{16a^4 p_0}{\pi^6 D} \sum_m \sum_n \frac{\sin \dfrac{m\pi x}{a} \sin \dfrac{n\pi y}{b}}{mn\left(m^2 + \dfrac{a^2}{b^2} n^2\right)^2} \qquad (m, n \text{ odd only}) \tag{4.25}$$

The maximum deflection is at $x = a/2$, $y = b/2$; for a square plate, $a = b$, this is:

$$w_{\max} = \frac{4a^4 p_0}{\pi^6 D}\left(1 - \frac{2}{75} + \frac{1}{729} + \frac{2}{845} \cdots\right) \approx 0{\cdot}00406 \frac{a^4 p_0}{D} \tag{4.26}$$

If only the first term had been used, the result would have been $0{\cdot}00416\, a^4 p_0/D$, about 2·5 percent too high.

In evaluating such a double series numerically, the order of importance of the terms can be indicated by the following scheme:

$$mn = \begin{array}{ccccc} \underline{11} & 12 & \underline{13} & 14 & \underline{15} \\ 21 & 22 & 23 & 24 & \\ \underline{31} & 32 & \underline{33} & \cdots & \\ 41 & 42 & \vdots & & \\ \underline{51} & & & & \end{array} \tag{4.27}$$

The most important term is that for $m, n = 1, 1$; for cases of symmetry between x and y the importance of the terms decreases with $m + n$, that is, the next most important terms (dropping the commas for simplicity) are 21 and 12, the next after that 31, 22, 13, etc., as shown separated by the diagonal lines in Eq. (4.27). If convergence is more rapid with respect to x or m than with respect to y or n, or vice versa, the importance would decrease with $m + cn$, where c is different from unity, corresponding to drawing the diagonal separators in (4.27) at a different angle. For example, if a/b in Eq. (4.24) is greater than unity, convergence would be faster in the y or n direction and the separators should be more nearly vertical.

In a case such as that considered above, where there is symmetry of structure and loading about the middle of a plate in both x and y directions, all terms in the series will be zero except those having odd values of both m and n, so that only the values of mn shown underlined in (4.27) will remain. Of the terms in the parentheses in Eq. (4.26) the first term is for $mn = 11$, the second term is the sum of the equal values for 31 and 13, while the third and fourth terms are for 33 and for the sum of the equal values for 51 and 15. It will be seen that the terms do decrease in importance in the manner predicted.

Using Eq. (4.25) in (4.16), the bending and twisting stresses are, ignoring nonlinear terms:

$$\sigma_x = \frac{192a^2p_0z}{\pi^4h^3}\sum_m\sum_n \frac{\left(m^2+\nu\frac{n^2a^2}{b^2}\right)\sin\frac{m\pi x}{a}\sin\frac{n\pi y}{b}}{mn\left(m^2+\frac{a^2}{b^2}n^2\right)^2}$$

$$\sigma_{xy} = -\frac{192(1-\nu)a^3p_0z}{\pi^4h^3b}\sum_m\sum_n \frac{\cos\frac{m\pi x}{a}\cos\frac{n\pi y}{b}}{\left(m^2+\frac{a^2}{b^2}n^2\right)^2} \tag{4.28}$$

The expression for σ_y is the same as for σ_x with a and b, m and n, x and y interchanged. If $a < b$ the σ_x stresses will be greater than the σ_y. For a square plate, $a = b$, the maximum bending stress will be at $z = h/2$, $x, y = a/2$, while the maximum twisting stress will be at $z = h/2$, $x, y = 0$:

$$\begin{aligned}\sigma_{x(\max)} &= \frac{24(1+\nu)a^2p_0}{\pi^4h^2}\left[1-\frac{2}{15}+\frac{1}{81}+\frac{2}{65}-\cdots\right]\\ &\approx 0{\cdot}224\frac{(1+\nu)a^2p_0}{h^2}\\ \sigma_{xy(\max)} &= -\frac{24(1-\nu)a^2p_0}{\pi^4h^2}\left[1+\frac{2}{25}+\frac{1}{81}+\frac{2}{169}+\cdots\right]\\ &\approx 0{\cdot}273\frac{(1-\nu)a^2p_0}{h^2}\end{aligned} \tag{4.29}$$

where again the terms are for $mn = 11,\ 31+13,\ 33,\ 51+15$.

These results are instructive in several ways. First, it will be seen that the maximum shear stress, which is caused by twisting, is nearly as large as (about 66 percent of) the maximum bending stress. Since the safe stress in shear may be only a half or so of that in tension the twisting stress is at least as important as the bending stress; however it is sometimes disregarded by engineers, perhaps because they are more familiar with beams than with plates and tend to think of the latter as if they were an array of beams.

Second, the convergence for these series is much poorer than it was for the deflection, as was discussed earlier. Nevertheless the value of σ_x is only about one percent too high and that for σ_{xy} about two percent too low, with the number of terms shown. With only the first term the result would have been about twelve percent too high for σ_x and too low for σ_{xy}.

An easy way to extend such series is to plot the results and extrapolate to get rough values for the less important terms beyond those calculated. In Fig. 4.14, for example, the sum of the terms of the series for $\sigma_{xy(\max)}$ between successive diagonal separators of Eq. (4.27) are plotted and the plot extrapolated (the dotted curve) to give rough values of the terms after the cut-off. More accuracy can be obtained by

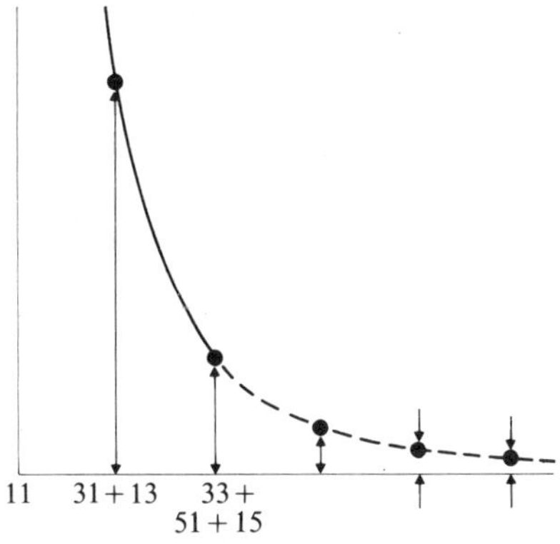

FIGURE 4.14

calculating every other value or every fifth value, etc., and interpolating. Instead of measuring the desired ordinates, their sum can be obtained approximately by measuring the area under the curve with a planimeter and dividing by the horizontal spacing (a simple graphical equivalent of the Fourier integral method). If the signs of terms alternate, as in the series for σ_x, two curves can be plotted, one for the sum of successive positive terms and the other for successive negative terms, or a plot of paired positive and negative terms can be made. More complicated analyses which give better convergence or other types of analysis can be used, but for practical numerical work such methods used with the simplest analysis usually save time.

Concentrated Load on Hinged-Edge Plate

For a concentrated load P acting at the point x_0, y_0, as in the derivation of Eq. (2.15) for beams, we replace the load by a distributed load P/Δ^2 acting uniformly over an infinitesimal square surface with sides Δ centered about the point x_0, y_0, with zero loading elsewhere. Then the integral $\int_0^a dx \int_0^b dy\, p(x, y) \sin m\pi x/a \sin n\pi y/b$ in Eq. (4.24) will be zero everywhere except in this small square, where $\sin m\pi x/a \sin n\pi y/b$ can be taken as $\sin m\pi x_0/a \sin n\pi y_0/b$, so that the integral will equal $P \sin m\pi x_0/a \sin n\pi y_0/b$. As with beams, solutions for different concentrated loads or of elemental loads forming a distributed loading can be superposed in such linear analyses.

Such series solutions for the deflection w converge well even at the point of application of the load. However, the corresponding series for the second derivatives of w in the expression for the bending stresses do not converge at the point of application of the concentrated load. Closed solutions for this case, such as that given in Sec. 4.7 for a round plate, show that the curvatures predicted by the classical plate and shell theories go to infinity at such a point. Actually this result is erroneous and is due to the limitations of the classical theory. The stress must of course be infinite at the point of the surface to which a truly concentrated load is applied, but it is actually finite everywhere else. This question will be discussed in Sec. 5.3, where the correct distribution of stress for a concentrated load on a plate or a thin shell will be derived.

Free Vibration of Hinged-Edge Plate

The function $w(x, y, t)$:

$$w = w_{mn} \sin\frac{m\pi x}{a} \sin\frac{n\pi y}{b} \sin 2\pi N_{mn}t \qquad (m, n = 1, 2, 3, \ldots) \tag{4.30}$$

describes a vibration of amplitude w_{mn} and frequency N_{mn} cycles per unit of time t, and of shape or mode consisting of m half waves in the x direction and n half waves in the y direction, separated by a rectangular grid of stationary nodes. The transverse loading is the 'inertia loading' $p = -\rho h \partial^2 w/\partial t^2$ where ρ is the mass density of the material. Substituting these values of w and p into Eq. (4.19) we obtain

$$\pi^4 D\left(\frac{m^2}{a^2} + \frac{n^2}{b^2}\right)^2 w_{mn} \sin\frac{m\pi x}{a} \sin\frac{n\pi y}{b} \sin 2\pi N_{mn}t$$
$$= \rho h 4\pi^2 N_{mn}^2 w_{mn} \sin\frac{m\pi x}{a} \sin\frac{n\pi y}{b} \sin 2\pi N_{mn}t \tag{4.31}$$

from which the frequency is:

$$N_{mn} = \frac{\pi}{2a^2}\sqrt{\frac{D}{\rho h}}\left(m^2 + \frac{a^2}{b^2}n^2\right) = \frac{\pi h}{4a^2}\sqrt{\frac{E}{3(1-\nu^2)\rho}}\left(m^2 + \frac{a^2}{b^2}n^2\right) \tag{4.32}$$

The analysis will be an excellent approximation for the lower modes of vibration of thin plates, that is when the half wave lengths are large compared to the thickness, or a/m, $b/n \gg h$. As in the case of beams, for the higher modes transverse shear strains and rotatory inertia have to be considered, as will be discussed in Sec. 5.6.

Stability of Hinged-Edge Plate Under Edge Compressions

Suppose a plate is under edge compressive stresses s_x uniformly distributed over the sides $x = 0, a$ and s_y over the sides $y = 0, b$, Fig. 4.15. Then the plane stress solution for this loading (which, as discussed before, applies before buckling and also during the infinitesimal buckling displacement assumed in this classical stability study) is that stresses $\sigma_{xm} = -s_x$, $\sigma_{ym} = -s_y$ act throughout the plate. Buckling can take place in the shape:

$$w = w_{mn} \sin\frac{m\pi x}{a} \sin\frac{n\pi y}{b} \qquad (m, n = 1, 2, 3, \ldots) \tag{4.33}$$

which satisfies the boundary conditions. Substituting these values into Eq. (4.18)

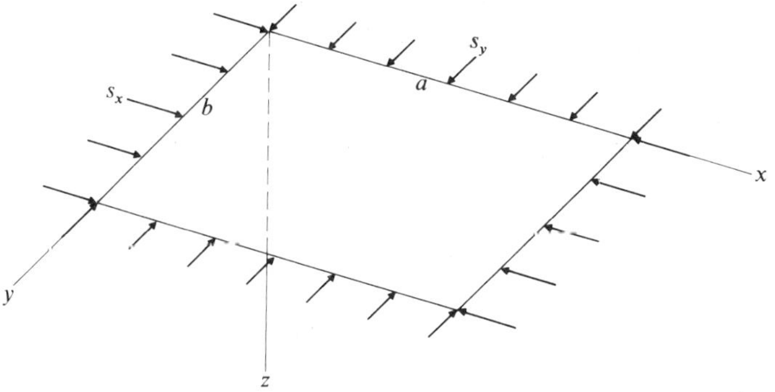

FIGURE 4.15

we obtain:

$$\pi^4 D\left(\frac{m^2}{a^2}+\frac{n^2}{b^2}\right)^2 w_{mn} \sin\frac{m\pi x}{a}\sin\frac{n\pi y}{b}$$
$$= \pi^2 h\left(\frac{m^2}{a^2}s_x+\frac{n^2}{b^2}s_y\right) w_{mn} \sin\frac{m\pi x}{a}\sin\frac{n\pi y}{b} \qquad (4.34)$$

from which, cancelling the $w_{mn} \sin m\pi x/a \sin n\pi y/b$:

$$m^2\left(\frac{b^2h}{\pi^2D}s_x\right)+n^2\left(\frac{a^2h}{\pi^2D}s_y\right) = m^2S_x+n^2S_y = \left(\frac{b}{a}m^2+\frac{a}{b}n^2\right)^2 \qquad (4.35)$$

Equation (4.35) defines the values of S_x and S_y and hence s_x and s_y for which the plate can be in equilibrium in a deflected state. There are many solutions, corresponding to different values of m and n; however, only those giving the lowest values of s_x and s_y are of practical interest to us. There are two stresses s_x, s_y, and only one condition, Eq. (4.35), from which to determine them; if one of the stresses (or perhaps a fixed ratio between them) is given, Eq. (4.35) can be used to determine the other. Figure 4.16(a) shows plots of S_x against S_y obtained from Eq. (4.35) for the case of a square plate $a = b$, for various values of mn. The full line, made up of the sections of the lines closest to the origin, represents the part of the plot of practical interest, because if the stresses are increased in a certain ratio, that is if we go away from the origin in a certain direction, buckling will occur at the first line crossed and the dotted lines further out will never be reached. It will be seen that buckling will occur in one half wave in each direction (that is for $mn = 11$) if both stresses are compression. This is true even if one of the stresses is a tension (that is if s_x or s_y has a negative value) as long as the tension does not exceed about half the compression in the other direction. If the tension exceeds this, then buckling will occur in two half waves in the direction of the compression (mn equals 21 or 12), until a point is reached where the 31 or 13 curve crosses the 21 or 12 curve at a very high tension, when there would be three half waves in the direction of the compression, etc. Figure 4.16(b) shows a similar plot for $a = 2b$.

If one of the stresses or the ratio between them is given, then all values of

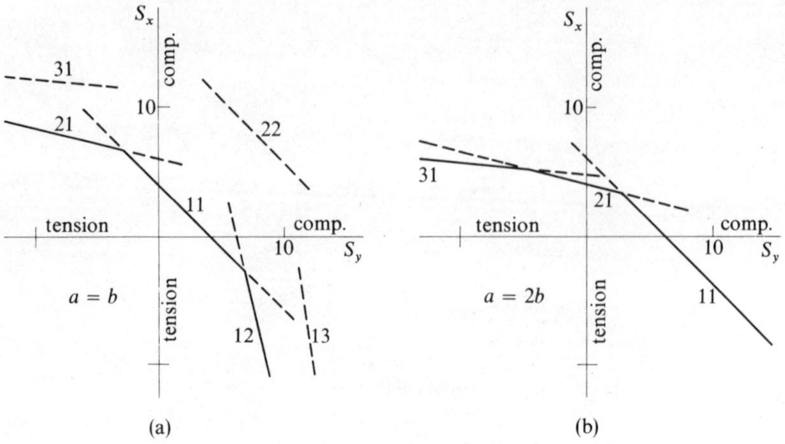

FIGURE 4.16

a/b can be covered in one plot. The most important case is that of uniaxial compression. For example if $s_y = 0$, n appears only on the right side of Eq. (4.35), so n must have its smallest possible value, unity, to give the smallest value of S_x or s_x; physically this means that the plate will always buckle in one half wave in the crosswise direction, at right angles to the compression. Equation (4.35) then reduces to:

$$S_x = \frac{b^2h}{\pi^2D}s_x = \left(\frac{mb}{a} + \frac{a}{mb}\right)^2 \tag{4.36}$$

Figure 4.17 shows plots of S_x vs a/b for values of $m = 1, 2, 3, \ldots$. Only the lowest parts of the curves, shown as a heavy full line, are of practical interest because if, for a given value of a/b, s_x and so S_x is increased, buckling will occur when the

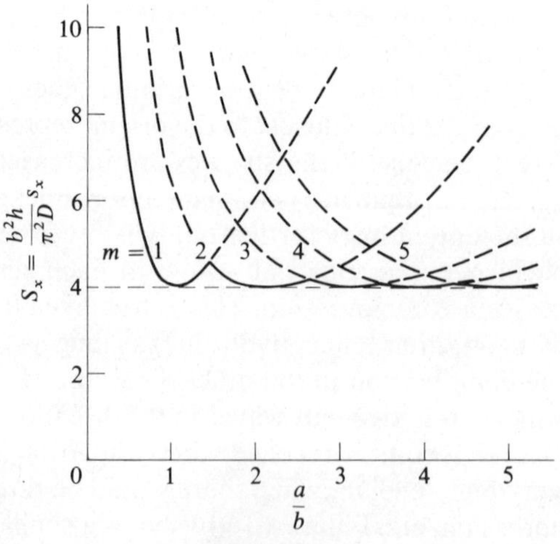

FIGURE 4.17

lowest curve is crossed, and curves above that (shown dashed) can never be reached. Up to a value of a/b represented by the intersection of the curves for $m = 1$ and $m = 2$ (which can be found by setting $(b/a + a/b)^2 = (2b/a + a/2b)^2$, from which $a/b = \sqrt{2}$) the plate will buckle in one wave in the x direction. Then until a/b reaches the intersection of the curves for $m = 2$ and $m = 3$, given by $(2b/a + a/2b)^2 = (3b/a + a/3b)^2$ the plate will buckle in two waves, etc.

For values of a/b greater than unity S_x is close to the value of 4 (which is always on the safe side), that is for:

$$\begin{aligned} a/b < 1: \qquad & S_x = \left(\frac{b}{a} + \frac{a}{b}\right)^2 \\ a/b \geq 1: \qquad & S_x \approx 4 \end{aligned} \tag{4.37}$$

The line $S_x = 4$, shown dotted, is the 'envelope' of the curves for different values of m. Such an envelope is the relation which we obtain if we ignore boundary conditions which limit the buckling shape to certain specific wave forms, in the present case to those described by the fact that m must be an integer. If we ignore this fact and minimize S_x by setting $dS_x/dm = 0$ in Eq. (4.36) we obtain $m = a/b$ (which means that the plate *tends* to buckle into squares) and $S_x = 4$.

The above situation is typical of many stability problems of plates and shells. It is similar to the case of an elastically supported strut discussed in Sec. 2.5; Eq. (2.28) for that case can be put in the form $P' = P(\beta EI)^{-1/2} = (m^2/L^2 + L^2/m^2)$, where $L = l(\beta/EI)^{1/4}/\pi$, and plotting P' against L or L^2 for different values of m gives a plot similar to Fig. 4.17. The initial deviations from flatness which always occur in actual plates can also be considered in the same manner that they were for beams in Sec. 2.5; this case is studied by the energy method in Sec. 4.6.

4.5 PLATES UNDER EDGE CONDITIONS OTHER THAN HINGED

Kirchhoff Boundary Condition

The methods of considering the flexural boundary conditions associated with w which were developed in Sec. 2.7 for beams can usually be applied without great change or difficulty to plate and shell problems. However there is one respect in addition to those mentioned in Sec. 4.1 in which plate and shell theory based on the Love–Kirchhoff approximation differs significantly from that for transversely loaded beams. In Fig. 4.1 we see that there are three flexural forces or moments on each side, for example F_{xz}, M_x, M_{xy} on the side normal to the x direction, whereas there are only two, F_{xz} and M_x, on a transversely loaded beam. But Eq. (2.4) for beams and the corresponding Eq. (4.18) for plates are both of fourth order, and complete solutions of them involve only enough constants of integration for the beams, and functions (of the distance along the side) of integration for the plates, to satisfy two conditions on each end or side.

This deficiency in the fourth order Eq. (4.18) is customarily disregarded (as we did in Sec. 4.4) for edge conditions which prevent any w displacement along the edge, such as hinged or clamped edges or elastic resistance to rotation. This is because the support along such an edge can obviously resist both the transverse shear forces such as F_{xz} and also the twisting moments such as M_{xy}, and the magnitudes of the forces which this resistance requires the support to exert on the edge are usually not of great practical importance. In such cases therefore it is usually sufficient to satisfy the edge condition $w = 0$ and the condition involving either M_x or the rotation corresponding to M_x.

On the other hand for an elastic instead of a rigid support against w displacements, we need an expression for the total transverse force which the support must exert to resist both the transverse shear and the twisting moment. And for a free edge all the forces and moments on the edge should be zero, for example we should have F_{xz}, M_x, $M_{xy} = 0$ along a free edge normal to the x direction, while the equations of the classical plate theory can only satisfy two such conditions on each edge. To meet this situation Kirchhoff calculated an 'effective' transverse shear force which is supposed to be equivalent to the combined effect of F_{xz} and M_{xy}; this effective shear force can be set equal to a constant times the deflection w in the case of elastic support against w displacements and set equal to zero in the case of a free edge, thus combining two edge conditions into one.

Figure 4.18(a) shows two adjacent elements each of width dx on an edge normal to the y direction. One is acted on by a twisting moment $M_{xy}\,dx$, due to σ_{xyf} stresses which vary linearly from the middle surface; the next element is an average of dx further on in the x direction and so is acted on by a twisting moment $[M_{xy} + (\partial M_{xy}/\partial x)\,dx]\,dx$ due to similar but slightly changed σ_{xyf} stresses. Kirchhoff replaces these moments, which are actually produced by distributed horizontal shear stresses, by couples which are composed of *transverse* shear forces (shown dashed) having a magnitude M_{xy} at each end of the first element (and hence having a moment $M_{xy}\,dx$, since they are a distance dx apart), and similar transverse shear forces $M_{xy} + (\partial M_{xy}/\partial x)\,dx$ at each end of the second element, as shown. The main parts of the two transverse shear forces at the dividing line between the two elements cancel, leaving a net downward force of $(\partial M_{xy}/\partial x)\,dx$. If we did the same thing with all the elements we would have similar (but gradually changing) transverse shear forces at intervals of dx, and these would be equivalent to a *distributed* force per unit length of section of $(\partial M_{xy}/\partial x)\,dx/dx = \partial M_{xy}/\partial x$ as indicated in Figs. 4.18(b), (c).

Combining this distributed transverse shear force which is required to replace the distributed twisting moment M_{xy} with the transverse shear force proper, F_{yz}, we obtain the 'effective' or Kirchhoff transverse shear force. Using Eqs. (4.14), (4.15), this is:

$$\begin{aligned} F'_{yz} &= F_{yz} + \frac{\partial M_{xy}}{\partial x} = -D\left[\frac{\partial^3 w}{\partial x^2\,\partial y} + \frac{\partial^3 w}{\partial y^3} + (1-\nu)\frac{\partial^3 w}{\partial x^2\,\partial y}\right] \\ &= -D\left[\frac{\partial^3 w}{\partial y^3} + (2-\nu)\frac{\partial^3 w}{\partial x^2\,\partial y}\right] \end{aligned} \tag{4.38}$$

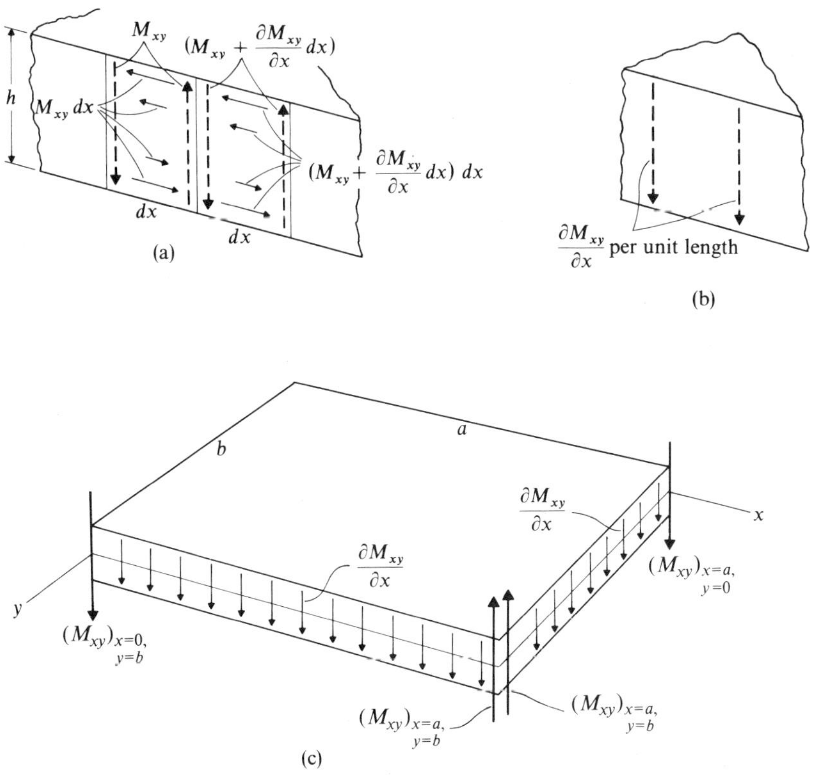

FIGURE 4.18

The condition for a free edge at $y = y_0$ is then:

$$y = y_0: \qquad \frac{\partial^2 w}{\partial y^2} + \nu \frac{\partial^2 w}{\partial x^2}, \qquad \frac{\partial^3 w}{\partial y^3} + (2 - \nu) \frac{\partial^3 w}{\partial x^2\, \partial y} = 0 \tag{4.39}$$

with similar conditions (with x and y interchanged) for a free edge normal to the x direction. The condition for elastic support against w displacement along an edge normal to the y direction is $\alpha w = \pm F'_{yz}$, where α is a measure of the stiffness of the support, etc.

Since we are replacing moments due to horizontal shear stresses which are distributed over a thickness h, it is sometimes argued that, by Saint-Venant's principle, any errors introduced by replacing F_{yz} and M_{xy} by F'_{yz} should be important only in a narrow edge zone of width equal to the thickness h. However, it is not quite as simple as this, because, while the main parts of the transverse shear forces M_{xy} which we have introduced are cancelled by opposite forces on each *interior* dividing line of the plate's edge, they are not balanced by anything at the *ends* of the edge. By using Kirchhoff's device we are thus substituting for the actual distributed M_{xy} moments not merely distributed transverse shear forces $\partial M_{xy}/\partial x$ but also concentrated transverse forces at the ends of the edge equal to the value which M_{xy} has at these points and with the directions indicated in Fig.

4.18(c). The concentrated forces corresponding to the two sides meeting at a corner add together, giving a total transverse force on the corner of twice the value which M_{xy} has at the corner, as indicated.

These concentrated forces are separated by distances of the length or width of the plate, and hence Saint-Venant's principle gives us little indication regarding the validity of what we are doing. These forces are not usually of much importance if the corners are supported or fixed, since they then could not affect the deflection of the plate, and the magnitude of the reactions is usually not critical. Kirchhoff's device gives a good approximation for calculating deflections or buckling loads, which depend upon average conditions over the whole plate.

However, if w displacements are not prevented at the corner of a plate, because the adjacent sides are free or elastically supported, and concentrated forces are introduced on the corner by the use of Kirchhoff's device, then these corner forces would obviously greatly affect the stresses and deflections. This brings up the question whether such corner forces actually exist in the physical problem, or whether they are fictitious, that is only the result of errors which we make in the analysis and not actually present.

This is perhaps a rather controversial question, but it seems safe to say that the answer depends upon the conditions of support which are actually present. In a fixed support such as that of a plate welded to a stiff wall or frame, the horizontal shear stresses which compose the twisting moment on internal sections are undoubtedly resisted directly by corresponding horizontal shear forces exerted by the support; in such a case therefore corner forces required by some different assumed disposition of the twisting moments are no doubt fictitious. They must also be fictitious for a free edge where *both* the transverse shear and the twisting moment must actually be zero, not balanced against each other as the Kirchhoff condition assumes. If a theoretical solution for adjacent free sides requires a fictitious force at the free corner, then we would presumably have to cancel this force by adding a balancing force to the loading system or, much better, use a more complete solution which gets away from the limitations imposed by the Love–Kirchhoff approximation.

On the other hand some types of support cannot resist horizontal flexural shear stresses, and for these the transverse forces must resist both the transverse shear and the twisting moment in the plate in somewhat the way that Kirchhoff assumed; there is evidence that high concentrations of distributed shear force, if not actual concentrated forces, do occur in such cases, as will be discussed in Sec. 5.5.

It should also be noted that we are considering here only resultant forces and moments, or average or middle surface displacements at the edges, which is all that can be expected of the classical theory. Actually some definite condition should be satisfied at every point of the edges; by Saint-Venant's principle the replacement of such conditions by conditions on resultants or on average displacements should only be important in an edge zone of width equal to the thickness and can be corrected by local stress fields such as are discussed for beams in Sec. 3.4 and for plates and shells in Secs. 5.2, 5.3.

Use of Normal Modes for Plate Problems

In Sec. 2.7 the normal modes of vibration of a beam with fixed, hinged or free ends were used to solve problems of transversely loaded beams having the same boundary conditions. General plate problems involving any combination of fixed and hinged edges can also be solved by the use of the normal modes of vibration of beams having the corresponding end conditions; however, this involves some added complications.

As an example, the expression given in Eq. (2.41) for a fixed end beam can be converted by analogy into the following double function for a plate having dimensions $2a$ by $2b$, Fig. 4.19, with all edges fixed, under symmetric loading:

$$w = \sum_m \sum_n w_{mn} X_m Y_n = \sum_m \sum_n w_{mn} \left(\frac{\cos c_m x/a}{\cos c_m} - \frac{\cosh c_m x/a}{\cosh c_m} \right) \left(\frac{\cos c_n y/b}{\cos c_n} - \frac{\cosh c_n y/b}{\cosh c_n} \right) \tag{4.40}$$

where X_m, Y_n are defined as indicated, and $\tan c_m + \tanh c_m = 0$ from which $c_m = c_1, c_2, c_3, \ldots = 2{\cdot}365, 5{\cdot}498, \ldots (m - \frac{1}{4})\pi$, with the same values for c_n. It can easily be checked that Eq. (4.40) satisfies the fixed-edge conditions:

$$x = \pm a: \quad w, \frac{\partial w}{\partial x} = 0; \qquad y = \pm b: \quad w, \frac{\partial w}{\partial y} = 0 \tag{4.41}$$

The fourth derivative of $X_m(x)$ and $Y_n(y)$ with respect to x and y respectively are functions of the same form; however, the second derivatives are functions of a different form, because $\cos c_m x/a$ changes sign while $\cosh c_m x/a$ does not change sign. This makes it impossible to obtain a simple solution of Eq. (4.18) or (4.19) with these functions, such as was obtained with sine functions. Iyengar and Narisimhan* get around this difficulty by taking advantage of the fact that the terms of series (4.40) are orthogonal to each other in the interval $-a < x < a$, $-b < y < b$, to replace each altered term with an infinite series of terms like the original ones.

* *Buckling of Rectangular Plates with Clamped and Simply Supported Edges*, Publications de L'Institut Mathématique, Nouvelle série, vol. 5, 1965, p. 31.

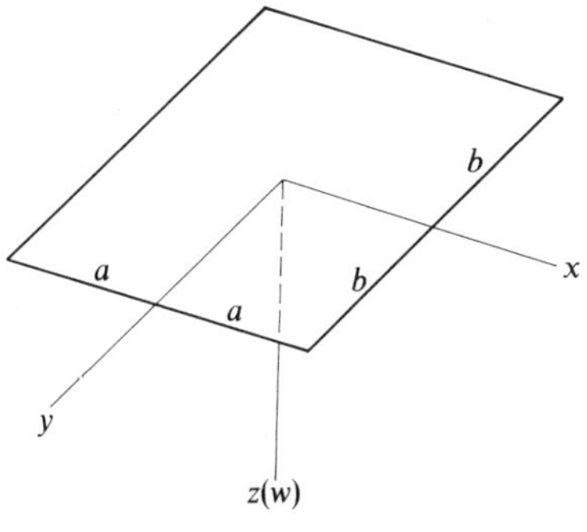

FIGURE 4.19

As an example, consider the symmetric buckling of the plate of Fig. 4.19 having all sides fixed, when it is subjected to unit compressive pressure s applied to all four edges. Substituting $\sigma_{xm} = \sigma_{ym} = -s$ into Eq. (4.18) and then using expression (4.40) for w we find:

$$\nabla^4 w + \frac{hs}{D}\nabla^2 w = 0$$

$$\sum_m \sum_n w_{mn}\left[\left(c_m^4 + \frac{a^4}{b^4}c_n^4\right)X_m Y_n + 2\frac{a^2}{b^2}c_m^2 c_n^2 X'_m Y'_n - S\left(c_m^2 X'_m Y_n + \frac{a^2}{b^2}c_n^2 X_m Y'_n\right)\right] = 0 \tag{4.42}$$

$$X'_m = \frac{\cos c_m x/a}{\cos c_m} + \frac{\cosh c_m x/a}{\cosh c_m}, \qquad Y'_n = \frac{\cos c_n y/b}{\cos c_n} + \frac{\cosh c_n y/b}{\cosh c_n} \tag{4.43}$$

where $S = a^2 hs/D$ and X_m, Y_n were defined in Eq. (4.40).
As discussed above, let:

$$X'_m = \sum_p d_{mp} X_p, \qquad Y'_n = \sum_q d_{nq} Y_q \tag{4.44}$$

where the coefficients d_{mp}, d_{nq} can be found in the same manner as in harmonic analysis:

$$\int_{-a}^{a} dx X'_m X_p = \int_{-a}^{a} dx\,(d_{m1}X_1 + \cdots d_{mp}X_p + \cdots)X_p = d_{mp}\int_{-a}^{a} dx X_p^2$$
$$d_{mp} = \int_{-a}^{a} dx X'_m X_p \Big/ \left(\int_{-a}^{a} dx X_p^2\right) \tag{4.45}$$

Carrying out these integrations, and using $\tanh c_m + \tan c_m = 0$ and $\sin^2\alpha \equiv 1 - \cos^2\alpha$, $\sinh^2\alpha \equiv \cosh^2\alpha - 1$, we find:

$$\int_{-a}^{a} dx X_p^2 = 2a,$$

$$\int_{-a}^{a} dx X'_m X_p = a[-2\tanh c_m/c_m - 1/\cosh^2 c_m + 1/\cos^2 c_m] \qquad (\text{for } p = m)$$
$$= 8c_p^2 a(c_m \tanh c_m - c_p \tanh c_p)/(c_p^4 - c_m^4) \qquad (\text{for } p \neq m)$$

from which, using the values of c_m given above:

$$\begin{array}{llllll} d_{mp}, d_{nq} = d_{11} & d_{12} & d_{13} = & 0{\cdot}550 & -0{\cdot}435 & -0{\cdot}340 \\ d_{21} & d_{22} & \cdots & -0{\cdot}0805 & 0{\cdot}818 & \cdots \\ d_{31} & \vdots & & -0{\cdot}0255 & \vdots & \end{array} \tag{4.46}$$

Using the values of X'_m, Y'_n from Eq. (4.44) in Eq. (4.42), the latter equation now contains only X_m, Y_n functions of x and y; to satisfy it for all values of x and y, the coefficients of each X_m, Y_n function (that is for each combination of m, n)

must add up to zero. Since there is also a value of w_{mn} for each combination of m, n, this will give us as many homogeneous, linear equations in the w_{mn}'s as there are unknown w_{mn}'s. Eliminating these unknowns simultaneously, we are left with an equation in S of the same degree as the number of combinations of m, n used; mathematically this is equivalent to setting the determinant of the array of coefficients of the w_{mn}'s in these equations equal to zero.

We obtain the coefficients by writing out Eq. (4.42), retaining only as many combinations of m, n as we propose to use. For the case of a square plate, $a/b = 1$, retaining the first three combinations of $mn = 11$, 21, 12, Eq. (4.42) becomes:

$$\begin{aligned}
&w_{11}\{2c_1^4[(1+d_{11}^2)X_1Y_1 + d_{11}\,d_{12}(X_2Y_1 + X_1Y_2)] \\
&- Sc_1^2[2\,d_{11}X_1Y_1 + d_{12}(X_2Y_1 + X_1Y_2)]\} \\
&+ w_{21}\{[2c_1^2c_2^2(d_{11}\,d_{21}X_1Y_1 + d_{11}\,d_{22}X_2Y_1 + d_{21}\,d_{12}X_1Y_2) \\
&+ (c_1^4 + c_2^4)X_2Y_1] - S[c_2^2\,d_{21}X_1Y_1 + (c_1^2\,d_{11} + c_2^2\,d_{22})X_2Y_1]\} \\
&+ w_{12}\{[2c_1^2c_2^2(d_{11}\,d_{21}X_1Y_1 + d_{21}\,d_{12}X_2Y_1 + d_{11}\,d_{22}X_1Y_2) \\
&+ (c_1^4 + c_2^4)X_1Y_2] - S[c_2^2\,d_{21}X_1Y_1 + (c_1^2\,d_{11} + c_2^2\,d_{22})X_1Y_2]\} = 0
\end{aligned} \tag{4.47}$$

From this, the coefficients of w_{11}, w_{21}, w_{12} in the simultaneous equations are given by Table 4.2.

Table 4.2

	($)w_{11}$ +	($)w_{21}$ +	($)w_{12} = 0$
X_1Y_1:	$2c_1^4(1+d_{11}^2)$ $- S2c_1^2\,d_{11}$	$2c_1^2c_2^2\,d_{11}\,d_{21}$ $- Sc_2^2\,d_{21}$	$2c_1^2c_2^2\,d_{11}\,d_{21}$ $- Sc_2^2\,d_{21}$
X_2Y_1:	$2c_1^4\,d_{11}\,d_{12}$ $- Sc_1^2\,d_{12}$	$(c_1^4+c_2^4)+2c_1^2c_2^2\,d_{11}\,d_{22}$ $- S(c_1^2\,d_{11}+c_2^2\,d_{22})$	$2c_1^2c_2^2\,d_{21}\,d_{12}$
X_1Y_2:	$2c_1^4\,d_{11}\,d_{12}$ $- Sc_1^2\,d_{12}$	$2c_1^2c_2^2\,d_{21}\,d_{12}$	$(c_1^4+c_2^4)+2c_1^2c_2^2\,d_{11}\,d_{22}$ $- S(c_1^2\,d_{11}+c_2^2\,d_{22})$

Using only one combination of $mn = 11$ and using the values of c_1, d_{11} previously given we have $2c_1^4(1+d_{11}^2) - S2c_1^2\,d_{11} = 0$, $S = a^2hs/D = c_1^2(1+d_{11}^2)/d_{11} = 2{\cdot}365^2(1+0{\cdot}550^2)/0{\cdot}550$, from which $s = 13{\cdot}24D/(a^2h)$. Similarly, equating to zero the determinant of the terms for $mn = 11$, 21, we can find in a similar manner $s = 13{\cdot}13D/(a^2h)$; equating to zero the determinant of all the terms shown we find $s = 13{\cdot}08D/(a^2h)$. These values check well with those obtained by other methods for this case.

Although this type of solution is not simple, it seems valuable because it is flexible, and no solution involving fixed edges can be very simple. It can be used in the same general manner for lateral loading and vibration problems, as well as for buckling under any combination of uniform compressions. However, it cannot be used with elastically constrained edges or with free edges, using the Kirchhoff transverse shear force, because such plate boundary conditions result in series with non-orthogonal terms.

Use of Particular and Homogeneous Solutions

For the case of transverse loading only, the homogeneous equation obtained from Eq. (4.19) by retaining only terms in w is $\nabla^4 w = 0$. Solutions for this type of equation were discussed in Sec. 3.3. For a particular solution of Eq. (4.19) for any type of loading we can use expression (4.21). As a typical application, Fig. 4.20 shows a plate with sides of length $a, 2b$, sides $x = 0, a$ being hinged, and sides $y = \pm b$ fixed, under a transverse loading $p(x, y)$. We will confine ourselves for simplicity to loads symmetrical about the x axis. Modifying equation (4.21) for the shift in axes, and taking the solution of the homogeneous equation $\nabla^4 w = 0$ as the symmetrical terms of the trigonometric–hyperbolic solution, Eq. (3.17b), we can take w in the form:

$$\begin{aligned} w = w_p + w_h &= \sum_m \sum_n w_{mn} \sin \frac{m\pi x}{a} \cos \frac{n\pi y}{2b} \\ &+ \sum_m \sin \frac{m\pi x}{a} \left(B_m \cosh \frac{m\pi y}{a} + C_m \frac{m\pi y}{a} \sinh \frac{m\pi y}{a} \right) \\ &(m = 1, 2, 3, \ldots, n = 1, 3, 5, \ldots) \end{aligned} \tag{4.48}$$

We can then satisfy Eq. (4.19) if we take:

$$p = \sum_m \sum_n p_{mn} \sin \frac{m\pi x}{a} \cos \frac{n\pi y}{2b} \tag{4.49}$$

where the coefficients p_{mn} can be found as previously by harmonic analysis:

$$p_{mn} = \frac{2}{ab} \int_0^a dx \int_{-b}^{b} dy\, p(x, y) \sin \frac{m\pi x}{a} \cos \frac{n\pi y}{2b} \tag{4.50}$$

Substituting expressions (4.48), (4.49) into Eq. (4.19), we obtain (remembering that

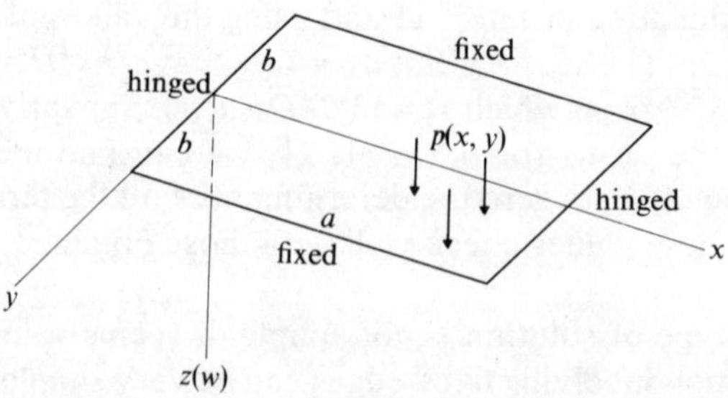

FIGURE 4.20

∇^4 applied to the trigonometric–hyperbolic part of expression (4.48) gives zero):

$$\pi^4 D \sum_m \sum_n \left(\frac{m^2}{a^2}+\frac{n^2}{4b^2}\right)^2 w_{mn} \sin\frac{m\pi x}{a}\cos\frac{n\pi y}{2b} = \sum_m \sum_n p_{mn} \sin\frac{m\pi x}{a}\cos\frac{n\pi y}{2b} \tag{4.51}$$

Equating the coefficients of corresponding terms and using Eq. (4.50), we find:

$$w_{mn} = \frac{2a^3}{\pi^4 bD\left(m^2+\dfrac{n^2a^2}{4b^2}\right)^2}\int_0^a dx \int_{-b}^{b} dy\, p \sin\frac{m\pi x}{a}\cos\frac{n\pi y}{2b} \tag{4.52}$$

The boundary conditions are

$$\begin{aligned} x = 0,\ a: &\qquad w,\ \frac{\partial^2 w}{\partial x^2}+\nu\frac{\partial^2 w}{\partial y^2}=0 \\ y = \pm b: &\qquad w,\ \frac{\partial w}{\partial y}=0 \end{aligned} \tag{4.53}$$

Substituting expression (4.48) into Eqs. (4.53), we find that the conditions at $x = 0, a$ are satisfied identically, while those at $y = \pm b$ yield the equations:

$$\begin{aligned} &\sum_m \sin\frac{m\pi x}{a}\left[B_m \cosh\lambda + C_m\lambda \sinh\lambda\right] = 0, \\ &\sum_m \sin\frac{m\pi x}{a}\left[\left(\sum_n \frac{n\pi}{2}\sin\frac{n\pi}{2} w_{mn}\right) - \lambda(B_m + C_m)\sinh\lambda \right. \\ &\left. - \lambda^2 C_m \cosh\lambda\right] = 0 \end{aligned} \tag{4.54}$$

where $\lambda = m\pi b/a$. In order for these conditions to be satisfied all along the edges $y = \pm b$, that is for all values of x, the expressions in the brackets of Eq. (4.54) must be equal to zero. Solving the resulting equations for B_m and C_m and using $\cosh^2\lambda - \sinh^2\lambda \equiv 1$, we find:

$$B_m = \frac{-\sinh\lambda \sum_n \dfrac{n\pi}{2}\sin\dfrac{n\pi}{2} w_{mn}}{\lambda + \sinh\lambda\cosh\lambda}, \qquad C_m = \frac{\cosh\lambda \sum_n \dfrac{n\pi}{2}\sin\dfrac{n\pi}{2} w_{mn}}{\lambda(\lambda + \sinh\lambda\cosh\lambda)} \tag{4.55}$$

Substituting expressions (4.52) and (4.55) into (4.48):

$$\begin{aligned} w = &\frac{2a^3}{\pi^4 bD}\sum_m \sum_n \frac{\displaystyle\int_0^a dx \int_{-b}^{b} dy\, p \sin\frac{m\pi x}{a}\cos\frac{n\pi y}{2b}}{\left(m^2 + n^2\dfrac{a^2}{4b^2}\right)^2} \sin\frac{m\pi x}{a} \\ &\times\left[\cos\frac{n\pi y}{2b} + \frac{\dfrac{n\pi}{2}\sin\dfrac{n\pi}{2}}{\lambda + \sinh\lambda\cosh\lambda}\left(\frac{y}{b}\cosh\lambda \sinh\lambda\frac{y}{b}\right.\right. \\ &\left.\left. - \sinh\lambda\cosh\lambda\frac{y}{b}\right)\right] \end{aligned} \tag{4.56}$$

Equation (4.56) applies to any distribution of p. Applying it to the case of uniformly distributed force $p = p_0$, the integral is equal to zero for m even, and to $8ab(\sin n\pi/2)p_0/(\pi^2 mn)$ for m odd. The maximum deflection for a square plate, $a/2b = 1$, $\lambda = m\pi/2$, would then be:

$$w_{max} = w_{(x=a/2,\, y=0)} = \frac{16a^4 p_0}{\pi^6 D} \sum_m \sum_n \frac{\sin\dfrac{m\pi}{2}\sin\dfrac{n\pi}{2}}{mn(m^2+n^2)^2}$$

$$\times\left[1 - \frac{\dfrac{n\pi}{2}\sinh\dfrac{m\pi}{2}\sin\dfrac{n\pi}{2}}{\dfrac{m\pi}{2}+\sinh\dfrac{m\pi}{2}\cosh\dfrac{m\pi}{2}}\right] \quad (m, n = 1, 3, 5, \ldots)$$

$$= \frac{16a^4 p_0}{\pi^6 D}\left[\frac{1}{4}\left(1 - \frac{3{\cdot}613}{7{\cdot}346}\right) - \frac{1}{300}\left(1 + \frac{10{\cdot}84}{7{\cdot}345}\right) - \frac{1}{300}\left(1 - \frac{87{\cdot}4}{3110}\right)\cdots\right] \approx \frac{a^4 p_0}{522D} \tag{4.57}$$

The three terms shown are for $mn = 11, 13, 31$. The correct value of the number in the denominator is about 521, so the result with three terms is about 0·2 percent too low. With only the first term, $mn = 11$, the result is about 10 percent too high, and with the first two terms it is about 3 percent too high.

Stability Under Compression of Plates with Opposite Sides Hinged

Consider a plate under uniform compressive stresses $\sigma_{xm} = -s_x$, $\sigma_{ym} = -s_y$ and having two opposite sides $x = 0, a$ hinged. The boundary conditions $x = 0, a$: w, $\partial^2 w/\partial x^2 + \nu\,\partial^2 w/\partial y^2 = 0$ are satisfied if we take $w = Y \sin m\pi x/a$, where $Y(y)$ is a function of y only. Substituting these values and $p = 0$, $\sigma_{xym} = 0$ into Eq. (4.18), and dividing through by $\sin m\pi x/a$, we obtain:

$$\frac{d^4Y^4}{dy^4} + 2d\frac{d^2Y}{dy^2} - cY = 0$$

where

$$c = \frac{hs_x}{D}e - e^2, \qquad d = \frac{hs_y}{2D} - e, \qquad e = \frac{m^2\pi^2}{a^2} \tag{4.58}$$

This ordinary homogeneous equation containing only even derivatives of Y can be satisfied by trigonometric or hyperbolic functions. Taking Y as a constant times $\sinh \alpha y$ or $\cosh \alpha y$ and substituting into Eq. (4.58) we can solve for α; similarly taking Y as a constant times $\sin \beta y$ or $\cos \beta y$ we can solve for β. Thus we find the general solution of the transverse equilibrium equation (4.18) for this case in the form:

$$w = \sin\frac{m\pi x}{a}[A \sinh \alpha y + A' \cosh \alpha y + B \sin \beta y + B' \cos \beta y]$$

$$\text{where} \qquad \alpha = \sqrt{-d + \sqrt{d^2 + c}}, \qquad \beta = \sqrt{d + \sqrt{d^2 + c}} \tag{4.59}$$

Applying the boundary conditions on the sides $y = 0, b$ to this expression for w, we obtain four equations from which we can eliminate A, A', B, B' (one of them being arbitrary, corresponding to the fact that the buckling stresses are independent of the magnitude of the displacement), leaving an equation relating α and β, from which the buckling stresses can be determined. This relation is the same as that given by equating to zero the determinant of the four equations. As a simple example, for the case of all sides hinged we have conditions $y = 0, b$: w, $\partial^2 w/\partial y^2 + \nu\,\partial^2 w/\partial x^2 = 0$, which are satisfied if $A = A' = B' = 0$, $\sin \beta b = 0$. From this we have $\beta = n\pi/b$, and setting this equal to $\sqrt{d + \sqrt{d^2 + c}}$ we obtain Eq. (4.35) previously found for this case.

Stability of Plate with Three Sides Hinged, One Side Free, Under Compression

For a plate with sides $x = 0, a$, $y = 0$ hinged, and $y = b$ free, the boundary conditions on $y = 0, b$ are, using Eq. (4.39):

$$\begin{aligned} y = 0: &\qquad w,\ \frac{\partial^2 w}{\partial y^2} + \nu \frac{\partial^2 w}{\partial x^2} = 0 \\ y = b: &\qquad \frac{\partial^2 w}{\partial y^2} + \nu \frac{\partial^2 w}{\partial x^2},\ \frac{\partial^3 w}{\partial y^3} + (2 - \nu)\frac{\partial^3 w}{\partial x^2\,\partial y} = 0 \end{aligned} \tag{4.60}$$

It can be seen that the first two conditions, at $y = 0$, will be satisfied if $A', B' = 0$ in Eq. (4.59). Substituting the remainder of expression (4.59) into the last two boundary conditions of Eq. (4.60), and cancelling $\sin m\pi x/a$, we obtain:

$$A(\alpha^2 - \nu e)\sinh \alpha b = B(\beta^2 + \nu e)\sin \beta b = 0$$
$$A\alpha[\alpha^2 - (2 - \nu)e]\cosh \alpha b = B\beta[\beta^2 + (2 - \nu)e]\cos \beta b = 0$$

Equating to zero the determinant of this set of equations, which in this case is the same as dividing one equation by the other, we obtain the following equation which relates s_x, s_y, m and the properties of the plate:

$$\frac{\alpha^2 - \nu e}{\alpha[\alpha^2 - (2 - \nu)e]}\tanh \alpha b = \frac{\beta^2 + \nu e}{\beta[\beta^2 + (2 - \nu)e]}\tan \beta b \tag{4.61}$$

The case $s_y = 0$, of a plate under uniform compression s_x in the direction of the free edge, was solved by Timoshenko* in 1907, using similar equations. He found that the lowest value of s_x at which buckling deflections can occur is for $m = 1$, that is the plate always buckles in one half sine wave in the direction of the compression, as shown in Fig. 4.21(a). Solving Eq. (4.61) for s_x, he found the critical value:

$$s_x = k\pi^2 D/(b^2 h) \tag{4.61a}$$

where k has the values shown by the full line in Fig. 4.21(b), for various values of a/b using $\nu = 0{\cdot}25$, apparently becoming asymptotic to the line $k = 1/2$ when a is

* *Theory of Elastic Stability*, 1st edn, pp. 337–44.

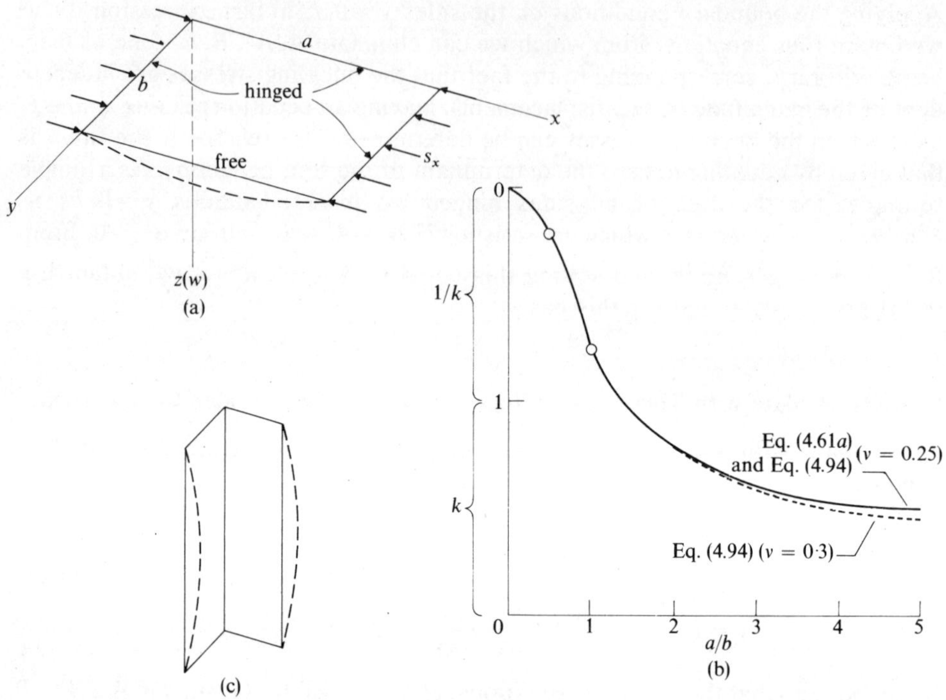

FIGURE 4.21

large compared to b. The dotted curve is for a more approximate solution of the same problem given in Sec. 4.6, using $\nu = 0{\cdot}30$. This solution has practical applications in such problems as the local buckling of outstanding flanges of columns. Thus, each leg of an equal-leg angle section used as a strut, Fig. 4.21(c), can be considered to be a plate with the outer edge free and the edge which is connected to the other leg hinged (since each can buckle as shown, without interfering with the other).* In general, outstanding flanges involve one edge free but other conditions on the three remaining edges.

Stability of Plate with Two Opposite Sides Free, Under Compression in Their Direction

This is the case of a column in the shape of a plate. It is of special interest because it throws light upon a subject which has come up frequently, the fact that in the bending of a wide plate the effective modulus of elasticity tends to be the modified modules $E/(1-\nu^2)$, whereas in a very narrow beam it is only E, because its material is free to expand or contract in the width direction. The present case should show what happens between these extremes.

* Bridget, Jerome, and Vosseller, 'Some new experiments of buckling of this wall construction,' *Trans. ASME* vol. 56, p. 569, 1934.

Consider a plate with the edges $x = 0, a$ hinged, and the edges $y = \pm b$ free, under a uniform compression s_x in the x direction, as shown in Fig. 4.22(a). Trial shows that in such a case buckling occurs under the lowest compression s_x if the buckling deflection is in the shape of a half sine wave in the x direction (as in the case of a simple strut), and is symmetrical about the x axis so that the antisymmetrical terms in expression (4.59) can be taken as zero. Then from Eqs. (4.58), (4.59), we can write w as:

$$w = \sin\frac{\pi x}{a}(A' \cosh \alpha y + B' \cos \beta y)$$

$$\alpha = \sqrt{-d + \sqrt{d^2 + c}}, \qquad \beta = \sqrt{d + \sqrt{d^2 + c}} \tag{4.62}$$

$$c = \frac{\pi^4}{a^4}\left(\frac{h s_x a^2}{\pi^2 D} - 1\right), \qquad d = -\frac{\pi^2}{a^2}$$

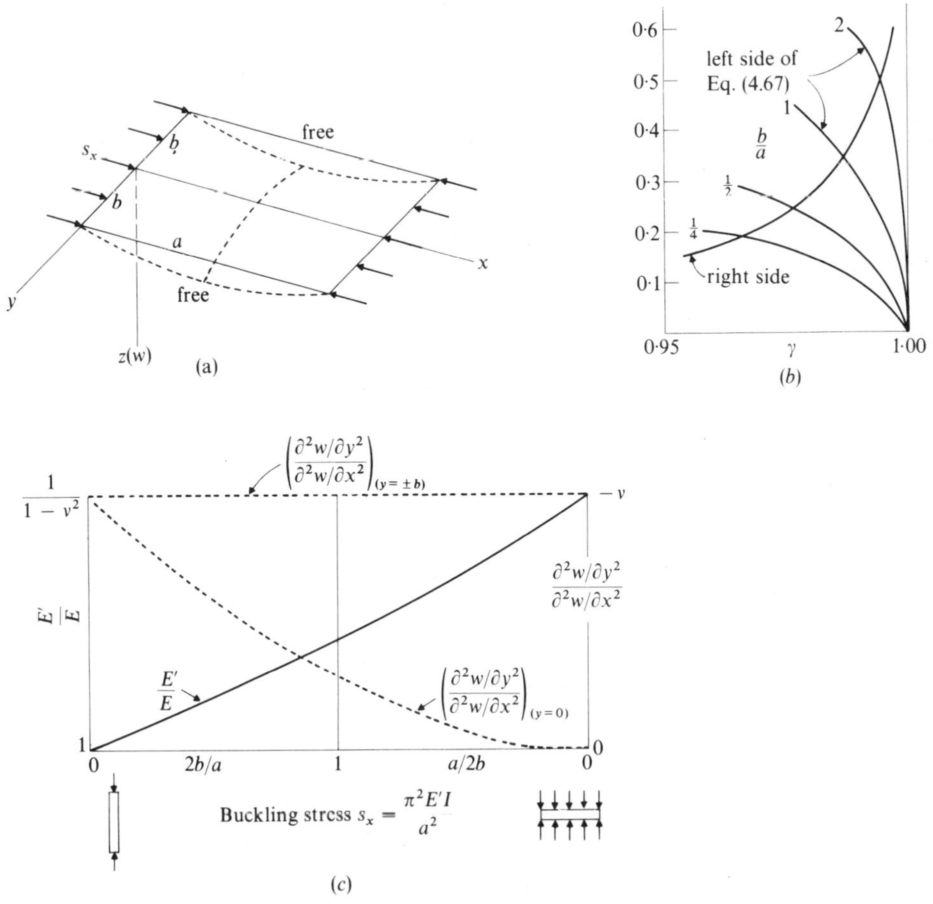

FIGURE 4.22

The expression $hs_x a^2/(\pi^2 D)$ has a simple physical meaning if we treat the plate like a simple Euler column with a length a, a moment of inertia of cross section $2bh^3/12$, and an 'effective' modulus of elasticity E' which is unknown, and related to s_x through the Euler column formula: $2bhs_x = \pi^2 E'(2bh^3/12)/a^2$. Using $D = Eh^3/12(1-\nu^2)$, this relation between E' and s_x can be written:

$$\frac{hs_x a^2}{\pi^2 D} = \frac{(1-\nu^2)E'}{E} = \gamma^2 \tag{4.63}$$

Using this in Eq. (4.62) we find $c = (\pi^4/a^4)(\gamma^2 - 1)$, $\alpha = (\pi/a)\sqrt{1+\gamma}$, $\beta = i(\pi/a)\sqrt{1-\gamma}$, where i is the imaginary $\sqrt{-1}$. We write β in this form because we expect E' to have a value somewhere between E and $E/(1-\nu^2)$, which would make γ^2 and therefore γ equal to or less than unity. Using $\cos i\theta = \cosh\theta$, Eq. (4.62) can then be written in the form:

$$w = \sin\frac{\pi x}{a}\left(A' \cosh\frac{\sqrt{1+\gamma}\,\pi y}{a} + B' \cosh\frac{\sqrt{1-\gamma}\,\pi y}{a}\right) \tag{4.64}$$

Expression (4.64) satisfies conditions on the hinged edges $x = 0, a$, while those on the free edges are, from Eq. (4.39):

$$y = \pm b: \qquad \frac{\partial^2 w}{\partial y^2} + \nu\frac{\partial^2 w}{\partial x^2}, \qquad \frac{\partial^3 w}{\partial y^3} + (2-\nu)\frac{\partial^3 w}{\partial x^2\,\partial y} = 0 \tag{4.65}$$

Substituting expression (4.64) into Eqs. (4.65) and simplifying by dividing by $\sin\pi x/a$ and by π^2/a^2 or π^3/a^3, we find:

$$[\gamma + (1-\nu)]A' \cosh\frac{\sqrt{1+\gamma}\,\pi b}{a} = [\gamma - (1-\nu)]B' \cosh\frac{\sqrt{1-\gamma}\,\pi b}{a}$$
$$\sqrt{1+\gamma}\,[\gamma - (1-\nu)]A' \sinh\frac{\sqrt{1+\gamma}\,\pi b}{a} = \sqrt{1-\gamma}\,[\gamma + (1-\nu)]B' \sinh\frac{\sqrt{1-\gamma}\,\pi b}{a} \tag{4.66}$$

Eliminating A' and B' by dividing the second equation by the first, we obtain an equation for determining γ and hence s_x:

$$\frac{\tanh\dfrac{\sqrt{1-\gamma}\,\pi b}{a}}{\tanh\dfrac{\sqrt{1+\gamma}\,\pi b}{a}} = \frac{\sqrt{1+\gamma}(\gamma - 1 + \nu)^2}{\sqrt{1-\gamma}(\gamma + 1 - \nu)^2} \tag{4.67}$$

We see that γ is a function of b/a and ν. While it seems hopeless to try to solve such an equation directly for γ, it can easily be solved by cross plotting. Figure 4.22(b) shows curves obtained by calculating the right-hand side of Eq. (4.67) for different values of γ, taking $\nu = 0{\cdot}3$, together with similar plots of the left-hand side for several values of b/a. The intersections of the curves give the values of γ for the corresponding b/a.

These results are shown by the full line in Fig. 4.22(c). At the left of this plot, for small values of $2b/a$, the plates are narrow and it can be seen that if the width $2b$ of a bar is small compared to its length it is correct to take the effective modulus as E, as we did in studying beams in Chap. 2. At the other extreme at the right-hand side of the plot, the plates are wide in comparison to their length and it is correct to use the modified modulus $E/(1-\nu^2)$ in calculations which treat the plate as a beam. However, it can be seen that there is no sudden transition from one case to the other and at the center of the plot, for $2b/a = 1$ (for a square plate), the proper modulus to use in treating the plate as a beam would be roughly half way between E and $E/(1-\nu^2)$. These results should be applicable to any problem in which we wish to treat a plate with free edges as a beam, if we take a as the 'primary half wave length of the deflection' as defined at the end of Sec. 3.5. It should be emphasized that these results do not in any way affect our studies of plates or shells *as* plates or shells, in this chapter and the rest of the book; in such studies we take everything properly into consideration in using $D = Eh^3/12(1-\nu^2)$, as we have shown in our derivations.

The shape of the deformation in this case is also of interest. This can be obtained by solving for A'/B' from either of Eqs. (4.66) and substituting into Eq. (4.64), which leaves one arbitrary factor corresponding to the arbitrary amplitude of the buckling deflection. The curvature $\partial^2 x/\partial y^2$ in the y direction is found to be opposite in sign from the curvature $\partial^2 w/\partial x^2$ in the x direction, as can be seen from the fact that w is a trigonometric function of x but a hyperbolic function of y. Thus if the plate buckles concave downward in the x direction it will be curved concave upward in the y direction as indicated by the dotted lines in Fig. 4.22(a). This type of curvature has been given names such as 'anticlastic' and 'saddle-shaped'. The ratio between the y curvature and the x curvature, indicated by the dotted lines in Fig. 4.22(c), is exactly minus Poisson's ratio ν at the free edges $y = \pm b$, as indicated by the horizontal dotted line. This is because there are no σ_y stresses at the free edges, and the σ_{xf} bending stresses produce strains in the y direction equal to $-\nu$ times the strains in the x direction, due to Poisson's ratio effect.

In the middle part of a wide plate, on the other hand, resistance to curving in the y direction builds up, probably due principally to the large amount of twist which such anticlastic curving produces. Along the center line, $y = 0$, the curvature in the y direction is therefore much less than $-\nu$ times the curvature in the x direction, ranging down to nothing for the infinitely wide plate, $a/2b = 0$, as indicated by the dotted curve in Fig. 4.22(c). These results are all for very small deflections at the start of bending, when the nonlinear membrane stresses are negligible, and this tendency would be greatly increased as these stresses build up. The anticlastic shape is of course a non-developable shape and there is great resistance to the membrane strains such a shape would involve, so that for wide plates under deflections large compared to the thickness, the y curvature is practically zero up to a short distance from the free edges, where it must have the value described above.

4.6 ENERGY METHODS APPLIED TO PLATES

Elastic Strain Energy in a Deformed Plate

As stated before, the elastic strain energy of a strained body is the sum of the work done in deforming each element of the body. We can take elements of a plate across the thickness of the plate, of volume $dx\,dy\,h$, and calculate the work done by the resultant forces and moments acting on the sides of the element as we did with beams; or, alternatively, we can take elements infinitesimal in the z direction also, of volume $dx\,dy\,dz$ located at the general point x, y, z in the plate, and calculate the work done by the stresses on its sides during the deformation of the element. Let us try the latter approach.

In accordance with the Love–Kirchhoff approximation, we ignore the transverse strains, and hence do not consider the transverse stresses σ_{xz}, σ_{yz}, and σ_z which act upon the element, since they would do no work on the zero relative displacements in their direction which this approximation assumes. Of the remaining stresses which act upon the element, σ_x, σ_y, σ_{xy}, the stress σ_x produces total forces $\sigma_x\,dy\,dz$ acting in opposite directions upon the two sides of the element normal to the x direction. These two sides suffer a relative displacement in the direction of the stresses of $\epsilon_x\,dx$. Since the forces on a body in equilibrium do no net work during a rigid body displacement, we can imagine one of the sides to be stationary so that the force on it does no work, while the other side moves the above distance, and the force on it therefore does the work $(\sigma_x\,dy\,dz)(\epsilon_x\,dx)/2$. The factor 1/2 is of course due to the fact that the force varies linearly with the displacement from zero to its final value $\sigma_x\,dy\,dz$, as discussed in Sec. 2.8 for beams. The effect of the change in σ_x is a small quantity of higher order compared to the stress itself, and disappears in the limit as dx approaches zero. The stress σ_y does a corresponding work $(\sigma_y\,dx\,dz)(\epsilon_y\,dy)/2$.

Figure 4.23 shows the four forces acting upon four sides of the element due to the shear stress σ_{xy}. Again since rigid body displacements involve no work we can imagine the top side in the figure to be stationary. The force $\sigma_{xy}\,dx\,dz$ on the bottom side will then do the work $(\sigma_{xy}\,dx\,dz)(\epsilon_{xy}\,dy)/2$ during the displacement $\epsilon_{xy}\,dy$ in its direction of the side on which it acts. The forces on the other sides do no work during the displacement since the displacement of the vertical sides is at

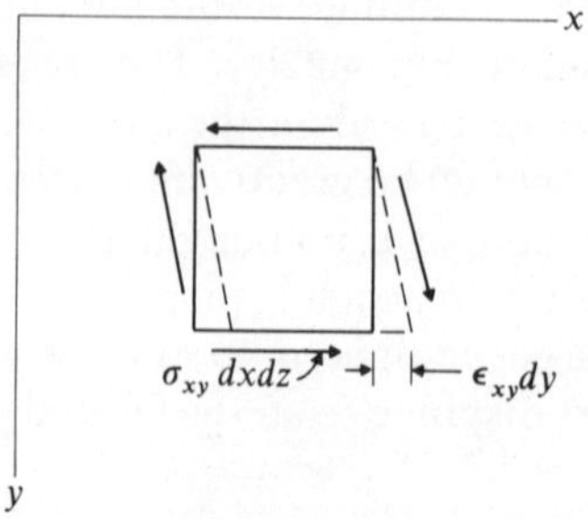

FIGURE 4.23

right angles to the forces on the sides (except for a higher order component, the work for which in any case cancels on the two sides).

The total strain energy in the plate is the sum of all the work done in deforming all the elements or:

$$\mathscr{E} = \frac{1}{2}\int dx \int dy \int_{-c}^{c} dz\ (\sigma_x\epsilon_x + \sigma_y\epsilon_y + \sigma_{xy}\epsilon_{xy}) \tag{4.68}$$

where the integrations are carried out over the whole plate in the x and y directions. Using the stress–strain relation given by Hooke's Law, Eq. (3.11a,b), to express the strains in terms of the stresses, or the stresses in terms of the strains, Eq. (4.68) becomes:

$$\mathscr{E} = \frac{1}{2E}\int dx \int dy \int_{-c}^{c} dz\ [\sigma_x^2 - 2\nu\sigma_x\sigma_y + \sigma_y^2 + 2(1+\nu)\sigma_{xy}^2]$$

or: (4.69)

$$\mathscr{E} = \frac{E}{2(1-\nu^2)}\int dx \int dy \int_{-c}^{c} dz\ \left(\epsilon_x^2 + 2\nu\epsilon_x\epsilon_y + \epsilon_y^2 + \frac{1-\nu}{2}\epsilon_{xy}^2\right)$$

Finally we can use Eq. (4.16) to express each stress as the sum of the corresponding membrane stress and flexural stress, for instance $\sigma_x = \sigma_{xm} + \sigma_{xf} = (\partial^2\phi/\partial y^2) - [Ez/(1-\nu^2)](\partial^2 w/\partial x^2 + \nu\partial^2 w/\partial y^2)$, and similarly for σ_y and σ_{xy}. When such a sum of membrane and flexural terms is squared or multiplied by a similar sum, we obtain squares or products of membrane terms, which are independent of z, cross products of membrane and flexural terms, which are proportional to z, and squares or products of flexural terms, which are proportional to z^2. The integral of the cross product terms is zero in this case, because $\int_{-h/2}^{h/2} z\,dz = 0$. This means that the strain energy due to the membrane stresses and the strain energy due to the flexural stresses are unconnected, or 'uncoupled'. Putting these expressions into Eq. (4.69) and separating the membrane and flexural energy terms, we can write the total strain energy as the sum of the flexural and membrane energies, $\mathscr{E} = \mathscr{E}_f + \mathscr{E}_m$, where:

$$\mathscr{E}_f = \frac{D}{2}\int dx \int dy \left\{(\nabla^2 w)^2 + 2(1-\nu)\left[\left(\frac{\partial^2 w}{\partial x\,\partial y}\right)^2 - \frac{\partial^2 w}{\partial x^2}\frac{\partial^2 w}{\partial y^2}\right]\right\} \tag{4.70}$$

$$\mathscr{E}_m = \frac{h}{2E}\int dx \int dy \left\{(\nabla^2 \phi)^2 + 2(1+\nu)\left[\left(\frac{\partial^2 \phi}{\partial x\,\partial y}\right)^2 - \frac{\partial^2 \phi}{\partial x^2}\frac{\partial^2 \phi}{\partial y^2}\right]\right\} \tag{4.71}$$

where as usual $\nabla^2 = \partial^2/\partial x^2 + \partial^2/\partial y^2$, $D = Eh^3/12(1-\nu^2)$.

Instead of expressing the membrane stresses in terms of the stress function ϕ, we might have used Eq. (4.16) to express them instead in terms of the displacements u, v, w. We would thus obtain the alternative expression for $\mathscr{E}_m$:

$$\begin{aligned}\mathscr{E}_m = \frac{Eh}{2(1-\nu^2)}\int dx \int dy \Bigg\{&\left(\frac{\partial u}{\partial x}\right)^2 + 2\nu\frac{\partial u}{\partial x}\frac{\partial v}{\partial y} + \left(\frac{\partial v}{\partial y}\right)^2 \\ &+ \frac{1-\nu}{2}\left(\frac{\partial u}{\partial y} + \frac{\partial v}{\partial x}\right)^2 + \left(\frac{\partial u}{\partial x} + \nu\frac{\partial v}{\partial y}\right)\left(\frac{\partial w}{\partial x}\right)^2 + \left(\frac{\partial v}{\partial y} + \nu\frac{\partial u}{\partial x}\right)\left(\frac{\partial w}{\partial y}\right)^2 \\ &+ (1-\nu)\left(\frac{\partial u}{\partial y} + \frac{\partial v}{\partial x}\right)\frac{\partial w}{\partial x}\frac{\partial w}{\partial y} + \frac{1}{4}\left[\left(\frac{\partial w}{\partial x}\right)^2 + \left(\frac{\partial w}{\partial y}\right)^2\right]^2\Bigg\}\end{aligned} \tag{4.71a}$$

For small deflections, which we will consider in the remainder of this chapter, we will need only the flexural strain energy $\mathscr{E}_f$ given by Eq. (4.70). As has been discussed in Sec. 2.8, the energy method gives an exact solution if we use exact displacements.

Hinged-Edge Plate Under Transverse Loads

Taking the sides of a rectangular plate as $x = 0, a, y = 0, b$, as shown in Fig. 4.13, the boundary conditions are satisfied if we take w as the sine–sine series:

$$w = \sum_m \sum_n w_{mn} \sin \frac{m\pi x}{a} \sin \frac{n\pi y}{b} \tag{4.72}$$

When we substitute this into Eq. (4.70) the part of the integral in the brackets cancels in this case, and we obtain:

$$\mathscr{E} = \frac{\pi^4 bD}{8a^3} \sum_m \sum_n \left(m^2 + \frac{a^2}{b^2} n^2\right)^2 w_{mn}^2 \tag{4.73}$$

Consider now the virtual displacement $dw_{mn} \sin m\pi x/a \sin n\pi y/b$ produced by a small change dw_{mn} in a typical coefficient w_{mn}. The change in the internal strain energy $\mathscr{E}$ during this virtual displacement will be the rate at which $\mathscr{E}$ changes with w_{mn}, $\partial\mathscr{E}/\partial w_{mn}$, times the amount dw_{mn} which w_{mn} changes, or $(\partial\mathscr{E}/\partial w_{mn})dw_{mn}$. The transverse loading $p(x, y)$ can be taken as constant during this small virtual displacement, so the work done by it will be $(p\,dx\,dy)dw_{mn} \sin m\pi x/y \sin n\pi y/b$, integrated over the area of the plate. By the law of virtual work, then:

$$\frac{\partial\mathscr{E}}{\partial w_{mn}} dw_{mn} = \frac{\pi^4 bD}{8a^3}\left(m^2 + \frac{a^2}{b^2}n^2\right)^2 2w_{mn}\,dw_{mn}$$

$$= dw_{mn} \int_0^a dx \int_0^b dy\, p(x, y) \sin \frac{m\pi x}{a} \sin \frac{n\pi y}{b},$$

from which:

$$w_{mn} = \frac{4a^3}{\pi^4 bD} \frac{1}{\left(m^2 + \frac{a^2}{b^2}n^2\right)^2} \int_0^a dx \int_0^b dy\, p(x, y) \sin \frac{m\pi x}{a} \sin \frac{n\pi y}{b} \tag{4.74}$$

Putting this expression back into Eq. (4.72), we obtain the same Eq. (4.24) which was derived by using the equilibrium equation and harmonic analysis.

Plate With Stiffening Ribs

The energy method is particularly useful for studying built-up structures made of several parts having different geometries. With equilibrium methods such structures would require separate equations for each part, together with continuity conditions which express the fact that the displacements of the parts where they are joined are continuous, and the forces which they exert upon each other are action and reaction. With the energy method, having assumed continuous expressions for the displacements, it is only necessary to use the total strain energy obtained by adding together the internal strain energy produced in each part by these displacements, and solve the problem in the same way as for a simple plate.

For example, consider the plate shown in Fig. 4.24(a) having reinforcing ribs or stiffeners in the y direction, located at distances $x_1, x_2, \ldots x_i \ldots$ from the y axis. It may also have ribs parallel to the x axis (not shown) at distances $y_1, y_2, \ldots y_j \ldots$ from the x axis. We will assume for the present that the ribs are symmetrical with respect to the middle surface of the plate taken as the x, y plane, as shown in Fig. 4.24(a), and that they are made of relatively thin material and of open cross section (such as a channel or 'I' or 'Z' section) in which case their torsional stiffness should be small compared to their bending stiffness and so can be

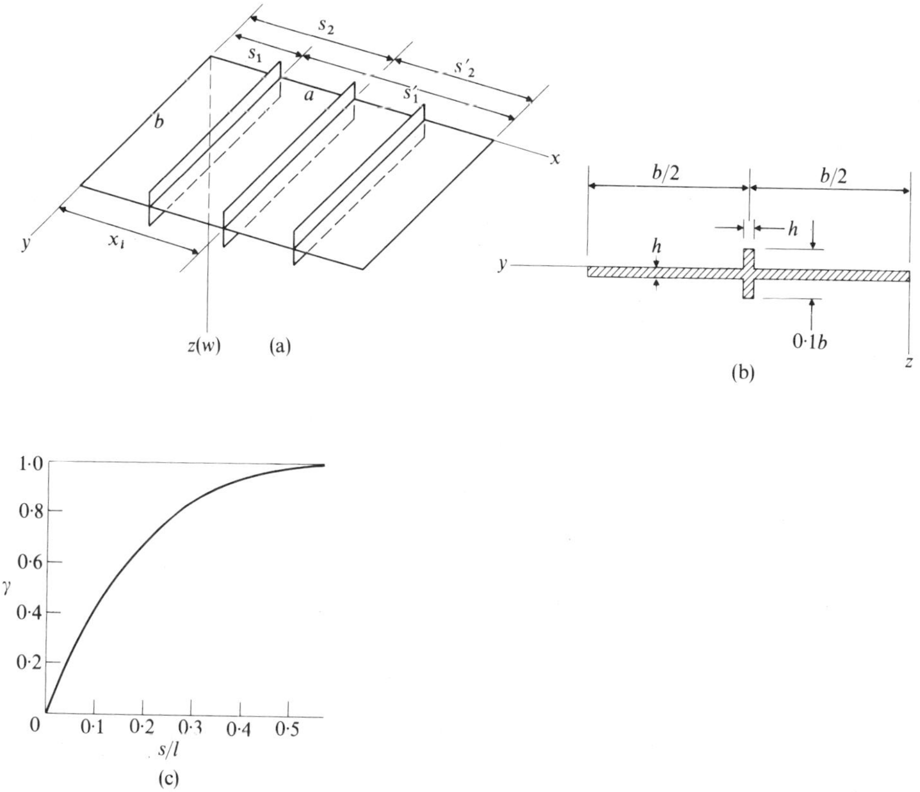

FIGURE 4.24

neglected. Then the total strain energy will be, using Eqs. (2.54) and (4.70):

$$\mathscr{E} = \frac{D}{2}\int_0^a dx \int_0^b dy \left\{(\nabla^2 w)^2 + 2(1-\nu)\left[\left(\frac{\partial^2 w}{\partial x\,\partial y}\right)^2 - \frac{\partial^2 w}{\partial x^2}\frac{\partial^2 w}{\partial y^2}\right]\right\}$$
$$+\sum_i \frac{EI_i}{2}\int_0^b dy \left(\frac{\partial^2 w}{\partial y^2}\right)^2_{(x=x_i)} + \sum_j \frac{EI_j}{2}\int_0^a dx \left(\frac{\partial^2 w}{\partial x^2}\right)^2_{(y=y_j)} \qquad (4.75)$$

where EI_i is the bending stiffness, about the middle surface, of the rib located at $x = x_i$, etc.

If the rib has a torsional stiffness which cannot be ignored, as will in general be the case if it forms a closed tube, then the expression for the torsional strain energy must be added to Eq. (4.75); this by analogy is:

$$\sum_i \frac{GJ_i}{2}\int_0^b dy \left(\frac{\partial^2 w}{\partial x\,\partial y}\right)_{(x=x_i)} + \sum_j \frac{GJ_j}{2}\int_0^a dx \left(\frac{\partial^2 w}{\partial x\,\partial y}\right)_{(y=y_j)} \qquad (4.75a)$$

where GJ_i is the torsional stiffness defined as the torsional moment required to twist a unit length of the rib through unit angle (one radian)* of the rib located at $x = x_i$, etc.

There is a third type of strain energy which is associated with twisting of the ribs although it is not strictly torsional. If the rib were subjected to a constant rate of twist then Eq. (4.75a), which describes the strain energy corresponding to torsional shear stresses and strains, would be sufficient. In practice the rate of twist is in general not constant and the parts of the rib at a distance from the plate will also be subjected to bending in the plane of the plate due to the variable rate of twist. While all parts of the rib take part in such bending it is generally sufficient to consider the flanges of the rib, since these are generally furthest from the plate and involve most of the stiffness in the plane of the plate. The moment of inertia I_{fl} of each flange of an I-beam used as a rib, for example, could be approximated as half the moment of inertia of the whole cross section about the web as an axis, which is given in structural handbooks.

Then for a rib running in the y direction and at a distance x_i from the y axis, the deflection of the flange in the x direction is $U = c\,\partial w/\partial x$, where c is the mean distance of the flange from the plate middle surface. From Eq. (2.54) the strain energy of bending of each flange in the x direction is then:

$$\frac{EI_{fl}}{2}\int_0^b dy \left(\frac{\partial^2 U}{\partial y^2}\right)^2_{(x=x_i)} = \frac{c^2 EI_{fl}}{2}\int_0^b dy \left(\frac{\partial^3 w}{\partial x \partial y^2}\right)^2_{x=x_i} \qquad (4.75b)$$

There will also be a shortening of the distance between the ends of the flange of:

$$\frac{1}{2}\int_0^b dy \left(\frac{\partial U}{\partial y}\right)^2 = \frac{c^2}{2}\int_0^b dy \left(\frac{\partial^2 w}{\partial x\,\partial y}\right)^2 \qquad (4.75c)$$

which must be taken into account in the work done by external forces in stability problems. Similar expressions apply to any other ribs in the x and y directions.

If the ribs are unsymmetrical with respect to the middle surface of the plate, for example if they are attached to one side of the plate (as is usually the case in

* See Timoshenko, *Strength of Materials* Part II, 2nd edn, pp. 265–93.

practice) then expression (4.75) for the bending strain energy will be somewhat too low, because the part of the plate close to each rib is attached to and so is strained with the rib, and hence acts like an enlargement of the rib and considerably increases the rib's effective stiffness. That is, this part of the plate will have, in addition to flexural stresses like those in other parts of the plate, also some membrane stresses corresponding to bending as a part of the rib.

This action of a plate as part of an attached rib (or of the plate-like flange of a wide-flanged beam) was studied by von Kármán.* The membrane strain in the plate must have the same value as the strain in the rib at the point where the plate is attached to the rib; the strain in the plate decreases exponentially on either side of the rib, slowly if the rib is bent in long waves and rapidly if it is bent in short waves. Kármán calculated 'effective widths' λ of the plate on each side of the rib which, if they were subjected to a uniform strain the same as that in the rib would have the same effect as the actually exponentially decreasing strain in an infinitely extended plate. In most applications the plate is wide enough so that the assumption that it is infinitely wide gives a good approximation, but to cover the case of a narrower plate the calculations were extended to give the effective width which, if subjected to the strain in the rib, would be equivalent to the exponentially decreasing strain in a plate of limited width.

The effective width on each side of the rib is thus found to be $\lambda = 2\gamma l/(3+2\nu-\nu^2)\pi \approx 0{\cdot}181\gamma l$, where l can be taken as the primary half wave length (the distance between nodes) of the bending deflection of the rib. The numerical factor γ has a value of unity for an infinitely wide plate and is shown in Fig. 4.24(c) as a function of s/l, where s is the width of plate on the side considered which can conceivably act with this rib. In Fig. 4.24(a) s_1, s_1' are the values of s on each side of the rib next to the y axis, s_2, s_2' of the next rib, etc., assuming that the effects on the plate of different ribs can superpose on each other.

If the rib is of the same material as the plate and is attached to the plate by a narrow weld or a row of closely spaced rivets, elementary calculations show that the moment of inertia I of the rib about its own centroid will be increased, due to the area added by these two effective widths of the plate, to an 'effective moment of inertia', I' about its new centroid of:

$$I' = I + \frac{Ae^2}{1 + \dfrac{A}{(\lambda + \lambda')h}} \approx I + \frac{Ae^2}{1 + \dfrac{5{\cdot}52\,A}{(\gamma + \gamma')lh}} \tag{4.76}$$

where A is the cross-sectional area of the rib, e is the distance from the center of gravity of the rib to the middle surface of the plate ($e = 0$ for symmetrical ribs), h is the thickness of the plate and λ, λ' and γ, γ' are the values of λ, γ for the two sides of the rib. If the rib and plate are made of different materials, then the second term in the denominators of Eq. (4.76) should be multiplied by the ratio of the modulus of elasticity of the rib to that of the plate; if the rib is attached by

* *Festschrift August Föppls*, p. 114, 1923. Or see Timoshenko, *Theory of Elasticity*, 1st edn, pp. 157–61.

widely spaced rivets, then there will be some slippage between rib and plate and this second term should be increased somewhat; however this would not be important if the rivet spacing is small compared to l, as will generally be the case.

The interaction between rib and plate could be considered in a different way by taking into account the u and v displacements of the middle surface, which will occur even under small displacements if non-symmetrical ribs are used. Expressions for u and v displacements with unknown coefficients would have to be assumed, the geometrically resulting strains in rib and plate calculated, and the unknown coefficients determined by the energy method along with the coefficients in the expression for w. This method would be more direct and potentially more complete, but would involve far more work (because of the tripling of the number of unknown coefficients) than the methods which have been presented for determining the strain energy of a stiffened plate, which consider all the factors which are likely to be important.

As a simple application of the above, consider a plate having one central symmetrical rib parallel to the x axis, having the cross section shown in Fig. 4.24(b) and under a uniform transverse loading p_0. Taking the deflection w in the form (4.72), with m, n odd only because of symmetry, the strain energy is given by Eq. (4.75). The strain energy in the rib is, neglecting energy due to twist:

$$\begin{aligned}\mathscr{E}_{\text{rib}} &= \frac{EI}{2}\frac{\pi^4}{a^4}\int_0^a dx\left(\sum_m\sum_n m^2 w_{mn}\sin\frac{m\pi x}{a}\sin\frac{n\pi}{2}\right)^2\\ &= \frac{EI}{2}\frac{\pi^4}{a^4}\frac{a}{2}\sum_m m^4\left(\sum_n w_{mn}\sin\frac{n\pi}{2}\right)^2 \qquad (4.77)\\ &= \frac{EI}{2}\frac{\pi^4}{a^4}\frac{a}{2}[(w_{11}-w_{13}+w_{15}\ldots)^2+81(w_{31}-w_{33}+w_{35}\ldots)^2\ldots]\end{aligned}$$

By the law of virtual work, during a virtual displacement $dw_{mn}\sin m\pi x/a \sin n\pi y/b$:

$$\left[\frac{\partial}{\partial w_{mn}}(\mathscr{E}_{\text{plate}}+\mathscr{E}_{\text{rib}})\right]dw_{mn} = p_0\,dw_{mn}\int_0^a dx\int_0^b dy\,\sin\frac{m\pi x}{a}\sin\frac{n\pi y}{b}$$

from which:

$$\frac{\pi^4 bD}{8a^3}\left(m^2+\frac{a^2}{b^2}n^2\right)^2 2w_{mn}\,dw_{mn}+\frac{EI}{2}\frac{\pi^4}{a^4}\frac{a}{2}m^4 2\left(\sum_{n'} w_{mn'}\sin\frac{n'\pi}{2}\right)dw_{mn}\sin\frac{n\pi}{2} = \frac{4ab}{\pi^2 mn}p_0\,dw_{mn} \qquad (4.78)$$

Thus in such cases, because the rib strain energy is integrated in one direction only, we do not obtain explicit formulas for each coefficient w_{mn}, but a set of simultaneous equations:

$$\left[bD\left(1+\frac{a^2}{b^2}\right)^2+2EI\right]w_{11}-2EI(w_{13}-w_{15}\ldots)=\frac{16a^4bp_0}{\pi^6}$$

$$\left[bD\left(1+9\frac{a^2}{b^2}\right)^2+2EI\right]w_{13}-2EI(w_{11}+w_{15}\ldots)=\frac{16a^4bp_0}{3\pi^6} \tag{4.79}$$

$$\left[bD\left(9+\frac{a^2}{b^2}\right)^2+162EI\right]w_{31}-162EI(w_{33}-w_{35}\ldots)=\frac{16a^4bp_0}{3\pi^6}, \text{ etc.}$$

Using only one term, for the coefficient w_{11}, we have:

$$w_{11}=w_{\max}=\frac{16a^4bp_0}{\pi^6\left[bD\left(1+\frac{a^2}{b^2}\right)^2+2EI\right]} \tag{4.80}$$

For a square plate, $a=b$, and the dimensions of the rib shown in Fig. 4.24(b), $I=hb^3/12000$ and:

$$w_{\max}=\frac{4a^4p_0}{\pi^6D}\frac{1}{1+\dfrac{(1-\nu^2)}{2000}\dfrac{b^2}{h^2}} \tag{4.81}$$

For a fairly thick plate, say for $b/h=20$ the effect of the rib, represented by the term in b^2/h^2, would not be great. But for thin plates, say for $b/h>100$, a rib having less than one-tenth of the weight of the plate can reduce the deflection by a factor of five and more; the use of a more efficient rib shape, such as a channel beam, would make the comparison even more striking. If the reader wishes to continue this problem using more terms, he will find that the convergence is good and the above conclusions still hold.

If the rib of Fig. 4.24(b) is nonsymmetrical, all on one side of the plate instead of being half above and half below, then for $a=b$ we would have $s/l=s'/l=1/2$, $\gamma=\gamma'\approx 1$, and the effective moment of inertia would, by Eq. (4.76) be about:

$$I'=\frac{hb^3}{12000}+\frac{(hb/10)(b/20)^2}{1+2{\cdot}76(hb/10)m/ah}=\frac{hb^3}{12000}\left(1+\frac{3}{1+0{\cdot}276\,mb/a}\right) \tag{4.82}$$

Replacing I in Eqs. (4.79) by I', we would change the $2EI$ in the first two equations to $2EI[1+3/(1+0{\cdot}276b/a)]$, the $162EI$ in the third equation to $162EI[1+3/(1+0{\cdot}828b/a)]$, etc. In Eq. (4.81) for $a=b$, using only one term, the term in b^2/h^2 would be increased by the large factor $[1+3/(1+0{\cdot}276)]=3{\cdot}35$. While this factor would be smaller for a rib of better shape, it is evident from this result that it is much more effective to put stiffening ribs on one side of the plate, as well as being more practical for manufacturing purposes. This is of course due to the greater effectiveness of the plate, an appreciable part of which performs a double function as both plate and part of the stiffener, greatly increasing the effective moment of inertia of the stiffener.

Stability Problems, Using Energy Method

In such problems during a virtual displacement the most important work done by external forces (and the *only* work except in cases of elastically supported edges) is that done by edge loads in the plane of the middle surface, which act over the u

and v displacements of the edges. To calculate these u, v displacements is rather a complex task involving solutions of the relations between ϕ and w given by Eq. (4.13) and the relations between ϕ and the displacements given by Eq. (4.11). For complex problems, such as those involving a different distribution of edge forces on opposite sides or large initial deviations from flatness, such a complete solution is probably needed. However, in the simpler cases of practical interest which we will consider, involving edge forces similarly distributed on opposite edges and zero or small initial deviations, it is sufficient and much easier to sum across the plate the contributions to the relative movements of the edges, in the direction of the edge forces, which are due to the tipping of the middle surface during buckling; the membrane stresses and strains *in* the middle surface can be assumed to remain unchanged during an infinitesimal virtual displacement, so that only the movement due to tipping has to be considered.

Buckling of Hinged-Edge Plates Under Compression

For example, consider a typical element $dx\,dy$ in a plate subjected to edge forces which produce σ_{xm} membrane stresses. During a displacement $w(x, y)$ the distance in the x direction between its sides dy will be decreased due to tipping of the element in the x direction. In Fig. 4.4(a) the hypotenuse $O'P'$ should now represent a tipped unstrained element whose length remains dx, the horizontal distance OP between its ends being reduced by the tipping from dx to $\{1-(\partial w/\partial x)^2\}^{1/2}\,dx$, that is by the amount $dx-\{1-(\partial w/\partial x)^2\}^{1/2}\,dx \approx (\partial w/\partial x)^2\,dx/2$. The total work done by the external forces producing σ_{xm} can then be taken as:

$$-\frac{1}{2}\int dx \int dy\, h\, \sigma_{xm} \left(\frac{\partial w}{\partial x}\right)^2 \tag{4.83a}$$

where the integration is over the plate and the work is negative because the direction of the displacement is opposite to that of the stress. The work done by forces producing a membrane stress σ_{ym} will be the same except that x and y are interchanged.

In studying the stability of a perfect specimen the actual displacement w is considered to be infinitesimal, and if desired can be taken as the virtual displacement. By the law of virtual work, then, we can equate expression (4.83a) for the external work directly to expression (4.73) for the change in strain energy. We can satisfy hinged boundary conditions by using expression (4.72) for w. Consider for example the important case of a uniform compressive loading s_x on the sides $x = 0$, a, Fig. 4.15. Then $\sigma_{xm} = -s_x$, corresponding to the Airy stress function $\phi = -s_x y^2/2$, and we obtain:

$$\frac{D}{2}\sum_m \sum_n \left(\frac{m^2\pi^2}{a^2}+\frac{n^2\pi^2}{b^2}\right)^2 w_{mn}\frac{ab}{4} = \frac{hs_x}{2}\sum_m \sum_n \frac{m^2\pi^2}{a^2}\, w_{mn}\frac{ab}{4}$$

$$s_x = \frac{\pi^2 D}{b^2 h}\sum_m \sum_n \left(\frac{mb}{a}+\frac{a}{mb}n^2\right)^2 \tag{4.83b}$$

Buckling will occur when s_x reaches the smallest value given by Eq. (4.83b). Since the terms in the summations are all positive, this will obviously be when the amplitudes w_{mn} of all terms except the smallest one are zero, and in this term n will have its smallest possible value, unity. Equation (4.83b) then coincides with Eq. (4.36) and the value of m to give the minimum s_x is given by Fig. 4.17.

In the case of an imperfect specimen having an initial deviation from flatness w_0, the displacement w from this initial position will increase steadily with the load and must therefore be considered to be finite at the instant studied. The decrease in the horizontal distance OP, Fig. 4.4(a), must therefore be found for an infinitesimal change Δw in w. This will be:

$$\left(\frac{\partial(w_0+w+\Delta w)}{\partial x}\right)^2\frac{dx}{2}-\left(\frac{\partial(w_0+w)}{\partial x}\right)^2\frac{dx}{2}$$

$$=\left[\left(\frac{\partial(w_0+w)}{\partial x}+\frac{\partial\Delta w}{\partial x}\right)^2-\left(\frac{\partial(w_0+w)}{\partial x}\right)^2\right]\frac{dx}{2}$$

$$=\left[2\frac{\partial(w_0+w)}{\partial x}\frac{\partial\Delta w}{\partial x}+\left(\frac{\partial\Delta w}{\partial x}\right)^2\right]\frac{dx}{2}=\frac{\partial(w_0+w)}{\partial x}\frac{\partial\Delta w}{\partial x}dx$$

in the limit, since Δw is infinitesimal compared to w or w_0. During the change Δw it can be assumed that σ_{xm} and the edge forces which cause it remain constant, so that the edge forces will do the work:

$$-\int dx\int dy h\sigma_{xm}\frac{\partial(w_0+w)}{\partial x}\frac{\partial\Delta w}{\partial x}\tag{4.84a}$$

If the initial deviation is small it can be assumed that the distribution of membrane stresses is the same as given by plane stress elasticity theory for a flat plate. Taking the same case of uniform compression $\sigma_{xm}=-s_x$ as studied above, and using expression (4.72) for w and a similar expression for w_0 with coefficients w_{0mn} in Eqs. (4.73) and (4.84a), the law of virtual work requires that during a virtual displacement $\Delta w = dw_{mn}\sin m\pi x/a \sin n\pi y/b$ (remembering that the strain energy depends only upon the *movement* w):

$$\frac{\partial\mathscr{E}}{\partial w_{mn}}dw_{mn}=\frac{\pi^4 bD}{4a^3}\left(m^2+\frac{a^2}{b^2}n^2\right)^2 w_{mn}\,dw_{mn}$$

$$=\int_0^a dx\int_0^b dy\,h\,s_x\,dw_{mn}\frac{m\pi^2}{a^2}\cos\frac{m\pi x}{a}\sin\frac{n\pi y}{b}\left[(w_{011}+w_{11})\cos\frac{\pi x}{a}\sin\frac{\pi y}{b}\right.$$

$$\left.+\cdots m(w_{0mn}+w_{mn})\cos\frac{m\pi x}{a}\sin\frac{n\pi y}{b}+\cdots\right]=\frac{\pi^2 hs_x}{a^2}\frac{ab}{4}m^2(w_{0mn}+w_{mn})\,dw_{mn},$$

$$w_{mn}=\frac{w_{0mn}}{\dfrac{\pi^2 D}{b^2 hs_x}\left(\dfrac{mb}{a}+\dfrac{a}{mb}n^2\right)^2-1}\tag{4.84b}$$

This equation corresponds to Eq. (2.24) for beams, and can be studied in a similar manner. If w_0 and hence w_{0mn} are zero the denominator of the right-hand side must be zero to give a buckling deflection, from which:

$$s_x = \frac{\pi^2 D}{b^2 h}\left(\frac{mb}{a} + \frac{a}{mb}\right)^2 \tag{4.84c}$$

the number of waves in the y direction n being assumed to be unity since this obviously gives the smallest buckling load. This checks with Eq. (4.36), obtained with the equilibrium method. It is assumed in such a solution that both w_0 and w are small compared to h. If either of these is of the order of magnitude of h we could no longer assume that σ_{xm} is uniform by ignoring the nonlinear membrane stresses produced by the buckling deflection, as we will see in Sec. 5.1.

Buckling of Hinged-Edge Plates Under Shear

As in the case of compression considered above, the calculation of the distance over which buckling forces act due to tipping of the middle surface is the reverse of the derivation of the nonlinear terms in the expressions for membrane strains of Eqs. (4.2). In that derivation we calculated the strains due to tipping alone without u, v displacements, while here we are interested in the u, v displacements due to tipping alone without strains; the expressions must be the same except for opposite signs.

Thus for shear, by reinterpreting Fig. 4.4(b) as we did Fig. 4.4(a), it can be shown that, considering one side dy of an unstrained element $dx\,dy$ as fixed for convenience, the movement in the y direction of the opposite side due to tipping produced by a displacement $w(x, y)$ is $(\partial w/\partial x)(\partial w/\partial y)dy$. If shear stresses σ_{xym} are acting, the external forces producing them will do work on the movement of this side which, summed up over the plate will be:

$$-\int dx \int dy\, h\, \sigma_{xym} \frac{\partial w}{\partial x}\frac{\partial w}{\partial y} \tag{4.85}$$

where the integration is over the plate and the work is negative because the direction of the displacement is opposite to that of the stress. There is no work due to the movement of the sides dx because their movement is normal to the shear stresses on them.

Similarly, if there is an initial deviation from flatness w_0 the movement of one side dy during a small change Δw in w will be:

$$\left[\frac{\partial(w_0 + w + \Delta w)}{\partial x}\frac{\partial(w_0 + w + \Delta w)}{\partial y} - \frac{\partial(w_0 + w)}{\partial x}\frac{\partial(w_0 + w)}{\partial y}\right] dy$$

$$= \left[\left(\frac{\partial(w_0 + w)}{\partial x} + \frac{\partial \Delta w}{\partial x}\right)\left(\frac{\partial(w_0 + w)}{dy} + \frac{\partial \Delta w}{\partial y}\right) - \frac{\partial(w_0 + w)}{\partial x}\frac{\partial(w_0 + w)}{\partial y}\right] dy$$

$$= \left[\frac{\partial(w_0 + w)}{\partial x}\frac{\partial \Delta w}{\partial y} + \frac{\partial(w_0 + w)}{\partial y}\frac{\partial \Delta w}{\partial x}\right] dy$$

resulting in a work done by the edge forces which cause σ_{xym} of

$$-\int dx \int dy\, h\, \sigma_{xym} \left[\frac{\partial \Delta w}{\partial x}\frac{\partial (w_0 + w)}{\partial y} + \frac{\partial \Delta w}{\partial y}\frac{\partial (w_0 + w)}{\partial x}\right] \tag{4.86}$$

Consider the case shown in Fig. 4.25(a) of a hinged-edge plate under a uniformly distributed shear forces per unit area s_{xy} on its sides $x = 0, a$ and $y = 0, b$. If the initial and final deflections w_0 and w are small compared to h, the membrane stress distribution can be assumed to be the uniform shear stresses $\sigma_{xym} = s_{xy}$ given by the Airy stress function $\phi = -s_{xy}xy$. We take expression (4.72) for w to satisfy the boundary conditions, with a similar expression for w_0 with coefficients w_{0mn}, which can be assumed to be known in any specific case. Then, using Eqs. (4.73) and (4.86), the law of virtual work requires that during a virtual displacement $\Delta w = dw_{mn} \sin m\pi x/a \sin n\pi y/b$:

$$\frac{\partial \mathscr{E}}{\partial w_{mn}} dw_{mn} = \frac{\pi^4 bD}{4a^3}\left(m^2 + \frac{a^2}{b^2}n^2\right)^2 w_{mn}\, dw_{mn} = -\frac{\pi^2 h s_{xy}}{ab}\, dw_{mn} \int_0^a dx \int_0^b dy$$

$$\times \sum_p \sum_q \left[(w_{0pq} + w_{pq})\left(mq \sin\frac{p\pi x}{a}\cos\frac{m\pi x}{a}\sin\frac{n\pi y}{b}\cos\frac{q\pi y}{b}\right.\right.$$

$$\left.\left. + np \sin\frac{m\pi x}{a}\cos\frac{p\pi x}{a}\sin\frac{q\pi y}{b}\cos\frac{n\pi y}{b}\right)\right]$$

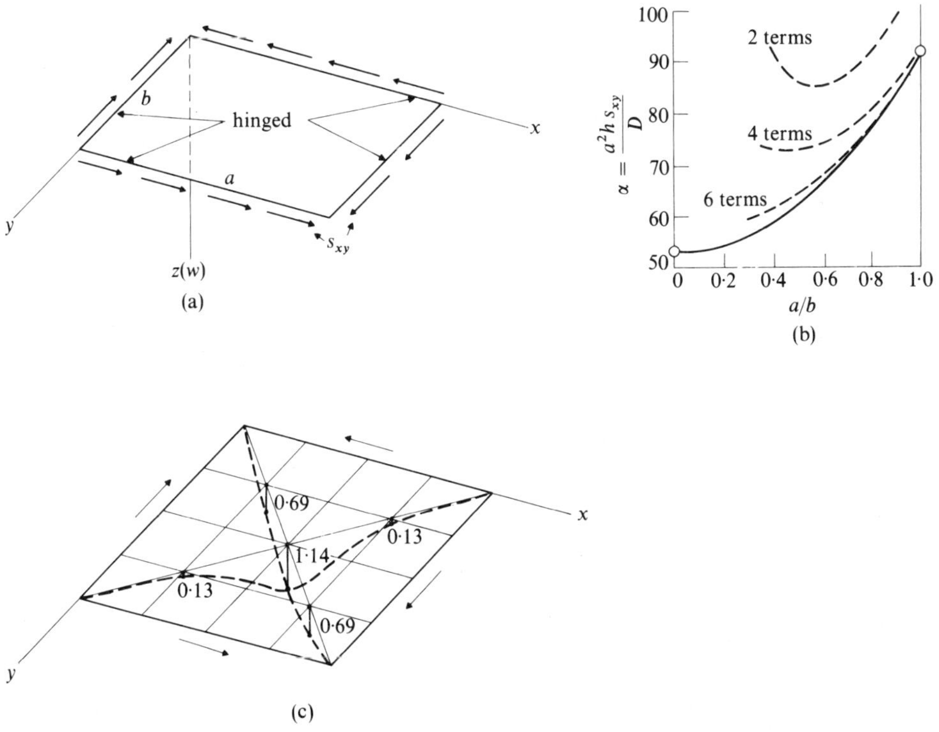

FIGURE 4.25

whence, using the integrals of products of sines and cosines:

$$S\left(m^2+\frac{a^2}{b^2}n^2\right)^2 w_{mn} - \sum_p \sum_q (w_{0pq}+w_{pq})\frac{mnpq}{(m^2-p^2)(n^2-q^2)} = 0 \tag{4.87}$$

where $m+p$ and $n+q$ are odd only, and where $S = \pi^4 bD/(32a^3hs_{xy})$. Writing out Eqs. (4.87) for each combination of m and n, we obtain two sets of simultaneous equations, one containing only terms with $(m+n)$ even and the other with $(m+n)$ odd. The first set is found by trial to yield smaller values of s_{xy}, and is:

$$\begin{aligned}
&S\left(1+\frac{a^2}{b^2}\right)^2 w_{11} - \frac{4}{9}(w_{022}+w_{22}) && \cdots = 0\\
-&\frac{4}{9}(w_{011}+w_{11}) + 16S\left(1+\frac{a^2}{b^2}\right)^2 w_{22} + \frac{4}{5}(w_{013}+w_{13}) + \frac{4}{5}(w_{031}+w_{31}) && \cdots = 0\\
&\qquad\frac{4}{5}(w_{022}+w_{22}) + S\left(1+9\frac{a^2}{b^2}\right)^2 w_{13} && \cdots = 0\\
&\qquad\frac{4}{5}(w_{022}+w_{22}) \qquad\qquad + S\left(9+\frac{a^2}{b^2}\right)^2 w_{31} && \cdots = 0\\
&\cdots \qquad \cdots \qquad \cdots \qquad \cdots && \cdots
\end{aligned} \tag{4.88}$$

Equating the determinant of this equation to zero for any shape ratio a/b, we obtain an equation for determining S and hence the critical s_{xy}. The dashed lines in Fig. 4.25(b) show values of $\alpha = \pi^4 b/(32aS) = a^2hs_{xy}/D$ for the case of a perfect plate, $w_0 = 0$, obtained by Timoshenko with different numbers of terms of such a set of equations.* The full line shows the most probably correct values of α, taking into consideration an exact (within the Love–Kirchhoff approximation) value of $\alpha = 52{\cdot}8$ for a long strip, $a/b = 0$, which was obtained by Southwell and Skan.**

The above general results for a perfect plate were obtained by the energy method, but they can also be obtained by using equilibrium considerations. Substituting expression (4.72) for w into Eq. (4.18), with σ_{xm}, σ_{ym}, $p = 0$, we obtain

$$\begin{aligned}
\pi^4 D \sum_m \sum_n \left(\frac{m^2}{a^2}+\frac{n^2}{b^2}\right)^2 w_{mn} \sin\frac{m\pi x}{a}\sin\frac{n\pi y}{b}\\
= \frac{2\pi^2 hs_{xy}}{ab}\sum_m \sum_n mnw_{mn}\cos\frac{m\pi x}{a}\cos\frac{n\pi y}{b}
\end{aligned} \tag{4.89}$$

This equation can be satisfied by replacing each cosine function by a sine series, as we did in using normal beam modes in Sec. 4.5. Let

$$\cos\frac{m\pi x}{a}\cos\frac{n\pi y}{b} = \sum_p \sum_q C_{mnpq}\sin\frac{p\pi x}{a}\sin\frac{q\pi y}{b} \tag{4.90}$$

* Timoshenko, *Theory of Elastic Stability*, 1st edn, pp. 357–62.

** R. V. Southwell and S. W. Skan, 'On the stability under shearing forces of a flat elastic strip', *Proc. Roy. Soc. Lond.*, vol. 105A, 1924, p. 582.

Determining the coefficients C_{mnpq} by harmonic analysis we find:

$$\begin{aligned} C_{mnpq} &= \frac{4}{ab}\int_0^a dx \int_0^b dy \sin\frac{p\pi x}{a}\cos\frac{m\pi x}{a}\sin\frac{q\pi y}{b}\cos\frac{n\pi y}{b} \\ &= \frac{16}{\pi^2}\frac{pq}{(m^2-p^2)(n^2-q^2)} \end{aligned} \tag{4.91}$$

where $m+p, n+q$ are odd only. Substituting this into Eq. (4.90), and putting it back into Eq. (4.89), we obtain an equation containing only sine-sine functions of x and y. For this to be satisfied for all values of x and y, the sum of the coefficients of each kind of sine-sine function must be zero, and this condition gives us the same set of equations as Eqs. (4.88).

The shape of the buckling deflection of a long narrow plate under shear is found by the theory and by experiment to consist of one half wave across the narrow dimension and waves of about the same wave length in the long direction. Unlike the simple sine-wave shape of buckling under compressive stresses, the waves are skewed, with the nodes at an angle, so that the plate is bent more sharply in the direction of the diagonal which is under compression than it is in the direction of the diagonal which is under tension. This trend increases greatly for thin plates when the deflection becomes large compared to the thickness, the tension diagonal becoming nearly straight and the compression diagonal bent in many waves, a shape similar to that produced by shearing a piece of thin paper or cloth between the hands. Such large deflection buckling is discussed in Chap. 5.

Equations (4.88) can easily be solved for the ratios of the deflection coefficients. For a square plate $a = b$ we find $S = 0{\cdot}031$, $w_{22} = 0{\cdot}28\, w_{11}$, $w_{13} = w_{31} = -0{\cdot}072\, w_{11}$. Using these in Eq. (4.72) we obtain the buckling shape for small deflections shown in Fig. 4.25(c), in which w_{11} has been taken as unity. The sharper curvature in the compression than in the tension diagonal direction is quite evident.

Approximate Solution for Plate with Three Sides Hinged, One Side Free, under Compression in the Direction of the Free Side

Let us consider the use with plates of the energy method as an approximate method, by assuming a deflection which 'looks about right'. If a piece of thin cardboard is buckled in the manner shown in Fig. 4.21(a), lines originally parallel to the y axis seem to remain straight, that is the cardboard seems to bend in the simple shape:

$$w = w_1 y \sin\frac{\pi x}{a} \tag{4.92}$$

This satisfies the boundary conditions on the sides $x = 0$, a and $y = 0$, but it does not satisfy those on the free edge $y = b$, Eq. (4.60). Nevertheless, from our findings in the case of the plate with two free edges, Fig. 4.22(a), we would expect that, at least for a wide plate, the plate will buckle nearly in the developable shape defined by Eq. (4.92) over most of its surface, only deviating much from it near the free edge. Hence it will be interesting to compare an energy solution using Eq. (4.92)* with the rather complex 'exact' solution represented by Eq. (4.61) for $s_y = 0$.

Using expression (4.92) for w in Eq. (4.70), the strain energy is:

$$\mathscr{E} = \frac{D}{2}\int_0^a dx \int_0^b dy\left[\left(\frac{\pi^2}{a^2} w_1 y \sin\frac{\pi x}{a}\right)^2 + 2(1-\nu)\left(\frac{\pi}{a} w_1 \cos\frac{\pi x}{a}\right)^2\right]$$
$$= \frac{\pi^2 bD}{4a}\left[\frac{\pi^2 b^2}{3a^2} + 2(1-\nu)\right] w_1^2 \tag{4.93}$$

By the law of virtual work during a virtual displacement $\Delta w = dw_1 y \sin\frac{\pi x}{a}$, using Eq. (4.84a) with $w_0 = 0$ for the work done by the external force s_x:

$$\frac{\partial \mathscr{E}}{\partial w_1} dw_1 = \frac{\pi^2 bD}{4a}\left[\frac{\pi^2 b^2}{3a^2} + 2(1-\nu)\right] 2w_1\, dw_1 = hs_x \int_0^a dx \int_0^b dy \frac{\partial \Delta w}{\partial x}\frac{\partial w}{\partial x}$$
$$= \frac{\pi^2 b^3 h s_x}{6a} w_1\, dw_1 \tag{4.94}$$

from which $s_x = \frac{\pi^2 D}{b^2 h}\left[\frac{b^2}{a^2} + \frac{6(1-\nu)}{\pi^2}\right]$, where the quantity in the brackets corresponds to the k plotted in Fig. 4.21(b). The dotted curve shows the plot obtained with this expression using $\nu = 0{\cdot}3$. If the same value of Poisson's ratio $\nu = 0{\cdot}25$ is used as was used before in the 'exact' solution shown by the full line, the curve obtained is almost indistinguishable from that given by the exact solution, being about 1 percent too high over the range shown. As was pointed out in Sec. 2.8, an approximate energy solution which satisfies the boundary conditions always gives higher than the correct buckling load, but this rule does not necessarily apply if the boundary conditions are not satisfied, as is the case here.

Stability of Stiffened Plates

This important subject is too specialized a field** to explore in its many ramifications in a book on general theory, but a few observations and examples can be cited. As with other problems involving built-up structures, an energy type solution is generally the most practical, and the work done by the part of the load, if any, which is taken by the stiffening ribs must of course be included in the work done by external forces during the virtual displacement. The possible buckling modes which must be investigated to determine the one which will occur at the smallest loading are more diverse than for a simple plate. They include: buckling across the ribs; buckling between ribs, in which the ribs influence the modes of the

* See Appendix written by L. H. Donnell, of Kármán's paper 'The strength of thin plates in compression', *Trans. ASME*, vol. 54, 1932, p. 54.

** For very approximate but very broad solutions in this field see L. H. Donnell, 'The stability of isotropic or orthotropic cylinders or flat or curved panels, between and across stiffeners, with any edge conditions between hinged and fixed, under any combination of compression and shear', *NACA* TN No. 918, 1943.

buckling deformation of the plate and undergo twisting but no bending out of the plane of the plate; and, finally, local buckling of parts of the rib such as outstanding flanges.

As an example, consider a hinged edge plate with sides $x = 0, a$, $y = 0, b$, having an I-beam stiffener with one flange attached to the plate at $y = b/2$. We will ignore initial deviation from perfect shape, and assume that the rib and stiffener are of the same material, and that the loading conditions are such that both are under the same compressive stresses $\sigma_x = -s_x$ during the buckling. We take the transverse deflection w as a sine-sine series given by Eq. (4.72). Then the strain energy of the rib is, using Eqs. (2.54), (4.75a), and (4.75b) and the symbols defined for these equations (with Eq. (4.76) for I′):

$$\frac{1}{2}\int_0^a dx\left[EI'\left(\frac{\partial^2 w}{\partial x^2}\right)^2 + GJ\left(\frac{\partial^2 w}{\partial x\,\partial y}\right)^2 + c^2 EI_{fl}\left(\frac{\partial^3 w}{\partial x^2\,\partial y}\right)^2\right]_{(y=b/2)} \tag{4.95}$$

During a virtual displacement $dw_{mn} \sin m\pi x/a \sin n\pi y/b$ the change in strain energy of the rib can then be written as:

$$\begin{aligned} &dw_{mn}\frac{\pi^4}{2}\left[\frac{m^4 EI'}{a^3}\left(\sum_{n'} w_{mn'} \sin\frac{n'\pi}{2}\right)\sin\frac{n\pi}{2}\right. \\ &\left.+\left(\frac{m^2 nGJ}{ab^2} + \frac{\pi^2 m^4 nc^2 EI_{fl}}{a^3 b^2}\right)\left(\sum_{n'} n' w_{mn'} \cos\frac{n'\pi}{2}\right)\cos\frac{n\pi}{2}\right] \end{aligned} \tag{4.96}$$

The work done by the external loading s_x on the rib during the virtual displacement is, using Eqs. (2.58a) and (4.75c):

$$\begin{aligned} &dw_{mn}\frac{\partial}{\partial w_{mn}}\frac{1}{2}\int_0^a dx\left[s_x A\left(\frac{\partial w}{\partial x}\right)^2 + \frac{s_x A}{2}c^2\left(\frac{\partial^2 w}{\partial x\,\partial y}\right)^2\right]_{(y=b/2)} \\ &= dw_{mn}\frac{\pi^2 s_x A}{2}\left[\frac{m^2}{a}\left(\sum_{n'} w_{mn'} \sin\frac{n'\pi}{2}\right)\sin\frac{n\pi}{2}\right. \\ &\left.+\frac{\pi^2 c^2 m^2 n}{2ab^2}\left(\sum_{n'} n' w_{mn'} \cos\frac{n'\pi}{2}\right)\cos\frac{n\pi}{2}\right] \end{aligned} \tag{4.97}$$

Equating expressions (4.96) and (4.97), and adding to Eq. (4.84b) with $w_{0mn} = 0$, to take care of the energy changes for the plate, we obtain, after dividing through by $\pi^2 b\, dw_{mn}/4a$ and rearranging:

$$\begin{aligned} &\left[\frac{\pi^2 D}{a^2}\left(m^2 + \frac{a^2}{b^2}n^2\right)^2 - m^2 h s_x\right]w_{mn} \\ &+\frac{2m^2}{b}\left[\frac{\pi^2 m^2 EI'}{a^2} - As_x\right]\left(\sum_{n'} w_{mn'} \sin\frac{n'\pi}{2}\right)\sin\frac{n\pi}{2} \\ &+\frac{2\pi^2 m^2 n}{b^3}\left[GJ + \left(\frac{\pi^2 m^2 EI_{fl}}{a^2} - \frac{As_x}{2}\right)c^2\right]\left(\sum_{n'} n' w_{mn'} \cos\frac{n'\pi}{2}\right)\cos\frac{n\pi}{2} = 0 \end{aligned} \tag{4.98}$$

The second of the three terms is zero for even values of n or n', while the third term is zero for odd values. Hence Eq. (4.98) represents two sets of simultaneous equations. In one set each equation contains all the w_{mn}'s for odd values of n; the determinant of this set gives the value of s_x which will produce buckling across the rib. The other set similarly involves even values of n and gives the value of s_x which will produce buckling between the rib and the sides of the plate. Each set involves a value of m which must be chosen to minimize s_x. If we take a square plate, $a = b$, and use only one term of each set, we obtain, for $n = 1$ and $n = 2$ respectively:

$$\frac{a^2 s_x}{\pi^2} = \frac{aD\left(m + \frac{1}{m}\right)^2 + 2m^2 EI'}{ah + 2A} \text{ (across)}$$

$$\frac{a^2 s_x}{\pi^2} = \frac{a^3 D\left(m + \frac{4}{m}\right)^2 + 8(a^2 GJ + \pi^2 m^2 c^2 EI_{fl})}{a^3 h + 4\pi^2 c^2 A} \text{ (between)} \qquad (4.99)$$

For buckling across the rib, $m = 1$ gives the lowest buckling stress, while for buckling between the ribs we must try $m = 1, 2,$ or 3. After finding the buckling mode which gives the lowest s_x we can go back to Eq. (4.98) and use more terms to refine the value of s_x for the mode or modes which are found to be important.

4.7 CIRCULAR PLATES UNDER AXISYMMETRIC DISPLACEMENT

Basic Theory

Up to now we have considered only a rectangular shaped plate, using rectangular coordinates and equilibrium and energy methods. While this is not only the simplest but also the most important type of plate, our discussion would not be complete without at least illustrating the study of other types of plate. The most important coordinate system for use in plate theory besides rectangular is the polar coordinate system, useful mainly for circular plates. For simplicity we will consider here only the case of axisymmetric deformation, produced by axisymmetric loading of circular plates or their axisymmetric modes of buckling and vibration; the general case can be derived from the general shell theories given in Chap. 6. The axisymmetric plate case is simpler than that of the rectangular plate in that there is variation along only one dimension, the radius. We will take the distance from the middle surface and the displacement normal to this surface as z and w respectively, as in rectangular coordinates, and the radial dimension and displacement of the middle surface as r and u, as shown in Fig. 4.26(a).

With the Love–Kirchhoff assumption and to the same degree of approximation as Eq. (4.2), the strain in the radial direction ϵ_r is evidently the same as ϵ_x, Eq. (4.2), with x replaced by r. A general point P, Fig. 4.26(a), has initially the radius r, and after displacement the radius $r + u - (dw/dr)z$. The unit circumferential strain ϵ_θ is therefore $\{2\pi[r + u - (dw/dr)z] - 2\pi r\}/2\pi r = u/r - (dw/dr)z/r$.

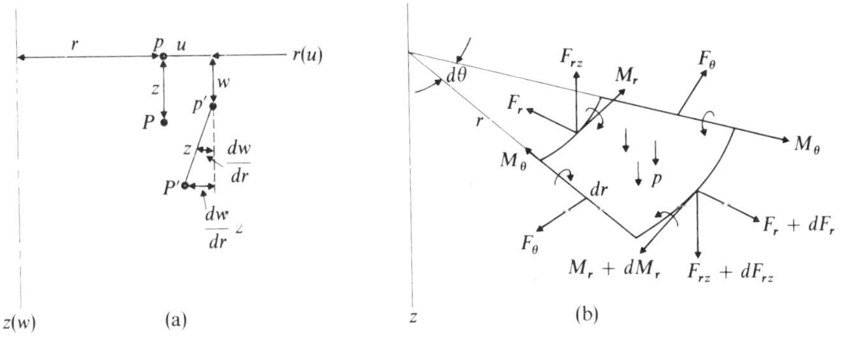

FIGURE 4.26

Hence:
$$\epsilon_r = \frac{du}{dr} + \frac{1}{2}\left(\frac{dw}{dr}\right)^2 - \frac{d^2w}{dr^2}z, \qquad \epsilon_\theta = \frac{u}{r} - \frac{dw}{dr}\frac{z}{r} \tag{4.100}$$

There is no nonlinear term for ϵ_θ because from symmetry there is no tipping of elements in this direction. Using Eq. (4.7) with x, y replaced by r, θ in the subscripts and applying Hooke's Law Eq. (3.11b) for the elastic case, the forces and moments per unit length of section are found to be:

$$F_r = \frac{Eh}{1-\nu^2}\left[\frac{du}{dr} + \frac{1}{2}\left(\frac{dw}{dr}\right)^2 + \nu\frac{u}{r}\right], \qquad F_\theta = \frac{Eh}{1-\nu^2}\left[\frac{u}{r} + \nu\frac{du}{dr} + \frac{\nu}{2}\left(\frac{dw}{dr}\right)^2\right] \tag{4.101a}$$

$$M_r = -D\left(\frac{d^2w}{dr^2} + \frac{\nu}{r}\frac{dw}{dr}\right), \qquad M_\theta = -D\left(\frac{1}{r}\frac{dw}{dr} + \nu\frac{d^2w}{dr^2}\right) \tag{4.101b}$$

In the present case $F_{r\theta}$ and $M_{r\theta}$ are zero because of symmetry. As in Eq. (4.16) the stresses $\sigma_{rm} = F_r/h$, $\sigma_{rf} = 12M_rz/h^3$, etc.

Figure 4.26(b) shows the forces and moments acting upon a typical element. In general F_r, M_r and the transverse shear force per unit length of section F_{rz} vary with r. From symmetry F_θ, M_θ do not vary with the angle θ, there is no $F_{\theta z}$, and the equations of equilibrium $\Sigma f_\theta = 0$ and $\Sigma m_r = 0$ are identically satisfied. Due to the angle $d\theta$ between the $F_\theta\, dr$ forces on opposite sides, these forces have a component $(F_\theta\, dr)\, d\theta$ in the radial direction, and similarly there is a moment $(M_0\, dr)\, d\theta$ about a circumferential axis. Due to the angles between opposite sides produced by the deflection there are $F_r r\, d\theta(d^2w/dr^2)\, dr$ (ignoring small quantities of higher order) and $(F_0\, dr)\, d\theta(dw/dr)$ (due to rotation of the above mentioned $F_0\, dr\, d\theta$ radial force through the angle dw/dr) forces in the transverse equilibrium equation which are important in stability problems and, as nonlinear terms, in large-deflection problems. As discussed in connection with Eq. (2.3b) and Eq. (4.8), we can ignore the components in the radial direction of the transverse shear forces F_{rz} due to the deflection.

Remembering that the products of infinitesimals disappear in the limit compared to the infinitesimals themselves, the equation of equilibrium of moments about the circumferential direction is

$$\sum m_\theta = (M_r + dM_r)(r + dr)\, d\theta - M_r r\, d\theta - (M_\theta\, dr)\, d\theta - (F_{rz} r\, d\theta)\, dr$$
$$= M_r\, dr\, d\theta + dM_r r\, d\theta - M_\theta\, dr\, d\theta - F_{rz} r\, d\theta\, dr = 0$$

and similarly for $\Sigma f_r = 0$ and $\Sigma f_z = 0$. Dividing through by $r\, d\theta\, dr$, we obtain:

$$\sum f_r = 0: \quad \frac{F_r - F_\theta}{r} + \frac{dF_r}{dr} = 0 \tag{4.102a}$$

$$\sum f_z = 0: \quad p + \frac{F_{rz}}{r} + \frac{dF_{rz}}{dr} + F_r \frac{d^2 w}{dr^2} + \frac{F_\theta}{r}\frac{dw}{dr} = 0 \tag{4.102b}$$

$$\sum m_\theta = 0: \quad \frac{M_r - M_\theta}{r} + \frac{dM_r}{dr} = F_{rz} \tag{4.102c}$$

Substituting expressions (4.101b) for M_r, M_θ in Eq. (4.102c) we obtain an expression for F_{rz}:

$$F_{rz} = -D\left(\frac{d^3 w}{dr^3} + \frac{1}{r}\frac{d^2 w}{dr^2} - \frac{1}{r^2}\frac{dw}{dr}\right) \tag{4.103}$$

and putting this expression for F_{rz} into the transverse equilibrium equation (4.102b) we find:

$$D\left(\frac{d^4 w}{dr^4} + \frac{2}{r}\frac{d^3 w}{dr^3} - \frac{1}{r^2}\frac{d^2 w}{dr^2} + \frac{1}{r^3}\frac{dw}{dr}\right) = p + F_r \frac{d^2 w}{dr^2} + \frac{F_\theta}{r}\frac{dw}{dr} \tag{4.104}$$

As was discussed for Eqs. (4.13) and (4.18), for small deflections we can ignore the nonlinear $(dw/dr)^2/2$ term in Eq. (4.101a), making F_r and F_θ independent of w. In a stability problem we can then solve the plane stress problem represented by Eqs. (4.101a), (4.102a) to find the distribution of F_r, F_θ, and consider these to be constants, independent of w, in solving Eq. (4.104); in the small deflection transverse loading problem we can take F_r and F_θ as zero. Only when the deflections are of the order of magnitude of the thickness must we use the nonlinear term and solve Eqs. (4.101a) and (4.104) simultaneously.

Small Deflections of Discs Under Transverse Loading

For this case Eq. (4.104) becomes

$$D\left(\frac{d^4 w}{dr^4} + \frac{2}{r}\frac{d^3 w}{dr^3} - \frac{1}{r^2}\frac{d^2 w}{dr^2} + \frac{1}{r^3}\frac{dw}{dr}\right) = p \tag{4.104a}$$

If we retain only the terms involving w in this equation, replacing p by zero, we have an ordinary homogeneous differential equation of the fourth order whose solution can be written as:

$$w = C_0 + C_1 r^2 + C_2 \log r + C_3 r^2 \log r \tag{4.105}$$

This can be added to any particular solution of Eq. (4.104a) to satisfy the two edge conditions on w and dw/dr, or on M_r and F_{rz}, at the outer edge, and similar conditions at an inner edge for a disc with a hole or for continuity at the center of a solid disc.

Disc Under Uniform Transverse Load

As a simple example of the application of the above theory, consider a disc under a uniform loading $p = p_0$. Taking $F_r, F_\theta = 0$, the particular solution of Eq. (4.104a) for this case is readily found as $p_0 r^4/64D$ and adding the solution (4.105) of the homogeneous equation we have:

$$w = \frac{p_0}{64D} r^4 + C_0 + C_1 r^2 + C_2 \log r + C_3 r^2 \log r \tag{4.106}$$

The condition of continuity at the center is that, from symmetry, using Eq. (4.103):

$$r = 0: \quad \frac{dw}{dr}, F_{rz} = 0, \qquad \text{or} \qquad \frac{dw}{dr}, \quad \frac{d^3w}{dr^3} + \frac{1}{r}\frac{d^2w}{dr^2} - \frac{1}{r^2}\frac{dw}{dr} = 0 \tag{4.107}$$

Putting expression (4.106) for w into these we find that to satisfy these conditions:

$$C_2, C_3 = 0, \qquad w = \frac{p_0}{64D} r^4 + C_0 + C_1 r^2 \tag{4.108}$$

Hinged Edge, Uniform Load

If the outer edge of the disc $r = a$ is hinged, then, using Eq. (4.101b):

$$r = a: \quad w, M_r = 0, \qquad \text{or} \qquad w, \quad \frac{d^2w}{dr^2} + \frac{\nu}{r}\frac{dw}{dr} = 0 \tag{4.109}$$

Substituting expression (4.108) for w into these, solving for C_0, C_1 and putting these values back into expression (4.108), we obtain:

$$C_0 = \frac{5+\nu}{1+\nu}\frac{p_0 a^4}{64D}, \qquad C_1 = -\frac{3+\nu}{1+\nu}\frac{p_0 a^2}{32D}$$
$$w = \frac{p_0}{64D}\left(\frac{5+\nu}{1+\nu}a^4 - \frac{6+2\nu}{1+\nu}a^2 r^2 + r^4\right), \qquad w_{(r=0)} = \frac{5+\nu}{1+\nu}\frac{p_0 a^4}{64D} \tag{4.110}$$

Using this expression for w in Eq. (4.101b), M_r is:

$$M_r = \frac{3+\nu}{16} p_0(a^2 - r^2), \qquad M_{r(r=0)} = \frac{3+\nu}{16} p_0 a^2 \tag{4.111}$$

M_θ is the same as M_r at the center $r = 0$, and less than M_r everywhere else.

Fixed Edge, Uniform Load

For this case:

$$r = a: \quad w, \frac{dw}{dr} = 0 \tag{4.112}$$

and substituting expression (4.108) for w into these we obtain:

$$C_0 = \frac{p_0 a^4}{64D}, \qquad C_1 = -\frac{p_0 a^2}{32D}$$

$$w = \frac{p_0}{64D}(a^2 - r^2)^2, \qquad w_{(r=0)} = \frac{p_0 a^4}{64D} \tag{4.113}$$

which from Eq. (4.101b) gives:

$$M_r = \frac{p_0}{16}[(1+\nu)a^2 - (3+\nu)r^2], \qquad M_{r(r=0)} = \frac{(1+\nu)p_0 a^2}{16},$$

$$M_{r(r=a)} = -\frac{p_0 a^2}{8} \tag{4.114}$$

Again M_θ is the same as M_r at $r = 0$ and smaller everywhere else. The maximum deflection for fixed edge is thus only about a quarter of that for hinged edge, and the maximum stress occurs at the rim and is about 0·6 that for hinged edge.

Uniform Load Over Part of Disc

As an example of a problem of this type, consider the case of a disc of radius a hinged at the outer edge and under a uniform compressive load per unit area $p = p_0$ which acts over an inner portion of the disc having a radius $r = b$, the outer part of the disc being unloaded. Then for the inner loaded portion Eqs. (4.108) apply. Using primes for the displacements, loads, moments, etc., of the outer unloaded region, we have for this region:

$$w' = C_0' + C_1' r^2 + C_2' \log\frac{r}{a} + C_3' r^2 \log\frac{r}{a} \tag{4.114a}$$

$$r = a: \quad w' = 0, M_r' = 0 \quad \text{or} \quad w', \quad \frac{d^2w'}{dr^2} + \frac{\nu}{r}\frac{dw'}{dr} = 0 \tag{4.114b}$$

Continuity at the dividing line between the two regions, $r = b$, requires that:

$$r = b: \quad w = w', \quad \frac{dw}{dr} = \frac{dw'}{dr}, \quad M_r = M_r' \quad \text{or} \quad \frac{d^2w}{dr^2} + \frac{\nu}{r}\frac{dw}{dr} = \frac{d^2w'}{dr^2} + \frac{\nu}{r}\frac{dw'}{dr} \tag{4.114c}$$

$$\text{and} \qquad F_{rz} = F_{zr}' \text{ or } \frac{d^3w}{dr^3} + \frac{1}{r}\frac{d^2w}{dr^2} - \frac{1}{r^2}\frac{dw}{dr} = \frac{d^3w'}{dr^3} + \frac{1}{r}\frac{d^2w'}{dr^2} - \frac{1}{r^2}\frac{dw'}{dr}$$

We now have six unknown coefficients C_0, C_1, C_0', C_1', C_2', C_3' to satisfy the six boundary conditions (4.114b, c). Substituting expressions (4.108) and (4.114a) for w and w' into these boundary conditions, solving them simultaneously, and using the symbol $P' = b^2 p_0/64D$, we obtain the values:

$$C_0 = \left[-(7+3\nu) + 4(3+\nu)\frac{a^2}{b^2} - 4(1+\nu)\log\frac{a}{b}\right]\frac{b^2 P'}{1+\nu}$$

$$
\begin{aligned}
C_1 &= \left[-4+(1-\nu)\frac{b^2}{a^2}-4(1+\nu)\log\frac{a}{b}\right]\frac{2P'}{1+\nu} \\
C_0' &= \left[2(3+\nu)-(1-\nu)\frac{b^2}{a^2}\right]\frac{2a^2P'}{1+\nu}, \qquad C_2' = 4b^2P' \\
C_1' &= \left[-2(3+\nu)+(1-\nu)\frac{b^2}{a^2}\right]\frac{2P'}{1+\nu}, \qquad C_3' = 8P'
\end{aligned}
\tag{4.114d}
$$

Whcn $b - a$ the load p_0 is distributed over the whole disc and the above values of C_0 and C_1 coincide with those given in Eqs. (4.110). At the other extreme when b approaches zero while the total load $\pi b^2 p_0 = P$, we obtain the case of a concentrated load P acting at the center of the disc. For this case $C_2' = 0$, $C_3' = P/8\pi D$, while C_0', C_1', w', etc., have the same values as C_0, C_1, w, etc., obtained below in a somewhat different manner in Eqs. (4.117), (4.118).

Disc Under Concentrated Load at Center

Consider a disc under a concentrated transverse load P at the center. There is no transverse loading over the surface of the disc except at a singular point at the center. Hence p and the particular solution of Eq. (4.104a) can be taken as zero, and w can be taken as the homogeneous solution (4.105).

The continuity condition that the slope dw/dr must be zero at the center from symmetry still holds. However, the transverse shear F_{rz} will not be zero. The condition on F_{rz} can be found by considering the transverse equilibrium of the portion of the disc inside a radius r. The only transverse forces on this portion are P and the constant force per unit length F_{rz} acting over the circumference $2\pi r$, Fig. 4.27, from which $2\pi r F_{rz} + P = 0$. Using Eq. (4.103) the two conditions at the center are:

$$
\begin{aligned}
r = 0: \quad & dw/dr = 0 \\
F_{rz} = -\frac{P}{2\pi r} &= D\left(\frac{1}{r^2}\frac{dw}{dr} - \frac{1}{r}\frac{d^2w}{dr^2} - \frac{d^3w}{dr^3}\right)
\end{aligned}
\tag{4.115}
$$

Substituting expression (4.105) for w in these conditions we find that they are satisfied if:

$$
C_2 = 0, \qquad C_3 = \frac{P}{8\pi D}, \qquad w = C_0 + C_1 r^2 + \frac{P}{8\pi D} r^2 \log r \tag{4.116}
$$

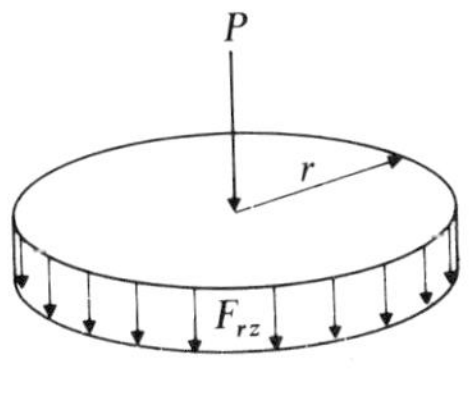

FIGURE 4.27

Hinged Edge, Concentrated Load

Substituting expression (4.116) for w into conditions (4.109) for a hinged edge at $r = a$, solving for C_0, C_1 and putting these back into (4.116) we find:

$$C_0 = \frac{3+\nu}{1+\nu}\frac{Pa^2}{16\pi D}, \qquad C_1 = -\frac{P}{16\pi D}\left(\frac{3+\nu}{1+\nu} + 2\log a\right)$$

$$w = \frac{P}{16\pi D}\left[\frac{3+\nu}{1+\nu}(a^2 - r^2) + 2r^2 \log\frac{r}{a}\right], \qquad w_{(r=0)} = \frac{3+\nu}{1+\nu}\frac{Pa^2}{16\pi D} \tag{4.117}$$

Using this expression for w in Eq. (4.101b) we find:

$$M_r = -\frac{(1+\nu)P}{4\pi}\log\frac{r}{a}, \qquad M_\theta = -\frac{(1+\nu)P}{4\pi}\left(\log\frac{r}{a} - \frac{1-\nu}{1+\nu}\right) \tag{4.118}$$

It can be seen that the values of the bending moments predicted by this classical plate theory solution become infinite at the point of application of the concentrated load, as has been mentioned in Sec. 4.4. As was discussed there, this result is erroneous, and is due to the limitations of the classical theory. Actually the stresses are finite everywhere except at the point of application of the concentrated load, as will be demonstrated in Sec. 5.3, where corrections will be developed for the classical stresses.

Fixed Edge, Concentrated Load

If we substitute expression (4.116) for w into the conditions (4.112) for a fixed edge at $r = a$, we obtain:

$$C_0 = \frac{Pa^2}{16\pi D}, \qquad C_1 = -\frac{P}{16\pi D}(1 + 2\log a)$$

$$w = \frac{P}{16\pi D}\left(a^2 - r^2 + 2r^2\log\frac{r}{a}\right), \qquad w_{(r=0)} = \frac{Pa^2}{16\pi D} \tag{4.119}$$

Using Eq. (4.101b):

$$M_r = -\frac{(1+\nu)P}{4\pi}\left(\log\frac{r}{a} + \frac{1}{1+\nu}\right), \qquad M_\theta = -\frac{(1+\nu)P}{4\pi}\left(\log\frac{r}{a} + \frac{\nu}{1+\nu}\right) \tag{4.120}$$

$$M_{r(r=a)} = -\frac{P}{4\pi}, \qquad M_{\theta(r=a)} = -\frac{\nu P}{4\pi}$$

The maximum deflection is thus about 0·4 of that for the hinged-edge plate under concentrated load. If the concentrated load P is the same as the total distributed load $\pi a^2 p_0$ in the uniform load case, the maximum deflection under the concentrated load is four times that for the corresponding uniform load for fixed edges, and about 2·5 times as much for hinged edges; the bending moment at an outer fixed edge for the concentrated load is twice that for the corresponding uniform load. The bending moments at the center given by the above solutions cannot be compared, as discussed above.

5

LARGE DEFLECTIONS OF PLATES, THICK PLATES

5.1 LARGE DEFLECTIONS OF PLATES

In the usual applications of the classical theories of bending of elastic beams and plates two important types of neglections are made: (a) neglection of the nonlinear effects of finite deformations, that is the effect of the changing geometry of the problem as the deformation progresses; and (b) the Love–Kirchhoff approximation, or neglection of the transverse stresses and strains, with associated simplification of boundary conditions, and ignoring of local stress conditions around concentrated loads, etc.

These two types of neglections can be studied separately, since cases in which both are important in the same problem are very unlikely and even if they do occur (as they might for a thick plate made of a rubber-like material) there is unlikely to be much 'cross effect', that is effect of one neglection upon the other, so that correction for each type could be made independently without serious error. Neglections (a) for nonlinear effects, which we will study in the present section, are usually important only in connection with the nonlocal, overall displacements of 'thin' beams and plates under loads generally in one direction, since otherwise the limitation to the elastic strains of ordinary engineering materials makes elastic deformation of a magnitude sufficient to produce important nonlinear effects impossible. In such cases the Love–Kirchhoff neglection gives an excellent approximation and we will use it here.

Neglections of type (b) will be studied in Sec. 5.2 and the remainder of this chapter. They are usually important only for 'thick' plates, in which the thickness is not very small compared to the other dimensions or compared to the cycle length of loads which alternate in direction, or where large local stresses would be important because of fatigue conditions or lack of material ductility.

Effects of Curvature

In Chap. 4 we considered problems involving thin flat plates whose middle surface deflections w were small compared to the thickness h; the membrane strains and stresses produced by the deflection, being proportional to a higher power than one of the deflection, are very small under these conditions. If finite membrane stresses are present before the deflection begins, due to edge forces in the plane of the plate, then these do not change appreciably during such small deflections.

In contrast, when *curved* plates, that is shells, are deflected, displacements u, v of the middle surface in its own plane and consequently membrane strains and stresses are in general produced, which are proportional to the *first* power of the deflection and are important even when the deflections are small; the same is true of nominally 'flat' plates which have initial deviations from flatness of the order of the thickness, because these are really shallow shells.

In summary, when the deflections of initially flat plates rise to the order of the thickness, the nonlinear u and v displacements and the resulting membrane stresses in general become important, and these are the cases which we will discuss in this section. When shells undergo deflections of the order of the thickness there are u, v displacements and resulting membrane stresses of *both* the linear variety due to the initial curvature, and of the nonlinear variety due to the deflections. If we like, we can think of the nonlinear displacements and membrane stresses as being also produced by curvature, but in this case not the initial curvature but the curvature produced by the deflection—the whole curvature for flat plates and the change in curvature for shells.

Membrane vs. Flexure Stresses

As has been discussed before, nonlinear effects on *flexure* (for example due to replacement of the slope of the deflection, that is the tangent of the angle of rotation, by the angle itself, etc.) are in general still negligible when deflections are of the same order of magnitude as the thickness, and would not become important until deflections reach the order of magnitude of other dimensions such as lengths or widths of plates. We will not consider problems of the latter type here, but a more complete, quantitative discussion of this subject will be given in Chap. 6.

Boundary Conditions

Whether the u, v displacements and resulting membrane stresses in plates and shells are of the linear or nonlinear variety, if they are important then we must consider their effects not only upon equilibrium or strain energy but also upon the boundary conditions. The flexural boundary conditions have been discussed before and can be resummarized as follows. For an edge normal to the x direction:

$$w=0, \quad M_x=\frac{\partial^2 w}{\partial x^2}+\nu\frac{\partial^2 w}{\partial y^2}=0 \qquad \text{for a hinged edge,}$$

$$w=0, \quad \frac{\partial w}{\partial x}=0 \qquad \text{for a fixed edge,} \tag{5.1}$$

$$M_x, F'_{xz}=0, \quad \text{or} \quad \frac{\partial^2 w}{\partial x^2}+\nu\frac{\partial^2 w}{\partial y^2}, \quad \frac{\partial^3 w}{\partial x^3}+(2-\nu)\frac{\partial^3 w}{\partial x\,\partial y^2}=0 \quad \text{for a free edge.}$$

For the same edge the membrane boundary conditions are:

$$F_x, F_{xy}=0 \text{ (or } \frac{\partial^2 \phi}{\partial y^2}, \frac{\partial^2 \phi}{\partial x\,\partial y}=0 \text{ if we use an Airy stress function } \phi$$
to represent the membrane stresses) if there is no restraint upon movement in the middle surface directions. (5.2)
$u, v=0$ if movements in the middle surface directions are prevented.

There are similar flexural and membrane conditions for an edge normal to the y direction obtained by interchanging x and y, and u and v in Eqs. (5.1) and (5.2). Depending upon the physical problem there may be various combinations of these conditions, as well as intermediate conditions such as elastic constraints, as was discussed for beams in connection with Eqs. (2.6).

For example, Fig. 5.1(a) shows in cross section a plate resting upon well lubricated rollers at the edges, while Fig. 5.1(b) shows a plate attached to rigid supports at the edges by well lubricated 'piano hinges'. In both cases the flexural boundary conditions can be considered to be hinged because there is negligible resistance to rotation of the edges. No distinction needs to be made between these two cases if the deflections are small as indicated by the dotted lines, because there is little tendency to 'pull in' at the edges and it therefore makes little difference whether this tendency is resisted or not. On the other hand if the deflections are large as indicated by the dashed lines the two cases are entirely different. In case (b), where the pulling-in is prevented, there will be large edge forces F required to prevent the pulling-in. In case (a) such forces will be absent

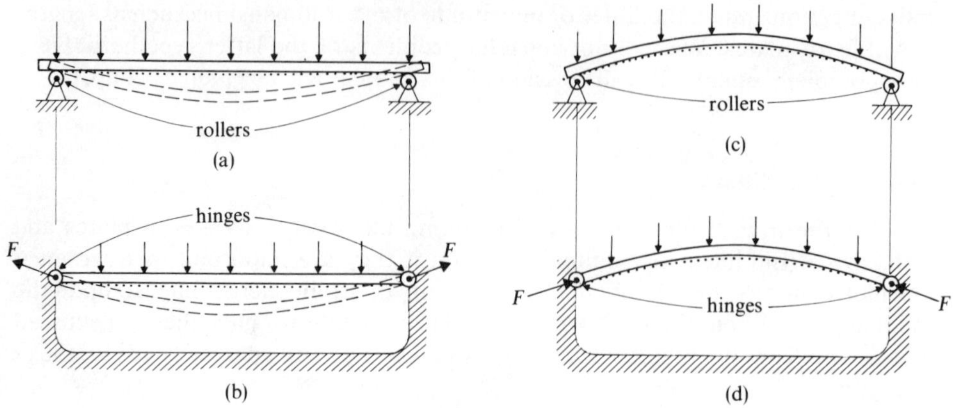

FIGURE 5.1

but, unlike the corresponding case of a beam, there will still be some membrane stresses if the plate is deformed into a nondevelopable shape.

In similar cases for the curved shell shown in Figs. 5.1(c) and (d), it can be appreciated that because of the initial curvature there will be an appreciable tendency to push out at the edges in the case shown, or to pull in if the concavity is upward and the load is still downward, even when the deflections are small. If these figures represent corresponding cases of beams, the beam shown at (d) will act as an arch as well as a beam (or in the case of concavity upward, as a cable as well as a beam) while the beam at (c) can resist the loading only by flexure.

Large Deflection Solutions by Equilibrium Method

When w is finite, that is not small compared to the thickness, conditions (4.13) and (4.18) must both be satisfied, and if the boundary conditions involve the displacements u, v then these displacements must be determined from Eq. (4.11) so as to satisfy these conditions. As has been mentioned, solutions of simultaneous nonlinear differential equations such as (4.13) and (4.18) are difficult. However, use can be made of the fact that the derivatives and the squares and products of such functions as power functions, as well as exponential functions and the trigonometric and hyperbolic functions related to them, are or can be converted to functions of the same general type as the original functions. For example:

$$\begin{aligned} \sin a \sin b &= [\cos(a-b) - \cos(a+b)]/2, \\ \cos a \cos b &= [\cos(a-b) + \cos(a+b)]/2, \\ \sin a \cos b &= [\sin(a-b) + \sin(a+b)]/2 \end{aligned} \tag{5.3}$$

Thus, by expressing w, ϕ, u, v as series of such functions we may be able to equate coefficients of similar terms and reduce the equations to nonlinear algebraic relations. While it is usually not easy to satisfy all possible boundary conditions, some interesting results have been obtained by using trigonometric functions in this way for rectangular plates by Levy,* and by using power functions for axisymmetric plates by Way.**

Large Deflections of Hinged-Edge Plates by Equilibrium Method

For rectangular hinged edge-plates we can start by assuming the deflection w and the lateral load p in the form previously used for the linear case, Eqs. (4.21), (4.22), that is as 'sine sine' functions of x and y:

$$w = \sum_m \sum_n w_{mn} \sin \frac{m\pi x}{a} \sin \frac{n\pi y}{b}, \qquad p = \sum_m \sum_n p_{mn} \sin \frac{m\pi x}{a} \sin \frac{n\pi y}{b} \tag{5.4}$$

This expression for w satisfies the flexural boundary conditions (5.1) on the edges $x = 0, a,\ \ y = 0, b$ of the plate shown in Fig. 4.13, and the coefficients p_{mn} can be found from Eq. (4.23) for any transverse loading $p(x, y)$. Then the Airy stress function ϕ can be taken as the sum $\phi = \phi_p + \phi_h$ of particular and homogeneous solutions ϕ_p and ϕ_h of Eq. (4.13). The function ϕ_p must be a cosine cosine function, with even values of m and n to satisfy Eq. (4.13), while any solution of $\nabla^4\phi_h = 0$ useful for satisfying the membrane boundary conditions may be used. To satisfy Eqs. (4.11) when u and v are involved, u can be taken as a sine cosine function and v as a cosine sine function of x and y respectively, with suitable functions of integration.

In using series solutions the series must be cut off somewhere and subsequent terms ignored. To illustrate this type of solution we will use about the simplest possible solution, consisting of the following terms, which include the most important terms required for many practical problems of this sort:

$$\begin{aligned} &w = w_{11} \sin \frac{\pi x}{a} \sin \frac{\pi y}{b}, \qquad p = p_{11} \sin \frac{\pi x}{a} \sin \frac{\pi y}{b} \\ &\phi = \frac{s_x y^2}{2} + \frac{s_y x^2}{2} + \phi_{20} \cos \frac{2\pi x}{a} + \phi_{02} \cos \frac{2\pi y}{b} \\ &u = u_0 + u_1 x + u_{20} \sin \frac{2\pi x}{a}, \qquad v = v_0 + v_1 y + v_{02} \sin \frac{2\pi y}{b} \end{aligned} \tag{5.5}$$

Substituting these expressions into Eqs. (4.13), using trigonometric transformations such as (5.3), and equating the coefficients of similar functions of x and y, we find:

$$\begin{aligned} 16\pi^4 \left(\frac{\phi_{20}}{a^4} \cos \frac{2\pi x}{a} + \frac{\phi_{02}}{b^4} \cos \frac{2\pi y}{b} \right) &= \frac{\pi^4 E w_{11}^2}{a^2 b^2} \left[\left(\cos \frac{\pi x}{a} \cos \frac{\pi y}{b} \right)^2 - \left(\sin \frac{\pi x}{a} \sin \frac{\pi y}{b} \right)^2 \right] \\ &= \frac{\pi^4 E w_{11}^2}{4a^2 b^2} \left[\left(1 + \cos \frac{2\pi x}{a}\right)\left(1 + \cos \frac{2\pi y}{b}\right) - \left(1 - \cos \frac{2\pi x}{a}\right)\left(1 - \cos \frac{2\pi y}{b}\right) \right] \\ &= \frac{\pi^4 E w_{11}^2}{2a^2 b^2} \left(\cos \frac{2\pi x}{a} + \cos \frac{2\pi y}{b} \right), \text{ whence } \phi_{20} = \frac{Ea^2}{32b^2} w_{11}^2, \quad \phi_{02} = \frac{Eb^2}{32a^2} w_{11}^2 \end{aligned} \tag{5.6}$$

* S. Levy, 'Bending of rectangular plates with large deflections', *NACA Tech. Note* No. 846, 1942.

** S. Way, *Trans. ASME*, vol. 56, p. 627, 1934.

Using these values of ϕ_{20}, ϕ_{02} and expressions (5.5) in Eq. (4.18) and in the first two equations of (4.11) (the linear part of the third equation is satisfied identically but satisfaction of its nonlinear part would require more terms in the series—this is one of the errors inherent in cutting off the series) and using trigonometric transformations, we obtain:

$$\left[\pi^4 D\left(\frac{1}{a^2}+\frac{1}{b^2}\right)^2 w_{11} - p_{11}\right]\sin\frac{\pi x}{a}\sin\frac{\pi y}{b} = h\left[\left(s_x - \frac{\pi^2 E w_{11}^2}{8a^2}\cos\frac{2\pi y}{b}\right)\right.$$

$$\times\left(-\frac{\pi^2 w_{11}}{a^2}\sin\frac{\pi x}{a}\sin\frac{\pi y}{b}\right) + \left(s_y - \frac{\pi^2 E w_{11}^2}{8b^2}\cos\frac{2\pi x}{a}\right)\left.\left(-\frac{\pi^2 w_{11}}{b^2}\sin\frac{\pi x}{a}\sin\frac{\pi y}{b}\right)\right] \qquad (5.7a)$$

$$= -\pi^2 h\left[\left(\frac{s_x}{a^2}+\frac{s_y}{b_2}\right) w_{11} + \frac{\pi^2 E}{16}\left(\frac{1}{a^4}+\frac{1}{b^4}\right) w_{11}^3\right]\sin\frac{\pi x}{a}\sin\frac{\pi y}{b} + \cdots,$$

$$\frac{s_x - \nu s_y}{E} - \frac{\pi^2 w_{11}^2}{8a^2}\cos\frac{2\pi y}{b} + \frac{\nu\pi^2 w_{11}^2}{8b^2}\cos\frac{2\pi x}{a} - u_1 - \frac{2\pi u_{20}}{a}\cos\frac{2\pi x}{a}$$

$$= \frac{\pi^2 w_{11}^2}{2a^2}\left(\cos\frac{\pi x}{a}\sin\frac{\pi y}{b}\right)^2 = \frac{\pi^2 w_{11}^2}{8a^2}\left(1 + \cos\frac{2\pi x}{a} - \cos\frac{2\pi y}{b} + \cdots\right), \qquad (5.7b)$$

$$\frac{s_y - \nu s_x}{E} - \frac{\pi^2 w_{11}^2}{8b^2}\cos\frac{2\pi x}{a} + \frac{\nu\pi^2 w_{11}^2}{8a^2}\cos\frac{2\pi y}{b} - v_1 - \frac{2\pi v_{02}}{b}\cos\frac{2\pi y}{b}$$

$$= \frac{\pi^2 w_{11}^2}{2b^2}\left(\sin\frac{\pi x}{a}\cos\frac{\pi y}{b}\right)^2 = \frac{\pi^2 w_{11}^2}{8b^2}\left(1 - \cos\frac{2\pi x}{a} + \cos\frac{2\pi y}{b} + \cdots\right) \qquad (5.7c)$$

where the dots indicate terms involving trigonometric functions (such as $\sin(\pi x/a)\sin(3\pi y/b)$ in the first equation) which are beyond the range of the terms which we decided to retain, and which we will ignore. These relations can only be satisfied for all values of x and y if the sums of the coefficients of similar functions of x and y on each side of the equations are equal. This gives us, using $D = Eh^3/12(1-\nu^2)$ and omitting identities:

$$\frac{p_{11}}{E} = \frac{\pi^4 h^3 w_{11}}{12(1-\nu^2)a^4}\left(1+\frac{a^2}{b^2}\right)^2 + \frac{\pi^2 h w_{11}}{Ea^2}\left(s_x + \frac{a^2}{b^2}s_y\right) + \frac{\pi^4 h w_{11}^3}{16a^4}\left(1+\frac{a^4}{b^4}\right) \qquad (5.8a)$$

$$u_1 = \frac{s_x - \nu s_y}{E} - \frac{\pi^2 w_{11}^2}{8a^2}, \qquad u_{20} = \frac{\pi a w_{11}^2}{16}\left(\frac{\nu}{b^2} - \frac{1}{a^2}\right)$$

$$v_1 = \frac{s_y - \nu s_x}{E} - \frac{\pi^2 w_{11}^2}{8b^2}, \qquad v_{02} = \frac{\pi b w_{11}^2}{16}\left(\frac{\nu}{a^2} - \frac{1}{b^2}\right) \qquad (5.8b)$$

Laterally Loaded Plates, Edges Prevented From Pulling In

From Eqs. (5.5) the condition that $x = 0, a: u = 0$, and $y = 0, b: v = 0$, Fig. 4.13, is satisfied if $u_0, v_0, u_1, v_1 = 0$. Setting $u_1, v_1 = 0$ in Eqs. (5.8b) and solving for s_x, s_y we find:

$$s_x = \frac{\pi^2 E w_{11}^2}{8(1-\nu^2)}\left(\frac{1}{a^2}+\frac{\nu}{b^2}\right), \qquad s_y = \frac{\pi^2 E w_{11}^2}{8(1-\nu^2)}\left(\frac{1}{b^2}+\frac{\nu}{a^2}\right) \qquad (5.9)$$

Putting these values into Eq. (5.8a) and simplifying, we obtain:

$$\frac{12(1-\nu^2)a^4}{\pi^4 Eh^4}p_{11} = \left(1+\frac{a^2}{b^2}\right)^2 \frac{w_{11}}{h} + \frac{3}{4}\left[(3-\nu^2)\left(1+\frac{a^4}{b^4}\right) + \frac{4\nu a^2}{b^2}\right]\left(\frac{w_{11}}{h}\right)^3 \quad (5.10)$$

For the case when a/b is zero, corresponding to a very wide plate of length a, the similarity between Eq. (5.10) and Eq. (2.36) for a narrow beam under similar loading and end conditions, Fig. 2.11, is evident.

The above is an approximate solution for a plate under a sinusoidally distributed load $p = p_{11} \sin(\pi x/a) \sin(\pi y/b)$. But we can also consider it as a still more approximate solution for other load distributions by taking it as the first (and for simple distributions the most important) harmonic component of these other distributions. Thus, from Eq. (4.23), $p_{11} = (4/ab) \int_0^a dx \int_0^b dy\, p_0 \sin(\pi x/a) \sin(\pi y/b) = 16p_0/\pi^2$, where $p = p_0$ is a uniformly distributed load. Substituting this for p_{11} in Eq. (5.10) and plotting $(1-\nu^2)a^4 p_0/(Eh^4)$ against $w_{11}/h = w_{max}/h$ for the case of a square plate $a = b$, we obtain the upper dashed curve in Fig. 5.2. The upper full line shows the solution for this case (of a uniform load on a square hinged-edge plate) which was obtained by Levy (op. cit.) in the same way but using enough terms to give a solution exact to the accuracy of plotting. It will be seen that the relatively simple solution (5.10) is nearly exact up to $w_{max}/h = 0{\cdot}8$ and is not a bad approximation for larger deflections. If the $(w_{11}/h)^3$ term in Eq. (5.10) were increased by the empirical factor 1·23 then it would give about the correct values in the range shown.

The linear solution for this case which was obtained in Chap. 4, Eq. (4.26), is shown by the dotted line; it is obviously not a good approximation beyond $w_{max} = 0{\cdot}3h$ or so. The linear solution obtained by neglecting the $(w_{11}/h)^3$ term in Eq. (5.10) is the same as Eq. (4.26) with only one term, and could hardly be distinguished from it in plotting. Equation (5.10) is no doubt an even better approximation for the sinusoidal loading case for which it was designed, but there is no exact solution for that case with which to compare it.

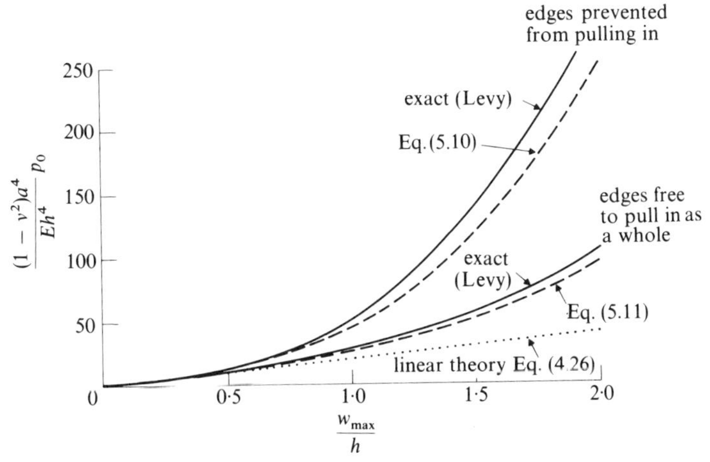

FIGURE 5.2

Laterally Loaded Plate, Edges Free to Pull In As a Whole

If we set s_x, $s_y = 0$ in Eq. (5.8a), we obtain:

$$\frac{12(1-\nu^2)a^4}{\pi^4 Eh^4} p_{11} = \left(1+\frac{a^2}{b^2}\right)^2 \frac{w_{11}}{h} + \frac{3(1-\nu^2)}{4}\left(1+\frac{a^4}{b^4}\right)\left(\frac{w_{11}}{h}\right)^3 \tag{5.11}$$

In this case the edges are not free to move in the plane of the plate at every point, because only the *average* edge membrane stresses s_x, s_y are zero. There is still a force per unit length on the edges $x = 0$, a of magnitude $h\sigma_{x(x=0,a)} = h(\partial^2\phi/\partial y^2) = -(4\pi^2 h/b^2)\phi_{02}\cos(2\pi y/b) = -\pi^2 E w_{11}/8a^2 \cos(2\pi y/b)$, with a similar force on the edges $y = 0$, b. But from the last of Eqs. (5.5) u does not vary in the y direction or v in the x direction, that is the edges remain straight. Hence the case described by Eq. (5.11) is that of a plate whose edges are hinged to rigid bars which are free to slide in the direction perpendicular to the edges.

Equation (5.11) is also plotted in Fig. 5.2 as the lower dashed line, using $p_{11} = 16p_0/\pi^2$ and $a = b$. Levy's results for the same case, obtained with many terms, are shown by the lower full line. Comparison of these results with the upper lines, which give the results for edges prevented from pulling in, shows the very great importance of the membrane boundary conditions when the displacements are large enough for the membrane strains and stresses to be important. The linearized case obtained by neglecting the $(w_{11}/h)^3$ term in Eq. (5.11) is the same as for Eq. (5.10), shown dotted in Fig. 5.2; this confirms the previous statements that the membrane boundary conditions are important only for large deflections.

Ultimate Compressive Buckling Load on Plate

Consider an initially flat hinged-edge thin plate which is subjected to a uniform shortening in the x direction. This condition is approximated if the edges of the plate are beveled and held vertically in a frame of bars with V-grooves, the horizontal top and bottom bars resting against the relatively rigid platens of a testing machine; the vertical bars at the sides are shortened enough so that they do not take any of the vertical load when the platens are brought closer together. This condition is also approximated in the panels between stiffening ribs of a large, thin plate which is stiffened by a grid of open section ribs in both the x and y directions, and subjected to a general compression in the x direction.

As the panel shortens in the x direction, which we take as the vertical direction, it is subjected to uniform compressive stress and strain in this direction which increases until the stress reaches the stability limit at which the panel buckles. If the shortening is continued after buckling has occurred, the stress and therefore the strain in the x direction remains nearly constant in the middle of the loaded side. But the vertical sides of the panel cannot buckle, being held straight by the grooved bars or ribs to which they are effectively hinged; thus the strain and hence the stress in the panel near these vertical sides continues to increase as the panel is shortened, and the total load taken by the panel can become much larger than the load at which the panel first buckles.

The ultimate load P_{ult} on the panel is reached when the stresses in the vertical edges of the panel reach some limiting value. This limiting value may be when the vertical membrane stress σ_{xm} in the panel edge reaches the buckling stress σ_r of the rib to which the edge is attached (assuming that the panel material has the same modulus of elasticity as that of the rib—otherwise, since the strains must be the same, the σ_{xm} in the panel edge would equal σ_r time the ratio between the moduli of the panel and of the rib).

The buckling stress of the rib σ_r must be determined as was discussed in Sec. 4.6. If the ribs are attached on one side of the sheet, then the effective widths of the sheet which must be considered to be acting with the rib in calculating the bending stiffness of the rib may be somewhat less than those prescribed in Sec. 4.6, due to the fact that the sheet has already buckled. While this is a complex subject which can only be touched upon here, it is probably safe to take the effective width of sheet acting with each vertical rib as the width b of each panel times the ratio of the critical stress at which the panel first buckles to the stress in the ribs when the ultimate load is reached, if this is less than the effective width which was used in calculating Eq. (4.76).

Alternatively, the limiting value of the stresses in the vertical edge of the panel may be reached when they result in yielding of the panel material at some point of this edge, since the buckling resistance of the panel material is greatly reduced after yielding occurs (because of the drastic reduction in the slope of the stress-strain curve, which can be regarded as the effective modulus of elasticity of the material for any increase in straining). Local buckling then inevitably occurs, followed by general collapse, with usually little or no increase in total load. Trial shows that yielding will first occur at the corner of the panel under the combined action of the membrane and flexural stresses. The actual condition which limits the ultimate load P_{ult} taken by the panel (buckling of the rib or yielding of the panel material) will of course be the one which results in the lowest value of P_{ult}.

Equation (5.8a) can be used to investigate this case. We assume that the vertical edges are free to move laterally in the plane of the panel, so that we can assume $s_y = 0$, and we take the transverse loading $p_{11} = 0$. We can then divide Eq. (5.8a) through by $\pi^2 h w_{11}/a^2$ (since w_{11} is not zero) and obtain:

$$\frac{\pi^2}{12(1-\nu^2)}\frac{h^2}{a^2}\left(1+\frac{a^2}{b^2}\right)^2 + \frac{s_x}{E} + \frac{\pi^2 w_{11}^2}{16a^2}\left(1+\frac{a^4}{b^4}\right) = 0 \tag{5.12}$$

Ultimate Buckling Load Limited by Rib Buckling

For this case, using Eqs. (4.16), (5.5), and (5.6) we have:

$$\sigma_{xm(y=0,b)} = \left(\frac{\partial^2\phi}{\partial y^2}\right)_{(y=0,b)} = s_x - \frac{\pi^2 E w_{11}^2}{8a^2} = -\sigma_r \tag{5.13}$$

where σ_r is the stress at which the rib buckles. Eliminating w_{11}^2 between Eqs.

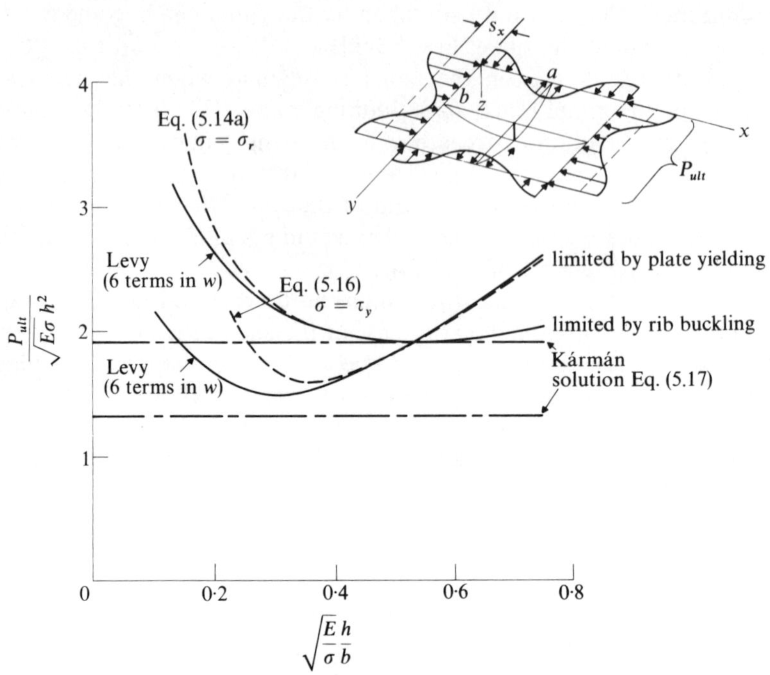

FIGURE 5.3
Square plate under edge load P_{ult}

(5.12) and (5.13), we can write the resulting relation in the form:

$$-\frac{s_x bh}{\sqrt{E\sigma_r h^2}} = \frac{P_{ult}}{\sqrt{E\sigma_r h^2}} = \frac{\pi^2}{6(1-\nu^2)} \frac{b^2 (a^2+b^2)^2}{a^2 \; a^4+3b^4} \sqrt{\frac{E}{\sigma_r}\frac{h}{b}} + \frac{a^4+b^4}{a^4+3b^4}\sqrt{\frac{\sigma_r}{E}\frac{b}{h}} \tag{5.14}$$

where $P_{ult} = -s_x bh$ is the ultimate compressive load taken by the panel, since s_x is the average value of σ_x across the width of the panel. For a square plate $a = b$ and $\nu = 0{\cdot}3$:

$$\frac{P_{ult}}{\sqrt{E\sigma_r h^2}} = 1{\cdot}81\sqrt{\frac{E}{\sigma_r}\frac{h}{b}} + 0{\cdot}5\sqrt{\frac{\sigma_r}{E}\frac{b}{h}} \tag{5.14a}$$

This relation between the dimensionless quantities $P_{ult}/(\sqrt{E\sigma_r h^2})$ and $\sqrt{E/\sigma_r}h/b$ given by Eq. (5.14a), which is derived by using one term for w, is shown in Fig. 5.3 by the upper dashed line, while the upper full line shows Levy's relatively exact solution obtained in the same way using six terms in w.

Ultimate Buckling Load Limited by Yielding

We again take σ_y, $p_{11} = 0$, Eq. (5.12). The yielding condition is given by Eq. (1.3), Sec. 1.6: $\sigma_x^2 - \sigma_x\sigma_y + \sigma_y^2 + 3\sigma_{xy}^2 = \tau_y^2$. Using Eqs. (4.16), (5.5) and (5.6) for the stresses and x, $y = 0$, $z = h/2$, we obtain:

$$\left(s_x - \frac{\pi^2 E w_{11}^2}{8a^2}\right)^2 + \left(s_x - \frac{\pi^2 E w_{11}^2}{8a^2}\right)\left(\frac{\pi^2 E w_{11}^2}{8b^2}\right) + \left(\frac{\pi^2 E w_{11}^2}{8b^2}\right)^2 + 3\left(\frac{\pi^2 E h w_{11}}{2(1+\nu)ab}\right)^2 = \tau_y^2 \tag{5.15}$$

We can then again eliminate w_{11} between Eqs. (5.12) and (5.15) without difficulty, but the resulting general expression is rather long, so we will consider only what it reduces to for a square plate $a = b$ and $\nu = 0{\cdot}3$. This can be put in the form:

$$A^2 + 8{\cdot}08AB - \left(38B^2 + \frac{1}{3B^2}\right) = 0, \text{ from which:}$$

$$A = -4{\cdot}04B + \{54{\cdot}3B^2 - 1/(3B^2)\}^{1/2}, \text{ where:} \tag{5.16}$$

$$A = -\frac{s_x hb}{\sqrt{E\tau_y}h^2} = \frac{P_{ult}}{\sqrt{E\tau_y}h^2}, \qquad B = \sqrt{\frac{E}{\tau_y}}\frac{h}{b}$$

If values of B are assumed A can easily be solved for. The plot of these quantities obtained from Eq. (5.16) is shown by the lower dashed line in Fig. 5.3, while Levy's solution obtained in similar manner by using six terms in w and a somewhat different yielding condition is shown by the lower full line.

Von Kármán's 'Effective Width' Solution

The problem of the ultimate buckling load of a plate is certainly a large-deflection plate problem, as it has been treated above. However, when experiments* on the compression of thin plates in V-grooves first showed that the ultimate strength of plates of a given material is nearly proportional to the square of the thickness and is relatively independent of the other dimensions, Kármán** derived a formula for the strength in quite another and much simpler manner, which checked extraordinarily well with the test results.

Kármán's reasoning was that for plates which are very thin compared to the other dimensions, the load carried by the plate at the stability limit is negligible compared to the load carried by two narrow 'effective widths' λ of the plate next to each side, which are artificially prevented from buckling until the compressive stress on them reaches a limiting values σ, which can be either the buckling stress $\sigma = \sigma_r$ of the ribs to which the sides were attached or the yield point $\sigma = \tau_y$ of the plate material, whichever was smaller.

The ultimate load taken by the plate P_{ult} is taken as the limiting stress σ times the area $2\lambda h$ on which it is assumed to act, or $P_{ult} = 2\lambda h\sigma$. The problem of determining the effective width λ is the reverse of the usual plate stability problem, in which we have a plate of a known width b and we find the compressive stress s_x under which it can be in equilibrium in a deflected state. Here we want to find the unknown width of a plate which will be in equilibrium in

* NACA Report No. 356, 1930.

** T. von Kármán, 'The strength of thin plates in compression' with an appendix written by L. H. Donnell, *Trans. ASME*, vol. 54, 1932, p. 53.

a deflected state under a known compressive stress σ. The relation between the width and the stress is the same in both cases, the only difference being in what is known and what is unknown.

If we make the rough assumption that the two effective widths on each side of the plate form together one plate of width 2λ under the stress σ, we can use Eq. (4.37), assuming that the length a is greater than the small width 2λ. We thus find $S_x = b^2hs_x/(\pi^2D) = 4$. Taking $b = 2\lambda$, and $s_x = \sigma$, with $D = Eh^3/12(1-\nu^2) \approx Eh^3/10{\cdot}9$, and solving for λ, we find $\lambda = 0{\cdot}952\sqrt{E/\sigma}h$. On the other hand, if we assume that the effective width on each side is in the condition of a plate hinged on three sides and free on the third side where it is attached to the middle part of the plate, then we can use Eq. (4.61a), taking $k = 0{\cdot}5$ since the length a of the plate can be assumed to be large compared to the small width. Taking $b = \lambda$, $s_x = \sigma$ we find: $s_x = 0{\cdot}5\pi^2D/(b^2h)$ or $\sigma = 0{\cdot}5\pi^2D/(\lambda^2h)$. Solving for λ we obtain $\lambda = 0{\cdot}676\sqrt{E/\sigma}h$. Using these values of λ to compute the ultimate load taken by the plate we find:

$$\frac{P_{ult}}{\sqrt{E\sigma}h^2} = 1{\cdot}91 \text{ (inner edges of effective widths continuous)}$$
$$\frac{P_{ult}}{\sqrt{E\sigma}h^2} = 1{\cdot}35 \text{ (inner edges of effective widths free)} \tag{5.17}$$

The values of $P_{ult}/(\sqrt{E\sigma}h^2)$ given by Eq. (5.17) are plotted in dash-dot lines in Fig. 5.3. In practice the inner edges of the strips of width λ are certainly not free but they are less highly constrained than they would be if the two strips were actually joined together along their inner edges. Thus the strips should be somewhere between these two conditions but probably closer to the free edge condition, say with $P_{ult}/(\sqrt{E\sigma}h^2) \approx 1{\cdot}5$. This is about the value shown by the tests on thin sheets in grooves. Tests of panels of a stiffened sheet seem to follow better the curves obtained with large-deflection plate theory.

However these checks may be to some extent fortuitous since actual plates have defects which have not been considered, and the edge conditions in the tests were certainly not exactly what was considered in even the 'exact' theories. There is always some pulling out of V-grooves at the middles of the vertical sides, and it is hard to tell exactly just what part of the load is taken by the panels and by the ribs in the stiffened sheet tests, etc. The above discussion of this important subject is obviously incomplete, but it is too specialized a field to devote more space to in a book such as this.

Ultimate Shear Buckling Load on Plate

The study of the buckling of a plate under shearing forces applied to the edges is not easy even when deflections are assumed to be infinitesimal, and naturally a rigorous study of this case becomes even more difficult when the deflections become large. However, as was mentioned in Sec. 4.6, some simple and

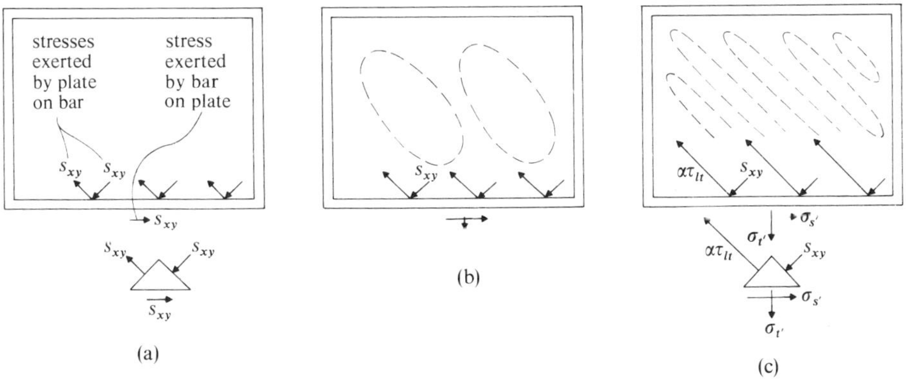

FIGURE 5.4

approximate but reasonable conclusions can be drawn about the ultimate conditions reached when thin panels are subjected to large shearing deformations.*

Figure 5.4 shows a plate attached to a frame of four bars, through which uniform shear forces are applied to the plate, together with the reactions of the plate on the bars (for simplicity only the reactions on the lower bar are shown). We will assume that the bars are hinged together or are too slender to take an appreciable part of the shearing load as a rigid frame. As the shear loading is increased, buckling of the panel will occur when the shear load per unit area reaches the stability limit s_{xy} given by Fig. 4.25(b). At this instant there will be membrane shear stresses $\sigma_{xym} = \sigma_s = s_{xy}$ uniformly distributed over the plate, and, as was shown in Fig. 3.3 and the development of Eq. (3.5a), there will be principal compressive and tensile stresses at 45° to the direction of the edges, of magnitudes σ_c, $\sigma_t = \sigma_s = s_{xy}$. These are shown in Fig. 5.4(a) as they act upon the bottom bar.

After the plate buckles conditions of course become complex, but in general the plate becomes more and more buckled in the compression direction while remaining relatively straight in the tension direction. For very thin plates it is probably not far from the truth to assume that the compressive stress in the plate remains substantially constant and equal to $\sigma_c = s_{xy}$ and acting at 45° to the horizontal, while the maximum tensile stresses also at 45° to the horizontal can increase as indicated at (b) until, as shown at (c) they reach some limiting value τ_{lt} which is determined by the strength of the plate material or its attachments. This limiting tensile stress can be taken as the yield point τ_y of the material if it is desired to design so that yielding will not occur, or, if the ultimate strength of the plate is desired, as the stress at which the plate would tear (which may be considerably different from the ultimate strength τ_u of the material as determined from a standard test specimen, because of the radically different shape), or the stress at which the plate can break loose from its attachment to the bar under the normal and shear loadings per unit length of bar calculated below. While the fibers under tension will be relatively straight compared to the compression fibers, they

* See paper by H. Wagner, *Z. Flugtech. Motorluftsch.*, vol. 20, 1929, p. 200.

will have some curvature which varies from place to place, and this will result in a varying tensile stress having an average value of $\alpha\tau_{lt}$, where α is a numerical factor which is probably close to unity for a very thin sheet.

Taking the area of the horizontal hypotenuse of an average triangular element of the plate such as shown in Fig. 5.4(c) as unity, its 45° sides will have areas of $1/\sqrt{2}$. If the stresses on these sides are $\alpha\tau_{lt}$ and s_{xy} as shown, the total forces on them will be $\alpha\tau_{lt}/\sqrt{2}$ and $s_{xy}/\sqrt{2}$. From horizontal and vertical equilibrium of the element, the shear stress σ'_s and tensile stress σ'_t on the horizontal side will be $\sigma'_s = (\alpha\tau_{lt}/\sqrt{2} + s_{xy}/\sqrt{2})/\sqrt{2} = (\alpha\tau_{lt} + s_{xy})/2$ and $\sigma'_t = (\alpha\tau_{lt}/\sqrt{2} - s_{xy}/\sqrt{2})/\sqrt{2} = (\alpha\tau_{lt} - s_{xy})/2$. The edge of the sheet can therefore exert on the bar a shear force per unit length of $(\alpha\tau_{lt} + s_{xy})h/2$, while there will also be a tensile force per unit length normal to the bar of $(\alpha\tau_{lt} - s_{xy})h/2$. If there is a similar panel on the other side of the bar, the normal forces due to the two panels will largely cancel each other; if not then the bar must be designed to withstand the bending due to this normal force. For a very thin panel s_{xy} may be negligible compared to τ_{lt}.

5.2 THICK PLATES—SERIES SOLUTIONS IN LOADING FUNCTIONS

A number of exact closed solutions of three-dimensional elasticity theory are known which represent practical plate problems except for details of the edge conditions; these, by Saint-Venant's principle, are usually important only near the edges, where approximate corrections can be applied as discussed later. As in the case of beams, most if not all of these solutions, as well as several general exact closed solutions involving no loads on the plate faces which are useful in satisfying the edge conditions as well as for other important purposes, are contained in series solutions in terms of the top and bottom loading, which are similar to Eqs. (3.28), (3.29) for beams. These series solutions converge to exact solutions for 'smooth' loading of any type, and provide at least the most important corrections to the results given by classical plate theory for the most general loading conditions. It is logical therefore to start our study of thick plates with such series solutions.

Series Solution in the Loading Functions for Plates, Normal Loading*

As before for beams, the first terms of the series represent the classical theory, the second terms represent the most important corrections to the classical theory and involve higher derivatives of the loading, and so on. All the terms are calculated to approach an exact three-dimensional elasticity solution in the limit.

* L. H. Donnell, 'A theory for thick plates', *Proc. 2nd US Natl. Congr. of Appl. Mech., ASME*, 1955, p. 369.

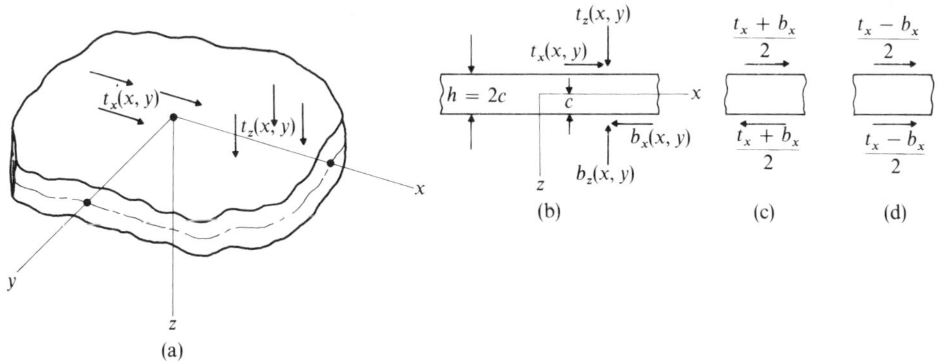

FIGURE 5.5

For simplicity we will consider the case of zero body forces. Unlike surface forces, body forces are usually uniform or vary in a simple way so that closed solutions like Eqs. (3.8o) can be used to allow for them, superposing them if necessary upon solutions for the surface forces.

Let the loading functions $t_z(x, y)$, $b_z(x, y)$ represent normal pressures, that is distributed forces per unit area in the z direction, on the top and bottom surfaces respectively of thc plate, as shown in Fig. 5.5. We will use the nondimensional coordinates $\mathbf{x} = x/c$, $\mathbf{y} = y/c$, $\mathbf{z} = z/c$, where c is the half thickness. We next define the functions $\mathbf{t}_z(x, y)$, $\mathbf{b}_z(x, y)$ by the relations:

$$\boldsymbol{\nabla}^4\mathbf{t}_z = t_z, \qquad \boldsymbol{\nabla}^4\mathbf{b}_z = b_z \tag{5.18a}$$

where operators $\boldsymbol{\nabla}^2$, $\boldsymbol{\nabla}^4$, etc., are the same as ∇^2, ∇^4, etc., with $\mathbf{x}, \mathbf{y}, \mathbf{z}$ substituted for x, y, z, that is $\boldsymbol{\nabla}^2 = \partial^2/\partial\mathbf{x}^2 + \partial^2/\partial\mathbf{y}^2$, etc. It should of course be kept in mind that operators such as ∇^2, $\boldsymbol{\Delta}^2$ when applied to the *displacements* as in Eqs. (3.8) include the second derivatives with respect to z or $\mathbf{z}$, because unlike the loading functions, the displacements are functions of z as well as of x and y. Comparison of Eqs. (5.18a) with Eq. (4.19) shows that the functions $\mathbf{t}_z$ or $-\mathbf{b}_z$ have the physical meaning of first approximations to the deflection w multiplied by the constant $2E/3(1-\nu^2)c$. This is also shown by the first term of the expression for u_z in Eqs. (5.19) calculated below.

For the first term in the series for u_z, in accord with the classical theory, we ignore the variation of u_z with z and take u_z the same as the deflection w of the middle surface. Using Eq. (5.18a) and the definition $D = Eh^3/12(1-\nu^2) = 2Ec^3/3(1-\nu^2)$ we can write Eq. (4.19) as $D\nabla^4 u_z = p$, or $[2Ec^3/3(1-\nu^2)]\nabla^4 u_z = t_z - b_z$, or $[2E/3(1-\nu^2)c]\boldsymbol{\nabla}^4 u_z = \boldsymbol{\nabla}^4(\mathbf{t}_z - \mathbf{b}_z)$. This will be satisfied if we take $u_z = [3(1-\nu^2)c/2E](\mathbf{t}_z - \mathbf{b}_z)$. The functions of integration involved in this integration process are of course important for satisfying the edge conditions and will be considered later. We take this as the first term of our series for u_z. As to the first terms in the series for u_x and u_y, since small-deflection classical theory takes such displacements as zero at the middle surface and ignores transverse shear strains, we can, as in Fig. 4.3 or Eq. (4.1), take:

$$u_x = -z\partial u_z/\partial x = [-3(1-\nu^2)c/2E]\mathbf{z}\partial(\mathbf{t}_z - \mathbf{b}_z)/\partial \mathbf{x},$$
$$u_y = -z\partial u_z/\partial y = [-3(1-\nu^2)c/2E]\mathbf{z}\partial(\mathbf{t}_z - \mathbf{b}_z)/\partial \mathbf{y}$$

The series as a whole must satisfy the three basic elasticity conditions of Eqs. (3.8) and the following boundary conditions on the upper and lower surfaces of the plate:

$$z = -c, c \text{ (or } \mathbf{z} = -1, 1\text{)}: \quad \sigma_{xz}, \sigma_{yz} = 0, \sigma_z = -t_z, -b_z \qquad (5.18b)$$

using Eqs. (3.7b) for the expressions for the stresses. The first terms of the series given above satisfy these conditions only if we neglect all terms in the conditions which contain $\boldsymbol{\nabla}^2\mathbf{t}_z$, $\boldsymbol{\nabla}^2\mathbf{b}_z$ and higher derivatives of these. By trial we assume expressions for the second terms of the series with unknown coefficients, and we find that we can solve for these coefficients so as to satisfy all the above conditions provided that we neglect terms which contain $\boldsymbol{\nabla}^4\mathbf{t}_z = t_z$, $\boldsymbol{\nabla}^4\mathbf{b}_z = b_z$ and higher derivatives. By adding third terms to the series we can satisfy all the conditions provided we neglect terms containing $\boldsymbol{\nabla}^6\mathbf{t}_z = \boldsymbol{\nabla}^2 t_z$, $\boldsymbol{\nabla}^6\mathbf{b}_z = \boldsymbol{\nabla}^2 b_z$ and higher derivatives, and so on. With each addition to the series the functions of the loading neglected increase in order by the operator $\boldsymbol{\nabla}^2$, that is all conditions are satisfied except for finer and finer detail in the variation of the loading, and hence in the displacements and stresses produced.

Trial shows that there is a simple pattern for the successive terms of our series, as will be seen from Eqs. (5.19). Each term is a product of a function of $t_z - b_z$ or $t_z + b_z$ by a polynomial in z, and while the functions increase in order by the operator ∇^2 the highest power in the polynomial in z increases by two. In calculating the unknown coefficients of the polynomial terms at each step, it is not necessary to put all the preceding terms into the conditions—all that is needed is to use the terms neglected in the preceeding step plus the new terms. However, as soon as the boundary condition $\sigma_z = -t_z, -b_z$ in Eqs. (5.18b) has been satisfied, that is as soon as σ_z has the proper value on the upper and lower surfaces, then any contributions to σ_z from additional terms must be zero, that is this boundary condition becomes $\mathbf{z} = -1, 1$: $\sigma_z = 0$.

Using this procedure we obtain the results shown in Eqs. (5.19). The expressions for the stresses are of course derived from those for the displacements by using Eqs. (3.7b), and the expressions for the resultant forces and moments per unit length of section F_x, M_x, etc., by integrating over the thickness the stresses or their moments about the middle surface. In deriving these expressions for forces and moments, the following formulas are convenient, in which $f = f(x, y)$ is a function of x and y only:

$$F_\alpha = \int_{-c}^{c} dz\,\sigma_\alpha = c\int_{-1}^{1} d\mathbf{z}\,\sigma_\alpha = c\int_{-1}^{1} d\mathbf{z}\sum \mathbf{z}^m f = 0 \text{ (for } m \text{ odd)}$$
$$= \sum \frac{2c}{m+1} f \text{ (for } m \text{ even)}$$

$$M_\alpha = \int_{-c}^{c} dz\,\sigma_\alpha z = c^2 \int_{-1}^{1} d\mathbf{z}\,\sigma_\alpha \mathbf{z} = c^2 \int_{-1}^{1} d\mathbf{z} \sum (\mathbf{z}^m f)\mathbf{z} = 0 \text{ (for } m \text{ even)}$$

$$= \sum \frac{2c^2}{m+2} f \text{ (for } m \text{ odd)}$$

Additional terms of the series (including a constant part of the $\nabla^2(t_z - b_z)$ term of u_z, which is not needed with the terms given) can be found by continuing the process outlined. Expressions for general terms for the series for stresses have been developed by C. W. Lee and are published in the paper cited below in the derivation of Eqs. (5.32).

$$u_x = \frac{(1+\nu)c}{E}\frac{\partial}{\partial \mathbf{x}}\left\{-\frac{3(1-\nu)\mathbf{z}}{2}(\mathbf{t}_z - \mathbf{b}_z) + \frac{5(2-\nu)\mathbf{z}^3 - 3(2+3\nu)\mathbf{z}}{20}\nabla^2(\mathbf{t}_z - \mathbf{b}_z) + \frac{\nu}{2}\nabla^2(\mathbf{t}_z + \mathbf{b}_z) + \left[\frac{-35(3-\nu)\mathbf{z}^5 - 70(1-3\nu)\mathbf{z}^3 + (87-157\nu)\mathbf{z}}{2800}(t_z - b_z) + \frac{(1-\nu)(3\mathbf{z}^2-1)}{12}(t_z + b_z)\ldots\right]\right\}$$

$$u_z = \frac{(1+\nu)c}{E}\left\{\frac{3(1-\nu)}{2}(\mathbf{t}_z - \mathbf{b}_z) + \frac{15\nu\mathbf{z}^2 - 3(8-3\nu)}{20}\nabla^2(\mathbf{t}_z - \mathbf{b}_z) + \frac{-175(1+\nu)\mathbf{z}^4 + 210(5-3\nu)\mathbf{z}^2 - 227 + 157\nu}{2800}(t_z - b_z) - \frac{(1-\nu)\mathbf{z}}{2}(t_z + b_z) + \left[\frac{35(2+\nu)\mathbf{z}^6 - 105(2-3\nu)\mathbf{z}^4 + 3(70-157\nu)\mathbf{z}^2 \cdots + 3(70+157\nu)\mathbf{z}^2 \cdots}{16\,800}\nabla^2(t_z - b_z) - \frac{\nu(\mathbf{z}^3 - \mathbf{z})}{12}\nabla^2(t_z + b_z)\ldots\right]\right\}$$

$$e = \frac{1}{c}\left(\frac{\partial u_x}{\partial \mathbf{x}} + \frac{\partial u_y}{\partial \mathbf{y}} + \frac{\partial u_z}{\partial \mathbf{z}}\right) = \frac{(1-2\nu)(1+\nu)}{E}\left\{-\frac{3\mathbf{z}}{2}\nabla^2(\mathbf{t}_z - \mathbf{b}_z) + \frac{5\mathbf{z}^3 + 9\mathbf{z}}{20}(t_z - b_z) - \frac{1}{2}(t_z + b_z) + \left[\frac{-35\mathbf{z}^5 - 210\mathbf{z}^3 + 157\mathbf{z}}{2800}\nabla^2(t_z - b_z) + \frac{3\mathbf{z}^2 - 1}{12}\nabla^2(t_z + b_z)\ldots\right]\right\}$$

$$\sigma_x = -\frac{3\mathbf{z}}{2}\left(\frac{\partial^2}{\partial \mathbf{x}^2} + \nu\frac{\partial^2}{\partial \mathbf{y}^2}\right)(\mathbf{t}_z - \mathbf{b}_z) + \frac{5\mathbf{z}^3 - 3\mathbf{z}}{10}\frac{\partial^2}{\partial \mathbf{x}^2}\nabla^2(\mathbf{t}_z - \mathbf{b}_z) + \frac{\nu(5\mathbf{z}^3 + 9\mathbf{z})}{20}\frac{\partial^2}{\partial \mathbf{y}^2}\nabla^2(\mathbf{t}_z - \mathbf{b}_z) - \frac{\nu}{2}\frac{\partial^2}{\partial \mathbf{y}^2}\nabla^2(\mathbf{t}_z + \mathbf{b}_z) + \left[\frac{-105\mathbf{z}^5 - 70\mathbf{z}^3 + 87\mathbf{z}}{2800}\frac{\partial^2}{\partial \mathbf{x}^2}(t_z - b_z) - \frac{\nu(35\mathbf{z}^5 + 210\mathbf{z}^3 - 157\mathbf{z})}{2800}\frac{\partial^2}{\partial \mathbf{y}^2}(t_z - b_z) + \frac{3\mathbf{z}^2 - 1}{12}\left(\frac{\partial^2}{\partial \mathbf{x}^2} + \nu\frac{\partial^2}{\partial \mathbf{y}^2}\right)(t_z + b_z)\ldots\right] \tag{5.19}$$

$$\sigma_z = \frac{-\mathbf{z}^3 + 3\mathbf{z}}{4}(t_z - b_z) - \frac{1}{2}(t_z + b_z) + \left[\frac{\mathbf{z}^5 - 2\mathbf{z}^3 + \mathbf{z}}{40}\nabla^2(t_z - b_z)\ldots\right]$$

$$\sigma_{xy} = \frac{\partial^2}{\partial \mathbf{x}\partial \mathbf{y}}\left\{-\frac{3(1-\nu)\mathbf{z}}{2}(\mathbf{t}_z - \mathbf{b}_z) + \frac{5(2-\nu)\mathbf{z}^3 - 3(2+3\nu)\mathbf{z}}{20}\boldsymbol{\nabla}^2(\mathbf{t}_z - \mathbf{b}_z) + \frac{\nu}{2}\boldsymbol{\nabla}^2(\mathbf{t}_z + \mathbf{b}_z)\right.$$
$$+\left[\frac{-35(3-\nu)\mathbf{z}^5 - 70(1-3\nu)\mathbf{z}^3 + (87-157\nu)\mathbf{z}}{2800}(t_z - b_z)\right.$$
$$\left.\left. + \frac{(1-\nu)(3\mathbf{z}^2-1)}{12}(t_z + b_z)\ldots\right]\right\} \qquad (5.19\text{ cont.})$$

$$\sigma_{xz} = \frac{\partial}{\partial \mathbf{x}}\left\{\frac{3(\mathbf{z}^2-1)}{4}\boldsymbol{\nabla}^2(\mathbf{t}_z - \mathbf{b}_z) + \left[\frac{-5\mathbf{z}^4 + 6\mathbf{z}^2 - 1}{40}(t_z - b_z)\right.\right.$$
$$\left.\left. + \frac{35\mathbf{z}^6 + 35\mathbf{z}^4 - 87\mathbf{z}^2 + 17}{5600}\boldsymbol{\nabla}^2(t_z - b_z) - \frac{\mathbf{z}^3 - \mathbf{z}}{12}\boldsymbol{\nabla}^2(t_z + b_z)\ldots\right]\right\}$$

$$F_x = -\nu c\frac{\partial^2}{\partial \mathbf{y}^2}\boldsymbol{\nabla}^2(\mathbf{t}_z + \mathbf{b}_z)\ldots, F_{xy} = \nu c\frac{\partial^2}{\partial \mathbf{x}\partial \mathbf{y}}\boldsymbol{\nabla}^2(\mathbf{t}_z + \mathbf{b}_z)\ldots, \; F_{xz} = -c\frac{\partial}{\partial \mathbf{x}}\boldsymbol{\nabla}^2(\mathbf{t}_z - \mathbf{b}_z)\ldots$$

$$M_x = c^2\left\{-\left(\frac{\partial^2}{\partial \mathbf{x}^2} + \nu\frac{\partial^2}{\partial \mathbf{y}^2}\right)(\mathbf{t}_z - \mathbf{b}_z) + \frac{2\nu}{5}\frac{\partial^2}{\partial \mathbf{y}^2}\boldsymbol{\nabla}^2(\mathbf{t}_z - \mathbf{b}_z) + \left[\frac{2\nu}{525}\frac{\partial^2}{\partial \mathbf{y}^2}(t_z - b_z)\ldots\right]\right\}$$

$$M_{xy} = c^2\frac{\partial^2}{\partial \mathbf{x}\partial \mathbf{y}}\left\{-(1-\nu)(\mathbf{t}_z - \mathbf{b}_z) - \frac{2\nu}{5}\boldsymbol{\nabla}^2(\mathbf{t}_z - \mathbf{b}_z) + \left[\frac{-2\nu}{525}(t_z - b_z)\ldots\right]\right\}$$

The expressions for u_y, σ_y, σ_{yz}, F_y, M_y are the same as those given for u_x, σ_x, σ_{xz}, F_x, M_x with x and y interchanged.

Plate With Harmonically Distributed Normal Loading

As a simple application of these results, we apply them to the case of a harmonically distributed load on the top surface of the plate shown in Fig. 5.6(a):*

$$t_z = \sum_m \sum_n p_{mn} \cos\frac{m\pi\mathbf{x}}{\mathbf{a}}\cos\frac{n\pi\mathbf{y}}{\mathbf{b}} \quad (m, n = 1, 3, 5, \ldots), \qquad b_z = 0 \qquad (5.20)$$

where $\mathbf{a} = a/c$, $\mathbf{b} = b/c$. Equations (5.18a) are satisfied if:

$$\mathbf{t}_z = \frac{\mathbf{a}^4}{\pi^4}\sum_m\sum_n \frac{p_{mn}}{(m^2 + n^2\mathbf{a}^2/\mathbf{b}^2)^2}\cos\frac{m\pi\mathbf{x}}{\mathbf{a}}\cos\frac{n\pi\mathbf{y}}{\mathbf{b}}, \qquad \mathbf{b}_z = 0 \qquad (5.21)$$

Substituting these values into Eqs. (5.19) we see that since we have only cosine functions of $m\pi\mathbf{x}/\mathbf{a}$ in the expression for u_y and of $n\pi\mathbf{y}/\mathbf{b}$ in the expression for u_x and of both in the expressions for σ_x, σ_y and u_z, the following boundary conditions are satisfied:

$$\mathbf{x} = \pm\frac{\mathbf{a}}{2}: \qquad \sigma_x,\, u_y,\, u_z = 0$$
$$\mathbf{y} = \pm\frac{\mathbf{b}}{2}: \qquad \sigma_y,\, u_x,\, u_z = 0 \qquad (5.22)$$

* C. W. Lee, 'A three-dimensional solution for simply supported thick rectangular plates', *Nucl. Engng and Design*, no. 6, 1967.

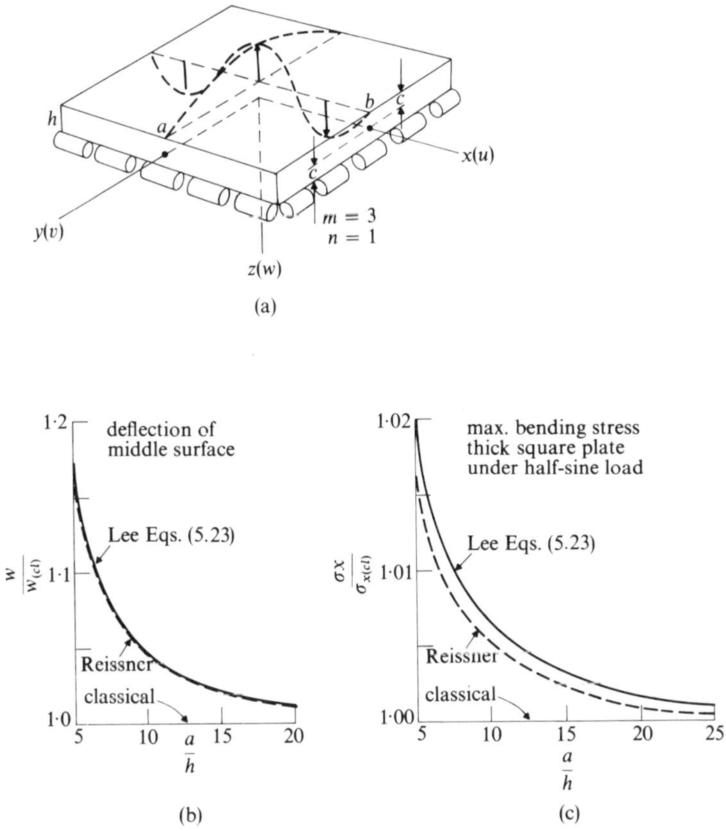

FIGURE 5.6

The absence of normal stresses on a boundary means of course that there are neither bending moments nor normal forces on these boundaries. Hence the boundary conditions are something like those suggested by Fig. 5.6(a) of a plate which rests on short rollers along the edges, friction being sufficient to prevent axial slippage.

However, the shear stresses σ_{xz}, σ_{yz}, σ_{xy} are not zero on the edges. The reactions are provided by parabolically distributed transverse shear stresses represented by the first terms in the expressions for σ_{xz}, σ_{yz}; these could be approximately transformed into reactions distributed along the bottoms of the edges by superposition of the plane strain version of the local stress field of Fig. 3.17, as discussed previously in Sec. 3.4. The remaining terms in the expressions for σ_{xz}, σ_{yz} are self balanced, and could similarly be eliminated to good approximation by superposition of the plane strain versions of the local stress fields of Eqs. (3.39), (3.40), and Fig. 3.16. A method for the elimination of the σ_{xy} stresses will be discussed later in Secs. 5.4 and 5.5.

The peak stresses and deflections are located at the center of the plate far from the edges and should be little affected by such edge conditions. Using Eqs.

(5.20), (5.21) in Eqs. (5.19) we obtain expressions for these in the form of infinite series such as:

$$\sigma_{x(x,y=0,z=1)} = \frac{3a^2}{2\pi^2} \sum_m \sum_n \frac{p_{mn}\left(m^2 + \nu \frac{n^2 a^2}{b^2}\right)}{\left(m^2 + \frac{n^2 a^2}{b^2}\right)^2}$$

$$\times \left[1 + \frac{2\pi^2}{15a^2}\left(m^2 + \frac{n^2 a^2}{b^2}\right) - \frac{142\pi^4}{1575a^4}\left(m^2 + \frac{n^2 a^2}{b^2}\right)^2 \cdots\right]$$

For the case of a square plate, $a = b$, under a simple half cosine load, $m = n = 1$, we obtain:

$$\sigma_{x(x,y=0,z=1)} = \frac{3(1+\nu)}{2\pi^2} p_{11} \frac{a^2}{h^2}\left(1 + \frac{\pi^2}{15}\frac{h^2}{a^2} - \frac{71\pi^4}{3150}\frac{h^4}{a^4} \cdots\right)$$

$$u_{z(x,y,z=0)} = w = \frac{6(1-\nu^2)}{\pi^4}\frac{p_{11}}{E}\frac{a^4 c}{h^4}\left(1 + \frac{\pi^2(8-3\nu)}{20(1-\nu)}\frac{h^2}{a^2} - \frac{\pi^4(227-157\nu)}{16\,800(1-\nu)}\frac{h^4}{a^4} \cdots\right) \tag{5.23}$$

where the first terms in the parentheses represent the values given by the classical plate theory.

Figures 5.6(b), (c) show the values of the parentheses (which represent the ratios of σ_x and w to their values given by the classical plate theory) as obtained by Lee from Eqs. (5.23). Also shown by dashed lines are results obtained by using the thick plate theory developed by Eric Reissner* which satisfies nearly the same boundary conditions. It will be seen that the two thick plate theories, which are derived in quite different ways, are in good agreement and give about 20 percent greater deflections and 2 percent greater stress than the classical theory, for a plate five times as wide as it is thick.

Convergence of Series in Loading Functions

The series of Eqs. 5.23 converge rapidly in the range covered in Figs. 5.6(b), (c), and solutions for harmonic loading with half wave lengths a/m, $b/n \geq h$ are easily obtained and can be combined to give other distributions of loading. However, it can be seen that, as in the analogous case for beams, the parts of the series involving derivatives of t_z or b_z, which are enclosed in square brackets in Eqs. (5.19), converge poorly or start to diverge when the wave lengths drop to the order of and less than the thickness. By analogy with the findings for beams, Eqs. (3.28), (3.29), it can be inferred that series (5.19) for normal loading as well as the other series in loading functions given later will give accurate results and will converge rapidly for loadings which do not vary rapidly, and they can safely be used in entirety in numerical applications if the terms are actually found to converge.

On the other hand at discontinuities in loading the derivatives of t_z or b_z required in the terms in square brackets usually do not exist, and, even when they

* *J. Appl. Mech.*, vol. 12, p. A68, 1945.

do exist, if the loading changes rapidly divergence of the series may make it possible to use only the first terms of the series, ignoring the terms in square brackets. The values thus obtained will always be an improvement on those given by classical theory alone, but the stresses especially may not be sufficiently accurate in the immediate neighborhood of discontinuities or near-discontinuities in the loading. As discussed above and previously for beams, such loading functions can be replaced by harmonic, that is trigonometric, series, using the whole of expressions (5.19) for the harmonic components with half wave lengths l for which l/h is considerably greater than unity, and the parts of the expressions up to the square brackets for the shorter components. Corrections for the stresses in the neighborhood of a concentrated load are derived later in Sec. 5.3 and corrections for other common types of discontinuities could be obtained by similar methods.

Closed Solutions Contained in Series

As in the case of beams, series solutions such as Eqs. (5.19) for plates contain exact closed solutions for loadings which can be expressed by power functions of $\mathbf{x}$ and $\mathbf{y}$, and also solutions for edge loading only. As an example of the former, if $\mathbf{t}_z$ is taken as proportional to $\mathbf{x}^4$, $\mathbf{x}^2\mathbf{y}^2$ or $\mathbf{y}^4$ in Eqs. (5.19) we obtain a closed solution for a plate with a uniform loading on the top surface; if $\mathbf{t}_z$ is taken proportional to a suitable fifth power function of $\mathbf{x}$, $\mathbf{y}$ we get solutions for linearly varying loading, etc. However, while each such solution will represent a definite physical case, the boundary conditions involved will not usually be simple or like those in important practical problems. To satisfy particular boundary conditions we can add to the values of $\mathbf{t}_z$ and $\mathbf{b}_z$ which satisfy Eqs. (5.18a) any biharmonic functions, that is functions satisfying the equations:

$$\nabla^4 \mathbf{t}_z = 0, \qquad \nabla^4 \mathbf{b}_z = 0 \tag{5.24}$$

As Lee has pointed out in the paper cited above, solutions of these equations are sufficiently general to permit the satisfaction of two conditions on each edge for each of the above equations, or four conditions in all on each edge. There are actually at least five conditions to be satisfied in the most general case, for instance on the three resultant forces and two resultant moments on each edge, Fig. 4.1, or corresponding displacements. Hence Kirchhoff's device may in general still have to be used even to satisfy only the 'gross' boundary conditions, unless supplementary solutions are superposed, as will be discussed in Sec. 5.5.

Uniformly Loaded, Simply Supported Thick Plate

Power function solutions of Eqs. (5.24) such as those of Eq. (3.17a) are insufficient to satisfy practical edge conditions. By using these in conjunction with trigonometric–hyperbolic solutions of Eqs. (5.24), and also solutions for edge loads obtained from both the normal and tangential load series solutions, Eqs. (5.19), (5.32) (given later), Lee, in the paper cited under Eq. (5.20), succeeded in satisfying the boundary conditions:

$$
\begin{aligned}
x &= 0, a: \; M_x, \; u_{z(z=0)}, \; F_x, \; u_{y(z=0)} = 0 \\
y &= \pm\frac{b}{2}: \; M_y, \; u_{z(z=0)}, \; F_y, \; u_{x(z=0)} = 0
\end{aligned}
\tag{5.25}
$$

for a plate under uniform loading $t_z = p_0$, $b_z = 0$, Fig. 5.7(a). These conditions are also nearly those of a plate supported on rollers at its edges as in Fig. 5.6(a), and the discussion of these conditions for that case applies here as well.

To save space we will give here a simplified version of this solution, satisfying only the first two of each of the four conditions (5.25); these involve the bending moments and deflections at the edges, which have the principal effect

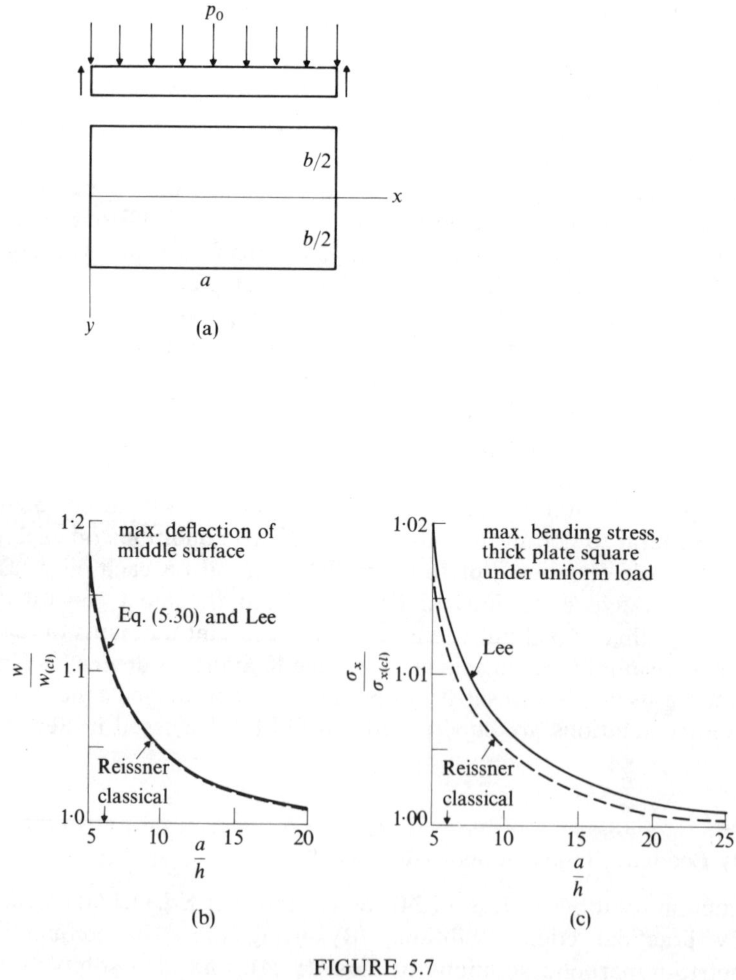

FIGURE 5.7

upon the maximum deflection and the stresses which occur in the middle part of the plate. Using only symmetrical functions of y, we can satisfy Eqs. (5.18a) if $\mathbf{b}_z = 0$ and:

$$\mathbf{t}_z = \frac{p_0}{24}\mathbf{x}^4 + C_3\mathbf{x}^3 + C_2\mathbf{x}^2 + C_1\mathbf{x} + C_0 \tag{5.26}$$

$$+\sum_m \sin\frac{m\pi\mathbf{x}}{\mathbf{a}}\left(A_m \cosh\frac{m\pi\mathbf{y}}{\mathbf{a}} + B_m\frac{m\pi\mathbf{y}}{\mathbf{a}}\sinh\frac{m\pi\mathbf{y}}{\mathbf{a}}\right) \qquad (m = 1, 3, 5, \ldots)$$

where $\mathbf{a} = a/c$. Substituting this into the expressions for M_x, M_y and $u_{z(z=0)}$ in Eqs. (5.19), and then into the edge conditions on the M's and w, Eqs. (5.25), we obtain:

$$2C_2 = 0, \qquad \frac{p_0}{2}\mathbf{a}^2 + 6C_3\mathbf{a} + 2C_2 = 0$$

$$\frac{\nu p_0}{2}\mathbf{x}^2 + 6\nu C_3\mathbf{x} + 2\nu C_2 + \sum_m \frac{m^2\pi^2}{\mathbf{a}^2}\sin\frac{m\pi\mathbf{x}}{\mathbf{a}}$$
$$\times\{[(1-\nu)A_m + 2B_m]\cosh\alpha_m + (1-\nu)B_m\alpha_m\sinh\alpha_m\}$$
$$-\frac{2\nu}{5}\left(p_0 - 2\sum_m \frac{m^4\pi^4}{\mathbf{a}^4}\sin\frac{m\pi\mathbf{x}}{\mathbf{a}}B_m\cosh\alpha_m\right) = 0$$

$$\frac{p_0}{24}\mathbf{a}^4 + C_3\mathbf{a}^3 + C_2\mathbf{a}^2 + C_1\mathbf{a} + C_0$$
$$-\frac{8-3\nu}{10(1-\nu)}\left(\frac{p_0}{2}\mathbf{a}^2 + 6C_3\mathbf{a} + 2C_2\right) - \frac{227-157\nu}{4200(1-\nu)}p_0 = 0 \tag{5.27}$$

$$C_0 - \frac{8-3\nu}{5(1-\nu)}C_2 - \frac{227-157\nu}{4200(1-\nu)}p_0 = 0$$

$$0 = \frac{p_0}{24}\mathbf{x}^4 + C_3\mathbf{x}^3 + C_2\mathbf{x}^2 + C_1\mathbf{x} + C_0 + \sum_m \sin\frac{m\pi\mathbf{x}}{\mathbf{a}}(A_m\cosh\alpha_m + B_m\alpha_m\sinh\alpha_m)$$
$$-\frac{8-3\nu}{10(1-\nu)}\left[\frac{p_0}{2}\mathbf{x}^2 + 6C_3\mathbf{x} + 2C_2 + 2\sum_m\frac{m^2\pi^2}{\mathbf{a}^2}\sin\frac{m\pi\mathbf{x}}{\mathbf{a}}B_m\cosh\alpha_m\right] - \frac{227-157\nu}{4200(1-\nu)}p_0$$

where $\alpha_m = m\pi b/2a$.

These conditions are satisfied if:

$$C_3 = -\frac{\mathbf{a}}{12}p_0, \qquad C_2 = 0, \qquad C_1 = \frac{\mathbf{a}^3}{24}p_0, \qquad C_0 = \frac{227-157\nu}{4200(1-\nu)}p_0$$

$$\sum_m \frac{m^2\pi^2}{\mathbf{a}^2}\sin\frac{m\pi\mathbf{x}}{\mathbf{a}}\left\{\left[(1-\nu)A_m + \left(2 + \frac{4\nu}{5}\frac{m^2\pi^2}{\mathbf{a}^2}\right)B_m\right]\cosh\alpha_m\right.$$
$$\left.+(1-\nu)B_m\alpha_m\sinh\alpha_m\right\} = -\frac{\nu p_0}{2}\left(\mathbf{x}^2 - \mathbf{a}\mathbf{x} - \frac{4}{5}\right) \tag{5.28}$$

$$\sum_m \sin\frac{m\pi\mathbf{x}}{\mathbf{a}}\left\{\left(A_m - \frac{8-3\nu}{5(1-\nu)}\frac{m^2\pi^2}{\mathbf{a}^2}B_m\right)\cosh\alpha_m + B_m\alpha_m\sinh\alpha_m\right\} =$$
$$-\frac{p_0}{24}\left[\mathbf{x}^4 - 2\mathbf{a}\mathbf{x}^3 + \mathbf{a}^3\mathbf{x} - \frac{6(8-3\nu)}{5(1-\nu)}(\mathbf{x}^2 - \mathbf{a}\mathbf{x})\right]$$

In the last two equations we now expand the right-hand sides into sine series in $\sin m\pi x/a$ by harmonic analysis. Each of these equations is then satisfied only if it is satisfied for each individual term of these sine series. Cancelling out the $\sin m\pi x/a$, and solving the resulting equations simultaneously for A_m and B_m we find:

$$B_m = \frac{2p_0\mathbf{a}^4}{m^5\pi^5\cosh\alpha_m}, \qquad A_m = -B_m(2+\alpha_m\tanh\alpha_m) \tag{5.29}$$

Putting the above values of C_0, C_1, C_2, C_3, A_m, B_m into the expression $u_{z(z=0)}$ in Eqs. (5.19), we find that the deflection w_{max} of the middle surface at the center of the plate is, for the case of a square plate, $a=b$ or $\alpha_m = m\pi/2$:

$$\begin{aligned} w_{max} &= \frac{12(1-\nu^2)a^4p_0}{Eh^3}\left\{\frac{5}{384} - \sum_m \frac{\sin\frac{m\pi}{2}\left(4+m\pi\tanh\frac{m\pi}{2}\right)}{m^5\pi^5\cosh\frac{m\pi}{2}}\right. \\ &\left. + \frac{8-3\nu}{10(1-\nu)}\left(\frac{h}{a}\right)^2\left(\frac{1}{32} - \sum_m \frac{\sin\frac{m\pi}{2}}{m^3\pi^3\cosh\frac{m\pi}{2}}\right)\right\} \approx 0{\cdot}004\,06\frac{a^4p_0}{D}\left[1+4{\cdot}6\left(\frac{h}{a}\right)^2\right] \end{aligned} \tag{5.30}$$

where the first term in the square brackets represents the value given by the classical theory, Eq. (4.26).

Figure 5.7(b) shows this deflection, which is practically the same as that obtained by Lee who satisfied all the boundary conditions of Eq. (5.25). The value obtained by using Reissner's thick plate theory satisfying three of these conditions is also nearly the same, and they all give about 20 percent greater deflection than the classical theory for a plate five times as wide as it is thick. Figure 5.7(c) shows the maximum bending stress obtained by Lee from Eqs. (5.19) and by using Reissner's theory for the same case.

Series Solution in Loading Function, Tangential Loading*

Again referring to Fig. 5.5, we let $t_x(x, y)$, $b_x(x, y)$ represent distributed forces per unit area acting in the x direction on the top and bottom surfaces respectively, as shown. (We take all loads opposite in direction to the positive directions of the corresponding stresses so as to be consistent with the case of normal loads, which are obviously more realistically taken as compressive than as tensile.) We next define the symbols $\mathbf{t}_x(x, y)$, $\mathbf{b}_x(x, y)$ by the relations:

$$\nabla^4\mathbf{t}_x = t_x, \qquad \nabla^4\mathbf{b}_x = b_x \tag{5.31a}$$

The boundary conditions on the upper and lower surfaces are:

$$\mathbf{z} = -1, 1:\ \sigma_z, \sigma_{yz} = 0,\ \sigma_{xz} = -t_x, -b_x \tag{5.31b}$$

* C. W. Lee and L. H. Donnell, 'A study of thick plates under tangential loads applied on the faces', *Proc. 3rd US Nat. Congr. of Appl. Mech.* p. 401, 1958.

This loading can be separated into antisymmetrical and symmetrical parts as shown in Figs. 5.5(c), (d). For the first, which represents distributed couples tending to bend the plate, we obtain the first terms of our series from the classical plate theory. This requires taking p, F_x, F_y, $F_{xy} = 0$ in the equations of equilibrium (4.8), and adding the term $c(t_x + b_x)$ (after dividing by $dx\,dy$) to the right side of $\Sigma\, m_y = 0$. Then, carrying out the development as before by eliminating F_{xz}, F_{yz} and using Eq. (4.14), we find that Eq. (4.19) is changed to $D\nabla^4 w = -c\,\partial(t_x + b_x)/\partial x$. From this, using Eqs. (5.31a) we obtain the first terms $2Eu_z = -3(1-\nu^2)c\,\partial(\mathbf{t}_x + \mathbf{b}_x)/\partial\mathbf{x}$, $2Eu_x = 3(1-\nu^2)c\mathbf{z}\ \partial^2(\mathbf{t}_x + \mathbf{b}_x)/\partial\mathbf{x}^2$, $2Eu_y = 3(1-\nu^2)c\mathbf{z}\ \partial^2(\mathbf{t}_x + \mathbf{b}_x)/\partial\mathbf{x}\,\partial\mathbf{y}$ in the same manner as was done for normal loading.

The symmetrical part of the loading, Fig. 5.5(d), can be taken, for the first approximation, to be distributed uniformly over the thickness as a body force B_x in two-dimensional elasticity theory. The displacements u_x, u_y can then be found from Eqs. (3.14c), taking $B_y = 0$ and substituting $(t_x - b_x)/2c$ for B_x. We can then eliminate u_y between these equations by applying the operators $-\partial^2/\partial x\,\partial y$ to the second and $[(1-\nu)/(1+\nu)]\partial^2/\partial x^2$ and $[2/(1+\nu)]\partial^2/\partial y^2$ successively to the first, and adding the resulting three equations. Similarly we eliminate u_x by applying $-\partial^2/\partial x\,\partial y$ to the first, $[(1-\nu)/(1+\nu)]\partial^2/\partial y^2$ and $[2/(1+\nu)]\partial^2/\partial x^2$ to the second and adding. Using Eqs. (5.31a), we thus obtain the first terms of u_x, u_y for symmetrical tangential loading:

$$2Eu_x = -(1-\nu^2)c\,\partial^2(\mathbf{t}_x - \mathbf{b}_x)/\partial\mathbf{x}^2 - 2(1+\nu)c\,\partial^2(\mathbf{t}_x - \mathbf{b}_x)/\partial\mathbf{y}^2,$$
$$2Eu_y = (1+\nu)^2 c\,\partial^2(\mathbf{t}_x - \mathbf{b}_x)/\partial\mathbf{x}\,\partial\mathbf{y}.$$

The lateral displacement u_z is only that due to Poisson's ratio effects. Taking u_z as zero at the middle surface and using Eqs. (3.6), (3.11a, b) and the above expressions for u_x, u_y, we find:

$$u_z = \int_0^z \epsilon_z\,dz,$$

$$\epsilon_z = -\nu(\sigma_x + \sigma_y)/E = -\nu(\epsilon_x + \epsilon_y)/(1-\nu) = -\nu(\partial u_x/\partial x + \partial u_y/\partial y)/(1-\nu)$$

from which: $2Eu_z = \nu(1+\nu)c\mathbf{z}\nabla^2\ \ \partial(\mathbf{t}_x - \mathbf{b}_x)/\partial\mathbf{x}$.

Starting with these first terms, we build up the series term by term so as to satisfy the elasticity conditions (3.8) and the boundary conditions (5.31b) in just the same manner as in the normal loading case. We thus find:

$$\begin{aligned}
u_x = \frac{(1+\nu)c}{E}\Bigg\{ & \frac{3(1-\nu)\mathbf{z}}{2}\frac{\partial^2(\mathbf{t}_x + \mathbf{b}_x)}{\partial\mathbf{x}^2} - \frac{1-\nu}{2}\frac{\partial^2(\mathbf{t}_x - \mathbf{b}_x)}{\partial\mathbf{x}^2} - \frac{\partial^2(\mathbf{t}_x - \mathbf{b}_x)}{\partial\mathbf{y}^2} \\
& - \frac{5(2-\nu)\mathbf{z}^3 - (6-\nu)\mathbf{z}}{20}\frac{\partial^2\nabla^2(\mathbf{t}_x + \mathbf{b}_x)}{\partial\mathbf{x}^2} - \mathbf{x}\frac{\partial^2\nabla^2(\mathbf{t}_x + \mathbf{b}_x)}{\partial\mathbf{y}^2} \\
& + \frac{3(2-\nu)\mathbf{z}^2 - (2+\nu)}{12}\frac{\partial^2\nabla^2(\mathbf{t}_x - \mathbf{b}_x)}{\partial\mathbf{x}^2} + \frac{3\mathbf{z}^2 - 1}{6}\frac{\partial^2\nabla^2(\mathbf{t}_x - \mathbf{b}_x)}{\partial\mathbf{y}^2} \\
& + \Bigg[\frac{105(3-\nu)\mathbf{z}^5 - 70(7-\nu)\mathbf{z}^3 + 3(53+17\nu)\mathbf{z}}{8400}\frac{\partial^2(t_x + b_x)}{\partial\mathbf{x}^2} + \frac{\mathbf{z}^3 - 3\mathbf{z}}{6}\frac{\partial^2(t_x + b_x)}{\partial\mathbf{y}^2} \\
& - \frac{15(3-\nu)\mathbf{z}^4 - 30(1+\nu)\mathbf{z}^2 + (1+29\nu)}{720}\frac{\partial^2(t_x - b_x)}{\partial\mathbf{x}^2} - \frac{15\mathbf{z}^4 - 30\mathbf{z}^2 + 7}{360}\frac{\partial^2(t_x - b_x)}{\partial\mathbf{y}^2}\cdot\cdot\Bigg]\Bigg\}
\end{aligned} \tag{5.32}$$

$$u_y = \frac{(1+\nu)c}{E}\frac{\partial^2}{\partial \mathbf{x}\,\partial \mathbf{y}}\left\{\frac{3(1-\nu)\mathbf{z}}{2}(\mathbf{t}_x+\mathbf{b}_x)+\frac{1+\nu}{2}(\mathbf{t}_x-\mathbf{b}_x)\right.$$
$$-\frac{5(2-\nu)\mathbf{z}^3-(26-\nu)\mathbf{z}}{20}\boldsymbol{\nabla}^2(\mathbf{t}_x+\mathbf{b}_x)-\frac{3\nu\mathbf{z}^2+\nu}{12}\boldsymbol{\nabla}^2(\mathbf{t}_x-\mathbf{b}_x)$$
$$+\left[\frac{105(3-\nu)\mathbf{z}^5-70(27-\nu)\mathbf{z}^3+3(1453+17\nu)\mathbf{z}}{8400}(t_x+b_x)\right.$$
$$\left.\left.-\frac{15(1-\nu)\mathbf{z}^4+30(1-\nu)\mathbf{z}^2-(13-29\nu)}{720}(t_x-b_x)\cdots\right]\right\} \qquad (5.32\text{ cont.})$$

$$u_z = \frac{(1+\nu)c}{E}\frac{\partial}{\partial \mathbf{x}}\left\{-\frac{3(1-\nu)}{2}(\mathbf{t}_x+\mathbf{b}_x)-\frac{15\nu\mathbf{z}^2-(4+\nu)}{20}\boldsymbol{\nabla}^2(\mathbf{t}_x+\mathbf{b}_x)+\frac{\nu\mathbf{z}}{2}\boldsymbol{\nabla}^2(\mathbf{t}_x-\mathbf{b}_x)\right.$$
$$+\left[\frac{175(1+\nu)\mathbf{z}^4-70(5+\nu)\mathbf{z}^2+(87-17\nu)}{2800}(t_x+b_x)-\frac{(1+\nu)\mathbf{z}^3-(3-\nu)\mathbf{z}}{12}(t_x-b_x)\right.$$
$$-\frac{35(2+\nu)\mathbf{z}^6-35(6+\nu)\mathbf{z}^4+3(70-17\nu)\mathbf{z}^2\ldots}{16\,800}\boldsymbol{\nabla}^2(t_x+b_x)$$
$$\left.\left.+\frac{3(2+\nu)\mathbf{z}^5-10(2-\nu)\mathbf{z}^3+(30-29\nu)\mathbf{z}}{720}\boldsymbol{\nabla}^2(t_x-b_x)\ldots\right]\right\}$$

$$e = \frac{(1+\nu)(1-2\nu)}{E}\frac{\partial}{\partial \mathbf{x}}\left\{\frac{3\mathbf{z}}{2}\boldsymbol{\nabla}^2(\mathbf{t}_x+\mathbf{b}_x)-\frac{1}{2}\boldsymbol{\nabla}^2(\mathbf{t}_x-\mathbf{b}_x)+\left[\frac{-5\mathbf{z}^3+\mathbf{z}}{20}(t_x+b_x)\right.\right.$$
$$\left.\left.+\frac{3\mathbf{z}^2+1}{12}(t_x-b_x)+\frac{105\mathbf{z}^5-70\mathbf{z}^3-51\mathbf{z}}{8400}\boldsymbol{\nabla}^2(t_x+b_x)-\frac{15\mathbf{z}^4+30\mathbf{z}^2-29}{720}\boldsymbol{\nabla}^2(t_x-b_x)\ldots\right]\right\}$$

To save space only the displacements and dilation are given. The stresses are easily obtained from these by using Eqs. (3.7b). For instance

$$\begin{aligned}\sigma_x &= E[(\partial u_x/\partial \mathbf{x})/c+\nu e/(1-2\nu)]/(1+\nu)\\ &= \partial^3(\mathbf{t}_x+\mathbf{b}_x)/\partial \mathbf{x}^3[3(1-\nu)\mathbf{z}/2]+\boldsymbol{\nabla}^2\,\partial(\mathbf{t}_x+\mathbf{b}_x)/\partial \mathbf{x}[3\nu\mathbf{z}/2]\cdots\\ &= \partial/\partial \mathbf{x}[\partial^2/\partial \mathbf{x}^2+\nu\,\partial^2/\partial \mathbf{y}^2](\mathbf{t}_x+\mathbf{b}_x)3\mathbf{z}/2\cdots,\ \text{etc.}\end{aligned}$$

Additional terms can be found by continuing the process as previously outlined, or from the general term expressions for the stresses given in the paper cited above. The expressions for tangential loadings t_y, b_y in the y direction are the same as those for tangential loadings t_x, b_x except that x and y are interchanged in all subscripts as well as in the derivatives.

Solutions for thick plates under tangential loading with specific boundary conditions can be found in the same manner as discussed above for normal loading. For example, if we take:

$$t_x = b_x = q_{mn}\sin\frac{m\pi\mathbf{x}}{\mathbf{a}}\cos\frac{n\pi\mathbf{y}}{\mathbf{b}} \qquad (m, n = 1, 2, 3, \ldots) \qquad (5.33a)$$

then Eqs. (5.31a) are satisfied if:

$$\mathbf{t}_x = \mathbf{b}_x = \frac{\mathbf{a}^4 q_{mn}}{\pi^4(m^2+n^2\mathbf{a}^2/\mathbf{b}^2)^2}\sin\frac{m\pi\mathbf{x}}{\mathbf{a}}\cos\frac{n\pi\mathbf{y}}{\mathbf{b}} \qquad (5.33b)$$

Substituting these values into Eqs. (5.32) and using Eqs. (3.7b) we find that the

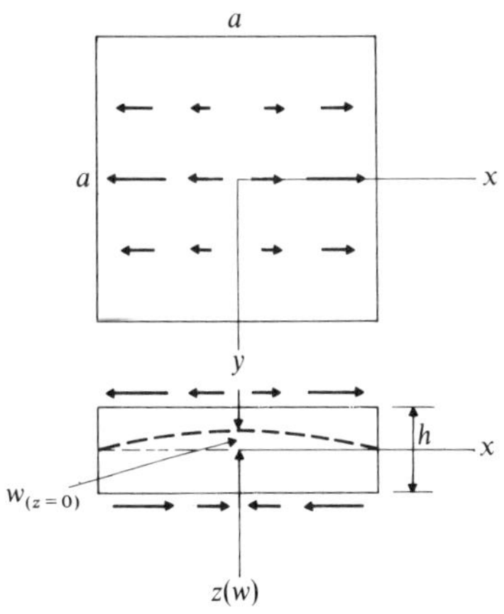

FIGURE 5.8

boundary conditions of Eqs. (5.22) are satisfied. For the case of a square plate $a = b$ with $m = n = 1$, we have a thick plate under the loading condition shown in Fig. 5.8 and the boundary conditions discussed before for the plate of Fig. 5.6(a). For this case the deflection of the center is:

$$u_{z(x,\,y,\,z=0)} = -\frac{6(1-\nu^2)q_{11}a^3c}{\pi^3Eh^3}\left\{1+\frac{\pi^2(4+\nu)}{60(1-\nu)}\left(\frac{h}{a}\right)^2-\frac{\pi^4(87-17\nu)}{16\,800(1-\nu)}\left(\frac{h}{a}\right)^4\cdots\right\} \tag{5.33c}$$

Series Solution in Loading Function in Cylindrical Coordinates, Normal Loading

Series solutions for normal and tangential loading in other coordinate systems can be derived in similar manner to those above in rectangular coordinates, or Eqs. (5.19), (5.32) can be converted to other coordinate systems as discussed at the end of Sec. 3.1. As an illustration let us consider the important practical case of cylindrical coordinates.

Taking the normal unit pressures on the top and bottom surfaces t_z, b_z as functions of r, θ, $t_z(r,\theta)$, $b_z(r,\theta)$, Eqs. (5.18a) still apply, with $\boldsymbol{\nabla}^4 = \boldsymbol{\nabla}^2\boldsymbol{\nabla}^2$ where $\boldsymbol{\nabla}^2 = \partial^2/\partial\mathbf{r}^2 + (1/\mathbf{r})\partial/\partial\mathbf{r} + (1/\mathbf{r}^2)\partial^2/\partial\theta^2$. Then using Eqs. (5.19) and (3.9e) in Eqs. (3.9b), we find that the expressions for u_r and u_θ are exactly the same as the expression for u_x in Eqs. (5.19) with $\partial/\partial\mathbf{r}$ and $(1/\mathbf{r})\partial/\partial\theta$ respectively substituted for $\partial/\partial\mathbf{x}$ (where $\mathbf{r} = r/c$). For example

$$u_r = (1+\nu)(c/E)\partial/\partial\mathbf{r}\{-3(1-\nu)\mathbf{z}(\mathbf{t}_z - \mathbf{b}_z)/2\ldots\}$$

The expressions for u_z and e are just the same as those in Eqs. (5.19) (but of course the symbols $\boldsymbol{\nabla}^2$, $\mathbf{t}_z$, etc., have different meanings, as described above). The stresses can then be found from the displacements and dilation by using Eqs. (3.9h).

Series Solution in Loading Function in Cylindrical Coordinates, Tangential Loading

It is much more difficult to convert the case of general tangential loading from rectangular to cylindrical coordinates because, unlike the normal load case, the directions of the loads are not constant. The relations between tangential unit loadings on the top and bottom surfaces $t_r(r, \theta)$, $b_r(r, \theta)$ in the radial direction and $t_\theta(r, \theta)$, $b_\theta(r, \theta)$ in the circumferential direction, and the tangential loadings in the x and y directions considered in Eqs. (5.32), are the same as the relations previously given by Eqs. (3.9c) between the tangential stresses σ_{zr}, $\sigma_{z\theta}$ and σ_{zx}, σ_{zy} of Fig. 3.5(c), that is:

$$t_x = t_r \cos\theta - t_\theta \sin\theta, \qquad t_y = t_r \sin\theta + t_\theta \cos\theta \tag{5.34a}$$

with similar relations between the bottom loadings, given by substituting b for t.

The presence of functions of θ in these expressions for t_x, etc., greatly complicates the solution in cylindrical coordinates, which we must find by adding to expressions (5.32) the similar terms for t_y, b_y, obtained by interchanging x and y in these expressions, and using Eqs. (3.9b), (3.9e), and (5.34a). However, it is found that we can obtain expressions in the form of Eqs. (5.35) which at least contain the desired solution.

To simplify the complex expressions we introduce the symbols s, s', etc., defined by the relations:

$$\begin{aligned} s &= \alpha(t_r + b_r) + \beta(t_\theta + b_\theta), & d &= \alpha(t_r - b_r) + \beta(t_\theta - b_\theta) \\ s' &= \beta(t_r + b_r) - \alpha(t_\theta + b_\theta), & d' &= \beta(t_r - b_r) - \alpha(t_\theta - b_\theta) \\ \mathbf{s} &= \alpha(\mathbf{t}_r + \mathbf{b}_r) + \beta(\mathbf{t}_\theta + \mathbf{b}_\theta), & \mathbf{d} &= \alpha(\mathbf{t}_r - \mathbf{b}_r) + \beta(\mathbf{t}_\theta - \mathbf{b}_\theta) \\ \mathbf{s}' &= \beta(\mathbf{t}'_r + \mathbf{b}'_r) - \alpha(\mathbf{t}'_\theta + \mathbf{b}'_\theta), & \mathbf{d}' &= \beta(\mathbf{t}'_r - \mathbf{b}'_r) - \alpha(\mathbf{t}'_\theta - \mathbf{b}'_\theta) \end{aligned} \tag{5.34b}$$

where α, β, γ (used later) are the operators:

$$\alpha = \frac{\partial}{\partial \mathbf{r}} + \frac{1}{\mathbf{r}}, \qquad \beta = \frac{1}{\mathbf{r}}\frac{\partial}{\partial \theta}, \qquad \gamma = \frac{\partial}{\partial \mathbf{r}} \tag{5.34c}$$

and the symbols $\mathbf{t}_r$, $\mathbf{t}_\theta$, $\mathbf{t}'_r$, $\mathbf{t}'_\theta$ are defined by:

$$\nabla^4 \alpha \mathbf{t}_r = \alpha t_r, \qquad \nabla^4 \beta \mathbf{t}'_r = \beta t_r, \qquad \nabla^4 \beta \mathbf{t}_\theta = \beta t_\theta, \qquad \nabla^4 \alpha \mathbf{t}'_\theta = \alpha t_\theta \tag{5.34d}$$

$\mathbf{b}_r$, $\mathbf{b}_\theta$, $\mathbf{b}'_r$, $\mathbf{b}'_\theta$ being defined by the same expressions with $\mathbf{t}$ replaced by $\mathbf{b}$. From (5.34b, d) it follows that $\nabla^4\mathbf{s} = s$, $\nabla^4\mathbf{d} = d$, $\nabla^4\mathbf{s}' = s'$, $\nabla^4\mathbf{d}' = d'$. Again it should be kept in mind that the order of application of operators should be strictly adhered to; thus in $\nabla^4\alpha\mathbf{t}_r$ the operator α must be applied first and then ∇^4.

Many relations exist between the above operators and ∇^2 (applied to t_r, etc.) such as $\alpha\beta = \beta\gamma$, $\nabla^2 = \alpha\gamma + \beta^2$. We will use in particular the following identities (which can readily be checked), f being any function of r, θ:

$$\left[\left(\nabla^2 - \frac{1}{\mathbf{r}^2}\right)\gamma - \gamma\nabla^2 - \frac{2}{\mathbf{r}}\beta^2\right]f = 0, \qquad \left[\left(\nabla^2 - \frac{1}{\mathbf{r}^2}\right)\beta - \beta\nabla^2 + \frac{2}{\mathbf{r}}\beta\gamma\right]f = 0,$$

$$(\beta\alpha - 2\alpha\beta + \gamma\beta)f = 0, \tag{5.34e}$$

The expressions for the displacements and dilation then become:

$$
\begin{aligned}
u_r = \frac{(1+\nu)c}{E}\Bigg\{ & \frac{3(1-\nu)\mathbf{z}}{2}\gamma\mathbf{s} - \frac{(1-\nu)}{2}\gamma\mathbf{d} - \beta\mathbf{d}' - \frac{5(2-\nu)\mathbf{z}^3-(6-\nu)\mathbf{z}}{20}\gamma\nabla^2\mathbf{s} - \mathbf{z}\beta\nabla^2\mathbf{s}' \\
& + \frac{3(2-\nu)\mathbf{z}^2-(2+\nu)}{12}\gamma\nabla^2\mathbf{d} + \frac{3\mathbf{z}^2-1}{6}\beta\nabla^2\mathbf{d}' \\
& + \Bigg[\frac{105(3-\nu)\mathbf{z}^5-70(7-\nu)\mathbf{z}^3+3(53+17\nu)\mathbf{z}}{8400}\gamma s + \frac{\mathbf{z}^3-3\mathbf{z}}{6}\beta s' \\
& - \frac{15(3-\nu)\mathbf{z}^4-30(1+\nu)\mathbf{z}^2+(1+29\nu)}{720}\gamma d - \frac{15\mathbf{z}^4-30\mathbf{z}^2+7}{360}\beta d' \ldots\Bigg]\Bigg\}
\end{aligned}
$$

$$
\begin{aligned}
u_\theta = \frac{(1+\nu)c}{E}\Bigg\{ & \frac{3(1-\nu)\mathbf{z}}{2}\beta\mathbf{s} - \frac{1-\nu}{2}\beta\mathbf{d} + \gamma\mathbf{d}' - \frac{5(2-\nu)\mathbf{z}^3-(6-\nu)\mathbf{z}}{20}\beta\nabla^2\mathbf{s} + \mathbf{z}\gamma\nabla^2\mathbf{s}' \\
& + \frac{3(2-\nu)\mathbf{z}^2-(2+\nu)}{12}\beta\nabla^2\mathbf{d} - \frac{3\mathbf{z}^2-1}{6}\gamma\nabla^2\mathbf{d}' \\
& + \Bigg[\frac{105(3-\nu)\mathbf{z}^5-70(7-\nu)\mathbf{z}^3+3(53+17\nu)\mathbf{z}}{8400}\beta s \\
& - \frac{\mathbf{z}^3-3\mathbf{z}}{6}\gamma s' - \frac{15(3-\nu)\mathbf{z}^4-30(1+\nu)\mathbf{z}^2+(1+29\nu)}{720}\beta d + \frac{15\mathbf{z}^4-30\mathbf{z}^2+7}{360}\gamma d' \ldots\Bigg]\Bigg\}
\end{aligned}
\tag{5.35}
$$

$$
\begin{aligned}
u_z = \frac{(1+\nu)c}{E}\Bigg\{ & -\frac{3(1-\nu)}{2}\mathbf{s} - \frac{15\nu\mathbf{z}^2-(4+\nu)}{20}\nabla^2\mathbf{s} + \frac{\nu}{2}\mathbf{z}\nabla^2\mathbf{d} \\
& + \Bigg[\frac{175(1+\nu)\mathbf{z}^4-70(5+\nu)\mathbf{z}^2+(87-17\nu)}{2800}s - \frac{(1+\nu)\mathbf{z}^3-(3-\nu)\mathbf{z}}{12}d \\
& - \frac{35(2+\nu)\mathbf{z}^6-35(6+\nu)\mathbf{z}^4+3(70-17\nu)\mathbf{z}^2\ldots}{16\,800}\nabla^2 s \\
& + \frac{3(2+\nu)\mathbf{z}^5-10(2-\nu)\mathbf{z}^3+(30-29\nu)\mathbf{z}}{720}\nabla^2 d\ldots\Bigg]\Bigg\}
\end{aligned}
$$

$$
\begin{aligned}
e = \frac{(1-2\nu)(1+\nu)}{E}\Bigg\{ & \frac{3\mathbf{z}}{2}\nabla^2\mathbf{s} - \frac{1}{2}\mathbf{d} + \Bigg[\frac{-5\mathbf{z}^3+\mathbf{z}}{20}s + \frac{3\mathbf{z}^2+1}{12}d \\
& + \frac{105\mathbf{z}^5-70\mathbf{z}^3-51\mathbf{z}}{8400}\nabla^2 s - \frac{15\mathbf{z}^4+30\mathbf{z}^2-29}{720}\nabla^2 d\ldots\Bigg]\Bigg\}
\end{aligned}
$$

It can be checked that these expressions satisfy the elasticity conditions Eqs. (3.9g) and, using Eqs. (3.9h) for the stresses, the boundary conditions:

$$
\mathbf{z} = -1, 1: \qquad \sigma_z = 0, \qquad \sigma_{rz} = -t_r, -b_r, \qquad \sigma_{\theta z} = -t_\theta, -b_\theta \tag{5.35a}
$$

The checks of Eqs. (3.9g) and of the boundary condition $\mathbf{z} = -1, 1$: $\sigma_z = 0$ are straightforward. However, checking the remaining boundary conditions requires some explanation and possible qualification. For example, consider the condition $\mathbf{z} = -1$: $\sigma_{\theta z} = -t_\theta$. Using expressions (5.35) in Eqs. (3.9h), we find that the expression for $\sigma_{\theta z}$ can be written as:

$$\sigma_{\theta z} = -\beta A + \gamma B - \beta C - \gamma D, \qquad \text{where:} \tag{5.35b}$$
$$A = \nabla^2 \alpha \mathbf{t}_r, \qquad B = \nabla^2 \beta \mathbf{t}'_r, \qquad C = \nabla^2 \beta \mathbf{t}_\theta, \qquad D = \nabla^2 \alpha \mathbf{t}'_\theta$$

Using these symbols A, B, C, D and $\nabla^4 = \nabla^2\nabla^2$, Eqs. (5.34d) can be written:

$$\nabla^2 A = \alpha t_r, \qquad \nabla^2 B = \beta t_r, \qquad \nabla^2 C = \beta t_\theta, \qquad \nabla^2 D = \alpha t_\theta \tag{5.35c}$$

Applying the operator β to the first and third of these equations and the operator $(2\alpha - \gamma)$ to the second and fourth, and using the identities (5.34e) and the operator $\lambda = 2\alpha^2 + \beta^2 - \gamma\alpha$, we obtain:

$$\lambda(\beta A) = \beta\alpha t_r, \quad \lambda(\gamma B) = \beta\alpha t_r, \quad \lambda(\beta C) = (2\alpha^2 - \gamma\alpha)t_\theta, \quad \lambda(\gamma D) = \beta^2 t_\theta \tag{5.35d}$$

When these expressions are substituted into Eq. (5.35b) to which the operator λ has been applied, this equation reduces to:

$$\lambda\sigma_{\theta z} = -\lambda t_\theta \tag{5.35e}$$

which is obviously satisfied by the required condition $\sigma_{\theta z} = -t_\theta$. Similar results can be obtained in the same manner for the other three boundary conditions.

Since equation of the type (5.35e) can conceivably also represent other boundary conditions, specific applications of Eqs. (5.35) should be checked to see that they satisfy the desired conditions. Consider for example the case of a round plate acted upon by uniformly distributed unit tangential forces $t_\theta = b_\theta = t$, t_r, $b_r = 0$, t being constant. With the addition of a uniform compression $\sigma_z = t/\mu$, where μ is a friction coefficient, this could represent a practical case of a dry disc pressed between a stationary surface and a surface which rotates about the central axis of the disc, or between two surfaces rotating about this axis at different speeds as in a disc clutch.

From Eqs. (5.34b) $s, d, d' = 0$, $s' = -2t/\mathbf{r}$. Using Eqs. (5.34d) $\mathbf{t}_r, \mathbf{t}'_r, \mathbf{b}_r, \mathbf{b}'_r, \mathbf{b}_\theta, \mathbf{t}_\theta, \mathbf{s}, \mathbf{d}, \mathbf{d}' = 0$, $\mathbf{t}'_\theta = \mathbf{b}'_\theta = t\mathbf{r}^4/45 + f$, $\mathbf{s}' = -2t r^3/9 - 2\alpha f$, where $f(r, \theta)$ is the most general solution of the homogeneous equation $\nabla^4\alpha\,\mathbf{t}'_\theta = 0$ or $\nabla^4\alpha\,\mathbf{b}'_\theta = 0$ and can be used for satisfying edge boundary conditions at the inner and outer edges of the disc. Ignoring f we have from Eqs. (5.35):

$$u_r, u_z, e = 0, \qquad u_\theta = -\frac{(1+\nu)ct}{E}\left(2\mathbf{z} + \frac{\mathbf{z}^3 - 3\mathbf{z}}{3\mathbf{r}^2}\cdots\right)$$
$$\sigma_z, \sigma_{rz} = 0, \qquad \sigma_{\theta z} = -t\left(1 + \frac{\mathbf{z}^2 - 1}{2\mathbf{r}^2}\cdots\right) \tag{5.35f}$$

It can be seen that the boundary conditions (5.35a) are satisfied. Unlike the previous series solutions in loading functions this solution for a uniform tangential loading is not a closed one. It is also not applicable to solid discs without a central hole, since it involves a singular point at the center $\mathbf{r} = 0$. For a disc with a central hole whose radius is large compared to the thickness it would converge rapidly. Solutions for loadings which are proportional to a power of the coordinates higher than one will be closed, as in the other series solution in the loading function, and they will be applicable to solid as well as to hollow discs.

Series Solution in Loading Function, Axisymmetric Normal Loading

The most important practical plate problems after those of rectangular plates are those in which the plate and loading are symmetrical about an axis such as the **z** axis, in which case it can be assumed (except as usual in stability and vibration problems) that the deflections and stresses are also axisymmetric. Taking u_θ, $\partial u_r/\partial\theta$, $\partial u_z/\partial\theta = 0$ in the equations described above for series in normal loading functions in cylindrical coordinates, we find, for axisymmetric normal loads t_z, b_z:

$$u_r = \frac{(1+\nu)c}{E}\frac{d}{d\mathbf{r}}\left\{-\frac{3(1-\nu)\mathbf{z}}{2}(\mathbf{t}_z-\mathbf{b}_z)+\frac{5(2-\nu)\mathbf{z}^3-3(2+3\nu)\mathbf{z}}{20}\nabla^2(\mathbf{t}_z-\mathbf{b}_z)+\frac{\nu}{2}\nabla^2(\mathbf{t}_z+\mathbf{b}_z)\right.$$
$$\left.+\left[\frac{-35(3-\nu)\mathbf{z}^5-70(1-3\nu)\mathbf{z}^3+(87-157\nu)\mathbf{z}}{2800}(t_z-b_z)+\frac{(1-\nu)(3\mathbf{z}^2-1)}{12}(t_z+b_z)\,.\,.\right]\right\}$$

$$u_z = \frac{(1+\nu)c}{E}\left\{\frac{3(1-\nu)}{2}(\mathbf{t}_z-\mathbf{b}_z)+\frac{15\nu\mathbf{z}^2-3(8-3\nu)}{20}\nabla^2(\mathbf{t}_z-\mathbf{b}_z)\right.$$
$$+\frac{-175(1+\nu)\mathbf{z}^4+210(5-3\nu)\mathbf{z}^2-227+157\nu}{2800}(t_z-b_z)-\frac{(1-\nu)\mathbf{z}}{2}(t_z+b_z) \tag{5.36}$$
$$\left.+\left[\frac{35(2+\nu)\mathbf{z}^6-105(2-3\nu)\mathbf{z}^4+3(70-157\nu)\mathbf{z}^2}{16\,800}\nabla^2(t_z-b_z)-\frac{\nu(\mathbf{z}^3-\mathbf{z})}{12}\nabla^2(t_z+b_z)\,.\,.\right]\right\}$$

$$e = \frac{(1-2\nu)(1+\nu)}{E}\left\{-\frac{3\mathbf{z}}{2}\nabla^2(\mathbf{t}_z-\mathbf{b}_z)+\frac{5\mathbf{z}^3+9\mathbf{z}}{20}(t_z-b_z)-\frac{1}{2}(t_z+b_z)\right.$$
$$\left.+\left[\frac{-35\mathbf{z}^5-210\mathbf{z}^3+157\mathbf{z}}{2800}\nabla^2(t_z-b_z)+\frac{3\mathbf{z}^2-1}{12}\nabla^2(t_z+b_z)\ldots\right]\right\}$$

$$\sigma_r = -\frac{3\mathbf{z}}{2}\left(\frac{d^2}{d\mathbf{r}^2}+\frac{\nu}{\mathbf{r}}\frac{d}{d\mathbf{r}}\right)(\mathbf{t}_z-\mathbf{b}_z)+\left(\frac{5\mathbf{z}^3-3\mathbf{z}}{10}\frac{d^2}{d\mathbf{r}^2}+\frac{5\mathbf{z}^3+9\mathbf{z}}{20}\frac{\nu}{\mathbf{r}}\frac{d}{d\mathbf{r}}\right)\nabla^2(\mathbf{t}_z-\mathbf{b}_z)$$
$$-\frac{1}{2}\frac{\nu}{\mathbf{r}}\frac{d}{d\mathbf{r}}\nabla^2(\mathbf{t}_z+\mathbf{b}_z)+\left[\left(\frac{-105\mathbf{z}^5-70\mathbf{z}^3+87\mathbf{z}}{2800}\frac{d^2}{d\mathbf{r}^2}\right.\right.$$
$$\left.\left.+\frac{-35\mathbf{z}^5-210\mathbf{z}^3+157\mathbf{z}}{2800}\frac{\nu}{\mathbf{r}}\frac{d}{d\mathbf{r}}\right)(t_z-b_z)+\frac{3\mathbf{z}^2-1}{12}\left(\frac{d^2}{d\mathbf{r}^2}+\frac{\nu}{\mathbf{r}}\frac{d}{d\mathbf{r}}\right)(t_z+b_z)\ldots\right]$$

$$\sigma_z = \frac{-\mathbf{z}^3+3\mathbf{z}}{4}(t_z-b_z)-\frac{1}{2}(t_z+b_z)+\left[\frac{\mathbf{z}^5-2\mathbf{z}^3+\mathbf{z}}{40}\nabla^2(t_z-b_z)\ldots\right]$$

$$\sigma_{rz} = \frac{d}{d\mathbf{r}}\left\{\frac{3(\mathbf{z}^2-1)}{4}\nabla^2(\mathbf{t}_z-\mathbf{b}_z)+\left[\frac{-5\mathbf{z}^4+6\mathbf{z}^2-1}{40}(t_z-b_z)\ldots\right]\right\}$$

where σ_θ is the same as σ_r with $d^2/d\mathbf{r}^2$ and $(1/\mathbf{r})\,d/d\mathbf{r}$ interchanged, $\nabla^2 = d^2/d\mathbf{r}^2+(1/\mathbf{r})\,d/d\mathbf{r}$, and $\mathbf{t}_z$, $\mathbf{b}_z$ are defined by:

$$\nabla^4\mathbf{t}_z = \nabla^2\nabla^2\mathbf{t}_z = t_z, \qquad \nabla^4\mathbf{b}_z = b_z \tag{5.37}$$

It should be kept in mind that the order of application of operators containing the independent variable can not be changed; thus in the expression for σ_r, ∇^2 must be applied to $(\mathbf{t}_z-\mathbf{b}_z)$ first and then $d^2/d\mathbf{r}^2$, etc.

Disc Under Uniform Transverse Load

As with other such solutions, Eqs. (5.36) contain exact closed solutions for loading functions which are in the form of power functions of $\mathbf{r}$. Thus for a disc acted upon by a uniform pressure p_0 on the top surface we have $t_z = p_0$, $b_z = 0$. The most general solutions of Eqs. (5.37) can then be written as:

$$\mathbf{t}_z = \frac{p_0 \mathbf{r}^4}{64} + C_0 + C_1 \mathbf{r}^2 + C_2 \log \mathbf{r} + C_3 \mathbf{r}^2 \log \mathbf{r}$$
$$\mathbf{b}_z = C_0' + C_1' \mathbf{r}^2 + C_2' \log \mathbf{r} + C_3' \mathbf{r}^2 \log \mathbf{r} \tag{5.38}$$

The terms involving C_0, C_0', etc., can be used to satisfy boundary conditions, and represent the complete solutions of the ordinary fourth degree homogeneous differential equations $\nabla^4 \mathbf{t}_z = 0$, $\nabla^4 \mathbf{b}_z = 0$. Actually in the following application the solution to $\nabla^4 \mathbf{b}_z = 0$ does not contribute to the satisfaction of boundary conditions, but conceivably there may be cases where it does.

For a solid disc without initial stresses, the coefficients C_2, C_3, C_2', C_3' must be taken as zero, since they produce infinite stresses at the center $r = 0$ of the disc when substituted into Eqs. (5.36). For a disc with a hole in the middle these coefficients could be used to satisfy conditions at the edge of the hole.

With the remaining coefficients and a uniform membrane stress $\sigma_r = \sigma_\theta = \sigma_0$, which can readily be shown to satisfy equilibrium and continuity conditions like the stresses of Fig. 3.8 (such stress conditions will be further discussed in Sec. 5.4) we can satisfy conditions such as $F_r = \int_{-1}^{1} \sigma_r \, d\mathbf{z} = 0$, $M_r = \int_{-1}^{1} \sigma_r \mathbf{z} \, d\mathbf{z} = 0$, $w = u_{z(z=0)} = 0$ at the outer edge $r = a$ of a simply supported disc. Such boundary conditions are no more complete than those satisfied in the classical plate theory solution of Eqs. (4.106) to (4.111) but we can now satisfy conditions far from the edges exactly. The resulting solution may be no more accurate than classical theory near the edges but if we satisfy the conditions for resultant forces and moments at the edges the errors should, by Saint-Venant's principle, be largely confined, at least as far as stresses are concerned, to a region about as wide as the thickness, where they can be corrected by the superposition of local stress fields which will be discussed in detail in Sec. 5.3.

For the case of a simply supported solid disc of radius a then, substituting expressions (5.38) with $C_2, C_3, C_0', C_1', C_2', C_3' = 0$ into the expressions for F_r, M_r, and w and setting these equal to zero for $\mathbf{r} = \mathbf{a} = a/c$, we obtain the boundary conditions:

$$\frac{1}{c} F_r = 2\sigma_0 - \frac{\nu}{2} p_0 = 0, \qquad \frac{1}{c^2} M_r = -\frac{(3+\nu)\mathbf{a}^2 p_0}{16} - 2(1+\nu)C_1 + \frac{\nu}{5} p_0 = 0$$
$$\frac{w}{c} = \frac{3(1-\nu^2)}{2E} \left\{ \frac{\mathbf{a}^4}{64} p_0 + C_0 + C_1 \mathbf{a}^2 - \frac{8-3\nu}{10(1-\nu)} \left(\frac{\mathbf{a}^2}{4} p_0 + 4C_1 \right) - \frac{227-157\nu}{4200(1-\nu)} p_0 \right\} = 0 \tag{5.39}$$

from which:

$$\sigma_0 = \frac{\nu p_0}{4}, \qquad C_1 = \frac{16\nu - 5(3+\nu)\mathbf{a}^2}{160(1+\nu)} p_0$$
$$C_0 = \left[\frac{5+\nu}{64} \mathbf{a}^4 - \frac{8+5\nu}{80} \mathbf{a}^2 + \frac{227 + 1414\nu - 661\nu^2}{4200(1-\nu)} \right] \frac{p_0}{1+\nu} \tag{5.40}$$

Using these values in Eqs. (5.36) the stresses are:

$$\sigma_r = \frac{p_0}{8}\left[(2+\nu)\mathbf{z}^3 - 3\mathbf{z}\left(\frac{(3+\nu)(\mathbf{r}^2-\mathbf{a}^2)}{4} + \frac{2+\nu}{5}\right)\right]$$

$$\sigma_\theta = \frac{p_0}{8}\left[(2+\nu)\mathbf{z}^3 - 3\mathbf{z}\left(\frac{(1+3\nu)\mathbf{r}^2-(3+\nu)\mathbf{a}^2}{4} + \frac{2+\nu}{5}\right)\right] \tag{5.41}$$

$$\sigma_z = -\frac{p_0}{4}(\mathbf{z}^3 - 3\mathbf{z} + 2), \qquad \sigma_{rz} = \frac{3p_0\mathbf{r}}{8}(\mathbf{z}^2 - 1)$$

while the maximum deflection is:

$$w_{(z=0)(r=0)} = \frac{5+\nu}{1+\nu}\frac{p_0 a^4}{64D}\left[1 + \frac{8(8+\nu+\nu^2)}{5(5+\nu)(1-\nu)}\left(\frac{c}{a}\right)^2\right] \tag{5.42}$$

where the first term in the brackets represents the value given by classical plate theory, Eq. (4.110). The correction factor given by the second term in the brackets is approximately $3{\cdot}6/\mathbf{a}^2$, which is about the same as that given for square plates, Eq. (5.30), if the width of the square plate is replaced by a little more than the diameter of the disc.

Series Solution in Loading Function, Axisymmetric Tangential Loading

We can obtain a solution for this case from Eqs. (5.35) by setting $\partial/\partial\theta$, β, t_θ, b_θ, $u_\theta = 0$ in Eqs. (5.34b, c, d), which gives $s = \alpha(t_r + b_r)$, $d = \alpha(t_r - b_r)$, s', $d' = 0$, defining $\mathbf{t}_r$, $\mathbf{b}_r$ by $\nabla^4\mathbf{t}_r = t_r$, $\nabla^4\mathbf{b}_r = b_r$; however, such a solution has the disadvantage of satisfying the boundary conditions for σ_{rz} ambiguously, as in Eq. (5.35).

Alternatively, it has been found possible, in this axisymmetric case, to derive a simpler and unambiguous series solution in the manner described before in the derivation of Eq. (3.29), by assuming expressions for successive terms so as to satisfy Eqs. (3.10a) and the boundary conditions:

$$\mathbf{z} = \pm 1: \ \sigma_z = 0, \ \sigma_{rz} = -b_r, \ -t_r \tag{5.43}$$

in steps; b_r and t_r are as shown in Fig. 5.5(b) with x replaced by r. In this way we find the following solution, in which the symbols $\mathbf{t}_r$, $\mathbf{b}_r$ are given the new definitions:

$$\Delta^2\mathbf{t}_r = t_r, \ \Delta^2\mathbf{b}_r = b_r, \qquad \text{where} \ \ \Delta^2 = \frac{d}{d\mathbf{r}}\left(\frac{d}{d\mathbf{r}} + \frac{1}{\mathbf{r}}\right) = \frac{d^2}{d\mathbf{r}^2} + \frac{1}{\mathbf{r}}\frac{d}{d\mathbf{r}} - \frac{1}{\mathbf{r}^2} \tag{5.44}$$

$$u_r = \frac{(1+\nu)c}{E}\Bigg(-\frac{1-\nu}{2}(\mathbf{t}_r - \mathbf{b}_r) + \frac{3(1-\nu)\mathbf{z}}{2}(\mathbf{t}_r + \mathbf{b}_r) + \frac{3(2-\nu)\mathbf{z}^2 - (2+\nu)}{12}(t_r - b_r)$$
$$- \frac{5(2-\nu)\mathbf{z}^3 - (6-\nu)\mathbf{z}}{20}(t_r + b_r) + \left[\frac{-15(3-\nu)\mathbf{z}^4 + 30(1+\nu)\mathbf{z}^2 - (1+29\nu)}{720}\Delta^2(t_r - b_r)\right.$$
$$\left.+\frac{105(3-\nu)\mathbf{z}^5 - 70(7-\nu)\mathbf{z}^3 + 3(53+17\nu)\mathbf{z}}{8400}\Delta^2(t_r + b_r)\ldots\right]\Bigg) \tag{5.45}$$

$$u_z = \frac{(1+\nu)c}{E}\left(-\frac{3(1-\nu)}{2}\int (\mathbf{t}_r + \mathbf{b}_r)\, d\mathbf{r} + \right. \qquad (5.45\ \text{cont.})$$

$$\left(\frac{d}{d\mathbf{r}} + \frac{1}{\mathbf{r}}\right)\left\{\frac{\nu\mathbf{z}}{2}(\mathbf{t}_r - \mathbf{b}_r) - \frac{15\nu\mathbf{z}^2 - (4+\nu)}{20}(\mathbf{t}_r + \mathbf{b}_r)\right.$$

$$-\frac{(1+\nu)\mathbf{z}^3 - (3-\nu)\mathbf{z}}{12}(t_r - b_r) + \frac{175(1+\nu)\mathbf{z}^4 - 70(5+\nu)\mathbf{z}^2 + (87-17\nu)}{2800}(t_r + b_r)$$

$$+\left[\frac{3(2+\nu)\mathbf{z}^5 - 10(2-\nu)\mathbf{z}^3 + (30-29\nu)\mathbf{z}}{720}\Delta^2(t_r - b_r)\right.$$

$$\left.\left.\left. - \frac{35(2+\nu)\mathbf{z}^6 - 35(6+\nu)\mathbf{z}^4 + 3(70-17\nu)\mathbf{z}^2 \ldots}{16\,800}\Delta^2(t_r + b_r)\ldots\right]\right\}\right)$$

$$e = \frac{(1+\nu)}{E}(1-2\nu)\left(\frac{d}{d\mathbf{r}} + \frac{1}{\mathbf{r}}\right)\left\{-\frac{1}{2}(\mathbf{t}_r - \mathbf{b}_r) + \frac{3\mathbf{z}}{2}(\mathbf{t}_r + \mathbf{b}_r) + \frac{3\mathbf{z}^2 + 1}{12}(t_r - b_r)\right.$$

$$\left. - \frac{5\mathbf{z}^3 - \mathbf{z}}{20}(t_r + b_r) + \left[\frac{-15\mathbf{z}^4 - 30\mathbf{z}^2 + 29}{720}\Delta^2(t_r - b_r) + \frac{105\mathbf{z}^5 - 70\mathbf{z}^3 - 51\mathbf{z}}{8400}\Delta^2(t_r + b_r)\ldots\right]\right\}$$

The stresses can be readily obtained from Eqs. (3.10b).

Unlike the previous series solutions in loading functions, Eqs. (5.45) do not contain closed solutions for loading functions which are in the form of power functions, because, due to the presence of the $1/r^2$ term in Eqs. (5.44), the terms of Eqs. (5.45) never become zero for such functions. However, series solutions for any type of smooth loading function should converge rapidly as in previous series solutions.

5.3 THICK PLATES—OTHER SOLUTIONS, LOCAL STRESS FIELDS

Harmonic Loading on Semi-Infinite Solid

The two-dimensional solutions (3.32), (3.33) give the displacements and stresses produced by normal or tangential loading on the edge of a semi-infinite plate, the loading being uniform across the width of the edge and varying harmonically along the length of the edge. These solutions were presented as potentially useful in studying the effects of loading on one face of a rectangular bar when the wave length of the harmonic loading is small compared to the thickness (depth) of the bar (but not small compared to the width of the face, since the two-dimensional elasticity theory would then be insufficiently accurate) in which case the stresses on the opposite face of the bar may be small enough to be neglected. Such solutions are obviously also useful for satisfying the edge boundary conditions for a plate; in this case the wave length of the harmonic loading need merely be small compared to the relatively large width of the plate in order for the stresses on the opposite edge to be negligible. Such applications of Eqs. (3.32), (3.33) and similar solutions such as antisymmetric analogs of them are discussed later in Secs. 5.4, 5.5.

Three-dimensional analogs of Eqs. (3.32), (3.33), giving the displacements and stresses produced by harmonically distributed loads on the surface of a semi-infinite solid, can also be derived. As in the case of solutions (3.32), (3.33) the stresses decrease exponentially with distance from the surface, becoming very small at distances from the surface large compared to the larger of the two wave lengths of the loading. Hence such a three-dimensional solution could be used for studying a loading on one face of a plate when the wave lengths are small compared to the thickness. Such solutions, while approximate, are simpler than exact solutions, just as in the two-dimensional case Eqs. (3.32), (3.33) are simpler than the more exact solutions of Eqs. (3.28), (3.29) or Table 3.3.

The following solutions were obtained by starting with expressions for u_x, u_y, u_z similar in form to those of Eqs. (3.32), (3.33) and with unknown coefficients, solving for the coefficients to satisfy Eqs. (3.8) and the conditions that two of the stresses σ_z, σ_{xz}, σ_{yz} given by Eqs. (3.7b) are zero at the surface. Taking the z axis normal to the surface and with the origin at the surface, the following solution is found for a normal pressure per unit area $p \cos Xx \cos Yy$ acting on the surface, where $X = \pi/l_x$, $Y = \pi/l_y$ (l_x, l_y being the half wave lengths in the x, y directions) and $Z^2 = X^2 + Y^2$:

$$
\begin{aligned}
u_x &= \frac{1+\nu}{E} \sin Xx \cos Yy\, e^{-Zz}(1-2\nu - Zz)\frac{X}{Z^2}p \\
u_y &= \frac{1+\nu}{E} \cos Xx \sin Yy\, e^{-Zz}(1-2\nu - Zz)\frac{Y}{Z^2}p \\
u_z &= \frac{1+\nu}{E} \cos Xx \cos Yy\, e^{-Zz}[-2(1-\nu) - Zz]\frac{1}{Z}p \\
\sigma_x &= \cos Xx \cos Yy\, e^{-Zz}(X^2 + 2\nu Y^2 - X^2 Zz)\frac{1}{Z^2}p \\
\sigma_y &= \cos Xx \cos Yy\, e^{-Zz}(2\nu X^2 + Y^2 - Y^2 Zz)\frac{1}{Z^2}p \\
\sigma_z &= \cos Xx \cos Yy\, e^{-Zz}(1 + Zz)p \\
\sigma_{xz} &= \sin Xx \cos Yy\, e^{-Zz} Xzp \\
\sigma_{yz} &= \cos Xx \sin Yy\, e^{-Zz} Yzp \\
\sigma_{xy} &= \sin Xx \sin Yy\, e^{-Zz}(-1 + 2\nu + Zz)\frac{XY}{Z^2}p
\end{aligned}
\tag{5.46a}
$$

Similarly the solution for a tangential loading per unit area in the x direction $q \sin Xx \cos Yy$ acting on the surface is:

$$
\begin{aligned}
u_x &= \frac{1+\nu}{E} \sin Xx \cos Yy\, e^{-Zz}[-2(1-\nu)X^2 - 2Y^2 + X^2 Zz]\frac{1}{Z^3}q \\
u_y &= \frac{1+\nu}{E} \cos Xx \sin Yy\, e^{-Zz}(2\nu + Zz)\frac{XY}{Z^3}q \\
u_z &= \frac{1+\nu}{E} \cos Xx \cos Yy\, e^{-Zz}(1 - 2\nu + Zz)\frac{X}{Z^2}q
\end{aligned}
\tag{5.46b}
$$

$$
\sigma_x = \cos Xx \cos Yy\, e^{-Zz}[-2X^2 - 2(1+\nu)Y^2 + X^2 Zz]\frac{X}{Z^3}q
$$

$$\sigma_y = \cos Xx \cos Yy\, e^{-Zz}(-2\nu X^2 + Y^2 Zz)\frac{X}{Z^3}\, q \qquad (5.46 \text{ cont.})$$

$$\sigma_z = \cos Xx \cos Yy\, e^{-Zz}(-Xz)q$$

$$\sigma_{xz} = \sin Xx \cos Yy\, e^{-Zz}\left(1 - \frac{X^2}{Z} z\right) q$$

$$\sigma_{yz} = \cos Xx \sin Yy\, e^{-Zz}\left(-\frac{XY}{Z} z\right) q$$

$$\sigma_{xy} = \sin Xx \sin Yy\, e^{-Zz}[(1-2\nu)X^2 + Y^2 - X^2 Zz)]\frac{Y}{Z^3}\, q$$

A similar solution for a harmonically distributed tangential loading in the y direction on the surface can of course be obtained by interchanging x, y, X, Y in Eqs. (5.46b), and the same solutions can be used on the bottom surfaces by taking the origin in that surface with the z axis upward. The distributions of the stresses are similar to those shown in Fig. 3.11 for the two-dimensional case. Equations (5.46a) can also be derived by using the biharmonic function $\phi = -[(1+\nu)/E] \cos Xx \cos Yy\, e^{-Zz}(2\nu + Zz)p/Z^3$ in the basic elasticity solution 14, Table 3.1, again with $Z^2 = X^2 + Y^2$, and similarly for Eqs. (5.46b). Unlike the approximate plain stress solutions (3.32), (3.33) solutions (5.46a, b) are exact three-dimensional elasticity solutions, so there is no limitation on the ratio of the loading wave lengths to dimensions other than the thickness.

Harmonically Distributed Pressures on Two Sides of Plates

Further possibilities are rather obvious. Thus in solutions (5.46a, b) Z can have values of $\pm(X^2 + Y^2)^{1/2}$, but only the negative value was used in the above. However, we can use both positive and negative exponentials, or, more conventionally and conveniently, the combination of these exponentials which we call hyperbolic sines and cosines, to get an exact solution for any desired magnitude of harmonically distributed pressures on both top and bottom surfaces of a plate. For example we can use the biharmonic function:

$$\phi = \cos X\mathbf{x} \cos Y\mathbf{y}(A \cosh Z\mathbf{z} + B \sinh Z\mathbf{z} + CZ\mathbf{z} \cosh Z\mathbf{z} + DZ\mathbf{z} \sinh Z\mathbf{z}) \qquad (5.47a)$$

in solution 14, Table 3.1. Calculating $\nabla^2\phi$ and $\nabla^4\phi$, we find that this function is biharmonic (and harmonic as well if $C, D = 0$) if we again take $Z^2 = X^2 + Y^2$. We can then calculate the displacements $u_x = \partial^2\phi/\partial z\,\partial x$, $u_y = \partial^2\phi/\partial y\,\partial z$, $u_z = \partial^2\phi/\partial z^2 - 2(1-\nu)\nabla^2\phi$, and the stresses from Eqs. (3.7b). Using the boundary conditions:

$$z = \pm c, \qquad \mathbf{z} = \pm 1: \qquad \sigma_{xz}, \sigma_{yz} = 0,$$

$$\sigma_z = -b_z \cos X\mathbf{x} \cos Y\mathbf{y}, -t_z \cos X\mathbf{x} \cos Y\mathbf{y} \qquad (5.47b)$$

with these conditions to solve for the coefficients A, B, C, D, we find that boundary conditions (5.47b) are satisfied if:

$$C = \frac{(1+\nu)(t_z + b_z)\sinh Z}{2EZ^3(Z + \sinh Z \cosh Z)}, \qquad B = -(2\nu + Z \coth Z)C$$
$$D = \frac{(1+\nu)(t_z - b_z)\cosh Z}{2EZ^3(Z - \sinh Z \cosh Z)}, \qquad A = -(2\nu + Z \tanh Z)D \tag{5.47c}$$

Solution (5.47a, b, c) can be regarded as a three-dimensional analog of the plane stress solution of Table 3.3, without any of the latter's approximations or limitations on the wave length of the loading function.

Both Eqs. (5.46a, b) and (5.47a, b, c) can be written in series form, substituting $X_m = m\pi/l_x$, $Y_n = n\pi/l_y$, $A = A_{mn}$, etc., and summing over m and n. Similar series can be added using sine instead of cosine functions for both x and y or for only one of them. In this way top and bottom loadings which are any desired functions of x, y can be represented in infinite series form, and the displacements and stresses due to them calculated. As in the case of Table 3.3, other solutions can be obtained by exchanging trigonometric for exponential functions or vice versa (but all three functions of x, y, z can not be trigonometric, nor all three exponential).

We have discussed three types of solution to the problem of thick plates under loading on their faces: the series solutions in loading functions (closed when these are power functions) such as Eqs. (5.19), (5.32); the quite different solutions of the type just discussed based on function (5.47a) or analogous functions, which are closed when the loadings are harmonic and infinite series for other loadings; and finally the simpler but approximate solutions of Eqs. (5.46a, b) which are good approximations only for loadings of short wave length. Like the corresponding solution for beams, Eqs. (3.28), (3.29), and Table 3.3 or Eqs. (3.26), (3.27), and Eqs. (3.32), (3.33), each of these types of solutions has its applications and limitations, advantages and disadvantages. Since all of them are based on fourth-order differential equations, they are capable by themselves of satisfying only two boundary conditions, at the ends in beams or on the edges in plates. More complete boundary conditions can be satisfied by superposing elasticity solutions involving unloaded faces, which are discussed in Sec. 3.4 for beams and in Secs. 5.4, 5.5 for plates.

Concentrated Normal Load on Semi-Infinite Solid

As in the corresponding two-dimensional case discussed in Sec. 3.4 of a concentrated load on the edge of a semi-infinite plate, this loading does not produce a true local stress field such as is produced by a balanced load system, although the stresses involved vary roughly inversely as the square of the distance from the point of application of the load. However, as in the two-dimensional case, important balanced systems of forces can be constructed using such concentrated loads.

We will take the point of application on the surface of the semi-infinite solid of the concentrated load P as the origin of cylindrical coordinates, Fig. 3.5(a), with the z axis normal to the surface of the solid and pointing into it, with the compressive load P acting in the z direction. Then the distance from any point of the solid to the origin is the square root of the quantity (r^2+z^2), and if the stresses are proportional to a negative power of this quantity they will satisfy the obvious requirement that they tend to infinity at this point and decrease everywhere as we go away from this point. Using the general solution 14, Table 3.1a (with $u_\theta=0$ for the axisymmetric case) it then seems logical to look for solutions in which the biharmonic function ϕ is such a negative power of (r^2+z^2) or some related function.

Substituting $\phi=(r^2+z^2)^n$ into the equations $\nabla^2\phi=[\partial^2/\partial r^2+(1/r)\partial/\partial r+\partial^2/\partial z^2]\phi=0$ and $\nabla^4\phi=\nabla^2\nabla^2\phi=0$, we obtain $2n(1+2n)(r^2+z^2)^{n-1}=0$ for the former and $2(1-2n)(n-1)(r^2+z^2)^{n-2}=0$ for the latter, which are satisfied by $n=0, -\frac{1}{2}$ and $n=\frac{1}{2}, 1$ respectively. The case of $n=0$ is trivial since it yields no stresses, but the other solutions $(r^2+z^2)^{-1/2}$, $(r^2+z^2)^{1/2}$, (r^2+z^2) are significant. Moreover, as pointed out in Sec. 3.1, we can derive an unlimited number of other related solutions from these solutions by applying to them any operator which gives zero when applied to zero and whose order of application with ∇^2 or ∇^4 is immaterial. Thus any derivative with respect to z, or conversely any integral with respect to z of these solutions is also a solution. Thus if we apply $\partial/\partial z$ to both sides of the relation $\nabla^2\phi=0$ we obtain $(\partial/\partial z)\nabla^2\phi=\nabla^2(\partial\phi/\partial z)=0$ (but the order of application would not be immaterial if we used $\partial/\partial r$ instead of $\partial/\partial z$ because of the $1/r$ factor in the second term of $\nabla^2\phi$).

In this manner we obtain from the first solution $(r^2+z^2)^{-1/2}$ the series of solutions: $z\log[z+(r^2+z^2)^{1/2}]-(r^2+z^2)^{1/2}$, $\log[z+(r^2+z^2)^{1/2}]$, $(r^2+z^2)^{-1/2}$, $z(r^2+z^2)^{-3/2}$, etc., each of which can be obtained by applying $\partial/\partial z$ to the preceding one. From the second solution $(r^2+z^2)^{1/2}$ we similarly obtain: $(r^2+z^2)\log[z+(r^2+z^2)^{1/2}]$, $(r^2+z^2)^{1/2}$, $z(r^2+z^2)^{-1/2}$, $z^2(r^2+z^2)^{-3/2}$, etc. The first series consists of harmonic functions; the second series consists of biharmonic functions and can also be obtained from the first series by multiplying its terms by z or (r^2+z^2). The third solution originally obtained above, (r^2+z^2), can also yield a series of solutions, but they are not useful for the present problem since they do not give stress expressions with the required negative exponents. All these solutions can of course by multiplied by any constant and combined as required.

In applying these solutions to the concentrated load problem we require only infinite stresses at the origin; their resultant can not be anything but a force in the z direction because of axisymmetry. We can fix the magnitude of this concentrated load P by using the equilibrium condition $P=-\int_0^\infty 2\pi r\,dr\,\sigma_z$, which can readily be satisfied for any value of z except zero (for which the presence of the singular point at the origin would unnecessarily complicate the mathematics). For the rest it is only required to satisfy the boundary conditions: $z=0, r\neq 0$: $\sigma_z, \sigma_{rz}=0$. If we calculate the expressions for σ_z and σ_{rz} for each of the above solutions it is not difficult to determine which ones can be combined and in what magnitudes to satisfy the above conditions.

In this way, by using general solution 14, Table 3.1a, and Eqs. (3.10b), we can find the following solution for this problem:

$$\phi = \frac{(1+\nu)P}{2\pi E}\{-(1-2\nu)z \log[z+(r^2+z^2)^{1/2}] - 2\nu(r^2+z^2)^{1/2}\}$$

$$u_r = \frac{(1+\nu)P}{2\pi E}\left[-\frac{1-2\nu}{r} + \frac{(1-2\nu)z}{r}(r^2+z^2)^{-1/2} + zr(r^2+z^2)^{-3/2}\right]$$

$$u_\theta = 0, \qquad u_z = \frac{(1+\nu)P}{2\pi E}[(3-2\nu)(r^2+z^2)^{-1/2} \quad r^2(r^2+z^2)^{-3/2}]$$

$$\sigma_r = \frac{(1-2\nu)P}{2\pi}\left\{\frac{1}{r^2} - \left[\frac{1}{r^2}(r^2+z^2)^{-1/2} + \frac{3r^2}{1-2\nu}(r^2+z^2)^{-5/2}\right]z\right\} \tag{5.48}$$

$$\sigma_\theta = \frac{(1-2\nu)P}{2\pi}\left\{-\frac{1}{r^2} + \left[\frac{1}{r^2}(r^2+z^2)^{-1/2} + (r^2+z^2)^{-3/2}\right]z\right\}$$

$$\sigma_z = -\frac{3P}{2\pi}z^3(r^2+z^2)^{-5/2}, \qquad \sigma_{rz} = -\frac{3P}{2\pi}z^2 r(r^2+z^2)^{-5/2}, \qquad \sigma_{r\theta}, \sigma_{\theta z} = 0$$

This solution was obtained by the French mathematician Boussinesq in 1885.

Solutions for balanced load systems consisting of several concentrated loads, for example a compressive load P with two tensile loads $P/2$ symmetrically placed on either side of it, can easily be obtained by superposition. Because the displacements and stresses at a general point due to different loads will be in different directions in cylindrical coordinates, it may be more convenient for such purposes to convert Eqs. (5.48) to rectangular coordinates. Using $r^2 = x^2 + y^2$ the expression for ϕ is then:

$$\phi = \frac{(1+\nu)P}{2\pi E}\{-(1-2\nu)z \log[z+(x^2+y^2+z^2)^{1/2}] - 2\nu(x^2+y^2+z^2)^{1/2}\} \tag{5.48a}$$

The displacements and stresses can then be found from this by using general solution 14, Table 3.1, and Eqs. (3.7b). A solution for a concentrated tangential load at a point on a semi-infinite solid can be derived in a similar fashion.

Distributed Normal Load on Surface of Semi-Infinite Solid

A solution for a distributed normal load can of course be obtained by superposition, using Eqs. (5.48) for an element of the load and summing up by integration over the loaded area. However, the integrals involved are not always easy to evaluate directly. As an example, consider a compressive force per unit area p_0 (that is a compressive stress) uniformly distributed over a circular area on the surface of a semi-infinite solid, with a radius b and center at the origin.

Because of symmetry the integrals present no difficulties for the displacements and stresses at points $z = z$, $r = 0$ on the z axis. Figure 5.9 shows an element $r\,d\theta\,dr$ on which a normal load $p_0 r\,d\theta\,dr$ acts. The u_r displacements and the σ_{rz} stresses are zero on the axis because of symmetry. The u_z displacements and σ_z stresses at points on the axis are in the same directions for all the elemental

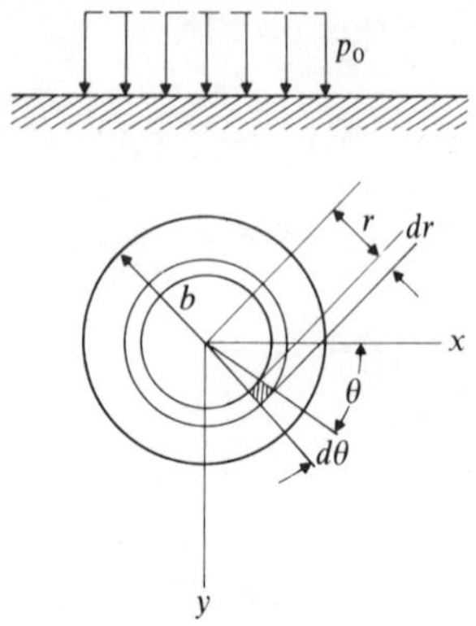

FIGURE 5.9

loads and so can be obtained by simple summing. Using Eqs. (5.48):

$$u_z = \frac{(1+\nu)p_0}{2\pi E}\int_0^b dr \int_0^{2\pi} d\theta\, r[(3-2\nu)(r^2+z^2)^{-1/2} - r^2(r^2+z^2)^{-3/2}]$$

$$= \frac{(1+\nu)p_0}{E}\{2(1-\nu)[(b^2+z^2)^{1/2} - z] - z^2(b^2+z^2)^{-1/2} + z\}, \qquad (5.49)$$

$$\sigma_z = -\frac{3p_0 z^3}{2\pi}\int_0^b dr \int_0^{2\pi} d\theta\, r(r^2+z^2)^{-5/2} = p_0[z^3(b^2+z^2)^{-3/2} - 1]$$

The σ_r and σ_θ stresses produced at points $z = z$, $r = 0$ on the axis by the pressures acting on elements located at different angles θ have different directions, and so the resultant σ_r and σ_θ stresses on the axis due to all the load elements cannot be found by simple summation. However, the magnitudes of these resultant σ_r and σ_θ stresses must be the same in all directions due to symmetry; they must therefore be equal to each other and to the corresponding stress in a particular direction, such as the resultant σ_x stress in the x direction, Fig. 5.10(a). Since the σ_x stresses due to all the load elements are in the same x direction, the resultant σ_x stress can be found by simple summation. From Eqs. (3.9d), remembering that $\sigma_{r\theta} = 0$: $\sigma_x = \sigma_r \cos^2\theta + \sigma_\theta \sin^2\theta$. Expressions (5.48) give the σ_r and σ_θ stresses at a point r, θ, z due to a load P on the axis, but we can use them also for the same stresses at a point on the axis due to a load $p_0 r\, d\theta\, dr$ located at point r, θ, z, merely by substituting $p_0 r\, d\theta\, dr$ for P and $\theta + \pi$ for θ. Considering that $\sin^2(\theta+\pi) = \sin^2\theta$, $\cos^2(\theta+\pi) = \cos^2\theta$, we then have at $z = z$, $r = 0$ the resultant stresses:

$$\sigma_r = \sigma_\theta = \sigma_x = \frac{(1-2\nu)p_0}{2\pi}\int_0^b dr\, r \int_0^{2\pi} d\theta \Big\{\cos^2\theta\Big[\frac{1}{r^2} - \frac{z}{r^2}(r^2+z^2)^{-1/2}$$

$$- \frac{3r^2 z}{1-2\nu}(r^2+z^2)^{-5/2}\Big] + \sin^2\theta\Big[-\frac{1}{r^2} + \frac{z}{r^2}(r^2+z^2)^{-1/2} + z(r^2+z^2)^{-3/2}\Big]\Big\}$$

$$= \frac{p_0 z}{2}\Big\{2(1+\nu)\Big[(b^2+z^2)^{-1/2} - \frac{1}{z}\Big] - z^2(b^2+z^2)^{-3/2} + \frac{1}{z}\Big\} \qquad (5.50)$$

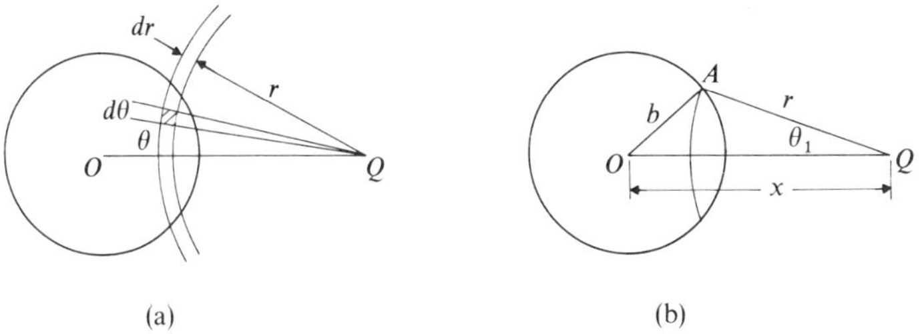

FIGURE 5.10

For $z \to 0$, that is at the center of the area of application of the loading, the above expressions converge to: $u_z = 2(1-\nu^2)p_0 b/E$, $\sigma_z = -p_0$, $\sigma_r = \sigma_\theta = -(1+2\nu)p_0/2$.

Distributed Normal Load on Circular Area of a Semi-Infinite Solid, General Case

Figure 5.10(a) shows a view looking in the z direction of the uniformly loaded circular area with a radius b and center at the origin O; also shown is a general point Q with coordinates z, x, at which we wish to determine the displacements and stresses due to this loading. There are actually two cases which require separate consideration, the case shown in the figure in which Q is outside the loaded area, and the case when it is inside. We will consider only the first case, since this is sufficient to calculate the results shown later in Fig. 5.11; the second case can be studied by similar methods.

The distributed load can be separated into elemental loads in various ways. In the figure the elemental load is $p_0 r\, d\theta\, dr$, where the distance r varies between $x-b$ and $x+b$ and the angle θ varies between $-\theta_1$ and θ_1, Fig. 5.10(b). From this figure, using the cosine law for the triangle OQA:

$$\theta_1 = \cos^{-1}\frac{x^2+r^2-b^2}{2xr} \tag{5.51}$$

Again the displacement u_z and stress σ_z are in the same direction for all elemental loads and require only simple summing up. Using Eqs. (5.48):

$$\begin{aligned} u_z &= \frac{(1+\nu)p_0}{2\pi E}\int_{x-b}^{x+b} dr \int_{-\theta_1}^{\theta_1} d\theta[(3-2\nu)r(r^2+z^2)^{-1/2} - r^3(r^2+z^2)^{-3/2}] \\ &= \frac{(1+\nu)p_0}{\pi E}\int_{x-b}^{x+b} dr\, \theta_1[(3-2\nu)r(r^2+z^2)^{-1/2} - r^3(r^2+z^2)^{-3/2}] \\ \sigma_z &= -\frac{3p_0 z^3}{2\pi}\int_{x-b}^{x+b} dr \int_{-\theta_1}^{\theta_1} d\theta\, r(r^2+z^2)^{-5/2} = -\frac{3p_0 z^3}{\pi}\int_{x-b}^{x+b} dr\, \theta_1 r(r^2+z^2)^{-5/2} \end{aligned} \tag{5.52}$$

The displacement u_r for each elemental load differs in direction by the angle θ from the resultant u_r desired, so we must sum up the components of the elemental displacements, obtained by multiplying them by $\cos\theta$, or:

$$u_r = \frac{(1+\nu)p_0}{2\pi E}\int_{x-b}^{x+b} dr \int_{-\theta_1}^{\theta_1} d\theta \cos\theta \left[-\frac{1-2\nu}{r} + \frac{(1-2\nu)z}{r}(r^2+z^2)^{-1/2} + zr(r^2+z^2)^{-3/2}\right]$$

$$= \frac{(1+\nu)p_0}{\pi E}\int_{x-b}^{x+b} dr \sin\theta_1 \left[-\frac{1-2\nu}{r} + \frac{(1-2\nu)z}{r}(r^2+z^2)^{-1/2} + zr(r^2+z^2)^{-3/2}\right] \quad (5.53)$$

The calculation of σ_r and σ_θ is similar to that previously given for points on the z axis except for the integration limits:

$$\sigma_r = \int_{x-b}^{x+b} dr \int_{-\theta_1}^{\theta_1} d\theta\, r\sigma_x, \qquad \sigma_\theta = \int_{x-b}^{x+b} dr \int_{-\theta_1}^{\theta_1} d\theta\, r\sigma_y$$

Taking σ_x as before and similarly $\sigma_y = \sigma_r \sin^2\theta + \sigma_\theta \cos^2\theta$, and using Eqs. (5.48), these yield the following expressions for σ_r, σ_θ:

$$\sigma_r = \frac{(1-2\nu)p_0}{2\pi}\int_{x-b}^{x+b} dr\left[A \sin 2\theta_1 + B\left(\theta_1 - \frac{1}{2}\sin 2\theta_1\right) - C\left(\theta_1 + \frac{1}{2}\sin 2\theta_1\right)\right]$$

$$(5.54)$$

$$\sigma_\theta = \frac{(1-2\nu)p_0}{2\pi}\int_{x-b}^{x+b} dr\left[-A \sin 2\theta_1 + B\left(\theta_1 + \frac{1}{2}\sin 2\theta_1\right) - C\left(\theta_1 - \frac{1}{2}\sin 2\theta_1\right)\right]$$

where $A = \frac{1}{r} - \frac{z}{r}(r^2+z^2)^{-1/2}$, $\quad B = zr(r^2+z^2)^{-3/2}$, $\quad C = \frac{3}{1-2\nu} zr^3(r^2+z^2)^{-5/2}$

Using Eqs. (3.9c) and remembering that $\sigma_{\theta z} = 0$, for the elementary loads: $\sigma_{xz} = \sigma_{rz}\cos\theta$, and the resultant σ_{rz} is then, using Eqs. (5.48):

$$\sigma_{rz} = \int_{x-b}^{x+b} dr \int_{-\theta_1}^{\theta_1} d\theta\, r\sigma_{xz} = -\frac{3p_0}{2\pi}\int_{x-b}^{x+b} dr \int_{-\theta_1}^{\theta_1} d\theta \cos\theta\, z^2r^2(r^2+z^2)^{-5/2}$$

$$= -\frac{3p_0}{\pi}\int_{x-b}^{x+b} dr \sin\theta_1 z^2r^2(r^2+z^2)^{-5/2} \quad (5.55)$$

Because of the fact that θ_1 is a rather complex function of r, direct integration over r of the above expressions seems impractical. However, numerical or graphical integrations can readily be carried out for any particular value of x. The results shown in Fig. 5.11 were obtained by graphical integration, by calculating the value of the integrand for eight values of r, plotting against r and measuring the area under the curve.

Figure 5.11(b) shows the local stress field due to the balanced load system shown in Fig. 5.11(a). This consists of a uniformly distributed tensile load per unit area $p_0 = -1000$ acting over a circular area of radius b on the surface of a semi-infinite solid, balanced by a concentrated compressive load $P = -\pi b^2 p_0 = 1000\,\pi b^2$ acting at the center of the area, the stresses due to P being found from Eqs. (5.48). The stresses are calculated for points on a plane of symmetry and separated by distances of $2b$. The stresses indicated by diagonal arrows are of course σ_θ. Figures 5.11(c), (d) shows similar results calculated for the balanced load system consisting of the same tensile loading $p_0 = -1000$ distributed over an

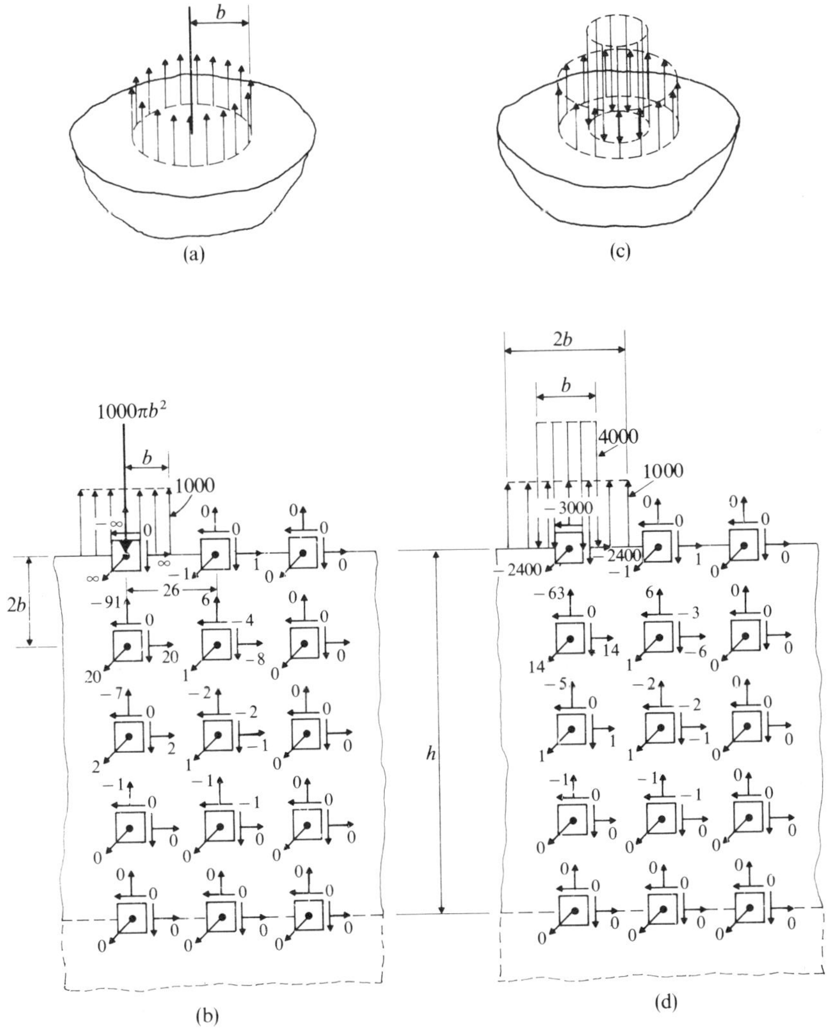

FIGURE 5.11

area of radius b balanced by another distributed compressive loading of intensity 4000 acting on a concentric circular area of radius $b/2$ (the net loading on this smaller area is of course a compression of $4000 - 1000 = 3000$). Other balanced load systems can be considered in a similar manner.

It will be seen that the stresses in such a balanced system die out rapidly as predicted by Saint-Venant's principle. Since the values of the stresses are shown to the nearest whole number they are less than 0·05 of one percent of the distributed loading at a distance from the surface of $h = 8b$. Hence the stress systems shown could be used without serious error for a plate of thickness h or even less, loaded as shown on one face.

Corrections to Classical Plate Theory for Concentrated Normal Load

As in the analogous case of a concentrated load on a beam, which was discussed in Sec. 3.4, the difference between the actual stresses produced by a concentrated load acting on one face of a plate and the stresses predicted by the classical theory for the same case form a local stress field which, when superposed upon the classical solution, gives the correct stresses.

We will derive such a corrective local stress field in the same general way as was suggested as a possible method for the corresponding beam case. We will consider a circular portion of the plate around the load, with a diameter $2a$ of four times the plate thickness, outside of which the local stresses will be negligible. The Boussinesq solution (5.48) can then be used for a concentrated normal load on one face of the plate, cancelling the normal and tangential stresses which this solution requires on the opposite face by using the series solutions (5.36) and (5.45). We can then subtract the classical solution (4.118) for the case of a concentrated load at the center of a disc with diameter four times the thickness; any remaining resultant forces on the edge of the circular portion of the disc studied can be removed by simple solutions for pure tension and pure bending.

This procedure is perfectly practical, but in the form stated it has one serious disadvantage. As was mentioned in the discussion following Eqs. (4.118), the classical theory predicts an infinite bending moment at the point of application of a concentrated load, and hence infinite bending stresses at every point of the line normal to the middle surface joining the point of application of the load and the corresponding point on the opposite face. Actually the stresses are finite everywhere except at the point of application of the load. The corrective local stress field must therefore provide infinite stresses of opposite sign for other points along this line. But when we come to superpose this corrective stress field on the classical solution, we will get an indeterminate quantity at these points, infinity minus infinity, which would give no clue to the correct finite values which the net stresses should actually have there.

To avoid this difficulty we will consider, instead of a concentrated load on one face, a uniformly distributed pressure p_0 acting over a small circular area having a diameter of one quarter of the thickness of the plate, or a radius $b = c/4$. The procedure for obtaining the corrective stress field for this case, which is shown in Fig. 5.12(a), is then the same as that outlined above except that we will subtract the stress field of Fig. 5.11(b) from the exact solution for a concentrated load, and use the classical solution of Eqs. (4.114a, . . . d) for the case of a distributed load on a disc, instead of solution (4.118). In this way we have finite stresses everywhere; if we wish to convert the final corrected result from a distributed to a concentrated load we need only add the field of Fig. 5.11(b) to this result.

Calculation of Corrective Stress Field

In deriving Eqs. (5.48) for a concentrated normal load P on a semi-infinite solid, the origin of coordinates was taken at the surface. For convenience in combining

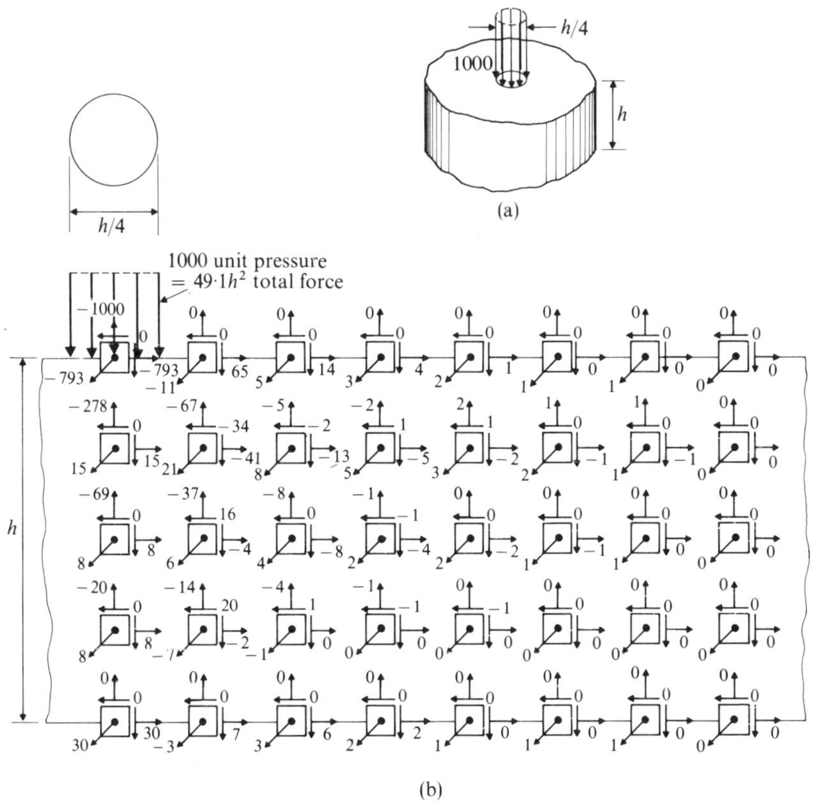

FIGURE 5.12
Correct stresses minus classical plate theory stresses ($\nu = 0{\cdot}3$)

with the other solutions we will convert Eqs. (5.48) so as to put the origin in the plate middle surface, by replacing z by $z + c$; at the same time we will make use of the dimensionless coordinates $\mathbf{r} = r/c$, $\mathbf{z} = z/c$, and replace P by $\pi c^2 p_0/16$. We thus obtain the expressions for the stresses:

$$\sigma_r = \frac{(1-2\nu)p_0}{32}\left\{\frac{1}{\mathbf{r}^2} - \left[\frac{1}{\mathbf{r}^2}(\mathbf{r}^2+\mathbf{z}^2+2\mathbf{z}+1)^{-1/2} + \frac{3\mathbf{r}^2}{1-2\nu}(\mathbf{r}^2+\mathbf{z}^2+2\mathbf{z}+1)^{-5/2}\right](\mathbf{z}+1)\right\}$$

$$\sigma_\theta = \frac{(1-2\nu)p_0}{32}\left\{-\frac{1}{\mathbf{r}^2} + \left[\frac{1}{r^2}(\mathbf{r}^2+\mathbf{z}^2+2\mathbf{z}+1)^{-1/2} + (\mathbf{r}^2+\mathbf{z}^2+2\mathbf{z}+1)^{-3/2}\right](\mathbf{z}+1)\right\}$$

$$\sigma_z = -\frac{3p_0}{32}(\mathbf{z}+1)^3(\mathbf{r}^2+\mathbf{z}^2+2\mathbf{z}+1)^{-5/2}, \qquad \sigma_{rz} = -\frac{\mathbf{r}}{(\mathbf{z}+1)}\sigma_z, \qquad \sigma_{r\theta}, \sigma_{\theta z} = 0 \tag{5.56}$$

On the bottom surface of the plate, $z = c$, $\mathbf{z} = 1$, we then have stresses $\sigma_z = 3p_0(\mathbf{r}^2+4)^{-5/2}/4$, $\sigma_{rz} = -3p_0\mathbf{r}(\mathbf{r}^2+4)^{-5/2}/8$. To free the bottom surface of loads we superpose solutions (5.36), (5.45) in which t_z, t_r, $\mathbf{t}_z$, $\mathbf{t}_r = 0$, and

$-b_z + 3p_0(\mathbf{r}^2+4)^{-5/2}/4 = 0$, $-b_r - 3p_0\mathbf{r}(\mathbf{r}^2+4)^{-5/2}/8 = 0$. To satisfy conditions (5.37), (5.44) it can readily be checked that:

$$\mathbf{b}_z = -\frac{p_0}{128}(\mathbf{r}^2+8)\operatorname{csch}^{-1}\frac{\mathbf{r}}{2} + \frac{p_0}{64}(\mathbf{r}^2+4)^{1/2} + A_1\log\mathbf{r} + A_2\mathbf{r}^2\log\mathbf{r} + A_3 + A_4\mathbf{r}^2$$

$$\nabla^2\mathbf{b}_z = -\frac{p_0}{32}\operatorname{csch}^{-1}\frac{\mathbf{r}}{2} + \frac{p_0}{16}(\mathbf{r}^2+4)^{-1/2} + 4A_2\log\mathbf{r} + 4(A_2+A_4) \qquad (5.57)$$

$$\nabla^4\mathbf{b}_z = b_z = -\frac{3}{4}p_0(\mathbf{r}^2+4)^{-5/2}, \quad \nabla^2 b_z = -\frac{15p_0}{4}(5\mathbf{r}^2-8)(\mathbf{r}^2+4)^{-9/2}, \text{ etc.}$$

$$\mathbf{b}_r = -\frac{p_0}{8\mathbf{r}}(\mathbf{r}^2+4)^{-1/2} + \frac{p_0}{16\mathbf{r}} + \frac{B_1}{\mathbf{r}} + B_2\mathbf{r}$$

$$\Delta^2\mathbf{b}_r = b_r = -\frac{3}{8}p_0\mathbf{r}(\mathbf{r}^2+4)^{-5/2} \qquad \Delta^2 b_r = -\frac{15}{8}p_0\mathbf{r}(3\mathbf{r}^2-16)(\mathbf{r}^2+4)^{-9/2}, \text{ etc.} \qquad (5.58)$$

where A_1, A_2, A_3, A_4, B_1, B_2 can have any values, and can be used to satisfy the boundary condition that the final stresses are zero on the boundary surface $\mathbf{r} = 4$ of the disc which we are studying. Actually σ_θ, σ_{rz} are found to be zero within the limits of accuracy of the results shown in Fig. 5.12 without using these constants; σ_r can be made equally close to zero by superposing simple tension and pure bending, which are included in the stresses defined by these constants, but which can be obtained more easily from Eqs. (3.10c, d). Hence we will take these constants as zero.

The classical solution for this case, using Eqs. (4.114a, . . . d) and taking $\nu = 0{\cdot}3$, $a = 4c$, $b = c/4$, $P' = b^2p_0/(64D) = 3(1-\nu^2)p_0/(2048Ec)$, gives, from Eq. (4.114d):

$$C_1 = -\left[1 - \frac{1-\nu}{1024} + (1+\nu)\log 16\right]\frac{c^2p_0}{128(1+\nu)D} = -0{\cdot}0277\frac{c^2p_0}{D}$$

$$C_1' = -\left[3+\nu-\frac{1-\nu}{512}\right]\frac{c^2p_0}{256(1+\nu)D} = -0{\cdot}009\,91\frac{c^2p_0}{D}, \qquad (5.58a)$$

$$C_2' = \frac{c^4p_0}{4096D}, \qquad C_3' = \frac{c^2p_0}{128D}$$

Using expression (4.108) for w and $D = 2Ec^3/3(1-\nu^2)$, the stresses at $r \leqq c/4$ are:

$$\sigma_r = -\frac{Ez}{1-\nu^2}\left(\frac{d^2w}{dr^2} + \frac{\nu}{r}\frac{dw}{dr}\right) = -\frac{Ez}{1-\nu^2}\left[\frac{3{\cdot}3p_0}{16D}r^2 + 2{\cdot}6C_1\right] = (0{\cdot}1079 - 1{\cdot}238\mathbf{r}^2)p_0\mathbf{z}$$

$$\sigma_\theta = -\frac{Ez}{1-\nu^2}\left(\nu\frac{d^2w}{dr^2} + \frac{1}{r}\frac{dw}{dr}\right) = -\frac{Ez}{1-\nu^2}\left[\frac{(1+3\nu)p_0}{16D}r^2 + 2(1+\nu)C_1\right] \qquad (5.59)$$

$$= (0{\cdot}1079 - 0{\cdot}7125\mathbf{r}^2)p_0\mathbf{z}$$

$$\sigma_{rz} = -\frac{E(c^2-z^2)}{2(1-\nu^2)}\left(\frac{d^3w}{dr^3} + \frac{1}{r}\frac{d^2w}{dr^2} - \frac{1}{r^2}\frac{dw}{dr}\right) = -\frac{E(c^2-z^2)p_0}{4(1-\nu^2)D}r = -0{\cdot}375p_0(1-\mathbf{z}^2)\mathbf{r}$$

Similarly, using expression (4.114a) for w' and corresponding formulas for the stresses, the stresses in the region $r \geqq c/4$ are:

$$\sigma_r = \frac{Ez}{1-\nu^2}\left[-2{\cdot}6C_1' + \frac{0{\cdot}7}{\mathbf{r}^2}C_2' - \left(3{\cdot}3 + 2{\cdot}6\log\frac{\mathbf{r}}{4}\right)C_3'\right] = \left(\frac{25{\cdot}63}{\mathbf{r}^2} - 1{\cdot}6 - 3047\log\frac{\mathbf{r}}{4}\right)\frac{p_0\mathbf{z}}{10^5}$$

$$\sigma_\theta = \frac{Ez}{1-\nu^2}\left[-2{\cdot}6C_1' - \frac{0{\cdot}7}{\mathbf{r}^2}C_2' - \left(1{\cdot}9 + 2{\cdot}6\log\frac{\mathbf{r}}{4}\right)C_3'\right] \tag{5.60}$$

$$= \left(-\frac{25{\cdot}63}{\mathbf{r}^2} + 1641 - 3047\log\frac{\mathbf{r}}{4}\right)\frac{p_0\mathbf{z}}{10^5}$$

$$\sigma_{rz} = -\frac{2E(c^2-z^2)C_3'}{1-\nu^2}\frac{C_3'}{r} = -0{\cdot}02344\,p_0\frac{(1-\mathbf{z}^2)}{\mathbf{r}}$$

This classical solution is then subtracted from the exact solution to obtain the correction to the classical solution (so that, when added to the classical solution, it gives the exact solution). The exact solution is obtained as outlined above, by superposing solution (5.48) and solutions (5.36) and (5.45) in which $\mathbf{b}_z$ and $\mathbf{b}_r$ are taken as given by Eqs. (5.57), (5.58), sufficient magnitude of solutions (3.10c, d) being added to bring the resultant σ_r stresses to zero at $\mathbf{r} = 4$; this gives the exact solution for a disc with a concentrated load, which is then converted to the same distributed load as was used for the classical solution by superposing the local stress field of Fig. 5.11(b).

Figure 5.12 shows the final results of those calculations in which p_0 is taken as 1000.

5.4 ELASTICITY SOLUTIONS FOR PLATES UNLOADED ON FACES

Introduction

By setting the loading equal to zero in any of the series solutions in loading functions which were presented in Sec. 5.2, we obtain exact closed solutions for plates with unloaded faces, which can be used for satisfying plate boundary conditions. Such solutions are of course also useful for problems which involve only edge loading, such as the 'plane stress' problems studied in Sec. 3.2 (for which only approximate general solutions were given there), as well as for corresponding flexural problems involving antisymmetrical edge loading.

Of course other solutions of the problem of plates with unloaded faces are possible, and could theoretically be derived directly from the general elasticity solutions of Table 3.1; we will find one such type of solution, which is given by Eqs. (5.68), (5.69) below, particularly useful. Even the solutions which we will obtain from series solutions in loading functions could theoretically be derived directly from the general solutions of Table 3.1, but this would be difficult unless one knew in advance how to go about it, except in such an obvious case as Eqs. (5.62) below.

Solutions Involving Harmonic Functions

For example, if we take t_z, b_z, $(\mathbf{t}_z - \mathbf{b}_z) = 0$ in Eqs. (5.19) and let $\nabla^2(\mathbf{t}_z + \mathbf{b}_z) = -(2/\nu)\psi(x, y)$, where $\nabla^2\psi = 0$ to satisfy Eqs. (5.18a), we obtain the previous exact plane stress solution Eqs. (3.12b, c). A more novel solution can be obtained from Eqs. (5.19) by taking t_z, b_z, $(\mathbf{t}_z + \mathbf{b}_z) = 0$, $(\mathbf{t}_z - \mathbf{b}_z) = -\{2/[3(1-\nu)]\} \int dx\, \psi$, where $\nabla^2\psi = 0$. Taking ψ as a harmonic function satisfies Eqs. (5.18a) (although these conditions would also be satisfied if ψ is taken as a biharmonic function, which leads to a more general solution, as will be given later). Using the relation $\partial^2\psi/\partial x^2 + \partial^2\psi/\partial y^2 = 0$, this gives us the solution:

$$\frac{E}{1+\nu} u_x = -z\frac{\partial\psi}{\partial x}, \quad \frac{E}{1+\nu} u_y = -z\frac{\partial\psi}{\partial y}, \quad \frac{E}{1+\nu} u_z = \psi;$$
$$\sigma_x = -\sigma_y = z\frac{\partial^2\psi}{\partial y^2}, \quad \sigma_{xy} = -z\frac{\partial^2\psi}{\partial x\,\partial y}, \quad \sigma_z, \sigma_{xz}, \sigma_{yz} = 0, \quad \nabla^2\psi(x, y) = 0. \tag{5.61}$$

This solution can also be obtained from Eqs. (5.32), by taking t_x, b_x, $(\mathbf{t}_x - \mathbf{b}_x) = 0$, and using $(\mathbf{t}_x + \mathbf{b}_x)$ in the same way as $(\mathbf{t}_z - \mathbf{b}_z)$ above.

The exact solution (5.61) is the antisymmetric (flexural) analog of the exact symmetric (membrane) solution (3.12b, c). Investigation also shows that, just as Eqs. (3.12a) give a slightly more general version of Eqs. (3.12b, c) by use of what we have called a 'plu-harmonic' function, Eqs. (5.61) can similarly be expanded to the more general form:

$$\frac{Eu_x}{1+\nu} = -z\frac{\partial\psi}{\partial x} + \mathbf{b}xyz + \frac{\mathbf{a}}{2}y^2z, \qquad \frac{Eu_y}{1+\nu} = -z\frac{\partial\psi}{\partial y} + \mathbf{a}xyz + \frac{\mathbf{b}}{2}x^2z$$
$$\frac{Eu_z}{1+\nu} = \psi - \frac{1}{2}(\mathbf{a}xy^2 + \mathbf{b}x^2y + \nu\mathbf{c}z^2),$$
$$\sigma_x = z\frac{\partial^2\psi}{\partial y^2} - \mathbf{a}zx + \mathbf{c}z, \qquad \sigma_y = z\frac{\partial^2\psi}{\partial x^2} - \mathbf{b}yz + \mathbf{c}z \tag{5.61a}$$
$$\sigma_{xy} = -z\frac{\partial^2\psi}{\partial x\,\partial y} + \mathbf{b}zx + \mathbf{a}yz, \quad \sigma_z, \sigma_{xz}, \sigma_{yz} = 0$$
$$\nabla^2\psi(x, y) = \mathbf{a}x + \mathbf{b}y - (1-\nu)\mathbf{c}$$

where **a**, **b**, **c** are arbitrary constants.

Alternative Solutions Involving Harmonic Functions

Taking the symmetric case first, let t_x, b_x, $(\mathbf{t}_x + \mathbf{b}_x) = 0$ in Eqs. (5.32), and take $(\mathbf{t}_x - \mathbf{b}_x) = -\int dx\, \phi$, where $\nabla^4\phi = 0$ to satisfy Eqs. (5.31a). This gives us expressions for the displacements which are 'unsymmetrical' with respect to x and y, meaning that they do not remain the same when x and y are interchanged in both the coordinates and the subscripts. The expressions are:

$$Eu_x/(1+\nu) = [(1-\nu)/2]\partial\phi/\partial x + \int dx\, \partial^2\phi/\partial y^2 + [\nu(3z^2+1)/12]\nabla^2\partial\phi/\partial x;$$

$$Eu_y/(1+\nu) = -[(1+\nu)/2]\partial\phi/\partial y + [\nu(3z^2+1)/12]\nabla^2\partial\phi/\partial y;$$

$$Eu_z/(1+\nu) = -(\nu/2)z\nabla^2\phi$$

We can get a similar but different solution by interchanging x and y in both the coordinates and subscripts (or this could be obtained from the series solution in loading functions t_y and b_y). These two solutions can be exchanged for two solutions in more convenient form by adding them together (that is adding the expressions for each displacement) and by similarly subtracting them from each other.

Thus if we subtract them from each other and substitute $\nabla^2\phi = \partial^2\psi/\partial x\partial y$, $\nabla^2\psi = 0$, we obtain the solution:

$$\frac{Eu_x}{1+\nu} = \frac{\partial\psi}{\partial y}, \quad \frac{Eu_y}{1+\nu} = -\frac{\partial\psi}{\partial x}, \quad u_z = 0; \quad \sigma_x = -\sigma_y = \frac{\partial^2\psi}{\partial x\,\partial y} \tag{5.62}$$

$$\sigma_{xy} = \frac{\partial^2\psi}{\partial y^2}, \quad \sigma_z, \sigma_{xz}, \sigma_{yz} = 0, \; \nabla^2\psi(x, y) = 0$$

Although it looks quite different, this solution is essentially the same as solution (3.12b, c), from which it can be derived by applying the operator $\partial/\partial x$, using $\partial^2\psi/\partial x^2 = -\partial^2\psi/\partial y^2$ (from $\nabla^2\psi = 0$) and then integrating with respect to y. However, it may be more convenient in this form for some purposes. This is also obviously essentially the same as general solution 1, Table 3.1, in which ψ is taken as a function of x and y only.

The antisymmetric solution (5.61) can be modified in the same way to:

$$\frac{E}{1+\nu}u_x = z\frac{\partial\psi}{\partial y}, \quad \frac{E}{1+\nu}u_y = -z\frac{\partial\psi}{\partial x}, \quad \frac{E}{1+\nu}u_z = \int dy \frac{\partial\psi}{\partial x};$$

$$\sigma_x = -\sigma_y = z\frac{\partial^2\psi}{\partial x\,\partial y}, \quad \sigma_{xy} = z\frac{\partial^2\psi}{\partial y^2}, \quad \sigma_z, \sigma_{xz}, \sigma_{yz} = 0, \; \nabla^2\psi(x, y) = 0 \tag{5.63}$$

as can be the 'expanded' Eqs. (3.12a) and (5.61a).

Solutions Involving Biharmonic Functions

If, instead of subtracting the above mentioned solutions, we add them together, multiply by $(1+\nu)$, and simplify the result by also adding solution (3.12b, c) in which we have substituted $\psi = (\nu/6)\nabla^2\phi$ where $\nabla^4\phi = 0$, we obtain the following exact symmetric (membrane) solution involving the biharmonic function ϕ:

$$Eu_x = \int dx \frac{\partial^2\phi}{\partial y^2} - \nu\frac{\partial\phi}{\partial x} + \frac{\nu}{2}z^2\nabla^2\frac{\partial\phi}{\partial x}, \quad Eu_y = \int dy \frac{\partial^2\phi}{\partial x^2} - \nu\frac{\partial\phi}{\partial y} + \frac{\nu}{2}z^2\nabla^2\frac{\partial\phi}{\partial y},$$

$$Eu_z = -\nu z\nabla^2\phi; \quad \sigma_x = \frac{\partial^2\phi}{\partial y^2} - \frac{\nu}{2(1+\nu)}z^2\nabla^2\frac{\partial^2\phi}{\partial y^2}, \tag{5.64}$$

$$\sigma_y = \frac{\partial^2\phi}{\partial x^2} - \frac{\nu}{2(1+\nu)} z^2\nabla^2 \frac{\partial^2\phi}{\partial x^2}, \quad \sigma_{xy} = -\frac{\partial^2\phi}{\partial x\,\partial y} + \frac{\nu}{2(1+\nu)} z^2\nabla^2 \frac{\partial^2\phi}{\partial x\,\partial y},$$

$$\sigma_z, \sigma_{xz}, \sigma_{yz} = 0, \quad \nabla^4\phi(x, y) = 0 \qquad (5.64 \text{ cont.})$$

If ϕ is also a harmonic function, that is if $\nabla^2\phi = \nabla^2\psi = 0$, then the last terms in each of the above expressions would drop out, $\int dx\, \partial^2\phi/\partial y^2 = -\int dx\, \partial^2\phi/\partial x^2 = -\partial\phi/\partial x$, etc., and the solution coincides with solution (3.12b, c).

For the antisymmetric (flexure) case, we can obtain an analogous exact solution involving a biharmonic function by taking t_z, b_z, $(\mathbf{t}_z + \mathbf{b}_z) = 0$ in Eqs. (5.19), and $(\mathbf{t}_z - \mathbf{b}_z) = -2\phi/3$, where $\nabla^4\phi = 0$ to satisfy Eqs. (5.18a). The resulting solution can be simplified by adding to it solution (5.61) in which we have substituted $\psi = -[(2+3\nu)/10]\nabla^2\phi$, where $\nabla^4\phi = 0$. The final exact antisymmetric solution is then:

$$\frac{Eu_x}{1+\nu} = (1-\nu)z\frac{\partial\phi}{\partial x} - \frac{2-\nu}{6} z^3\nabla^2\frac{\partial\phi}{\partial x}, \qquad \frac{Eu_y}{1+\nu} = (1-\nu)z\frac{\partial\phi}{\partial y} - \frac{2-\nu}{6} z^3\nabla^2\frac{\partial\phi}{\partial y}$$

$$\frac{Eu_z}{1+\nu} = -(1-\nu)\phi + \left(c^2 - \frac{\nu}{2} z^2\right)\nabla^2\phi; \qquad (5.65)$$

$$\sigma_x = z\left(\frac{\partial^2\phi}{\partial x^2} + \nu\frac{\partial^2\phi}{\partial y^2}\right) - \frac{2-\nu}{6} z^3\nabla^2\frac{\partial^2\phi}{\partial x^2}, \quad \sigma_y = z\left(\frac{\partial^2\phi}{\partial y^2} + \nu\frac{\partial^2\phi}{\partial x^2}\right) - \frac{2-\nu}{6} z^3\nabla^2\frac{\partial^2\phi}{\partial y^2},$$

$$\sigma_{xy} = (1-\nu)z\frac{\partial^2\phi}{\partial x\,\partial y} - \frac{2-\nu}{6} z^3\nabla^2\frac{\partial^2\phi}{\partial x\,\partial y}, \quad \sigma_{xz} = \frac{c^2 - z^2}{2}\nabla^2\frac{\partial\phi}{\partial x}, \quad \sigma_{yz} = \frac{c^2 - z^2}{2}\nabla^2\frac{\partial\phi}{\partial y},$$

$$\sigma_z = 0; \quad \nabla^4\phi(x, y) = 0$$

Again if ϕ is a harmonic as well as a biharmonic function, letting $\nabla^2\phi = -\nabla^2\psi/(1-\nu) = 0$ and using $\partial^2\psi/\partial x^2 + \partial^2\psi/\partial y^2 = 0$, this reduces to solution (5.61).

So far we have discussed only exact solutions to the problem of plates with unloaded faces. The related equations (3.12b, c) and (5.62) for the membrane case, and (5.61) and (5.63) for the flexural case are somewhat restricted in range of application because they are based on harmonic functions. These are contained in (but are simpler and hence more convenient to use when they suffice) the more general equations (5.64) and (5.65) which are based upon biharmonic functions.

Approximate Solutions Based on Biharmonic Functions

The qualities of generality and convenience can be combined, at the expense of introducing approximations which in many cases can be tolerated, by merely dropping the terms containing $\nabla^2\phi$ in the expressions for the displacements and σ_x, σ_y, σ_{xy} in Eqs. (5.64), (5.65), while continuing to assume that ϕ is a biharmonic function. We must retain the expressions for σ_{xz} and σ_{yz} in Eqs. (5.65) in order to satisfy equilibrium conditions (3.4); these conditions are satisfied in the following equations and must be satisfied in even an approximate solution. We thus obtain:

$$Eu_x = \int dx \frac{\partial^2 \phi}{\partial y^2} - \nu \frac{\partial \phi}{\partial x}, \quad Eu_y = \int dy \frac{\partial^2 \phi}{\partial x^2} - \nu \frac{\partial \phi}{\partial y}; \quad u_z, \sigma_z, \sigma_{zx}, \sigma_{yz} = 0,$$

$$\sigma_x = \frac{\partial^2 \phi}{\partial y^2}, \quad \sigma_y = \frac{\partial^2 \phi}{\partial x^2}, \quad \sigma_{xy} = -\frac{\partial^2 \phi}{\partial x \, \partial y}, \quad \nabla^4 \phi = 0. \tag{5.64a}$$

$$\frac{Eu_x}{1+\nu} = (1-\nu) z \frac{\partial \phi}{\partial x}, \quad \frac{Eu_y}{1+\nu} = (1-\nu) z \frac{\partial \phi}{\partial y}, \quad \frac{Eu_z}{1+\nu} = -(1-\nu)\phi;$$

$$\sigma_x = z\left(\frac{\partial^2 \phi}{\partial x^2} + \nu \frac{\partial^2 \phi}{\partial y^2}\right), \quad \sigma_y = z\left(\frac{\partial^2 \phi}{\partial y^2} + \nu \frac{\partial^2 \phi}{\partial x^2}\right), \quad \sigma_{xy} = (1-\nu) z \frac{\partial^2 \phi}{\partial x \, \partial y},$$

$$\sigma_z = 0, \quad \sigma_{xz} = \frac{c^2 - z^2}{2} \nabla^2 \frac{\partial \phi}{\partial x}, \quad \sigma_{yz} = \frac{c^2 - z^2}{2} \nabla^2 \frac{\partial \phi}{\partial y}, \quad \nabla^4 \phi = 0. \tag{5.65a}$$

It will be seen that the solution (5.64a) is essentially the traditional Airy-function plane stress elasticity solution (3.16a, b, c), while the solution (5.65a) is its antisymmetric analog involving similar approximations. It can also be seen that the errors in the traditional solution as well as in its flexural analog consist in ignoring most of the nonlinear (in z), higher order (in x, y) terms involving $\nabla^2\phi$. This is consistent with the discussion of errors in Sec. 3.2. We now have, in Eqs. (5.64) and (5.65) a means of eliminating these errors at the cost of some added complexity both in their application and in the satisfaction of edge boundary conditions. While the approximate equations (5.64a) and (5.65a) satisfy equilibrium conditions, they only partially satisfy the continuity conditions between the stresses and the displacements expressed by Eqs. (3.7a) or (3.7b).

Exact Solutions Derived from General Solutions of Table 3.1

In Table 3.1 there is only one solution in which the displacement u_z in the z direction (which we are taking as normal to the plate middle surface) is zero; this is solution number 1, which can be written as: $Eu_x/(1+\nu) = \partial\psi/\partial y$, $Eu_y/(1+\nu) = -\partial\psi/\partial x$, $u_z = 0$, where $\nabla^2\psi(x, y, z) = 0$. The fact that u_z is zero and the associated fact that $e \propto \partial^2\psi/\partial x\,\partial y - \partial^2\psi/\partial x\,\partial y + 0 = 0$ means that, from Eq. (3.7b), $\sigma_z = 0$. Moreover σ_{xz} and σ_{yz} are both proportional to the first derivative of ψ with respect to z. The latter fact makes it easy to satisfy the boundary conditions (with the origin of coordinates in the middle surface):

$$z = \pm c: \quad \sigma_z, \sigma_{xz}, \sigma_{yz} = 0 \tag{5.66}$$

The simplest and probably most useful way to satisfy these boundary conditions is to take ψ as a trigonometric function of z, for example

$$\psi(x, y, z) = \sum_m \psi_m(x, y) \cos \frac{m\pi z}{2c} \tag{5.67a}$$

The condition that ψ is a harmonic function then requires that:

$$\nabla^2\psi(x, y, z) = \sum_m \left(\frac{\partial^2 \psi_m}{\partial x^2} + \frac{\partial^2 \psi_m}{\partial y^2} - \frac{m^2\pi^2}{4c^2}\psi_m\right) \cos \frac{m\pi z}{2c} = 0,$$

from which, since $\cos m\pi z/2c$ is not in general zero:

$$\frac{\partial^2 \psi_m}{\partial x^2} + \frac{\partial^2 \psi_m}{\partial y^2} - \frac{m^2\pi^2}{4c^2}\psi_m = 0$$

$$\nabla^2 \psi_m(x, y) = \frac{m^2\pi^2}{4c^2}\psi_m \tag{5.67b}$$

The complete solution is then:

$$\frac{Eu_x}{1+\nu} = \sum_m \frac{\partial \psi_m}{\partial y}\cos\frac{m\pi z}{2c}, \quad \frac{Eu_y}{1+\nu} = -\sum_m \frac{\partial \psi_m}{\partial x}\cos\frac{m\pi z}{2c}, \quad u_z, e, \sigma_z = 0$$

$$\sigma_x = -\sigma_y = \sum_m \frac{\partial^2 \psi_m}{\partial x\,\partial y}\cos\frac{m\pi z}{2c}, \quad \sigma_{xy} = \frac{1}{2}\sum_m \left(\frac{\partial^2 \psi_m}{\partial y^2} - \frac{\partial^2 \psi_m}{\partial x^2}\right)\cos\frac{m\pi z}{2c} \tag{5.68}$$

$$\sigma_{xz} = -\frac{\pi}{4c}\sum_m m\frac{\partial \psi_m}{\partial y}\sin\frac{m\pi z}{2c}, \quad \sigma_{yz} = \frac{\pi}{4c}\sum_m m\frac{\partial \psi_m}{\partial x}\sin\frac{m\pi z}{2c},$$

$$\nabla^2 \psi_m(x, y) = \frac{m^2\pi^2}{4c^2}\psi_m$$

It is evident that the boundary conditions (5.66) will be satisfied if m is an even integer, $m = 2, 4, 6, \ldots$. They are of course also satisfied if $m = 0$, but then solution (5.68) becomes identical with the simpler previous solution (5.62) since, by (5.67b), $\nabla^2\psi_m = 0$ in this case, so $(\partial^2\psi_m/\partial y^2 - \partial^2\psi_m/\partial x^2)/2 = \partial^2\psi_m/\partial y^2$ in the expression for σ_{xy}.

Equations (5.68) define symmetric (membrane) solutions. By using $\sin m\pi z/2c$ in Eq. (5.67a) in place of $\cos m\pi z/2c$ we obtain the analogous antisymmetric (flexural) case. This gives

$$\frac{Eu_x}{1+\nu} = \sum_m \frac{\partial \psi_m}{\partial y}\sin\frac{m\pi z}{2c}, \quad \frac{Eu_y}{1+\nu} = -\sum_m \frac{\partial \psi_m}{\partial x}\sin\frac{m\pi z}{2c}, \quad u_z, e, \sigma_z = 0$$

$$\sigma_x = -\sigma_y = \sum_m \frac{\partial^2 \psi_m}{\partial x\,\partial y}\sin\frac{m\pi z}{2c}, \quad \sigma_{xy} = \frac{1}{2}\sum_m \left(\frac{\partial^2 \psi_m}{\partial y^2} - \frac{\partial^2 \psi_m}{\partial x^2}\right)\sin\frac{m\pi z}{2c}$$

$$\sigma_{xz} = \frac{\pi}{4c}\sum_m m\frac{\partial \psi_m}{\partial y}\cos\frac{m\pi z}{2c}, \quad \sigma_{yz} = -\frac{\pi}{4c}\sum_m m\frac{\partial \psi_m}{\partial x}\cos\frac{m\pi z}{2c},$$

$$\nabla^2 \psi_m(x, y) = \frac{m^2\pi^2}{4c^2}\psi_m \tag{5.69}$$

where, in order to satisfy the boundary conditions (5.66), m must be an odd integer, $m = 1, 3, 5, \ldots$.

Solutions of the basic relation of Eqs. (5.68) and (5.69), $\nabla^2\psi_m(x, y) = C\psi_m$, where C is a positive constant, must be exponential or exponential-based functions of x and y, since these are the only functions which have the same form as their second derivatives. Thus we can have $\psi_m(x, y) = g(x)h(y)$, where g, h can be exponential, hyperbolic or trigonometric functions. In order for C to be positive no more than one of these functions can be trigonometric. By making one of them an exponential function such as $g = e^{-\alpha x}$, where x is the distance into the plate from one edge, we can obtain a localized solution which is important only in the neighborhood of that edge.

Solutions for Plates with Unloaded Faces in Cylindrical Coordinates

All the above solutions for edge loaded plates with unloaded faces can of course be converted to other coordinate systems. For example, using Eqs. (3.9b, . . . , h) we find the following solutions in cylindrical coordinates. From Eqs. (3.12a) we obtain the symmetric (membrane) solution in terms of a harmonic function:

$$\frac{E}{1+\nu} u_r = -\frac{\partial \psi}{\partial r}, \quad \frac{E}{1+\nu} u_\theta = -\frac{1}{r}\frac{\partial \psi}{\partial \theta}, \quad \frac{E}{1+\nu} u_z = -\nu \mathbf{c} z; \tag{5.70}$$

$$\sigma_r = -\frac{\partial^2 \psi}{\partial r^2} + \nu \mathbf{c}, \quad \sigma_\theta = \frac{\partial^2 \psi}{\partial r^2} + \mathbf{c}, \quad \sigma_{r\theta} = \frac{1}{r^2}\frac{\partial \psi}{\partial \theta} - \frac{1}{r}\frac{\partial^2 \psi}{\partial r\, \partial \theta}, \quad \sigma_z, \sigma_{rz}, \sigma_{\theta z} = 0$$

$$\nabla^2 \psi(r, \theta) = -(1-\nu)\mathbf{c}$$

or, from Eqs. (5.62):

$$\frac{E}{1+\nu} u_r = \frac{1}{r}\frac{\partial \psi}{\partial \theta}, \quad \frac{E}{1+\nu} u_\theta = -\frac{\partial \psi}{\partial r}, \quad u_z = 0; \tag{5.71}$$

$$\sigma_r = -\sigma_\theta = \frac{1}{r}\frac{\partial^2 \psi}{\partial r\, \partial \theta} - \frac{1}{r^2}\frac{\partial \psi}{\partial \theta}, \quad \sigma_{r\theta} = -\frac{\partial^2 \psi}{\partial r^2}, \quad \sigma_z, \sigma_{rz}, \sigma_{\theta z} = \theta, \quad \nabla^2 \psi(r, \theta) = 0$$

From Eqs. (5.61a) we obtain the antisymmetric (flexure) solution in terms of a harmonic function:

$$\frac{E}{1+\nu} u_r = -z\frac{\partial \psi}{\partial r}, \quad \frac{E}{1+\nu} u_\theta = -\frac{z}{r}\frac{\partial \psi}{\theta}, \quad \frac{E}{1+\nu} u_z = \psi - \frac{\nu}{2}\mathbf{c} z^2;$$

$$\sigma_r = -z\frac{\partial^2 \psi}{\partial r^2} + \nu \mathbf{c} z, \quad \sigma_\theta = z\frac{\partial^2 \psi}{\partial r^2} + \mathbf{c} z; \quad \sigma_z, \sigma_{rz}, \sigma_{\theta z} = 0, \tag{5.72}$$

$$\sigma_{r\theta} = z\left(\frac{1}{r^2}\frac{\partial \psi}{\partial \theta} - \frac{1}{r}\frac{\partial^2 \psi}{\partial r\, \partial \theta}\right), \quad \nabla^2 \psi(r, \theta) = -(1-\nu)\mathbf{c}$$

or, from Eqs. (5.63):

$$\frac{E}{1+\nu} u_r = \frac{z}{r}\frac{\partial \psi}{\partial \theta}, \quad \frac{E}{1+\nu} u_\theta = -z\frac{\partial \psi}{\partial r}, \quad \frac{E}{1+\nu} u_z = r\int d\theta \frac{\partial \psi}{\partial r},$$

$$\sigma_r = -\sigma_\theta = z\left(\frac{1}{r}\frac{\partial^2 \psi}{\partial r\, \partial \theta} - \frac{1}{r^2}\frac{\partial \psi}{\partial \theta}\right), \quad \sigma_{rz}, \sigma_{\theta z} = 0, \quad \nabla^2 \psi(r, \theta) = 0 \tag{5.73}$$

The more general exact symmetric solution in terms of a biharmonic function, Eqs. (5.64), becomes:

$$Eu_r = \int dr \left(\nabla^2 \phi - \frac{\partial^2 \phi}{\partial r^2}\right) - \nu \frac{\partial \phi}{\partial r} + \frac{\nu}{2} z^2 \frac{\partial (\nabla^2 \phi)}{\partial r}$$

$$Eu_\theta = r\int d\theta \frac{\partial^2 \phi}{\partial r^2} - \int dr \int d\theta \left(\nabla^2 \phi - \frac{\partial^2 \phi}{\partial r^2}\right) - \frac{\nu}{r}\frac{\partial \phi}{\partial \theta} + \frac{\nu z^2}{2\, r} \nabla^2 \frac{\partial \phi}{\partial \theta} \tag{5.74}$$

$$Eu_z = -\nu z \nabla^2\phi, \quad \sigma_r = \nabla^2\phi - \frac{\partial^2\phi}{\partial r^2} + \frac{\nu z^2}{2(1+\nu)} \frac{\partial^2(\nabla^2\phi)}{\partial r^2} \qquad (5.74\text{ cont.})$$

$$\sigma_\theta = \frac{\partial^2\phi}{\partial r^2} - \frac{\nu z^2}{2(1+\nu)} \frac{\partial^2(\nabla^2\phi)}{\partial r^2}. \quad \sigma_{r\theta} = -\frac{\partial}{\partial r}\left(\frac{1}{r}\frac{\partial\phi}{\partial\theta}\right) + \frac{\nu z^2}{2(1+\nu)} \frac{\partial}{\partial r}\left(\frac{1}{r}\nabla^2\frac{\partial\phi}{\partial\theta}\right)$$

$$\sigma_z, \sigma_{rz}, \sigma_{\theta z} = 0, \quad \nabla^4\phi(r,\theta) = \nabla^2\nabla^2\phi = 0$$

The exact antisymmetric solution in terms of a biharmonic function, Eqs. (5.65), becomes:

$$\frac{E}{1+\nu}u_r = (1-\nu)z\frac{\partial\phi}{\partial r} - \frac{2-\nu}{6}z^3\frac{\partial(\nabla^2\phi)}{\partial r}, \quad \frac{E}{1+\nu}u_\theta = (1-\nu)\frac{z}{r}\frac{\partial\phi}{\partial\theta} - \frac{2-\nu}{6}\frac{z^3}{r}\nabla^2\frac{\partial\phi}{\partial\theta}$$

$$\frac{E}{1+\nu}u_z = -(1-\nu)\phi + \left(c^2 - \frac{\nu}{2}z^2\right)\nabla^2\phi$$

$$\sigma_r = z\left[\nu\nabla^2\phi + (1-\nu)\frac{\partial^2\phi}{\partial r^2}\right] - \frac{2-\nu}{6}z^3\frac{\partial^2(\nabla^2\phi)}{\partial r^2}$$

$$\sigma_\theta = z\left[\nabla^2\phi - (1-\nu)\frac{\partial^2\phi}{\partial r^2}\right] + \frac{2-\nu}{6}z^3\frac{\partial^2(\nabla^2\phi)}{\partial r^2}, \quad \sigma_z = 0 \qquad (5.75)$$

$$\sigma_{r\theta} = (1-\nu)z\frac{\partial}{\partial r}\left(\frac{1}{r}\frac{\partial\phi}{\partial\theta}\right) - \frac{2-\nu}{6}z^3\frac{\partial}{\partial r}\left(\frac{1}{r}\nabla^2\frac{\partial\phi}{\partial\theta}\right)$$

$$\sigma_{rz} = \frac{1}{2}(c^2 - z^2)\frac{\partial(\nabla^2\phi)}{\partial r}, \quad \sigma_{\theta z} = \frac{1}{2}(c^2 - z^2)\frac{1}{r}\nabla^2\frac{\partial\phi}{\partial\theta}, \quad \nabla^4\phi(r,\theta) = 0$$

For transforming Eqs. (5.68) to cylindrical coordinates we write solution 1, Table 3.1a, as $Eu_r/(1+\nu) = (1/r)\partial\psi/\partial\theta$, $Eu_\theta/(1+\nu) = -\partial\psi/\partial r$, $u_z = 0$, and take $\psi(r,\theta,z) = \sum_m \psi_m(r,\theta)\cos(m\pi z/2c)$. From the condition $\nabla^2\psi(r,\theta,z) = 0$ we find that $\nabla^2\psi_m(r,\theta) = m^2\pi^2\psi_m/(4c^2)$, and using this and Eqs. (3.9h):

$$\frac{Eu_r}{1+\nu} = \frac{1}{r}\sum_m \frac{\partial\psi_m}{\partial\theta}\cos\frac{m\pi z}{2c}, \quad \frac{Eu_\theta}{1+\nu} = -\sum_m \frac{\partial\psi_m}{\partial r}\cos\frac{m\pi z}{2c}, \quad u_z, e, \sigma_z = 0$$

$$\sigma_r = -\sigma_\theta = \sum_m \frac{\partial}{\partial r}\left(\frac{1}{r}\frac{\partial\psi_m}{\partial\theta}\right)\cos\frac{m\pi z}{2c}, \quad \sigma_{r\theta} = \sum_m \left(\frac{m^2\pi^2}{8c^2}\psi_m - \frac{\partial^2\psi_m}{\partial r^2}\right)\cos\frac{m\pi z}{2c}$$

$$\sigma_{rz} = -\frac{\pi}{4r}\sum_m m\frac{\partial\psi_m}{\partial\theta}\sin\frac{m\pi z}{2c}, \quad \sigma_{\theta z} = \frac{\pi}{4}\sum_m m\frac{\partial\psi_m}{\partial r}\sin\frac{m\pi z}{2c}, \qquad (5.76)$$

$$\nabla^2\psi_m(r,\theta) = \frac{m^2\pi^2}{4c^2}\psi_m$$

where $m = 2, 4, 6, \ldots$. Similarly Eqs. (5.69) become:

$$\frac{Eu_r}{1+\nu} = \frac{1}{r}\sum_m \frac{\partial\psi_m}{\partial\theta}\sin\frac{m\pi z}{2c}, \quad \frac{Eu_\theta}{1+\nu} = -\sum_m \frac{\partial\psi_m}{\partial r}\sin\frac{m\pi z}{2c}, \quad u_z, e, \sigma_z = 0$$

$$\sigma_r = -\sigma_\theta = \sum_m \frac{\partial}{\partial r}\left(\frac{1}{r}\frac{\partial\psi_m}{\partial\theta}\right)\sin\frac{m\pi z}{2c}, \quad \sigma_{r\theta} = \sum_m \left(\frac{m^2\pi^2}{8c^2}\psi_m - \frac{\partial^2\psi_m}{\partial r^2}\right)\sin\frac{m\pi z}{2c} \qquad (5.77)$$

$$\sigma_{rz} = \frac{\pi}{4r}\sum_m m\frac{\partial\psi_m}{\partial\theta}\cos\frac{m\pi z}{2c}, \qquad \sigma_{\theta z} = -\frac{\pi}{4}\sum_m m\frac{\partial\psi_m}{\partial r}\cos\frac{m\pi z}{2c},$$

$$\nabla^2\psi_m(r, \theta) = \frac{m^2\pi^2}{4c^2}\psi_m$$

where $m = 1, 3, 5, \ldots$.

Solutions of the basic equations of (5.76), (5.77) $\nabla^2\psi_m(r, \theta) = m^2\pi^2\psi_m/(4c^2)$ can be of the type $\psi_m = \sum_n a_{mn}I_n(m\pi r/2c)\sin n\theta$ or $\psi_m = \sum_n a_{mn} K_n(m\pi r/2c)\sin n\theta$, where a_{mn} is a numerical coefficient, and I_n, K_n are modified Bessel functions of the first and second kind, respectively, and of the nth order, with the argument $m\pi r/2c$. We can of course replace $\sin n\theta$ by $\cos n\theta$, n being an integer. I_n is finite for $r = 0$ and tends to infinity for $r \to \infty$, while K_n does the reverse; they are thus useful for discs of finite size without a hole, and for infinite plates with a hole respectively, while they can both be used to satisfy the boundary conditions on the inner and outer boundaries of rings.

Applications of Solutions for Plates With Unloaded Faces

The solutions discussed above are obviously useful both for problems involving plates which are under edge loads only, and for satisfying edge boundary conditions for other plate problems. We considered some membrane ('plane stress') problems of these types in Secs. 3.3 and 3.4, and we are now in a position to obtain solutions of the analogous antisymmetric (flexural) problems by using solutions such as (5.61) and (5.65a). We can also use Eqs. (5.64), (5.65) or (5.74), (5.75) to upgrade symmetric or antisymmetric solutions which are not already exact to exact solutions; this is not always a simple matter, because the presence of terms nonlinear in z in the exact solutions may complicate the satisfaction of boundary conditions. We will give some illustrations of all these applications below.

Stress Fields Due to Harmonically Distributed Edge Loads

Solutions of this type, such as Eqs. (3.32), (3.33), Fig. 3.11, are particularly useful because such harmonic distributions can be easily combined to give practically any desired distribution of edge loads or displacements. Changing x to y and z to x for application to plates loaded on the edge $x = 0$, and using Eqs. (5.64a) and the symbols $x' = \pi x/l$, $y' = \pi y/l$, Eqs. (3.32) and (3.33) can be rewritten and expanded in the form:

$$\phi = \frac{l^2}{\pi^2}\cos y' e^{-x'}(a + bx'), \quad \frac{Eu_x}{1+\nu} = \frac{l}{\pi}\cos y' e^{-x'}\left(a + \frac{1-\nu}{1+\nu}b + bx'\right),$$

$$\frac{Eu_y}{1+\nu} = \frac{l}{\pi}\sin y' e^{-x'}\left(a - \frac{2}{1+\nu}b + bx'\right); \quad \sigma_x = \cos y' e^{-x'}(-a - bx'), \qquad (5.78a)$$

$$\sigma_y = \cos y' e^{-x'}(a - 2b + bx'), \quad \sigma_{xy} = \sin y' e^{-x'}(-a + b - bx'),$$

$$u_z, \sigma_z, \sigma_{xz}, \sigma_{yz} = 0.$$

The values of a and b in Eqs. (5.78a) can be chosen to give various loadings or displacements on the edge. For example, using $F_x = \int_{-c}^{c} dz\, \sigma_x$ etc., we can have at $x = 0$: for $a = b = -F_{xo}/(2c)$: $F_x = F_{xo} \cos y'$, $F_{xy} = 0$; for $a = 0$, $b = F_{xyo}/(2c)$: $F_{xy} = F_{xyo} \sin y'$, $F_x = 0$; for $a = 2b/(1+\nu)$, $b = \pi E u_{xo}/[(3-\nu)l]$: $u_x = u_{xo} \cos y'$, $u_y = 0$; for $a = -(1-\nu)b/(1+\nu)$, $b = -\pi E u_{yo}/[(3-\nu)l]$: $u_y = u_{yo} \sin y'$, $u_x = 0$.

Similarly, for the corresponding flexural case we use Eqs. (5.65a) and $w = u_{z(z=0)}$ and obtain:

$$\phi = \frac{l^2}{(1-\nu)\pi^2} \cos y' e^{-x'}(a + bx'), \quad \frac{Eu_x}{1+\nu} = \frac{l}{\pi} z \cos y' e^{-x'}(-a + b - bx'),$$

$$\frac{Eu_y}{1+\nu} = \frac{l}{\pi} z \sin y' e^{-x'}(-a - bx'), \quad \frac{Ew}{1+\nu} = \frac{l^2}{\pi^2} \cos y' e^{-x'}(-a - bx'),$$

$$\sigma_x = z \cos y' e^{-x'}\left(a - \frac{2}{1-\nu} b + bx'\right), \quad \sigma_y = z \cos y' e^{-x'}\left(-a - \frac{2\nu}{1-\nu} b - bx'\right),$$

$$\sigma_{xy} = z \sin y' e^{-x'}(a - b + bx'), \quad \sigma_z = 0, \tag{5.78b}$$

$$\sigma_{xz} = \frac{\pi(c^2 - z^2)}{l} \cos y' e^{-x'}(b), \quad \sigma_{yz} = \frac{\pi(c^2 - z^2)}{l} \sin y' e^{-x'}(b)$$

As examples of its application, the values of a and b in Eqs. (5.78b) can be chosen to give the following loadings or displacements at $x = 0$, using $M_x = \int_{-c}^{c} dz\, \sigma_x z$, etc.: for $a = b = -3(1-\nu)M_{xo}/[2(1+\nu)c^3]$: $M_x = M_{xo} \cos y'$, $M_{xy} = 0$; for $a = 3M_{xyo}/[(1+\nu)c^3]$, $b = (1-\nu)a/2$: $M_{xy} = M_{xyo} \sin y'$, $M_x = 0$; for $a = 0$, $b = -\pi E(\partial w/\partial x)_o/[(1+\nu)l]$: $w = 0$, $\partial w/\partial x = (\partial w/\partial x)_o \cos y'$. The last case should be very useful for superposing upon various kinds of solutions for hinged edges, to bring the net slope to zero at the edge and thus convert such solutions to the fixed edge condition.

Exact Solutions for Harmonically Distributed Edge Loads

Like solutions (3.32), (3.33) the above solutions are good approximations only if l is large compared to c. To obtain an exact solution which is not subject to any such limitation we can use the same stress functions ϕ that we did above in Eqs. (5.64a) and (5.65a), in Eqs. (5.64) and (5.65) respectively. For the symmetric case, using the symbol $\lambda = \nu\pi^2 c^2/[3(1+\nu)l^2]$ and $u'_x = u_{x(ave)} = (\int_{-c}^{c} dz\, u_x)/(2c)$, etc., we obtain from Eq. (5.64):

$$\frac{Eu'_x}{1+\nu} = \frac{l}{\pi} \cos y' e^{-x'}\left[a + \left(\frac{1-\nu}{1+\nu} + \lambda\right) b + bx'\right],$$

$$\frac{Eu'_y}{1+\nu} = \frac{l}{\pi} \sin y' e^{-x'}\left[a - \left(\frac{2}{1+\nu} - \lambda\right) b + bx'\right], \tag{5.78c}$$

$$\sigma_x = \frac{F_x}{2c} = \cos y' e^{-x'}(-a - \lambda b - bx'), \sigma_y = \frac{F_y}{2c} = \cos y' e^{-x'}[a - (2-\lambda)b + bx'],$$

$$\sigma_{xy} = \frac{F_{xy}}{2c} = \sin y' e^{-x'}[-a + (1-\lambda)b - bx'], \quad \sigma_z, \sigma_{xz}, \sigma_{yz} = 0$$

This solution gives on the edge $x=0$: for $b=a/(1-\lambda)=-F_{xo}/(2c)$: $F_x=F_{xo}\cos y'$, $F_{xy}=0$; for $b=-a/\lambda=F_{xyo}/(2c)$: $F_{xy}=F_{xyo}\sin y'$, $F_x=0$; for $b=-a/[\lambda+(1-\nu)/(1+\nu)]=-\pi Eu'_{yo}/[(3-\nu)l]$: $u'_y=u'_{yo}\sin y'$, $u'_x=0$; for $b=a/[2/(1+\nu)-\lambda]=\pi Eu'_{xo}/[(3-\nu)l]$: $u'_x=u'_{xo}\cos y'$, $u'_y=0$.

For the antisymmetric case, using the symbol $\mu=(2-\nu)\pi^2c^2/[5(1-\nu)l^2]$ and $M_x=\int_{-c}^{c}dz\,\sigma_x z$, $F_{xz}=\int_{-c}^{c}dz\,\sigma_{xz}$, etc., we find from Eqs. (5.65):

$$\frac{Ew}{1+\nu}=\frac{l^2}{\pi^2}\cos y' e^{-x'}\left(-a-\frac{2\pi^2c^2}{l^2}b-bx'\right),$$

$$\frac{3M_x}{2c^3}=\cos y' e^{-x'}\left[a-\left(\frac{2}{1-\nu}-\mu\right)b+bx'\right],$$

$$\frac{3M_y}{2c^3}=\cos y' e^{-x'}\left[-a-\left(\frac{2\nu}{1-\nu}+\mu\right)b-bx'\right], \tag{5.78d}$$

$$\frac{3M_{xy}}{2c^3}=\sin y' e^{-x'}[a-(1-\mu)b+bx'],$$

$$\frac{3F_{xz}}{4c^3}=\frac{\pi}{l}\cos y' e^{-x'}(b),\quad \frac{3F_{yz}}{4c^3}=\frac{\pi}{l}\sin y' e^{-x'}(b),\quad \sigma_z=0.$$

which gives on the edge $x=0$: for $b=a/(1-\mu)=-3(1-\nu)M_{xo}/[2(1+\nu)c^3]$: $M_x=M_{xo}\cos y'$, $M_{xy}=0$; for $b=a/[2/(1-\nu)-\mu]=3(1-\nu)M_{xyo}/[2(1+\nu)c^3]$: $M_{xy}=M_{xyo}\sin y'$, $M_x=0$; for $b=-l^2a/(2\pi^2c^2)=-\pi E(\partial w/\partial x)_o/[(1+\nu)l]$: $\partial w/\partial x=(\partial w/\partial x)_o\cos y'$, $w=0$.

It will be seen that the exact solutions (5.78c, d) differ from (5.78a, b) only by numbers such as λ or μ which are proportional to c^2/l^2 and are negligible when the half wavelength l is fairly large compared to the thickness $2c$. Even when the exact and approximate solutions give the same resultant forces or moments per unit length on the edge $x=0$, the distribution of unit forces on the edge are slightly different, but this would make no difference if the solutions are used for satisfying gross boundary conditions; if they are used for satisfying complete boundary conditions then they would merely require somewhat different additional corrective stress fields than those which would usually be required in any case.

As mentioned previously, solutions for different wavelengths l can be combined in infinite series, using the exact solutions for the shorter wavelength components and the simpler approximate solutions for the components of longer wavelength. If $\sinh x'$ and $\cosh x'$ are used instead of the exponential function $e^{-x'}$ any of the above solutions can be extended to give any desired loadings on *two* surfaces, such as the two opposite sides of a rectangular plate, in the same manner that the trigonometric–hyperbolic solutions tabulated in Table 3.3 did for the symmetric case. Similar independent solutions of this type can also be obtained from Eqs. (5.68) and (5.69), and this would enable more conditions to be satisfied on the edges.

Complete Solutions

The solutions discussed above are primarily supplementary solutions for use in satisfying edge boundary conditions. As an example of the use of solutions for plates with unloaded faces for complete problems involving edge loading only, consider the case of a disc with a concentric hole (or a tube, since the solution will be exact, and not limited to a small plate thickness) with inside radius a and outside radius b. We will consider the case of uniform unit pressures p_i, p_o on the inside and outside surfaces respectively. For this axisymmetric case Eqs. (5.70) become:

$$\frac{E}{1+\nu}u_r = -\frac{d\psi}{dr}, \quad \frac{E}{1+\nu}u_z = -\nu \mathbf{c} z; \quad \sigma_r = -\frac{d^2\psi}{dr^2} + \nu\mathbf{c}, \quad \sigma_\theta = \frac{d^2\psi}{dr^2} + \mathbf{c} \tag{5.79a}$$

$$u_\theta, \sigma_z, \sigma_{r\theta}, \sigma_{rz}, \sigma_{\theta z} = 0, \quad \nabla^2\psi(r) = \frac{d^2\psi}{dr^2} + \frac{1}{r}\frac{d\psi}{dr} = -(1-\nu)\mathbf{c}$$

The general solution of the latter second degree ordinary differential equation is:

$$\psi = -(1-\nu)\mathbf{c}\frac{r^2}{4} + \mathbf{d}\log r + \mathbf{e} \tag{5.79b}$$

where **d**, **e** are arbitrary constants of integration. Substituting these expressions into the boundary conditions:

$$r = a, b: \quad \sigma_r = -p_i, -p_o \tag{5.79c}$$

and solving for the constants **c**, **d** (we can take **e** as zero since it would yield no displacements or stresses) we find:

$$\mathbf{c} = \frac{2(a^2p_i - b^2p_o)}{(1+\nu)(b^2-a^2)}, \qquad \mathbf{d} = \frac{a^2b^2(p_o - p_i)}{b^2-a^2} \tag{5.79d}$$

The solution for this case is then:

$$u_r = \frac{[(1-\nu)r^2 + (1+\nu)b^2]a^2p_i - [(1-\nu)r^2 + (1+\nu)a^2]b^2p_o}{E(b^2-a^2)r}, \quad u_z = \frac{2\nu z(b^2p_o - a^2p_i)}{E(b^2-a^2)}; \tag{5.79e}$$

$$\sigma_r = \frac{a^2b^2(p_o - p_i)}{r^2(b^2-a^2)} + \frac{a^2p_i - b^2p_o}{b^2-a^2}, \quad \sigma_\theta = \frac{a^2b^2(p_i - p_o)}{r^2(b^2-a^2)} + \frac{a^2p_i - b^2p_o}{b^2-a^2}, \quad \sigma_z, \sigma_{rz} = 0$$

This case was solved by Lamé in 1852. The use of the constant **c** in Eqs. (5.79a) permitted a simpler solution here than is otherwise required.

In the corresponding antisymmetrical (flexural) case, the boundary conditions can be written as:

$$r = a, b: \quad \sigma_r = -p_i z, \quad -p_o z \tag{5.80a}$$

where p_i, p_o represent unit pressures at unit distance from the middle surface, producing the moments per unit length $m_i = 2c^3p_i/3$, $m_o = 2c^3p_o/3$ on the inner and outer surfaces (c being the half thickness). This case can be solved in the same manner as the symmetrical (membrane) case above. We still have symmetry about the axis of revolution and Eqs. (5.72) become for this case of axisymmetry:

$$\frac{E}{1+\nu} u_r = -z\frac{d\psi}{dr}, \quad \frac{E}{1+\nu} u_z = \psi - \frac{\nu}{2}\mathbf{c}z^2; \quad \sigma_r = -z\frac{d^2\psi}{dr^2} + \nu\mathbf{c}z,$$

$$\sigma_\theta = z\frac{d^2\psi}{dr^2} + \mathbf{c}z, \quad u_\theta, \sigma_{r\theta} = 0, \quad \nabla^2\psi(r) = -(1-\nu)\mathbf{c} \tag{5.80b}$$

Using expression (5.79b) for ψ and solving for **c**, **d**, **e** as before, we obtain the same expressions (5.79d) as in the membrane case above. Using these results in Eqs. (5.80b), we find that the values of u_r, σ_r, σ_θ are the same as given in Eqs. (5.79e) except for an added factor of z in each case, and again $\sigma_z, \sigma_{rz} = 0$. The value of u_z is:

$$u_z = \frac{[(1-\nu)r^2 + 2\nu z^2](b^2 p_o - a^2 p_i) + 2(1+\nu)a^2 b^2 (p_o - p_i)\log r}{2E(b^2 - a^2)} \tag{5.80c}$$

The arbitrary constant **e** is again taken as zero, since it would only yield a rigid body displacement in this case.

Equations (3.10c, d) are special cases of the above problems for the case when $a = 0$, that is for a disc without a hole, and $p_o = -s$.

5.5 MORE EXACT PLATE EDGE BOUNDARY CONDITIONS

Some discussion of plate edge boundary conditions has been given before in Secs. 4.4 and 4.5. In this section we will cover the subject more completely, as well as recapitulate what was said before. In applying classical plate theory we are usually (and quite properly) content if we can satisfy what we found it convenient to call 'gross edge boundary conditions', defined as conditions on the resultant forces and moments per unit length of edge, or on the displacements of some particular surface such as the middle surface. In fact even these conditions are usually satisfied only approximately, because of the use of Kirchhoff's device.

In applying second-approximation 'thick plate' theories, or even more exact solutions which satisfy three-dimensional elasticity theory, we may be tempted, because of the difficulties involved in going further, to also satisfy no more than gross edge boundary conditions. This practice may possibly give us accurate stresses and displacements in the middle part of the plate at distances from the edges large compared to the thickness but it is not likely to give very accurate results at or near the edges, where the stresses are often critical.

Moreover the errors involved in this practice can conceivably have an important effect upon stresses and displacements in the middle part of the plate also. This is especially true of the use of Kirchhoff's device, since it can replace distributed shear stresses due to twist by a combination of distributed transverse shear stresses and concentrated transverse forces at the ends of each edge—forces which are separated by distances of the order of the largest dimensions of the plate. In the practical case of an edge support which is constructed in such a way that it cannot directly resist shear stresses parallel to the edge due to twist, it would seem reasonable to assume that the transverse supporting forces resist both the transverse shear and the twisting moment at the edge in about the way that Kirchhoff assumed, and we will prove later that this is roughly correct. The transverse forces probably approach those assumed by Kirchhoff as the plate thickness approaches zero, hence for thin plates under such conditions the use of his device is usually justified.

On the other hand for a free edge, or for a fixed edge in which the plate is welded to a relatively rigid wall, the shear stresses parallel to the edge must in the first case be zero and in the second case must be directly resisted by reactive shear stresses exerted by the wall; in these cases the concentrated corner loads which are usually required by Kirchhoff's device are obviously fictitious and it is possible for their presence to seriously distort the results. This would obviously be the case, for instance, if the use of Kirchhoff's device should require a concentrated load on a free corner of a plate, that is a corner at the intersection of two adjacent free edges.

It is certainly safer, and in using the more accurate theories much more consistent, to satisfy as far as possible what it is convenient to call 'complete edge boundary conditions', defined as the actual conditions on the edge stresses or displacements at every point of the edges. Thus, for the above mentioned extreme cases, a free edge parallel to the xz plane must obviously have zero σ_y, σ_{xy}, and σ_{yz} stresses at *every* point of the edge and not merely zero resultant forces and moments per unit length of edge, and the edge of a soft rubber plate cemented to a heavy steel wall must have substantially zero displacements in the x, y, z directions at *every* point of the edge.

Mathematical Requirements for Satisfying Plate Edge Boundary Conditions

Figure 5.13(a) illustrates the three quantities, stresses or displacements, whose values must be specified at every point of each edge of a plate if 'complete edge boundary conditions' are to be satisfied. Figure 5.13(b) shows the corresponding quantities which must be specified to satisfy 'gross edge boundary conditions',

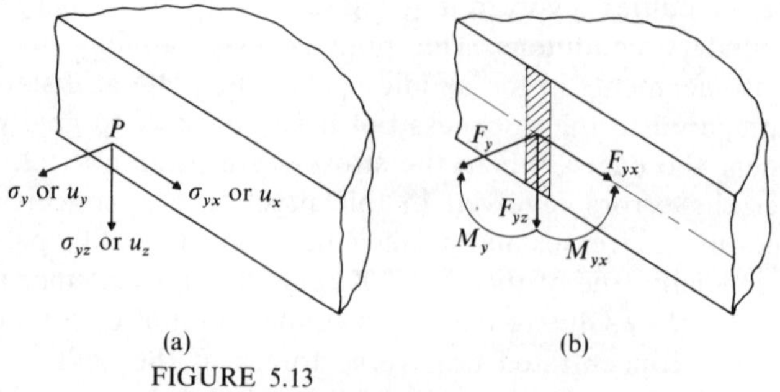

FIGURE 5.13
(a) Complete edge boundary conditions
(b) Gross edge boundary conditions

namely the three resultant forces per unit length, obtained by integrating over the thickness the stresses on the narrow strip shown shaded, and the two moments similarly obtained from the moments of two of the stresses about the midpoint of the strip (the transverse shear stress has no such moment); or the arrows shown can represent the three displacements u, v, w, or two rotations $\partial w/\partial x$, $\partial w/\partial y$, of the middle surface of the plate at the point considered.

At first glance the five quantities which have to be specified to satisfy 'gross' conditions may seem more complex than the three specified for 'complete' conditions. Actually of course they are far simpler, because they are functions of only one coordinate, the distance along the edge, in this case the coordinate x; the three quantities in the complete conditions, on the other hand, must be specified over the whole area of the edge, that is they are functions of two coordinates, in this case x and z.

Gross Edge Boundary Conditions

Let us consider first the satisfaction of the simpler gross edge boundary conditions. We have available for this purpose the general solutions (5.64), (5.65) and (5.68), (5.69) described in Sec. 5.4. The former, solutions (5.64), (5.65) in terms of biharmonic functions, include as special cases or approximations all the other solutions in rectangular coordinates given in Sec. 5.4 and earlier in Secs. 3.2 and 3.3 on two-dimensional elasticity; they also include the solutions given by the homogeneous equations derived from any of the various solutions for lateral loading of plates by setting the lateral loading equal to zero (as is indicated by the fact that they themselves were derived in this way from the general series solutions in loading functions). Hence in the following discussion we will let Eqs. (5.64), (5.65) take the place of all these other conditions contained in them. For simplicity we will also, in the following discussion, speak only of specified stresses, or forces and moments per unit length, it being understood that any of these may be replaced in the specifications by the corresponding displacements or rotations of the middle surface.

Since Eqs. (5.64), (5.65) are in terms of biharmonic functions ϕ defined by the fourth order partial differential equation $\nabla^4\phi(x, y) = 0$, each of them can satisfy the conditions on two quantities varying with the distance along the side on which they act, on all four sides of a rectangular plate. The antisymmetric Eqs. (5.65) can thus satisfy the conditions on the flexural quantities M_y and F_{yz} on the sides parallel to the xz plane and the two corresponding quantities on the other two sides. Similarly the symmetrical Eqs. (5.64) can satisfy the conditions on the membrane forces F_y and F_{yx} and the corresponding forces on the other two sides; as has been discussed in previous chapters, these can be taken as zero when the deflections are very small compared to the thickness ('small-deflection' case), in which case Eqs. (5.64) will not be needed.

This leaves only the twisting moments M_{yx} and M_{xy} to be taken care of. In classical plate theory these are taken care of by using Kirchhoff's device, which combines them with the transverse shear forces F_{yz} and F_{xz} to obtain 'effective' transverse shear forces F'_{yz} and F'_{xz}, Eq. (4.38); as we will show below, this is in general an approximation. To take care of M_{yx}, M_{xy} without using Kirchhoff's device, we now have available the independent antisymmetrical solution (5.69). Since this is in terms of a function $\psi(x, y, z)$ defined by the second order differential equation $\nabla^2\psi_m(x, y) = (m^2\pi^2/4c^2)\psi_m$, one term of it only (which should obviously be the first term, taking $m = 1$) can satisfy the conditions on one quantity which varies with the distance along the side on which it acts, on all four sides of a plate, in this case M_{yx} on the sides parallel to the xz plane and M_{xy} on the other two sides. E. Reissner and others have used a solution somewhat similar to this first term of the general solution (5.69) for similar purposes.

Complete Edge Boundary Conditions

A. E. Green* has devised a general exact method for satisfying complete plate edge boundary conditions, but we will consider here a simpler general method for accomplishing this purpose, which is in part an approximation but which should be sufficiently accurate for most purposes. We will discuss first the small-deflection case, in which F_y and F_{yx} (and F_x and F_{xy} on the sides parallel to the yz plane) can be taken as zero. As was discussed at the beginning of this chapter and elsewhere, it is only in such an extreme case as a thick plate made of a rubber-like material that large deflection effects, such as these membrane forces, would have to be considered at the same time as thick-plate considerations such as complete boundary conditions.

The proposed method consists in first satisfying the gross conditions on the stresses σ_y and σ_{yz}, Fig. 5.13, and the complete conditions on σ_{yx} (or the corresponding displacements) and corresponding conditions on the sides parallel to the yz plane. This can be done exactly by using the antisymmetric general solutions (5.65) and (5.69). The only difference between this and satisfying the gross edge boundary conditions as described above is that *all* the terms of Eqs. (5.69) are used, enabling us to satisfy the conditions on σ_{yx} all over the side instead of merely the conditions on its moment M_{yx}. This leaves to be satisfied only the differences between the stresses σ_y and σ_{yz} and their resultants M_y and F_{yz}. Since these differences will form a self-balanced system, they can be corrected for by using local stress fields, such as the plane strain version of Horvay's end corrections, as discussed in Sec. 3.4, Eqs. (3.39), (3.40), and also in the paragraphs headed 'Application of beam end corrections to plate and shell edge correction' which is given on p. 142. Only the odd values of n, $n = 3, 5, \ldots$, Fig. 3.16(b), (c) will be needed, since the corrections will be antisymmetric.

For the large deflection case, when F_y and F_{yx} can not be ignored, the symmetric general solutions (5.64) and (5.68) (all terms) and the even values of n in Horvay's end corrections would have to be used also. It seems probable that

* *Proc. Roy. Soc. A*, vol. 195, p. 533, 1949.

this symmetrical system of conditions will be independent of the antisymmetrical system discussed above and can be treated separately.

The above descriptions of theoretical methods for satisfying gross and complete edge boundary conditions may seem somewhat complex and confusing. To make their application clearer we will now give examples of each type of calculation.

Satisfaction of Gross Edge Conditions Without Use of Kirchhoff Approximation; Square Plate Under Uniform Lateral Load

To illustrate the exact calculation of gross edge boundary conditions we will take the case of a thin plate with edges $x = \pm a$, $y = \pm a$, having the top and bottom surfaces $z = \pm c$, under a uniform unit loading p_o in the z direction. We will use classical plate theory to take care of the lateral loading and we will take the boundary conditions on the edge $x = a$ as:

$$w = 0, \quad M_x = 0, \quad M_{xy} = 0 \tag{5.81a}$$

with similar conditions on the other three sides.

It will be seen that these conditions differ from previous 'hinged' or 'simply supported' gross edge boundary conditions used with classical plate theory, such as Eqs. (4.20), in the inclusion of the condition that twisting moments M_{xy} at the edges are zero; we previously took care of these by using Kirchhoff's device.

It would not be much more difficult to satisfy conditions (5.81a) using an exact lateral loading theory such as Eq. (5.30) instead of the approximate classical plate theory. But, aside from the inconsistency of satisfying approximate edge conditions using exact plate theory, this would not serve our purpose, which is to check the validity of using Kirchhoff's device with classical plate theory, especially with regard to the concentrated transverse corner loads which this device predicts. As a matter of fact, there is evidence that we would get practically the same transverse shear loads if we used the exact lateral loading theory, but then there would be no way of telling whether our results were due to non-use of Kirchhoff's device or to the consideration of transverse strains in the lateral loading theory.

Since we will study for simplicity a square plate, which is symmetrical not only about the x and y axes but also with respect to x and y (that is, so that all expressions and equations are unchanged if we interchange x and y in both coordinates and subscripts), if we satisfy conditions (5.81a) on the edge $x = a$ then similar conditions will be automatically satisfied on the other three sides (provided all the expressions which we use in satisfying the edge conditions have a similar symmetry, as those below do).

In order to satisfy conditions (5.81a) we will superpose three solutions, the classical plate theory solution given by using expression (5.81c) below for w in Eqs. (4.14), solution (5.65), and the first term (using $m = 1$) of solution (5.69). The last two involve zero loads on the faces. We take $w = u_{z(z=0)}$, and:

$$M_x = \int_{-c}^{c} \sigma_x z dz, \quad M_{xy} = \int_{-c}^{c} \sigma_{xy} z dz, \quad F_{yz} = \int_{-c}^{c} \sigma_{yz} dz \tag{5.81b}$$

To satisfy the conditions of symmetry discussed above, we will take the expression for w in the classical plate theory, derived in the same way as expression (4.25) for other boundaries, and functions ϕ and ψ_1 involved in Eqs. (5.65) and (5.69) respectively as:

$$w = \frac{256a^4 p_o}{\pi^6 D} \sum_m \sum_n \frac{\sin\frac{m\pi}{2} \sin\frac{n\pi}{2}}{mn(m^2+n^2)^2} \cos\frac{m\pi x}{2a} \cos\frac{n\pi y}{2a} \tag{5.81c}$$

$$\phi = \sum_m \left[A_m \left(\cos\frac{m\pi x}{2a} \cosh\frac{m\pi y}{2a} + \cosh\frac{m\pi x}{2a} \cos\frac{m\pi y}{2a} \right) \right.$$

$$\left. + D_m \left(\cos\frac{m\pi x}{2a} \frac{m\pi y}{2a} \sinh\frac{m\pi y}{2a} + \frac{m\pi x}{2a} \sinh\frac{m\pi x}{2a} \cos\frac{m\pi y}{2a} \right) \right] \tag{5.81d}$$

$$\psi_1 = \sum_m E_m \left(\sin\frac{m\pi x}{2a} \sinh \gamma_m y - \sinh \gamma_m x \sin\frac{m\pi y}{2a} \right)$$

$$\text{where } \gamma_m = \frac{\pi}{2c}\left(1 + \frac{m^2 c^2}{a^2}\right)^{1/2}, \quad m, n = 1, 3, 5, \ldots \tag{5.81e}$$

It can readily be checked that these expressions satisfy the required conditions that $\nabla^4 \phi = 0$ and $\nabla^2 \psi_1 = (\pi^2/4c^2)\psi_1$ and the required conditions of symmetry for w, M_x, M_{xy}.

It might be argued that in using Eqs. (5.65) it would be more consistent to ignore the parts of the expressions involving ∇^2 applied to derivatives of ϕ (or use Eqs. (5.61), taking ψ as a biharmonic function), since its accuracy should still be comparable to that of the classical plate theory (or traditional two-dimensional elasticity theory). To avoid any uncertainty about the accuracy of the results these terms were retained in making the calculations given below, but these calculations did show that for the value of the width–thickness ratio used in the calculations, $a/c = 20$, these terms were very small compared to the other terms present, and had a negligible effect upon the final results.

The displacement $w_{(x=a)}$ in the classical theory and $w = u_{z(z=0)}$ in Eqs. (5.69) are zero. Substituting expression (5.81d) into Eqs. (5.65) and considering that $\cos m\pi/2 = 0$, we find the first edge condition $w_{(x=a)} = 0$:

$$\sum_m \left\{ \left[-(1-\nu)A_m + \frac{m^2\pi^2c^2}{2a^2} D_m \right] \cosh\frac{m\pi}{2} - (1-\nu)D_m \frac{m\pi}{2} \sinh\frac{m\pi}{2} \right\} \cos\frac{m\pi y}{2a} = 0$$

which is satisfied if:

$$A_m = \left(-\frac{m\pi}{2} \tanh\frac{m\pi}{2} + \frac{m^2\pi^2c^2}{2(1-\nu)a^2} \right) D_m \tag{5.81f}$$

The bending moment $M_{x(x=a)}$ in the classical theory is zero. Using condition (5.81d) in (5.65), and (5.81e) in (5.69) we again find that all the terms in the second edge condition $M_{x(x=a)} = 0$ are proportional to $\cos(m\pi y/2a)$, and using (5.81f), the condition is satisfied if:

$$E_m = \frac{\pi^3 c}{12} \frac{m \cosh \frac{m\pi}{2}}{(\gamma_m a)\cosh(\gamma_m a)} \left(1 + \frac{(8+\nu)m^2\pi^2c^2}{40a^2}\right) D_m \qquad (5.81g)$$

The third and last edge condition, $M_{xy(x=a)} = 0$, is more complex and more difficult to satisfy because its terms do not involve the same functions of y. We use relations (5.81c) in (4.14), (5.81d) in (5.65), and (5.81e) in (5.69) to write $M_{xy(x=a)} = 0$ (and the expression for the transverse shear reaction at the edge $F_{xz(x=a)}$, given later). The resulting relation is then simplified somewhat by taking $\nu = 0{\cdot}3$, $\sin^2(n\pi/2) = 1$ (since n is odd), and dividing through by $64(1-\nu)$ $m[\cosh(m\pi/2)]c^3/(\pi^4 a)$. Finally, to eliminate the functions of y, we multiply through by $dy \sin m'\pi y/2a$, where m' is taken successively as 1, 3, 5, ..., and integrate between $-a$ and a. Using relations (5.81f, g) we thus obtain a set of equations, one for each value of m', which can be written in the form:

$$\sum_m \{a_{m'm} e_m (ap_o) - [2{\cdot}504m(c_{m'm} - b_{m'm}f_m) + 2{\cdot}275 d_{m'm} g_m] D_m\} = 0 \quad (5.81h)$$

where the dimensionless numbers $a_{m'm}$, e_m, etc., are:

$$a_{m'm} = \frac{1}{a}\int_{-a}^{a} \sin\frac{m'\pi y}{2a} \sin\frac{m\pi y}{2a}\, dy = 1 \text{ (when } m' = m),$$
$$= 0 \text{ (when } m' \neq m)$$

$$b_{m'm} = \frac{\sin\frac{m\pi}{2}}{a\cosh\frac{m\pi}{2}} \int_{-a}^{a} \sin\frac{m'\pi y}{2a} \sinh\frac{m\pi y}{2a}\, dy + \tanh\frac{m\pi}{2}\, a_{m'm}$$

$$= \frac{4m}{\pi(m'^2+m^2)} \sin\frac{m'\pi}{2} \sin\frac{m\pi}{2} + \tanh\frac{m\pi}{2}\, a_{m'm}$$

$$c_{m'm} = \frac{\sin\frac{m\pi}{2}}{a\cosh\frac{m\pi}{2}} \int_{-a}^{a} \sin\frac{m'\pi y}{2a} \frac{m\pi y}{2a} \cosh\frac{m\pi y}{2a}\, dy + \frac{m\pi}{2} a_{m'm}$$

$$= \frac{4m}{\pi(m'^2+m^2)} \sin\frac{m'\pi}{2} \sin\frac{m\pi}{2}\left(\frac{m\pi}{2}\tanh\frac{m\pi}{2} + \frac{m'^2-m^2}{m'^2+m^2}\right) + \frac{m\pi}{2} a_{m'm}$$

$$d_{m'm} = \frac{\sin\frac{m\pi}{2}}{a\cosh(\gamma_m a)} \int_{-a}^{a} \sin\frac{m'\pi y}{2a} \sinh(\gamma_m y) + \tanh(\gamma_m a) a_{m'm}$$

$$= \frac{2}{(\gamma_m a) + \frac{m'^2\pi^2}{4(\gamma_m a)}} \sin\frac{m'\pi}{2} \sin\frac{m\pi}{2} + \tanh(\gamma_m a) a_{m'm}$$

$$e_m = \frac{a^3}{c^3} \frac{\sin\frac{m\pi}{2}}{m\cosh\frac{m\pi}{2}} \sum_n \frac{1}{(m^2+n^2)^2}$$

$$f_m = \frac{m\pi}{2}\tanh\frac{m\pi}{2} - 1 - \frac{0{\cdot}593m^2\pi^2c^2}{a^2}, \quad g_m = \left(1 + \frac{0{\cdot}208m^2\pi^2c^2}{a^2}\right)\left[(\gamma_m a) + \frac{m^2\pi^2}{4(\gamma_m a)}\right]$$

Since each of the equations represented by (5.81h) contains all the D_m's, the latter must be found by solving the equations simultaneously. Since a computer was not available, the following results were obtained by using only the first three equations, for $m' = 1, 3, 5$, and the first three D_m's. In this way we find D_1, D_3, $D_5 = 11{\cdot}32(ap_o)$, $0{\cdot}00069(ap_o)$, $-0{\cdot}00013(ap_o)$ respectively, for $a/c = 20$.

Similarly the expression for the transverse shear reaction $F_{xz(x=a)} = R$ can be written as:

$$R = \sum_m \left\{-1{\cdot}032\frac{\sin m\pi/2}{m}\left(\sum_n \frac{1}{m^2+n^2}\right)(ap_o)\cos\frac{m\pi y}{2a}\right. \tag{5.81i}$$

$$+ D_m\left[2{\cdot}584\frac{mc}{a}\cosh\frac{m\pi}{2}\left(1 + \frac{m^2\pi^2c^2}{4{\cdot}82a^2}\right)\left(\frac{\sin m\pi/2}{\cosh(\gamma_m a)}\cosh(\gamma_m y)\right.\right.$$

$$\left.\left.\left. - \frac{m\pi/2\tanh(\gamma_m a)}{(\gamma_m a)}\cos\frac{m\pi y}{2a}\right) + \frac{m^3\pi^3c^3}{6a^3}\left(\sinh\frac{m\pi}{2}\cos\frac{m\pi y}{2a} - \sin\frac{m\pi}{2}\cosh\frac{m\pi y}{2a}\right)\right]\right\}$$

The full line in Fig. 5.14 shows the distribution of the transverse reaction given by this analysis, using the first three D_m's. The dotted line shows the nearly identical results obtained, also for $a/c = 20$, by A. Kromm,* who likewise satisfied the gross boundary conditions without using Kirchhoff's approximation, but who used a different method for doing this and also considered the principal effects of transverse strains. The dashed line with the concentrated reaction at the corner shows the results predicted by the usual classical plate analysis using Kirchhoff's approximation, which is given by the first term of expression (5.81i). The similarity between the present results and those of Kromm indicates that the

* *Ingr. Arch.*, vol. 21, p. 266, 1953. Also *Z. Ang. Math. Mech.*, vol. 35, p. 231, 1955.

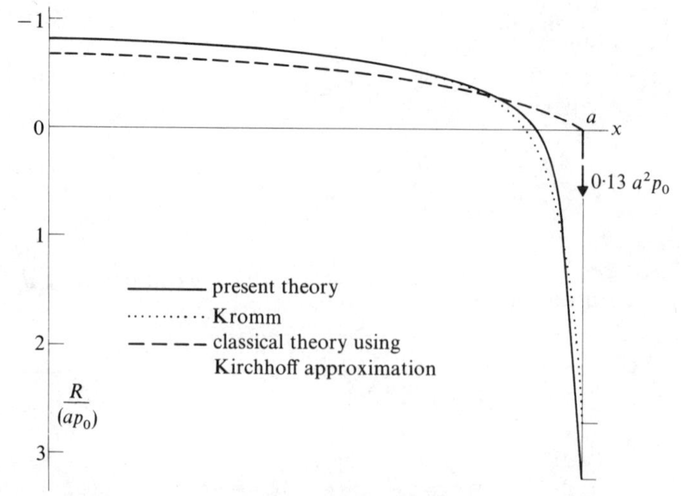

FIGURE 5.14

difference between his results and the usual classical results was due principally to his satisfying $M_{xy(x=a)} = 0$ without using Kirchhoff's approximation, rather than to his consideration of transverse strains.

It can be seen that the Kirchhoff approximation gives a rough approximation to the exact distribution of reactions, with the part of the reaction curves above the axis in Fig. 5.14 (corresponding to negative values of F_{xz}) of somewhat similar area and shape. The part of the reaction curve in the exact distribution below the axis (corresponding to positive F_{xz}) is quite close to the corner of the plate and has roughly an area corresponding to the concentrated load at the corner given by the Kirchhoff approximation (the total concentrated corner reaction has twice the value shown, the other half corresponding to the reaction distribution on the adjacent side). It seems probable that the classical Kirchhoff result is exact in the limit as $c/a \rightarrow 0$.

Satisfaction of Complete Edge Boundary Conditions; Round Hole in Plate Under Pure Bending

To illustrate the method of satisfying complete plate edge boundary conditions which has been outlined above, and to test its accuracy in a case where the edge of the plate has curvature of any magnitude, we will consider the case of a plate with a circular cylindrical hole, subjected to a uniformly distributed bending moment $M_x = M_o$ per unit length of section on edges parallel to the y axis and far from the hole. The plate has a thickness $h = 2c$, and the origin of rectangular or cylindrical coordinates, Fig. 3.5, is taken on the middle surface at the center of the hole, which has a radius a or diameter $d = 2a$.*

This case is particularly suitable for our purpose because it is one of the few cases (possibly the only case) of flexure of plates for which an exact analysis is available for comparison; several less exact analyses of this case are also available, which enable the effect of various types of approximate methods to be demonstrated. The study will be confined primarily to the calculation of the factor of stress concentration, that is the ratio between the maximum normal stresses which would occur with the hole and without the hole, since this is the main feature of interest in practical applications.

Measuring the angle θ from the x axis, the complete edge boundary conditions can readily be shown to be, using Fig. 3.5 and Eqs. (3.9d):

$$r \rightarrow \infty: \quad M_r = \int_{-c}^{c} z dz\, \sigma_r = \frac{M_o}{2}(1 + \cos 2\theta)$$

$$M_{r\theta} = \int_{-c}^{c} z dz\, \sigma_{r\theta} = -\frac{M_o}{2}\sin 2\theta, \quad \int_{-c}^{c} dz\, \sigma_r, \sigma_{rz}, \sigma_{r\theta} = 0 \tag{5.82a}$$

$$r = a: \quad \sigma_r, \sigma_{rz}, \sigma_{r\theta} = 0 \tag{5.82b}$$

* L. H. Donnell, 'On the satisfaction of complete plate edge boundary conditions' *Advances in Mechanics of Deformable Solids*, 100th anniversary volume in memory of B. G. Galerkin, Publishing house 'Science', Moscow, 1975, p. 227 (in Russian).

The detailed distribution of stresses at $r \to \infty$ is not significant for our purposes because it would not affect the conditions in which we are interested in the region of the hole. We will use solutions (5.75) and (5.77) to satisfy the conditions (5.82a) at $r \to \infty$, and instead of (5.82b), the conditions:

$$r = a: \quad \int_{-c}^{c} dz\, \sigma_r = 0, \quad \int_{-c}^{c} z dz\, \sigma_r = 0, \quad \int_{-c}^{c} dz\, \sigma_{rz} = 0, \quad \sigma_{r\theta} = 0 \tag{5.82c}$$

We can then use the plane strain version of Horvay's local stress fields to add corrective stresses σ'_r, σ'_z, σ'_{rz}, σ'_θ (the original plane stress version of course does not involve any $\sigma_{r\theta}$ or $\sigma_{\theta z}$ stresses and neither would the simple plane strain version of it proposed here; however, changing conditions along the edge would produce such stresses, which might be taken into consideration in a further refinement of the method) so as to reduce the remaining balanced σ_r and σ_{rz} stresses to zero. However, it will not be necessary for our present purposes to calculate these corrective stresses in detail, because as shown below they will not be needed for calculating the factor of stress concentration.

As would be expected, trial confirms that the maximum stress with the hole will be the tangential stress σ_θ at the point $r = a$, $\theta = \pi/2$, $z = c$. The local stress field correction for σ_{rz} will not affect σ_θ at this point, but that for σ_r will affect it. As has been stated, we will use the 'plane strain version' of these local stress fields. This merely means superposing upon the plane stress fields of Horvay (which involve in the present case only σ'_r, σ'_z, σ'_{rz} stresses normal to the direction of the edge) sufficient stress parallel to the direction of the edge, in the present case σ'_θ, to make the strain parallel to the edge zero, that is to make $\epsilon'_\theta = (\sigma'_\theta - \nu\sigma'_r - \nu\sigma'_z)/E = 0$. This assumption, which represents the chief approximation in our calculations, is justified by the highly localized and balanced nature of the stress field added. At $z = c$, $\sigma'_z = 0$ and hence $\sigma'_\theta = \nu\sigma'_r = -\nu\sigma_r$, where σ_r is the radial stress given by solutions (5.75) and (5.77), which the local stress field must counteract. The total tangential stress after all corrections is then $\sigma_\theta + \sigma'_\theta = \sigma_\theta - \nu\sigma_r$, where σ_θ and σ_r are the tangential and radial stress given by solutions (5.75) and (5.77). The factor of stress concentration k is therefore:

$$k = \frac{\sigma_\theta - \nu\sigma_r}{3M_0/2c^2} \tag{5.82d}$$

To accomplish our purposes we take the value of $\phi(r, \theta)$ in solution (5.75) as the most general solution of $\nabla^4\phi(r, \theta) = 0$ for the cases where ϕ is independent of θ and proportional to $\cos 2\theta$:

$$\phi = (a_0 + b_0 r^2 + c_0 \log r + d_0 r^2 \log r) + \left(\frac{a_2}{r^2} + b_2 + c_2 r^2 + d_2 r^4\right) \cos 2\theta \tag{5.82e}$$

where a_0, a_2, ... are coefficients to be determined.

Similarly, in solution (5.77) we will take the function $\psi_m(r, \theta)$ as the following product of functions of r and of θ:

$$\psi_m(r,\theta)=a_m K_2\left(\frac{m\pi r}{2c}\right)\sin 2\theta \tag{5.82f}$$

where a_m represents the numerical coefficient of the typical term in the series in m, and $K_2(m\pi r/2c)$ is the modified Bessel function of the second kind (since it must remain finite as $r\to\infty$) and second order, with the argument $m\pi r/2c$. The general solution (5.77) then becomes for this case:

$$\psi(r,\theta,z)=\sum_m \psi_m(r,\theta)\sin\frac{m\pi z}{2c}=\sum_m a_m K_2\left(\frac{m\pi r}{2c}\right)\sin 2\theta\sin\frac{m\pi z}{2c}$$

$$\frac{Eu_r}{1+\nu}=\frac{2}{r}\sum_m a_m K_2\cos 2\theta\frac{m\pi z}{2c},\quad \frac{Eu_\theta}{1+\nu}=-\frac{\pi}{2c}\sum_m a_m K_2'\sin 2\theta\sin\frac{m\pi z}{2c}$$

$$u_z,\sigma_z=0,\ \sigma_r=-\sigma_\theta=-\frac{1}{r}\sum_m a_m\left(\frac{2}{r}K_2-\frac{m\pi}{c}K_2'\right)\cos 2\theta\sin\frac{m\pi z}{2c}$$

$$\sigma_{r\theta}=\frac{\pi^2}{8c^2}\sum_m m^2 a_m(K_2-2K_2'')\sin 2\theta\sin\frac{m\pi z}{2c} \tag{5.82g}$$

$$\sigma_{rz}=\frac{\pi}{2cr}\sum_m m a_m K_2\cos 2\theta\cos\frac{m\pi z}{2c},\quad \sigma_{\theta z}=-\frac{\pi^2}{8c^2}\sum_m m^2 a_m K_2'\sin 2\theta\cos\frac{m\pi z}{2c}$$

where $K_2'(m\pi r/2c)$ and $K_2''(m\pi r/2c)$ are the first and second derivatives of K_2 with respect to its argument $m\pi r/2c$, and $m=1,3,5\ldots.$

Using the expressions for σ_r, σ_{rz}, $\sigma_{r\theta}$ given by (5.82e) in (5.75) and by (5.82g) in the boundary conditions (5.82c), we find that these require that:

$$d_0,d_2=0,\quad b_0=\frac{3M_0}{8(1+\nu)c^3},\quad c_2=\frac{3M_0}{8(1-\nu)c^3} \tag{5.82h}$$

Substituting the expressions for σ_r, σ_{rz}, $\sigma_{r\theta}$ with these values into the boundary conditions (5.82c), we find that we can cancel out the factor $\cos 2\theta$ (thus satisfying them for all values of θ if the remaining conditions are satisfied) if the constant part of the expression for σ_r is zero, which requires that:

$$c_0=\frac{3M_0a^2}{4(1-\nu)c^3} \tag{5.82i}$$

The coefficient a_0 does not enter the present problem. The condition $\int_{-c}^{c} dz\sigma_r=0$ is satisfied identically, while from the condition $\int_{-c}^{c} dz\sigma_{rz}=0$ we find, using (5.82h, i) and considering the fact that m is odd:

$$cb_2=\frac{3a^2}{8c^2}\sum_m a_m K_2\left(\frac{m\pi a}{2c}\right) \tag{5.82j}$$

Using this in the condition $\int_{-c}^{c} zdz\sigma_r=0$ we find:

$$\frac{1}{c}a_2=-\frac{M_0a^4}{8(1-\nu)c^4}+\frac{a^3}{(1-\nu)c^3}\sum_m a_m\left[\frac{2}{m\pi}K_2'\left(\frac{m\pi a}{2c}\right)\right.$$

$$\left.+\left(\frac{\nu a}{4c}-\frac{3(2-\nu)c}{20a}-\frac{4c}{m^2\pi^2a}\right)K_2\left(\frac{m\pi a}{2c}\right)\right] \tag{5.82k}$$

Finally, we can satisfy the condition $\sigma_{r\theta} = 0$:

$$\left(\frac{6(1-\nu)}{a^4}a_2 + \frac{2(1-\nu)}{a^2}b_2 - \frac{3M_0}{4c^3}\right)z + \frac{4(2-\nu)}{a^4}b_2 z^3$$
$$-\frac{1}{c^2}\sum_m \left(\frac{m\pi}{2}\right)^2 a_m\left[K_2''\left(\frac{m\pi a}{2c}\right) - \frac{1}{2}K_2\left(\frac{m\pi a}{2c}\right)\right]\sin\frac{m\pi z}{2c} = 0 \qquad (5.82l)$$

by using (5.82j, k) and making a harmonic analysis, that is by multiplying (5.82l) through by $dz \sin m'\pi z/2c$ and integrating from $-c$ to c, which gives us the conditions for determining the a_m's. This final result can be simplified somewhat by using the standard relations between K_2', K_2'' and K_2 and K_0. We thus obtain:

$$\left(\frac{m'\pi}{2}\right)^4 a_{m'}\left[K_2\left(\frac{m'\pi a}{2c}\right)\left(\frac{24c^2}{(m'\pi a)^2}+1\right) - \frac{1}{2}K_0\left(\frac{m'\pi a}{2c}\right)\right]$$
$$-\sum_m a_m\left\{K_2\left(\frac{m\pi a}{2c}\right)\left[\left(\frac{144}{(m\pi)^2} + \frac{72(2-\nu)}{(m'\pi)^2} - \frac{36(2-\nu)}{5}\right)\left(\frac{c}{a}\right)^2 + \frac{9-3\nu}{2}\right] \right. \qquad (5.82m)$$
$$\left. -6K_0\left(\frac{m\pi a}{2c}\right)\right\} = 3M_0$$

where m' is a particular value of m. This gives us a set of simultaneous equations, one for each value of m', for determining the a_m's, which can then be put into expressions (5.82j, k). Using these and (5.82h, i) and the expressions for σ_θ and σ_r from solutions (5.75) and (5.77), we can find the factor of stress concentration k from (5.82d). The latter expression reduces to the form:

$$k = 1 + \frac{(1+\nu)}{M_0}\sum_m m^2 a_m K_2\left(\frac{m\pi a}{2c}\right)\left[\left(\frac{2(12-\nu)}{5m^2} - \frac{48}{m^4\pi^2}\right)\left(\frac{c}{a}\right)^2 + \frac{\pi^2}{6} - \frac{1}{m^2}\right.$$
$$\left. + \left(\frac{K_0}{K_2}\right)\left(\frac{m\pi a}{2c}\right)\left(\frac{2}{m^2} - \frac{\pi^2}{6}\right)\right] \qquad (5.83)$$

Values of k so obtained, using $\nu = 1/4$, are plotted in the diagonal full line in Fig. 5.15 against the ratio $c/a = h/d$ of the plate thickness to the hole diameter. The convergence is rapid for the smaller values of h/d and is fairly good up to the largest value shown, as can be judged from the dotted line which was obtained by using only the first three terms, for $m = 1, 3, 5$. It is not very necessary to use computer calculations, since it is found that the contributions of further terms can apparently be calculated to the accuracy of the plot by using the first three terms so obtained in the fourth equation, the four terms so obtained in the fifth equation and so on, so that the results shown were obtained without solving more than three equations simultaneously. While this book was being typeset the author's attention was called to a solution of this problem by E. Reissner, *Int. J. Solids Struct.*, vol. 11, 1975, p. 569, which arrives at much the same results in a different way.

Figure 5.15 also shows the horizontal full line indicating the constant value 1·769 found by Goodier* by using the antisymmetric equivalent of classical

* *Phil. Mag.*, ser. 7, vol. 22, 1936, p. 69.

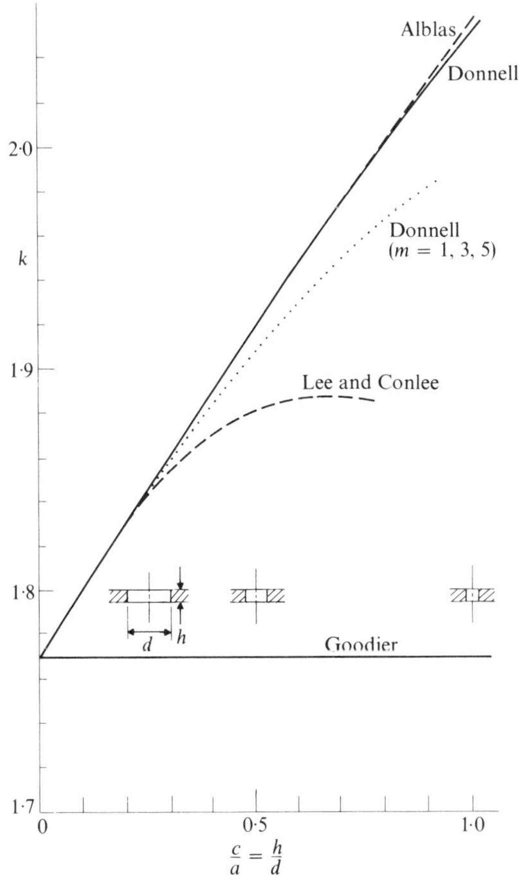

FIGURE 5.15 (from *Advances in Mechanics of Deformable Solids* (B. G. Galerkin Memorial Volume) p. 227, 1975, courtesy of Publishing House 'Science', Moscow)

two-dimensional elasticity theory; this result is correct only for very large holes in very thin plates. The figure also shows by dashed lines two previous three-dimensional elasticity solutions of this case. That of Lee and Conlee,* who used the equivalent of solution (5.75) and the first term of solution (5.77), satisfies only the gross edge boundary conditions. It is a decided improvement on the classical two-dimensional solution, giving substantially correct values for k for h/d up to 1/4 and fair approximations up to more than one half. The solution of Alblas** satisfies the complete edge conditions exactly by using the method of Green, previously mentioned, and, as can be seen, differs only slightly from the present simpler solution for values of h/d up to unity, which is as far as Alblas carried his calculations.

* *J. of the Franklin Inst.*, vol. 285, no. 5, May 1968, p. 377.

** *Theory of the Three-Dimensional State of Stress in a Plate with a Hole*, Dissertation, Tech. Hogeschool, Delft, published by H. J. Paris, Amsterdam, 1957.

Conclusions

This result, for a representative stress at a representative and critical point, lends quite strong confirmation to the validity of the general approximate method for satisfying complete edge boundary conditions which has been presented. The excellent results obtained even when the edge has the extremely sharp radius of curvature of only half the plate thickness (there is also a rapid variation of conditions along the edge, changing from maximum values to zero in a distance of less than half the sheet thickness, but the rate of change at the point studied is zero) suggests that this method should be usable with reasonable accuracy not only for plates but also for cases of curvature *out* of the middle surface as well as in the surface, that is for shells, at least when nothing better is available.

As mentioned, calculations have not been carried out for holes of thickness–diameter ratio h/d greater than unity; such calculations become increasingly difficult, and in the case of the present method probably increasingly inexact as this ratio increases. However, it is of interest to note that a fair idea of how the factor of stress concentration k must vary for larger values of h/d can be gained by noting that in the limit, as this ratio approaches infinity, that is when the diameter of the hole becomes extremely small, the variation in the thickness direction of the bending stress in the plate must become less and less important.

In the limit the stress concentration should therefore be the same as it is for a plate under uniform tension or compression. The classical theory for this case gives a factor of stress concentration of $k = 3$. This classical theory is based on the approximate two-dimensional elasticity theory; however, it would be exact for a material with a Poisson's ratio of zero, and the factor probably does not differ greatly from three for usual materials (an exact theory exists, but again calculations have apparently not been carried out for very small holes). If we take this

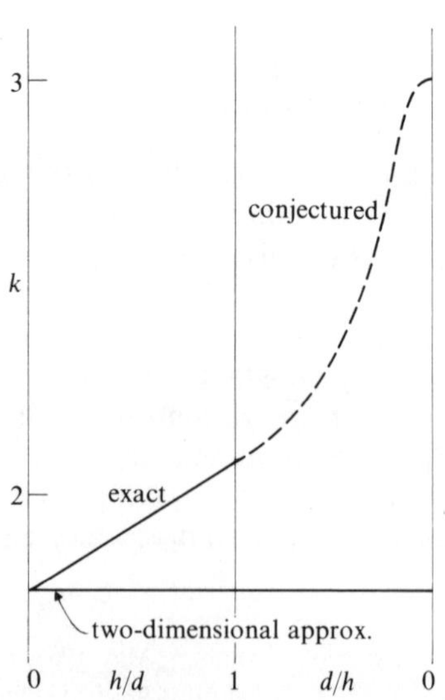

FIGURE 5.16

factor as three, Fig. 5.16, in which the plot has been extended to an h/d of infinity by plotting d/h from 1 to 0 in the right-hand side, shows the general way in which the factor k probably varies over the whole range of h/d from zero to infinity.

5.6 THICK PLATES—CORRECTIONS TO CLASSICAL PLATE DEFLECTIONS

General Introduction

As was discussed in Sec. 3.5 in connection with beams, the Love–Kirchhoff neglection of the effect of transverse strains and stresses, which is used in all classical beam, plate and shell theories, causes errors not only in stresses but also in strains and hence in displacements such as deflections. The errors in stresses are seldom important for steady loading of constructions made of ductile materials but should be considered when there are fatigue conditions or brittle materials; these errors can be corrected by using elasticity methods, which were considered for beams in Secs. 3.3, 3.4 and for plates in Secs. 5.2–5.5.

The errors in strains and displacements are important under somewhat different circumstances. They are generally unimportant in the usual loading of slender or thin constructions. They are likely to be important when the distances between nodes, or points where the loading comes to zero and changes sign, are not very large compared to the thickness, either because the loading varies rapidly (for example reverses at short intervals) or because the thickness is not very small compared to the other dimensions, that is for 'thick' constructions.

The errors in the strains and displacements predicted by classical theories are also likely to be important for constructions in which the middle part, which resists most of the transverse shear but little of the bending moment, has been lightened so that it is no stronger to resist transverse shear than the outer part is to resist bending stresses; in such a case the shear strains and the deflections due to shear strains must be expected to be of the same order of magnitude as those due to bending, which are all that are considered in classical theories. Common examples are beam trusses, and beams, plates or shells made of 'sandwich' material. Actual deflections are always greater, and so buckling strengths and vibration frequencies are always less, than those predicted by the classical theories.

The refinements introduced by application of elasticity theory discussed in Secs. 3.3, 3.4, 5.2, 5.3, 5.4, 5.5 of course give correct, or more nearly correct, values for the strains and displacements as well as for the stresses. However, these methods are generally difficult or impossible to apply to such constructions as trusses and sandwich materials, and in any case a much simpler correction to the classical theory can be used when the errors in deflections are the prime concern; this is the subject of the present section. Such corrections are based upon adding the deflections due to transverse strains (principally transverse shear strains) to the flexural deflections due to bending considered in the classical theories. This type of correction was first used by Timoshenko for beams, and apparently by the author for plates.*

Derivation of Correction to Classical Deflections

As in the beam analysis of Sec. 3.5, we take the total deflection w_t of the middle surface of a general plate as:

$$w_t = w_f + w_s \tag{5.84a}$$

We define w_f as the flexural deflection considered in classical plate theory, which is due to the bending and twisting of the plate. The formulas for the flexural moments M_x, M_y, M_{xy}, Eqs. (4.14), can be considered to be exact if we replace w by w_f (at least for small displacements, far from discontinuities in load or shape).

We will calculate w_s as the deflection due to transverse shear, using the approximate assumption that the shear stresses are uniformly distributed across the thickness of the plate instead of parabolically distributed as they usually are. However, as we did in the case of beams, we will multiply the value of w_s so obtained by the numerical factor α, which should have a value of the order of unity but which we can adjust to allow for the error made not only in assuming an incorrect distribution of transverse shear stresses but also for neglecting additional deflections due to transverse normal stresses, which it was shown in Sec. 3.5 vary in the same way as those due to shear and hence can be taken into consideration by use of such a single factor.

From moment equilibrium considerations the total transverse shearing forces F_{xz} and F_{yz} are given by Eqs. (4.15) with w replaced by w_f, and are:

$$\begin{aligned} F_{xz} &= -D\frac{\partial}{\partial x}\nabla^2 w_f = -\frac{Eh^3}{12(1-\nu^2)}\frac{\partial}{\partial x}\nabla^2 w_f \\ F_{yz} &= -\frac{Eh^3}{12(1-\nu^2)}\frac{\partial}{\partial y}\nabla^2 w_f \end{aligned} \tag{5.84b}$$

Strictly speaking there should be some additional terms on the right-hand side of these equations, similar to the term $F_x\, dw_s/dx$ which was included in Eq. (3.55a) and illustrated in Fig. 3.20(a). These terms are: $F_x\, \partial w_s/\partial x + F_{yx}\, \partial w_s/\partial y$ in the expression for F_{xz} and $F_y\, \partial w_s/\partial y + F_{xy}\, \partial w_s/\partial x$ in the expression for F_{yz}. We will omit these minor terms because, as was shown in Eq. (3.57) for the corresponding term in the case of beams, all these terms can be shown to be very small compared to the F_{xz} and F_{yz} terms, having a ratio to them of an elastic unit strain to unity.

Turning from equilibrium to displacement–strain–stress relations, again as in the beam case, since the shear displacement w_s leaves cross sections parallel to each other, the slopes $\partial w_s/\partial x$ and $\partial w_s/\partial y$ at any point would be equal to the shear

* See the author's discussion of paper by E. Reissner, *Trans. ASME*, vol. 67, p. A69, 1945; see also lecture 'Shell Theory', *Proc. of 4th Midwestern Conference on Solid Mechanics*, published by University of Texas, p. 1, 1959.

strains ϵ_{zx} and ϵ_{yz} at that point if the cross sections also remain vertical. As shown in Fig. 3.20(b), (c), for a cantilever or a simply supported beam, cross sections in w_s displacements *do* remain vertical. And the argument given after Eq. (3.59) that this is a general rule (any deviation from which would require rigid body translations and rotations which are prevented in all ordinary cases by boundary constraints) applies equally well to plates.

Hence, using Eqs. (3.5a) and (5.84b) we take:

$$\frac{\partial w_s}{\partial x} = \epsilon_{xz} = \frac{\alpha \sigma_{xz}}{G} = \frac{\alpha F_{xz}}{Gh} = -\frac{\alpha h^2}{6(1-\nu)} \frac{\partial}{\partial x} \nabla^2 w_f$$
$$\frac{\partial w_s}{\partial y} = \epsilon_{yz} = \frac{\alpha \sigma_{yz}}{G} = \frac{\alpha F_{yz}}{Gh} = -\frac{\alpha h^2}{6(1-\nu)} \frac{\partial}{\partial y} \nabla^2 w_f \tag{5.84c}$$

The two equations (5.84c) can only be satisfied in general if:

$$w_s(x, y) = -\frac{\alpha h^2}{6(1-\nu)} \nabla^2 w_f \tag{5.84d}$$

from which, using Eq. (5.84a):

$$w_t = \left[1 - \frac{\alpha h^2}{6(1-\nu)} \nabla^2\right] w_f \tag{5.84e}$$

This relationship between the total deflection w_t and the deflection w_f due to flexure alone can then be used with the transverse equilibrium equation (4.18) to find the total deflection produced by a lateral loading $p(x, y)$ and, if they are present, buckling normal loadings per unit length of section in the x and y directions F_x, F_y and shear loading F_{xy}. Equation (4.18) should now be written:

$$D\nabla^4 w_f = p + F_x \frac{\partial^2 w_t}{\partial x^2} + F_y \frac{\partial^2 w_t}{\partial y^2} + 2F_{xy} \frac{\partial^2 w_t}{\partial x \, \partial y} \tag{5.84f}$$

In the term on the left-hand side, w has been changed to w_f because this term represents the transverse internal resistance calculated by considering flexural displacements only. On the other hand the transverse components of F_x, F_y, F_{xy} obviously are proportional to the *total* curvatures and twist, so w has been replaced by w_t in the last three terms.

The two equations (5.84e, f) can be solved together for any particular loading to eliminate w_f, and yield the total deflection w_t, or buckling loads or vibration frequencies corrected for the effects of transverse strains and stresses, in the same manner as Eqs. (3.60) and (3.56) were used in the corresponding case of beams.

Direct Relation Between Loading and w_t

Alternatively, as in the development of Eq. (3.67) for beams, we can eliminate w_f between the general equations (5.84e and f) to obtain a general direct relation between the loading and w_t. To do this we expand w_f in the series:

$$w_f = w_t + a\nabla^2 w_t + b\nabla^4 w_t + \cdots \tag{5.84g}$$

Substituting expression (5.84e) for w_t into (5.84g) we obtain:

$$w_f = w_f + \left(a - \frac{\alpha h^2}{6(1-\nu)}\right)\nabla^2 w_f + \left(b - a\frac{\alpha h^2}{6(1-\nu)}\right)\nabla^4 w_f + \cdots$$

which can only be satisfied in general if:

$$a = \frac{\alpha h^2}{6(1-\nu)}, \qquad b = a\frac{\alpha h^2}{6(1-\nu)} = \left(\frac{\alpha h^2}{6(1-\nu)}\right)^2, \text{ etc.}$$

or, from Eqs. (5.84g) and (5.84f)

$$w_f = w_t + \frac{\alpha h^2}{6(1-\nu)}\nabla^2 w_t + \left(\frac{\alpha h^2}{6(1-\nu)}\right)^2 \nabla^4 w_t + \cdots,$$

$$D\left[1 + \frac{\alpha h^2}{6(1-\nu)}\nabla^2 + \cdots\right]\nabla^4 w_t = p + F_x\frac{\partial^2 w_t}{\partial x^2} + F_y\frac{\partial^2 w_t}{\partial y^2} + 2F_{xy}\frac{\partial^2 w_t}{\partial x\,\partial y} \tag{5.84h}$$

This correction to the classical plate equation (4.18), in the form of a direct relation between the loading and the correct total deflection, is exactly equivalent to the two equations (5.84e, f) if we use the entire series represented by the parenthesis of Eq. (5.84h). If we consider Eq. (4.18) to be a first approximation, then Eq. (5.84h) with only the first two terms of the series $[1 + \alpha h^2\nabla^2/6(1-\nu)]$ represents a second approximation, which should be sufficient for practical purposes if the correction to be made is small.

Evaluation of Factor α

Equations (5.19) provide an exact elasticity-based relation between w_t and the lateral loading p, which can be put in a series form comparable to Eq. (5.84h) for lateral load only. Setting $u_{z(z=0)} = w_t$, $t_z - b_z = p$, $t_z + b_z = 0$, $\boldsymbol{\nabla}^2 = c^2\nabla^2 = h^2\nabla^2/4$, and applying the operator ∇^4 and using Eqs. (5.18a) the second of Eqs. (5.19) becomes:

$$D\nabla^4 w_t = p - \frac{(8-3\nu)h^2}{40(1-\nu)}\nabla^2 p \ldots \tag{5.84i}$$

Equation (5.84i), like Eq. (3.67a) for beams, is in itself a direct relation between the lateral loading p and w_t, which may be convenient to use in some problems. To put Eq. (5.84i) in the form of (5.84h) we can expand p in the series:

$$p = D(1 + A\nabla^2 + B\nabla^4 + \cdots)\nabla^4 w_t \tag{5.84j}$$

Substituting (5.84i) into (5.84j) we find:

$$p = p + \left(A - \frac{(8-3\nu)h^2}{40(1-\nu)}\right)\nabla^2 p + \left(B - A\frac{(8-3\nu)h^2}{40(1-\nu)}\right)\nabla^4 p + \cdots$$

which in general requires that:

$$A = \frac{(8-3\nu)h^2}{40(1-\nu)}, \qquad B = \left(\frac{(8-3\nu)h^2}{40(1-\nu)}\right)^2 \cdots$$

or, from Eq. (5.84j)

$$D\left[1+\frac{(8-3\nu)h^2}{40(1-\nu)}\nabla^2+\left(\frac{(8-3\nu)h^2}{40(1-\nu)}\right)^2\nabla^4+\cdots\right]\nabla^4 w_t = p \tag{5.84k}$$

Comparing Eq. (5.84k) to Eq. (5.84h) for transverse load only it is apparent that to correct for the approximation which were made in deriving Eqs. (5.84c, d) we should take

$$\frac{\alpha h^2}{6(1-\nu)}=\frac{(8-3\nu)h^2}{40(1-\nu)}, \qquad \text{or} \qquad \alpha=\frac{3(8-3\nu)}{20}\approx 1{\cdot}065 \tag{5.84l}$$

It seems reasonable to use this value of α as the best possible estimate for all types of loading. Hence we can rewrite Eq. (5.84e) as:

$$w_t=\left[1-\frac{(8-3\nu)h^2}{40(1-\nu)}\nabla^2\right]w_f \tag{5.84m}$$

which can be solved simultaneously with Eq. (5.84f) to obtain the most accurate solution for any particular loading condition.

In applying the above result to get the best direct relation between the loading and w_t, we can of course use the above value of α in Eq. (5.84h). However, in truncating the series in the brackets to get the best and simplest second approximation we may also remember that in truncating the similar series of Eq. (3.67) for beams, it was found that we could allow approximately for the terms cut off by slightly modifying the last term retained; thus, if we retain only two terms, reducing the second term by about 10 percent (although the terms are all positive they typically alternate in sign when applied to an actual case) enabled the truncated series to closely duplicate the entire series in the practical range of interest. Making the reasonable assumption that this finding applies to plates, we can rewrite Eq. (5.84h) in the form:

$$D\left[1+\frac{(8-3\nu)h^2}{44(1-\nu)}\nabla^2\right]\nabla^4 w_t = p+F_x\frac{\partial^2 w_t}{\partial x^2}+F_y\frac{\partial^2 w_t}{\partial y^2}+2F_{xy}\frac{\partial^2 w_t}{\partial x\,\partial y} \tag{5.85}$$

Application to Sandwich Plates

We will make the assumption, which is certainly a very close approximation for practical sandwich plates, that transverse shear is taken entirely by the core, with transverse shear stresses distributed uniformly over the core, and that transverse normal stresses play a negligible role in this case. Then, in developing Eqs. (5.84c) for the case of sandwich plates we would have no need for the factor α, and can express the strain ϵ_{xz} by $F_{xz}/G'h'$, where G' is the shear modulus of the core material for transverse shear strains, and h' is the core thickness. Calling the flexural stiffness of the sandwich plate D' (for instance, if we ignore the contribution of the core, $D'=E(h^3-h'^3)/[12(1-\nu^2)]$), then $F_{xz}=-D'(\partial/\partial x)\nabla^2 w_f$ and the first of Eqs. (5.84c)

$$\frac{\partial w_s}{\partial x}=\frac{F_{xz}}{G'h'}=-\frac{D'}{G'h'}\frac{\partial}{\partial x}\nabla^2 w_f \tag{5.86a}$$

If we continue the development as previously, then we obtain equations the same as (5.84e, f, h) except that D is replaced by D' and $\alpha h^2/[6(1-\nu)] = (8-3\nu)h^2/[40(1-\nu)]$ is replaced by $D'/(G'h')$. If we wish to truncate the series in the brackets of Eq. (5.84h) to two terms, then the second term can be reduced by 10 percent to allow for the terms cut off.

Application to Stability of Hinged-Edge Plate Under Edge Compression

To illustrate the application of corrections to classical plate deflections, consider the case studied by classical plate theory in Sec. 4.4, Eqs. (4.36, 4.37), of the hinged plate shown in Fig. 4.15 under a uniaxial compressive stress s_x (taking $s_y = 0$). This case is not of great practical importance because, as will be seen, limitations to the elastic stresses possible for ordinary engineering materials make corrections to the buckling resistance predicted by classical theory for homogeneous plates of such materials relatively minor and unimportant. However, this case is very suitable for illustration, and a relatively simple extension of it to sandwich plates may have great practical importance.

Using the finding of classical theory (which certainly applies here also) that the easiest buckling mode for uniaxial compression is in one half wave in the direction normal to the compression, in this case the y direction, we take:

$$w_f = W_f \sin\frac{m\pi x}{a}\sin\frac{\pi y}{b} \qquad (m = \text{an integer}) \tag{5.86b}$$

$$w_t = W_t \sin\frac{m\pi x}{a}\sin\frac{\pi y}{b} \tag{5.86c}$$

Substituting these expressions into Eq. (5.84m) and dividing through by the sine functions, we find

$$W_t = \left[1 + \frac{\pi^2(8-3\nu)h^2}{40(1-\nu)a^2}\left(m^2 + \frac{a^2}{b^2}\right)\right] W_f \tag{5.86d}$$

Substituting expressions (5.86b, c) into Eq. (5.84f) with $F_x = \sigma_x h$ and p, F_y, $F_{xy} = 0$, using expressions (5.86d) for W_t and dividing through by $W_t \sin m\pi x/a \sin \pi y/b$, we obtain:

$$\frac{b^2 h}{\pi^2 D}\sigma_x = \left(\frac{mb}{a} + \frac{a}{mb}\right)^2 \left[\frac{1}{1 + \dfrac{\pi^2(8-3\nu)h^2}{40(1-\nu)a^2}\left(m^2 + \dfrac{a^2}{b^2}\right)}\right] \tag{5.86e}$$

Comparing this result with Eq. (4.36) we see that the correction for transverse strains is given by the quantity in the brackets. For $a/h = 20$ and $m = 2, 4$ for example, the classical theory would give buckling strengths too high by factors of 1·03 and 1·11 respectively. However, such a number of waves would require $a/b = 2, 4$ respectively and buckling stresses of about $0{\cdot}04E$, $0{\cdot}15E$ respectively. Such stresses could only be reached within the elastic range with soft plastics or rubber-like materials; the correction for buckling of solid core, that is homogeneous, metallic or hard plastic plates loaded within the elastic limit can never be great.

This case could have been analysed even more easily by substituting expression (5.86c) for w_t into Eq. (5.85), which gives the result:

$$\frac{b^2h}{\pi^2D}s_x = \left(\frac{mb}{a}+\frac{a}{mb}\right)^2\left[1-\frac{\pi^2(8-3\nu)h^2}{44(1-\nu)a^2}\left(m^2+\frac{a^2}{b^2}\right)\right] \tag{5.86f}$$

This formula gives the same correction factors as those listed above for $a/h=20$, $m=2, 4$.

These results can readily be extended to the similar and practically important case of buckling of a sandwich plate. For the same case of uniaxial compression F_x per unit width of section, where $F_x = hs_x$, setting $D = D'$ and $(8-3\nu)h^2/[40(1-\nu)] = D'/(G'h')$ in Eq. (5.86f), we obtain:

$$\frac{b^2}{\pi^2D'}F_x = \left(\frac{mb}{a}+\frac{a}{mb}\right)^2\left[1-\frac{0{\cdot}9\pi^2D'}{G'h'a^2}\left(m^2+\frac{a^2}{b^2}\right)\right] \tag{5.86g}$$

Frequency of Lateral Vibration of a Plate

Like the corresponding case of beam vibration, this case is complicated by the fact that when the effects of transverse strains are important, then the effect of acceleration of the outer fibers of the plate in the plane of the plate (which was described as 'rotary' or 'rotatory inertia' in beams) is also important. Since the w_s deflection due to transverse strains produces no rotations of cross sections under the approximate assumption of uniform distribution of transverse shear stresses (there are some negligible movements in the plane of the plate corresponding to the 'warping' of the cross sections due to the actual parabolic distribution) only the flexural displacements w_f of classical plate theory need be considered in calculating this inertia effect.

From Eq. (4.1) the displacement of an element $dx\,dy\,dz$ in the x direction due to flexure is $-z\,\partial w_f/\partial x$, giving rise to a resisting 'inertia force' of $\rho\,dx\,dy\,dz\,z\,\partial^3 w_f/\partial x\,\partial t^2$, where ρ is the mass per unit volume of the material and t is time. The total moment about the y axis of these resistances in an element $dx\,dyh$, obtained by integrating the above expression multiplied by its moment arm z from $-h/2$ to $h/2$, is $\rho h^3\,dx\,dy(\partial^3 w/\partial x\,\partial t^2)/12$. This, divided by $dx\,dy$, must be added to the left-hand side of the moment equilibrium equation $\Sigma\,m_y = 0$ in Eqs. (4.8). Using Eqs. (4.14) and solving for F_{xz}, with a similar calculation for F_{yz}, we obtain the following modifications of Eqs. (4.15) and (5.84b):

$$F_{xz} = -\frac{\partial}{\partial x}\left(D\nabla^2 w_f - \frac{\rho h^3}{12}\frac{\partial^2 w_f}{\partial t^2}\right)$$

$$F_{yz} = -\frac{\partial}{\partial y}\left(D\nabla^2 w_f - \frac{\rho h^3}{12}\frac{\partial^2 w_f}{\partial t^2}\right) \tag{5.87a}$$

We could now express the shearing forces F_{xz}, F_{yz} in terms of the shearing strains $\partial w_s/\partial x$ and $\partial w_s/\partial y$ respectively, as we did in Eqs. (5.84c), and derive new modified expressions for w_s and w_t, similar to Eqs. (5.84d, e or m), with $D\nabla^2 w_f$ replaced by the parenthesis in Eqs. (5.87a). However, as was argued in the discussion after Eq. (3.65) for beams, this would not be justified because it would represent a minor correction of the minor correction w_s, and would have no importance except for wave lengths so short that they are out of the range of applicability of the present essentially second-approximation theory.

The important application of the above new expressions for F_{xz}, F_{yz} is in the derivation of the transverse equilibrium equation $\Sigma f_z = 0$, Eq. (4.8). Using expressions (5.87a) and taking $F_x, F_y, F_{xy} = 0$, $p = -\rho h\, \partial^2 w_t/\partial t^2$, we obtain:

$$-\nabla^2\left(D\nabla^2 w_f - \frac{\rho h^3}{12}\frac{\partial^2 w_f}{\partial t^2}\right) = \rho h \frac{\partial^2 w_t}{\partial t^2} \tag{5.87b}$$

Assuming the expressions:

$$\begin{aligned} w_f &= W_f \sin\frac{m\pi x}{a}\sin\frac{n\pi y}{b}\sin 2\pi N_{mn}t \\ w_t &= W_t \sin\frac{m\pi x}{a}\sin\frac{n\pi y}{b}\sin 2\pi N_{mn}t \end{aligned} \tag{5.87c}$$

where N_{mn} is the number of free vibrations per unit time, and substituting them into Eqs. (5.87b), (5.84m), eliminating W_f between the two equations, dividing through by $W_t \sin(m\pi x/a)\sin(n\pi y/b)\sin 2\pi N_{mn}t$ and solving for N_{mn}, we find for $\nu = 0{\cdot}3$:

$$\begin{aligned} N_{mn} &= \frac{\pi}{2a^2}\sqrt{\frac{D}{\rho h}}\left(m^2 + \frac{a^2}{b^2}n^2\right)\left[1 + \frac{\pi^2 h^2}{a^2}\left(\frac{8-3\nu}{40(1-\nu)} + \frac{1}{12}\right)\left(m^2 + \frac{a^2}{b^2}n^2\right)\right]^{-1/2} \\ &\approx N_{mn(class.)}\left[1 - \frac{5}{3}\frac{h^2}{a^2}\left(m^2 + \frac{a^2}{b^2}n^2\right)\right] \end{aligned} \tag{5.87d}$$

where $N_{mn(class.)}$ is the frequency given by classical theory, Eq. (4.32). In the above expression the term $(8-3\nu)/40(1-\nu)$ is due to the transverse strains and the term 1/12 to the rotary inertia.

Range of Application of Various Approximations

As in the discussion of this subject at the end of Chapter 3 for beams, we can summarize the range of application of classical plate theory, and of the various improvements upon it which have been discussed, in the relations between the three quantities h, w_{max}, and l defined in that previous discussion, and the relations and limitations presented there apply equally well to plates, with the following supplementary definitions. Thus the nonlinear term $\sigma_{xm} h\, \partial^2 w/\partial x^2$ should be changed and expanded to $F_x\, \partial^2 w/\partial x^2$, $F_y\, \partial^2 w/\partial y^2$, and $F_{xy}\, \partial^2 w/\partial w\, \partial y$, and the half wave length l should be interpreted as the *smaller* of the half wave lengths in the x and y directions.

6

CLASSICAL SHELL THEORY

6.1 INTRODUCTION TO THIN SHELL THEORY

The Love–Kirchhoff Approximation

As in the case of bars and plates, we use the term 'classical' shell theory for all theories based upon the Love–Kirchhoff approximate assumption, described in the first paragraph of Chap. 2, that straight lines normal to the middle surface before deformation remain straight, normal to the middle surface and unchanged in length after deformation. Actually, of course, such lines after deformation are in general no longer exactly normal to the middle surface because of transverse shear strains, no longer straight because these strains vary with the distance from the middle surface, and no longer have the same length because of transverse normal strains.

As has been emphasized before, the errors due to this approximation are negligible for 'thin shells' of homogeneous material under usual loading conditions, that is shells whose wall thickness is small compared to the radii of curvature and the corresponding 'radius of twist', and compared to the half wavelength of the loading, defined as the distance between nodes or points where the loading reverses in sign; this means in practice a thickness which is also small compared to dimensions such as the widths and lengths of curved panels. Most shell problems of practical interest fall into this category of thin shell problems, which is the subject of this and much of the next chapter.

As also discussed previously, the simplification which can be made in such cases by use of the Love–Kirchhoff approximation is very great, as it permits the displacement of every point in the shell wall, and hence the strains and stresses at every point, to be defined in terms of the displacement of one surface such as the middle surface of the shell wall. This represents in effect the reduction of the problem from a three- to a two-dimensional one.

General Effects of Curvature

Some consideration was given to these effects in the introduction to Sec. 5.1 and the discussion of Fig. 5.1 in connection with boundary conditions. Figure 6.1 shows how the initial curvature of a curved surface, such as the middle surface of a thin cylindrical shell, affects the conditions over the whole surface. The figure shows the cylinder wall deflected harmonically in both the axial and circumferen-

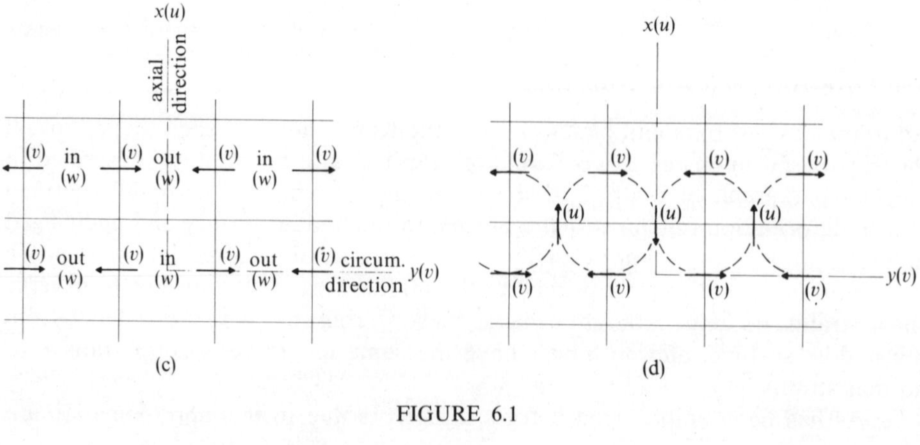

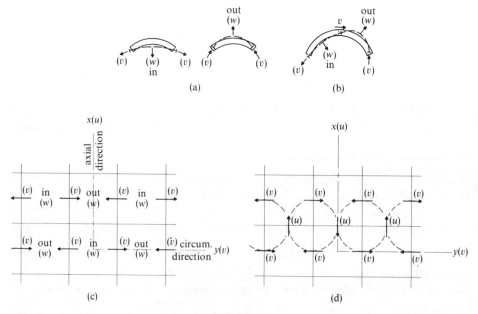

FIGURE 6.1

tial directions as pictured later in Fig. 7.5. It would undergo such a deflection under similarly distributed transverse loading (which may represent only one harmonic component of a more general loading distribution), or, as experiments and theory discussed later in Sec. 7.2 show, when it buckles under an axial compression.

Figure 6.1(a) shows how a circumferential section of the wall between nodes tends to 'push out' tangentially at the nodes when it is deflected inward (like the curved plate in Fig. 5.1(c)) and to 'pull in' at the nodes when it is deflected outward. Figure 6.1(b) shows how these tendencies reinforce one another at the common node due to the alternate inward and outward deflections of a harmonic deflection pattern. Calling the transverse displacements or deflections w and circumferential displacements v, the side view of the wall, Fig. 6.1(c), shows the pattern of v displacements thus caused by 'in' and 'out' w displacements. Finally, Fig. 6.1(d) shows how in turn the circumferential displacements v tend to cause axial displacements u, by tending to rotate the sections of the wall outlined by dots in the direction of the arrows. The u, v, w displacements shown are at the peak positions of these harmonically varying displacements.

Thus, the w displacements are caused directly by the loading. These w displacements tend to produce v displacements which are in general much smaller than the w displacements. In turn the v displacements tend to produce u displacements which are much smaller than the v displacements and hence even smaller compared to the w displacements. However, this relationship depends upon the wave size, and these statements are strictly true only for wavelengths which are smaller than the circumference and other overall shell dimensions, as is typical of harmonic components of usual shell deformations. In an extreme case such as the pure bending of a cylindrical tube, given later in Eq. (7.3d), where there is only one wave around the circumference and the longitudinal wavelength is infinite, the v displacement may have the same size as the w displacement.

This displacement pattern corresponds to the formulas of Eq. (6.12) given later, which were calculated for buckling of a thin cylinder under axial compression. Also the u, v, w amplitudes found, given in Table 6.4 (for which $pq = 11$ represents the primary or 'fundamental' harmonic component), confirm the above statements concerning relative magnitudes of these displacements in a cylindrical shell. In other types of shell the u displacements may not be smaller than the v displacements, but they will both be much smaller than the w displacements in common cases such as lateral loading or buckling.

Differentiation of Theories

In the above discussion we referred to classical shell 'theories', rather than a classical shell 'theory'. Even in the simpler case of flat plates we found it convenient to separate solutions based upon the Love–Kirchhoff assumption into special cases such as small- and large-displacement theories. In the case of general shells the broad variety possible, as well as the serious complications introduced by curvature, require simplifications which are valid only in certain ranges, and make advisable differentiation of numerous classes of shells.

More than this, theories designed by different investigators for the same applications frequently differ in details, presumably because simplifications are introduced at different stages in the development and so affect the final results differently, or because the decision as to what kind of simplification is legitimate under given conditions is made intuitively and intuitions differ. To be sure we will find, as Koiter* and others have argued, that differences between various theories designed for the same purpose are usually unimportant because the terms involved are unimportant. But even if this is true, the question remains as to what terms *are* necessary, and what are the simplest valid forms for solutions designed for various purposes.

Because of the complexities introduced by curvature and the consequent difficulties discussed above when simplifications are introduced on an intuitive basis or at various stages we will start our study of thin shells by developing a general theory** without any simplifications except for the use of the Love–Kirchhoff assumption. Even though we will subsequently find that many of its complexities can safely be eliminated even in the quite general case, it is felt that the only rational and safe way to make these simplifications is to start with the whole picture, so that we can compare all the terms retained and all those eliminated. As mentioned at the beginning of the book, this process turns out to be no more difficult than attempting to set up many special theories on an intuitive basis.

While such an exact general theory can be and has been set up more compactly by using tensor notation and analysis, doing this with more elementary mathematical tools presents no difficulty if no attempt is made to eliminate the intermediate quantities involved while the theory is in its general form. Moreover this is probably easier than doing it in tensor form if we include the task of interpreting it physically and of translating it back into a form where it can be applied to practical problems.

6.2 GENERAL THIN SHELL DISPLACEMENT–STRAIN RELATIONS

Coordinates

As coordinate lines we will use the 'lines of curvature' of the undisplaced middle surface of the shell wall, together with normals to this surface. These lines of

* W. T. Koiter, *A Consistent First Approximation in the General Theory of Thin Elastic Shells*, Technological University, Delft, 1959.

** L. H. Donnell, 'General thin shell displacement–strain relations', *Proc. of the 4th US Nat. Congr. of Appl. Mech.*, 1962, p. 529.

curvature are defined as lines along which the twist is zero, and it is shown in the theory of continuous surfaces that there are always at least two such systems of lines, and that these systems are orthogonal to each other, that is tangents to two such lines at the point where they intersect will be at right angles to each other. It is evident that the intersection of a plane of symmetry of the shell with its middle surface is a line of curvature, because any twist along such a line would violate the condition of symmetry; in most of the types of shells of practical interest this fact determines the systems of lines of curvature.

In most practical problems the shell boundaries follow or are normal to these lines of symmetry and hence coincide with the lines of curvature and the coordinate lines. As has been emphasized in the past, edge boundary conditions are most easily satisfied when edges follow coordinate lines, and in the case of flat plates we tried to choose the coordinate systems to make this true. In the case of shells the use of lines of curvature as coordinate lines has such great advantages in simplifying the theory that it pays to use such coordinates even in those few cases where they cannot be made to coincide with edge boundary lines.

Lines of Curvature Through a Point

Because simple geometric considerations suffice for determining the positions and curvatures of lines of curvature in the cases which we will study and in most practical cases, we will not develop here a complete mathematical theory of surfaces. However, we can learn a good deal of interest just by considering the conditions at a general point of a curved surface.

Figure 6.2(a) shows a point O of a continuous curved surface such as the middle surface of a shell, and rectangular coordinates x, y, z with origin at O. We take the x, y plane tangent to the surface at the point O. This means, if we define the surface by the distances $z(x, y)$ from the x, y plane to the surface, that at O:

$$x, y = 0: \qquad z, \frac{\partial z}{\partial x}, \frac{\partial z}{\partial y} = 0 \tag{6.1a}$$

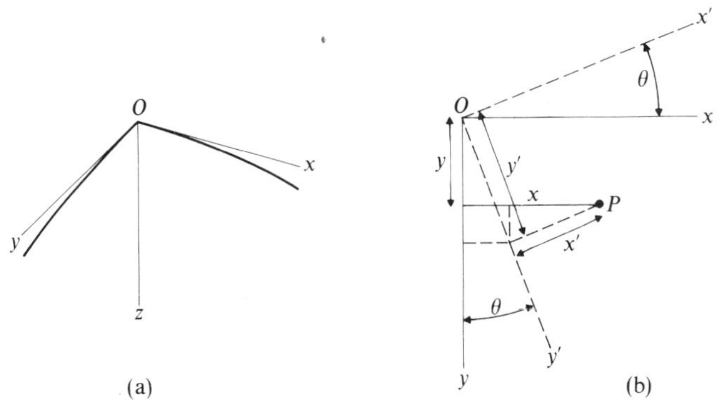

FIGURE 6.2

We define the rate of change in the x direction, $\partial/\partial x$, of the slope of the surface in the x direction, $\partial z/\partial x$, as the curvature of the surface in the x direction, k_x, and similarly for the y direction:

$$\frac{\partial}{\partial x}\left(\frac{\partial z}{\partial x}\right)=\frac{\partial^2 z}{\partial x^2}=k_x, \qquad \frac{\partial}{\partial y}\left(\frac{\partial z}{\partial y}\right)=\frac{\partial^2 z}{\partial y^2}=k_y \tag{6.1b}$$

We also define the rate of change in the x direction, $\partial/\partial x$, of the slope of the surface in the y direction, $\partial z/\partial y$, as the twist of the surface in the x direction, k_{xy}:

$$\frac{\partial}{\partial x}\left(\frac{\partial z}{\partial y}\right)=\frac{\partial^2 z}{\partial x \partial y}=k_{xy} \tag{6.1c}$$

Since the order of differentiation is immaterial, the twist in the y direction, defined as the rate of change in the y direction, $\partial/\partial y$, of the slope in the x direction, $\partial z/\partial x$, has the same expression $\partial^2 z/\partial y \partial x = \partial^2 z/\partial x \partial y$ and the same numerical value at the same point O. However, this equality applies only to perpendicular directions such as x and y; the twists of the surface in other directions through the point O may be quite different. The reciprocals of k_x, k_y and k_{xy} are called the radii of curvature and of twist in the corresponding directions.

Consider now new axes x', y' through O and in the x, y plane but at an angle θ to the x, y axes, as specified in Fig. 6.2(b), which is a view looking in the z direction. From this figure it can be seen that the coordinates of a point P in the neighborhood of O with respect to the two sets of axes are related by:

$$x = y' \sin\theta + x' \cos\theta, \qquad y = y' \cos\theta - x' \sin\theta \tag{6.1d}$$

We wish to determine the curvatures and twist of the surface in the x', y' directions, k'_x, k'_y, k'_{xy}, defined in the same way as above, in terms of k_x, k_y, k_{xy}, and θ. This can be done by relating derivatives of $z(x, y)$ with respect to x,y to derivatives of $z(x', y')$ with respect to x', y' by using Eqs. (6.1d) and the basic laws of differentiation.

However, we can see more clearly what we are doing physically in another way. We can replace a continuous surface function $z(x, y)$, such as we are considering, in the vicinity of the origin O by a double power series:

$$z=\sum_m \sum_n C_{mn} x^m y^n \tag{6.1e}$$

in the same way that a Maclaurin series is used to represent a function of one variable. In this series expression the zero and unit power terms, that is those for $(m+n)=0, 1$, must be absent in order to satisfy Eq. (6.1a), while the second power terms, for $(m+n)=2$, must be

$$z=\frac{k_x}{2}x^2+\frac{k_y}{2}y^2+k_{xy}xy+\cdots \tag{6.1f}$$

in order to satisfy Eqs. (6.1b,c). Terms of higher power than the second, that is for $(m+n)>2$, can be added to make the series expression approximate the actual surface at greater and greater distances from O, but the second derivatives of all such terms will contain x or y and hence will be zero at the origin $x, y = 0$. Therefore expression (6.1f) without further terms completely defines the curvatures and twist of the actual surface at the point O.

Substituting the values of x, y given by Eq. (6.1d) into Eq. (6.1f), equating the resulting expression for z to the expression $z = (k'_x/2)x'^2 + (k'_y/2)y'^2 + k'_{xy}x'y'$, which can be derived in the same manner as Eq. (6.1f), and separating out the terms involving the independent variables x'^2, y'^2, $x'y'$, we find that:

$$\begin{aligned} k'_x &= k_x \cos^2\theta + k_y \sin^2\theta - 2k_{xy} \sin\theta \cos\theta \\ k'_y &= k_y \cos^2\theta + k_x \sin^2\theta + 2k_{xy} \sin\theta \cos\theta \\ k'_{xy} &= k_{xy}(\cos^2\theta - \sin^2\theta) - (k_y - k_x) \sin\theta \cos\theta \end{aligned} \tag{6.1g}$$

Setting the twist in the x', y' directions, k'_{xy}, equal to zero in the third of Eqs. (6.1g) and using the relations between trigonometric functions, we find the angle θ, defined by Fig. 6.2(b), which lines of curvature make at any point with lines such as x, y for which the curvature and twist k_x, k_y, k_{xy} are known:

$$k_{xy} \cos 2\theta = (k_y - k_x)\frac{1}{2} \sin 2\theta, \qquad \theta = \frac{1}{2} \tan^{-1} \frac{2k_{xy}}{k_y - k_x} \tag{6.1h}$$

The values of the curvatures k'_x and k'_y of the surface in the directions of the lines of curvature can then be found by substituting the value of θ given by Eq. (6.1h) into the first two equations of (6.1g).

Since an angle π (or any multiple of it) can be added to an angle without changing the value of its tangent, the angle $\theta + \pi/2$ must also define a line of curvature; this confirms the existence of two systems of lines of curvature and their orthogonality.

There is obviously a close analogy between the above analysis and the calculation of the principal axes of inertia of an area or a body, that is the axes for which the product of inertia is zero, as well as the calculation of the principal stresses at a point in a stressed body, on planes on which the shear stress is zero.

Geometric Relation of Points in a Shell Wall

Figure 6.3 shows, in original position, a point o (the projection on the middle surface of a general point O) with the orthogonal lines of curvature of the middle surface, labelled α, β, passing through o. We assume α, β to be independent, continuously varying parameters, having constant values along the β, α lines respectively, and regard the values of these parameters at any point as the coordinates of that point. We take the coordinates of point o as α, β, and of the points p, q, adjacent to o in the directions of increase of α, β, as $\alpha + d\alpha, \beta$ and $\alpha, \beta + d\beta$ respectively, as indicated.

Coordinate systems such as rectangular and the corresponding cylindrical have the same scale throughout, so that distances measured along a coordinate line are proportional to, or for suitable units equal to, the change in the coordinate. But this is not true for polar coordinates or for our orthogonal curvilinear

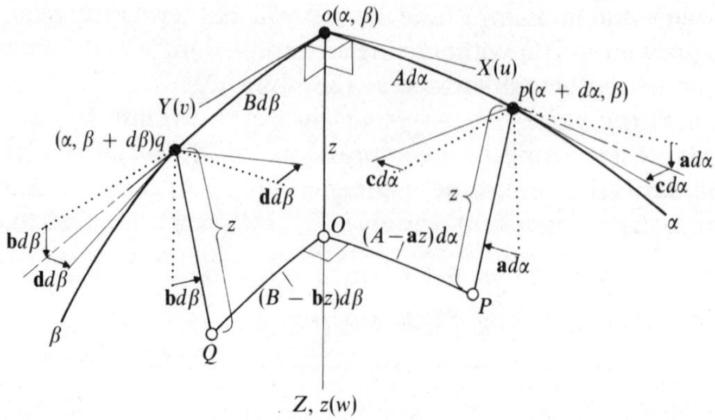

FIGURE 6.3
Original positions

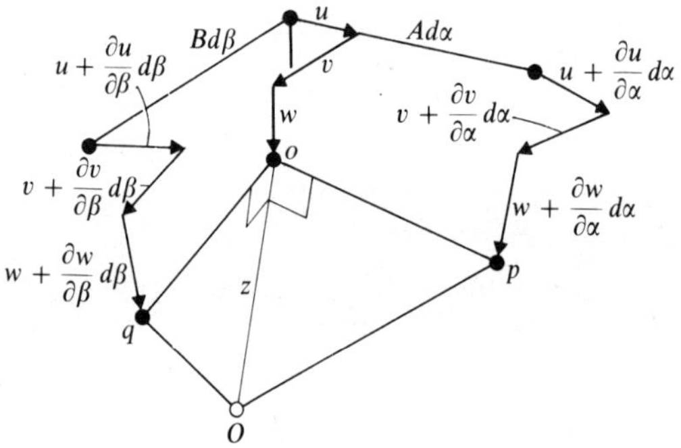

FIGURE 6.4
Displaced positions

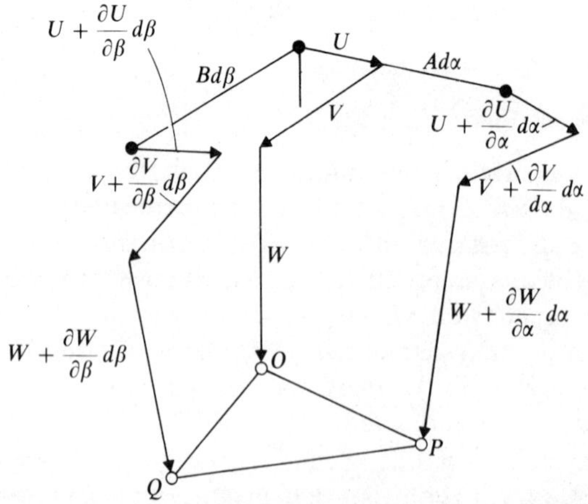

FIGURE 6.5
Displaced positions

coordinates in general. We are concerned with the computation of strains, and these must be studied by considering the actual distances between points. We therefore introduce variable scale factors A, B, defined so that $A\,d\alpha$ and $B\,d\beta$ are the distances measured along the curves between o and p and between o and q; we assume A, B and their first derivatives to be continuous functions of α, β.

We next erect right-handed rectangular coordinate axes X, Y, Z, with origin at o, taking the X and Y axes tangent respectively to the orthogonal α, β lines at o. Their positive directions are in the direction of increase of α, β, and the right handedness of the system defines the positive direction of the Z axis. They are fixed axes because the α, β lines are fixed in the undisplaced middle surface.

Similar triads of axes are erected at points p and q; they are not labelled, but are indicated in Fig. 6.3 by the full straight lines through p and q. Due to the curvature of the surface and of the α, β coordinate lines in the surface, these triads will in general be rotated relative to the XYZ directions, that is relative to the dotted lines shown in the figure through p and q, which are parallel to the XYZ axes.

These angles of rotation about the XYZ directions are shown in Fig. 6.3. The angles for the triad at p are evidently continuous functions of $d\alpha$, and so can be expanded in power series in $d\alpha$. The constant terms in these series are absent, because the angles are zero when $d\alpha$ is zero. The first power term for the angle of rotation about the X axis is also absent, because the α line is a line of curvature along which there is no twist. Now, if terms containing higher powers than one or products of $d\alpha$ and $d\beta$ are retained until the end of the calculations, they are all found to tend to zero, compared to other terms present, as $d\alpha$ and $d\beta$ approach zero, and they can therefore be ignored as small quantities of higher order. For simplicity we will ignore all such terms from the start.

We therefore take the rotations of the triad at p about the XYZ directions to be 0, $\mathbf{a}\,d\alpha$, $\mathbf{c}\,d\alpha$ respectively, as shown; $\mathbf{a}$, $\mathbf{c}$ and their first derivatives are assumed to be continuous functions of α, β, having the physical meanings of a curvature of the surface in the α direction and of the α line in the surface, each multiplied by A. We take $\mathbf{a}, \mathbf{c}$ as positive when the axis tangent to the α line at p is brought by the rotation into the first quadrant of the XYZ directions, as shown in Fig. 6.3. Similarly, interchanging pq, $\alpha\beta$, AB, XY, $\mathbf{ab}$, $\mathbf{cd}$, the triad at q is rotated through the angles $\mathbf{b}\,d\beta$, 0, $\mathbf{d}\,d\beta$, about the XYZ directions, as shown. Calculations of A, B, $\mathbf{a}$, $\mathbf{b}$, $\mathbf{c}$, $\mathbf{d}$ for important types of shell will be given later and are tabulated in Table 6.2. The physical dimensions of these geometric functions of course depend upon the definitions of the coordinates α, β, which may for instance be lengths or angles.

XYZ Coordinates of Middle Surface Points

We will use the fixed XYZ axes at o as 'master axes', and determine the XYZ coordinates of the points o, p, q in the middle surface and of the corresponding general points O, P, Q, in their original and displaced positions. The exact

Table 6.1 *XYZ* **COORDINATES OF DISPLACED POINTS**

Point	X	Y	Z
o	u	v	w
p	$u+(1+f)A\,d\alpha$	$v+g\,A\,d\alpha$	$w+h\,A\,d\alpha$
q	$u+i\,B\,d\beta$	$v+(1+j)B\,d\beta$	$w+k\,B\,d\beta$
O	U	V	W
P	$U+(1+F)A\,d\alpha$	$V+G\,A\,d\alpha$	$W+H\,A\,d\alpha$
Q	$U+I\,B\,d\beta$	$V+(1+J)B\,d\beta$	$W+K\,B\,d\beta$

distances between any of these points are simply related, by the theorem of Pythagoras, to their coordinates in such a set of fixed rectangular axes. The expressions found for the XYZ coordinates of the displaced points are listed in Table 6.1. We will now proceed to derive these coordinates.

Before displacement the XYZ coordinates of point o are 0, 0, 0. For point p, the Y, Z coordinates, and the difference between the X coordinate and $A\,d\alpha$, can be expressed as power series in $A\,d\alpha$, of which both the constant and first power terms are absent because the α line is tangent to the X axis. Hence, ignoring higher powers of $d\alpha$ than one for the same reason as discussed above, the XYZ coordinates of p before displacement can be taken as $A\,d\alpha$, 0, 0, and similarly of q as 0, $B\,d\beta$, 0. Figures 6.4 and 6.5 have been drawn accordingly.

Figure 6.4 shows the points o, p, q in their displaced positions, and unlabelled in their original positions. The displacement components of the point o tangential to the α, β lines and normal to the original middle surface, that is in the XYZ directions, will be designated by u, v, w respectively. We assume that u, v, w, and their first and second derivatives are continuous functions of α, β (in dynamical problems they may also be functions of time, but we are concerned here only with their relation to α, β). Then the displacement components of point p tangential to the α, β lines and normal to the original middle surface, that is in the direction of the unlabelled full-line triad shown at p in Fig. 6.3, must be $u+(\partial u/\partial\alpha)\,d\alpha$, $v+(\partial v/\partial\alpha)\,d\alpha$, $w+(\partial w/\partial\alpha)\,d\alpha$, as shown in Fig. 6.4, since p differs from o only by the small change $d\alpha$ in α.

After displacement, then, the XYZ coordinates of o are u, v, w, and the coordinates of p in the directions of the unlabelled full line triad at p, Fig. 6.3, are $u+(\partial u/\partial\alpha)\,d\alpha$, etc. To obtain the XYZ coordinates of p we must consider that the triad at p is displaced relative to the XYZ triad a distance $A\,d\alpha$ in the X direction, and rotated through angles $\mathbf{a}\,d\alpha$ and $\mathbf{c}\,d\alpha$ about the Y, Z directions. Ignoring powers of $d\alpha$ higher than one as small quantities of higher order, the sines of these angles can be taken as $\mathbf{a}\,d\alpha$ and $\mathbf{c}\,d\alpha$, and the cosines as unity. Again ignoring terms containing higher powers of $d\alpha$, the X coordinate of p after displacement is therefore:

$$A\,d\alpha + \left(u + \frac{\partial u}{\partial \alpha}\,d\alpha\right) - \left(v + \frac{\partial v}{\partial \alpha}\,d\alpha\right)\mathbf{c}\,d\alpha - \left(w + \frac{\partial w}{\partial \alpha}\,d\alpha\right)\mathbf{a}\,d\alpha$$

$$= A\,d\alpha + u + \left(\frac{\partial u}{\partial \alpha} - \mathbf{c}v - \mathbf{a}w\right)d\alpha = u + (1+f)A\,d\alpha$$

and similarly for the other coordinates of p, and the coordinates of q, which are all listed in Table 6.1. In these expressions f, g, h, i, j, k are functions having the following values (we will continue to use the symbol h also for the thickness, defining it when there might be confusion):

$$f = \left(\frac{\partial u}{\partial \alpha} - \mathbf{c}v - \mathbf{a}w\right)/A, \qquad g = \left(\frac{\partial v}{\partial \alpha} + \mathbf{c}u\right)/A, \qquad h = \left(\frac{\partial w}{\partial \alpha} + \mathbf{a}u\right)/A$$

$$i = \left(\frac{\partial u}{\partial \beta} + \mathbf{d}v\right)/B, \qquad j = \left(\frac{\partial v}{\partial \beta} - \mathbf{d}u - \mathbf{b}w\right)/B, \qquad k = \left(\frac{\partial w}{\partial \beta} + \mathbf{b}v\right)/B \tag{6.2}$$

Coordinates of General Points

Having found the XYZ coordinates of the adjacent points o, p, q in the middle surface, we are ready to consider the corresponding general points O, P, Q, which are at a common distance $z(-c \leqq z \leqq c)$, where c is the constant half thickness, from o, p, q, measured normal to the middle surface. [Actually, of course, points which are at a common distance from middle surface points before displacement are not at quite the same distance after displacement, because of lateral strains; these are produced by the stresses in the α, β directions and by the usually very small stresses in the lateral direction. The effect of assuming the distances to be the same will be very small in thin shells under the usual conditions. For shells of varying thickness, z can be taken as $z(\alpha, \beta) = \zeta c(\alpha, \beta)$, where ζ has values between ± 1, and c is the variable half thickness. This gives no great difficulty, but it complicates the final results considerably, so the results shown here are derived for z and c constant.]

We consider first the general point O, which is shown labelled in Fig. 6.4 in its displaced position, and in Fig. 6.3 in its original position. The XYZ coordinates of O in its original position are 0, 0, z. The coordinates in the displaced position can be found from the coordinates of o, p, q and the fact that, by the Love–Kirchhoff assumption, $\overline{oO} = z$ and the angles Oop and Ooq are right angles. Then, since Oop and Ooq are right triangles, we have the three equations:

$$\overline{Op}^2 = \overline{op}^2 + z^2, \qquad \overline{Oq}^2 = \overline{oq}^2 + z^2, \qquad \overline{Oo}^2 = z^2 \tag{6.3}$$

in which $\overline{Oo}^2$, $\overline{Op}^2$, etc., can be expressed as the sum of the squares of the differences between the XYZ coordinates of O, o, p, and q. All these coordinates are now known except for the three unknown coordinates of O, which we will call U, V, W, as indicated in Fig. 6.5. Writing Eqs. (6.3) in terms of these coordinates and solving the three equations for the three unknowns, we obtain:

$$U = u + smz, \qquad V = v + snz, \qquad W = w + s(1+l)z \tag{6.4}$$

where:

$$m = -h(1+j) + gk, \qquad n = -k(1+f) + hi$$
$$l = f + j + fj - gi \tag{6.5}$$
$$s = [m^2 + n^2 + (1+l)^2]^{-1/2}$$

Since the coordinates of O in the XYZ triad at o are U, V, W, and the points P, p differ from the points O, o only by the small change $d\alpha$ in α, the coordinates of P in the triad at p must be $U + (\partial U/\partial \alpha)\, d\alpha$, $V + (\partial V/\partial \alpha)\, d\alpha$, $W + (\partial W/\partial \alpha)\, d\alpha$, as indicated in Fig. 6.5. The XYZ coordinates of P can then be found by considering the displacement and rotation of the triad at p relative to the XYZ triad at o, in precisely the same way that the XYZ coordinates of p were found. We see that the XYZ coordinates of P and Q are related to those of O in the same way that those of p and q are related to those of o. To bring out and make use of this correspondence, symbols relating to the general points are taken the same as those for the middle surface points except that they are capitalized. Thus the XYZ coordinates given in Table 6.1 are the same for OPQ as for opq except for this capitalization, and the definitions:

$$F = \left(\frac{\partial U}{\partial \alpha} - \mathbf{c}V - \mathbf{a}W\right)/A, \quad G = \left(\frac{\partial V}{\partial \alpha} + \mathbf{c}U\right)/A, \quad H = \left(\frac{\partial W}{\partial \alpha} + \mathbf{a}U\right)/A$$
$$I = \left(\frac{\partial U}{\partial \beta} + \mathbf{d}V\right)/B, \quad J = \left(\frac{\partial V}{\partial \beta} - \mathbf{d}U - \mathbf{b}W\right)/B, \quad K = \left(\frac{\partial W}{\partial \beta} + \mathbf{b}V\right)/B \tag{6.6}$$

are similarly related to those of Eq. (6.2).

Strain Expressions

We now have expressions, exact within the limitations of the Love–Kirchhoff assumption, for the coordinates in a common fixed rectangular system, of three general points located at the same distances from the upper and lower surfaces of the shell and differing only by small changes in the α, β coordinates. The coordinates for the original positions of these points can be found from these expressions for the displaced positions by setting $u = v = w = 0$. From these values it is easy to calculate the Lagrangian strains (that is the strains at a point and in directions which are fixed in the shell) at the general point O in the plane of the three points.

The strains normal to this plane are zero or very small (depending upon our definition of strain) according to the Love–Kirchhoff assumption used. Although they are not actually zero or so very small, they should have negligible effect in practical problems whenever this assumption is valid.

The strain in a plane can be completely described by three independent quantities which may be defined in various ways. If the strains are finite it is mathematically convenient to take them as two 'normal' strains and a 'shear' strain defined as follows: Let δx, δy, δs be the sides of a small triangle whose sides

δx and δy were initially at right angles and had the lengths δx_0, δy_0. Then the three strains in the plane of the triangle at the apex of the right angle are defined as the values approached by:

$$\epsilon_x = \frac{\delta x^2 - \delta x_0^2}{2\delta x_0^2}, \qquad \epsilon_y = \frac{\delta y^2 - \delta y_0^2}{2\delta y_0^2}, \qquad \epsilon_{xy} = \frac{\delta x^2 + \delta y^2 - \delta s^2}{2\delta x_0 \delta y_0} \tag{6.7}$$

as δx_0 and δy_0 approach zero.

The three points O, P, Q are the vertices of such an initially right-angled triangle since it can be checked that $\overline{PQ}^2 = \overline{OP}^2 + \overline{OQ}^2$ for the initial state, that is for $u, v, w = 0$. [This is true in general only because we took the lines of curvature as our coordinate lines. It may also be noted that points such as O and P are actually points on a curved line, that OP is actually not quite tangent to this curved line at O, etc. The errors we make by ignoring such facts are proportional to higher powers of $d\alpha$ or $d\beta$ and can be ignored as small quantities of higher order, as in previous similar cases.] We take the x, y directions as the α, β directions, δx, δy, δs as $\overline{OP}$, $\overline{OQ}$, $\overline{PQ}$ and δx_0, δy_0 as the values of $\overline{OP}$, $\overline{OQ}$ when $u, v, w = 0$. Taking $\overline{OP}^2$ as the sum of the squares of the differences between the X, Y, Z coordinates of O and P, etc., and substituting in Eqs. (6.7), we obtain the 'normal' strains in the α, β directions and the 'shear' strain:

$$\begin{aligned} \epsilon_\alpha &= \frac{F + (F^2 + G^2 + H^2)/2 + \mathbf{a}z/A - (\mathbf{a}z/A)^2/2}{(1 - \mathbf{a}z/A)^2} \\ \epsilon_\beta &= \frac{J + (I^2 + J^2 + K^2)/2 + \mathbf{b}z/B - (\mathbf{b}z/B)^2/2}{(1 - \mathbf{b}z/B)^2} \\ \epsilon_{\alpha\beta} &= \frac{G + I + FI + GJ + HK}{(1 - \mathbf{a}z/A)(1 - \mathbf{b}z/B)} \end{aligned} \tag{6.8}$$

The strains at the middle surface, obtained from Eqs. (6.8) by setting $z = 0$ (or by taking δx, δy, δs as $\overline{op}$, $\overline{oq}$, $\overline{pq}$, etc.), are relatively simple:

$$\left.\begin{aligned} \epsilon_\alpha &= f + (f^2 + g^2 + h^2)/2 \\ \epsilon_\beta &= j + (i^2 + j^2 + k^2)/2 \\ \epsilon_{\alpha\beta} &= g + i + fi + gj + hk \end{aligned}\right\} (z = 0) \tag{6.8a}$$

If we ignore strain components which are proportional to z^2 and still higher even powers of z (which a study presented in Sec. 6.3 indicates to be of negligible importance) these strains can be taken as the 'membrane' strains $\epsilon_{\alpha m}$, $\epsilon_{\beta m}$, $\epsilon_{\alpha\beta m}$.

The linear relations obtained by neglecting higher powers than one or products of the displacements or their derivatives are useful for problems in which the displacements can be considered to be infinitesimal. For this case $m = -h$, $n = -k$, $l = f + j$, $s = 1 - l$, $U = u - hz$, $V = v - kz$, $W = w + z$, $F = f - (\partial h/\partial\alpha - \mathbf{c}k)z/A - \mathbf{a}z/A$, $G = g - (\partial k/\partial\alpha + \mathbf{c}h)z/A$, etc., $F^2/2 = -[f - (\partial h/\partial\alpha - \mathbf{c}k)z/A]\mathbf{a}z/A + (\mathbf{a}z/A)^2/2$, $G^2 = 0$, etc.

We thus obtain the following fairly simple linear relations, exact within the

Love–Kirchhoff approximation:

$$\epsilon_\alpha = \left[f - \left(\frac{\partial h}{\partial \alpha} - \mathbf{c}k\right) z/A\right] \bigg/ (1 - \mathbf{a}z/A)$$

$$\epsilon_\beta = \left[j - \left(\frac{\partial k}{\partial \beta} - \mathbf{d}h\right) z/B\right] \bigg/ (1 - \mathbf{b}z/B) \tag{6.8b}$$

$$\epsilon_{\alpha\beta} = \left[g - \left(\frac{\partial k}{\partial \alpha} + \mathbf{c}h\right) z/A\right] \bigg/ (1 - \mathbf{a}z/A) + \left[i - \left(\frac{\partial h}{\partial \beta} + \mathbf{d}k\right) z/B\right] \bigg/ (1 - \mathbf{b}z/B)$$

The commonly used 'engineering' strains, defined as

$$\epsilon'_x = \frac{\partial x - \partial x_0}{\partial x_0}, \qquad \epsilon'_y = \frac{\partial y - \partial y_0}{\partial y_0}, \qquad \epsilon'_{xy} = \sin^{-1} \frac{\partial x^2 + \partial y^2 - \partial s^2}{2\, \partial x\, \partial y} \tag{6.9}$$

yield somewhat more complex expressions. As is well known Eqs. (6.7) and (6.9) reduce to the same thing if neglections are made which are legitimate when the strains are small compared to unity.

Determination of *A*, *B*, a, b, c, d for Specific Shell Types

These geometric functions are not entirely independent of each other. For instance, the first two of the following relations can be derived by equating the difference between the lengths of opposite sides of an element of the middle surface due to the variation of the scale factors with that due to the relative rotations of the other two sides, as indicated later in Fig. 6.11. Altogether it can be shown* that:

$$\frac{\partial A}{\partial \beta} = -B\mathbf{c}, \qquad \frac{\partial B}{\partial \alpha} = -A\mathbf{d}$$

$$\frac{\partial \mathbf{a}}{\partial \beta} = -\mathbf{bc}, \qquad \frac{\partial \mathbf{b}}{\partial \alpha} = -\mathbf{ad} \tag{6.10}$$

$$\frac{\partial \mathbf{c}}{\partial \beta} + \frac{\partial \mathbf{d}}{\partial \alpha} = \mathbf{ab}$$

These relations, particularly the first two, will be found useful and can be used as checks, but it is generally quite easy to determine all the functions from simple geometric considerations.

Table 6.2 shows the most common way of defining the parameters α, β, and the resulting values of the functions A, B, **a**, **b**, **c**, **d** for the types of shells of usual practical interest: flat plates with rectangular and polar coordinates, cylinders using circumferential or angular coordinates and general, and conical and spherical shells. In all these cases there is no problem about selecting lines of curvature, because at least one of the two systems of lines of curvature can be taken as the intersections of planes of symmetry with the middle surface, the other system of course being at right angles to these intersections. Figures 6.6–6.9, which show

* Love, *Mathematical Theory of Elasticity*, 4th edn, p. 517.

Table 6.2

		Coordinates		Scale factors		Curvature factors of surface		Curvature factors in surface		
	Shell type	α	β	A	B	**a**	**b**	**c**	**d**	Fig.
1	Plate, rectangular coordinates	x	y	1	1	0	0	0	0	6.6(a)
2	Plate, polar coordinates	r	θ	1	r	0	0	0	-1	6.6(b)
3	Rt. circ. cylinder const. radius R	x	y	1	1	0	$\frac{1}{R}$	0	0	6.7(a)
4	Rt. circ. cylinder const. radius R	x	θ	1	R	0	1	0	0	6.7(b)
5	Cone, cone angle λ	x	θ	$\frac{1}{\cos\lambda}$	$x\tan\lambda$	0	$\cos\lambda$	0	$-\sin\lambda$	6.8
6	Sphere, const. radius R	φ	θ	R	$R\sin\varphi$	1	$\sin\varphi$	0	$-\cos\varphi$	6.9
7	General cylindrical shell	x	y	1	1	0	$\mathbf{b}(y)$	0	0	6.15

(Rows 3–6: $z = Z$ and w taken positive inward)

these cases, are drawn so as to resemble the middle surface part of Fig. 6.3, so as to make the physical relation of each special case to the general case as clear as possible, although this requires a rather awkward viewpoint in some cases.

Figure 6.6(a) shows the simplest case, of a flat plate using rectangular coordinates, for which the scale factors A, B can obviously be taken as unity and the curvature functions **a, b, c, d** as zero. At (b), using polar coordinates, the same is true in the radial, r (taken as the α), direction. In the angular direction, a small change $d\theta$ in the parameter $\beta = \theta$ produces an arc length of $r\,d\theta = B\,d\theta$, and a rotation of magnitude $d\theta = -\mathbf{d}\,d\theta$ of the coordinate line direction at point q

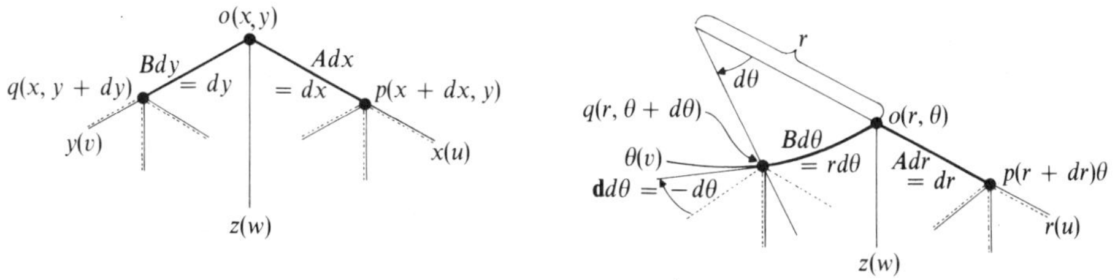

(a) (b)

(a) Flat plate, $\alpha = x$, $\beta = y$ (b) Flat plate, $\alpha = r$, $\beta = \theta$

FIGURE 6.6

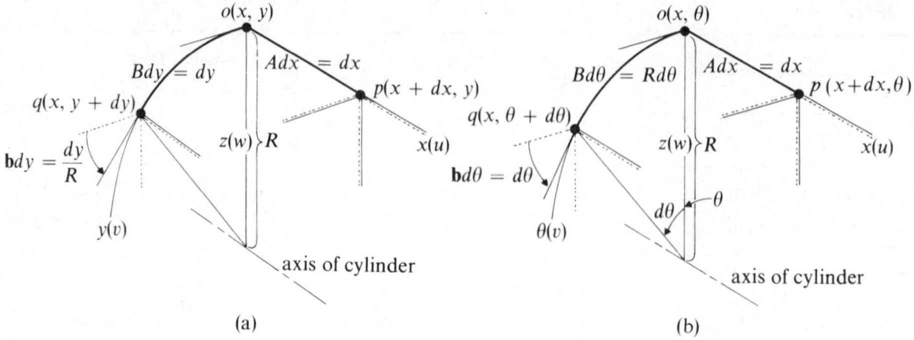

(a) Right circular cylinder, $\alpha = x$, $\beta = y$

(b) Right circular cylinder, $\alpha = x$, $\beta = \theta$

FIGURE 6.7

relative to that at point *o in* the middle surface, and no rotation *of* the surface, giving $B = r$ and $\mathbf{b} = 0$, $\mathbf{d} = -1$ (negative since the rotation is *out* of the first quadrant of the *XYZ* triad). Figures 6.7(a), (b) for the right circular cylinder, using a circumferential and an angular coordinate respectively, seem equally clear. In this case, as well as in the following two cases of a conical and spherical shell, the coordinate system is selected so that the coordinates Z, z and the displacement w are positive inward, as in the general case of Fig. 6.3.

Figure 6.8(a) shows the case of a conical shell, using distances along the axis $x = \alpha$, and rotations about the axis $\theta = \beta$ (measured clockwise as viewed from the

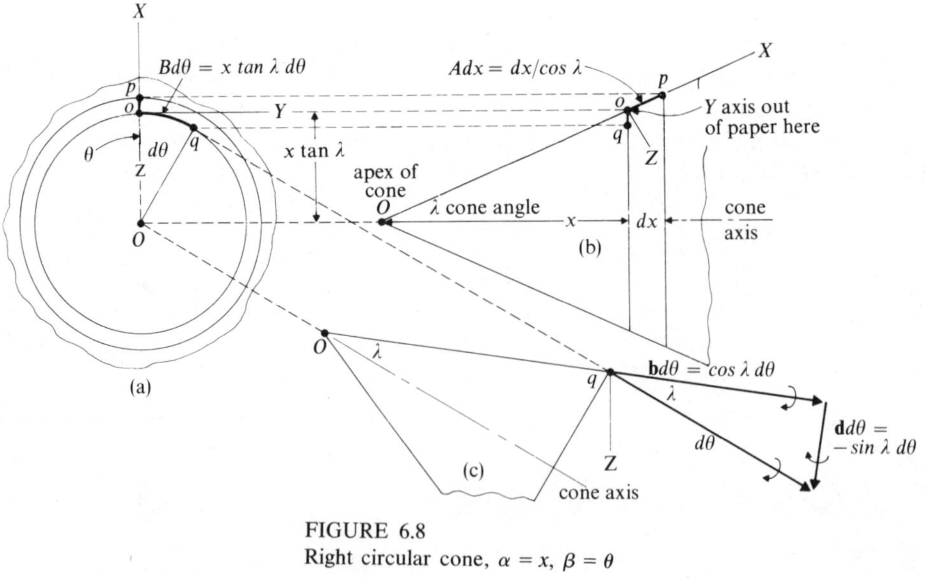

FIGURE 6.8
Right circular cone, $\alpha = x$, $\beta = \theta$

apex), as coordinates. In the x direction the actual distances along the middle surface, shown in true view at (b), are $x/\cos\lambda$, where λ is the cone angle; hence the scale factor A is $1/\cos\lambda$. There are evidently no rotations either *in* or *of* the middle surface at point p in this direction, so **a** and **c** are zero.

In the θ direction things are a little more complicated. The section taken through the axis including o and p, which is shown at (b), indicates that the radius of the arc oq (shown in true view in (a)) swept out by the change $d\theta$ in the angular coordinate θ is $x\tan\lambda$, so the length of the arc is $x\tan\lambda\, d\theta$ and the scale factor B is $x\tan\lambda$. There is obviously a resultant rotation $d\theta$ of the q end of the coordinate line segment oq; this must be separated into its components *in* the middle surface and *of* the middle surface, which can be done vectorially since the rotations are infinitesimal.

The vector diagram for making this separation is shown in true view in Fig. 6.8(c). Each rotational vector has the line of action of its axis of rotation, with a length proportional to the magnitude of the rotation, its sense being given by the right-hand rule, that is it points in the direction of the thumb of the right hand when the fingers point in the direction of rotation. The line of action of the axis of the resultant rotation $d\theta$ is of course paralled to the cone axis. Referring to Fig. 6.3, the axis of rotation of the component $\mathbf{d}\, d\beta = \mathbf{d}\, d\theta$, due to the curvature of the coordinate line oq *in* the middle surface, is normal to the surface at q. The axis of rotation for the other component $\mathbf{b}\, d\beta = \mathbf{b}\, d\theta$, due to the curvature *of* the surface, is in the middle surface at right angles to oq at q. The vector diagram thus gives $\mathbf{b} = \cos\lambda$ and $\mathbf{d} = -\sin\lambda$ (the negative sign because this component of rotation leads out of the first quadrant of the XYZ triad, opposite to the direction shown in Fig. 6.3).

In the case of the sphere, shown in Fig. 6.9, we take the radius of the middle surface as R, and the coordinates as two angles, an angle of latitude $\varphi = \alpha$, measured from the pole P_1, and an angle of longitude $\theta = \beta$ about the diameter P_1P_2, measured clockwise looking from P_1 towards P_2. At (b) is shown a section

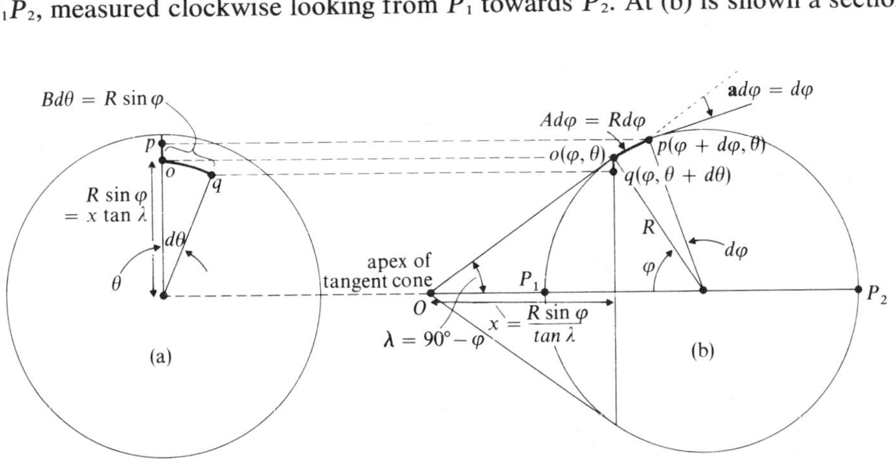

FIGURE 6.9
Spherical shell, $\alpha = \varphi$, $\beta = \theta$

through the diameter P_1P_2 and the points op. The length of the element op is $A\,d\varphi = R\,d\varphi$, so $A = R$. There is a rotation at p *of* the middle surface of $\mathbf{a}\,d\varphi = d\varphi$, and no rotation of the line *in* the middle surface so $\mathbf{a}, \mathbf{c} = 1, 0$. If we draw a cone with axis OP_1P_2 tangent to the spherical surface along oq we see from the figure at (b) that its cone angle is $\lambda = 90° - \varphi$, and the distance along the cone axis is $x = R \sin \varphi / \tan \lambda$. The values of B, $\mathbf{b}$, $\mathbf{d}$ for the sphere must then be the same as was just found for a cone with these values of λ and x.

The case of the general cylindrical shell is discussed on p. 359.

6.3 SIMPLIFICATION OF DISPLACEMENT–STRAIN RELATIONS

General Considerations

Expressions (6.8), together with (6.2), (6.4), (6.5), and (6.6), define the relation between the strains at all points of a shell and the displacements of the middle surface. Since our theory is limited to cases for which the Love–Kirchhoff assumption is a good approximation, we might as well make use of any simplification which this implies. Negligibility of transverse strains implies inequalities such as $(\partial^3 w / \partial \alpha^3) z / A \ll \partial^2 w / \partial \alpha^2$, but these do not seem to lead to useful simplification by themselves.

Use of the Love–Kirchhoff assumption also in general limits application of our theory to 'thin shells', for which $\mathbf{a}z/A, \mathbf{b}z/B \ll 1$, and this presents possibilities for simplifying expression (6.8) for the strains. The $\mathbf{a}z/A$, $\mathbf{b}z/B$ terms in the numerators of ϵ_α and ϵ_β are important for small displacements, and if they were dropped we would not obtain zero strains for the basic case when $u, v, w = 0$. However, if we neglect the $\mathbf{a}z/B$, $\mathbf{b}z/B$ in the denominators of the strain expressions, setting the denominators equal to unity, we only make errors in the strains at specific points of the order of the thickness–radius ratio. For calculating things such as deflections and buckling loads, which depend on average conditions, these errors nearly cancel between the two sides of the shell wall. The error in strain energy is about the square of the thickness–radius ratio, that is the error is about a tenth of a percent for a thickness one thirtieth of the radius. Hence it might seem that for thin shells, and for calculating deflections, buckling loads, etc., of fairly thin shells, the denominators in (6.8) could be taken as unity without serious error. However, although we will find that this is true, it requires justification, since seemingly unimportant terms may prove significant in later stages; this is considered in detail later in deriving Eq. (6.36).

The present theory is applicable to thin shells subjected to unlimited displacements and strains, for example displacements large compared to the radii of curvature and strains of the order of unity. For problems involving both large displacements and large strains probably no simplifications except those just discussed are possible, and calculations must be made with Eqs. (6.2), (6.4), (6.5), (6.6) and the modified (6.8), or equivalent relations. It is not at all impractical to use such indirect relationships, involving intermediate quantities such as f, m, U, and F, in numerical calculations, as will be demonstrated in the latter part of this section. For the more usual problems in which the displacements may be large but the strains are relatively small, a further and quite drastic simplication seems possible even for the quite general case, and this will be the principal subject of the remainder of this section.

The strains are necessarily small compared to unity in the elastic straining of materials such as metals, concrete, and the harder plastics. But the strains are also usually small in thin shells even when plastic flow occurs or if the material is rubber or rubber-like. This is because, for cases where the Love–Kirchhoff assumption applies, flexural strains are small even for displacements of the order of the thickness, and membrane strains involving compression in any direction are limited because of buckling. Only in rather unusual cases such as the inflation of rubber balloons, where the membrane stresses are entirely or almost entirely tensile, are large strains likely to occur in thin shells of any material.

As mentioned above, if strains are small, even though displacements may be unlimited, expressions (6.7), (6.8), and (6.8a, b) can be used for the engineering strains. In the elastic case they can then be used with the elementary definitions of Hooke's Law and internal strain energy for application of the law of virtual work.

It may also be of interest to note that $s = [(1+2\epsilon_{\alpha m})(1+2\epsilon_{\beta m}) - \epsilon^2_{\alpha\beta m}]^{-1/2}$, as can readily be checked from Eqs. (6.5) and (6.8a). If the strains are everywhere small, then the parts of the strains which are constant with z, which vary with z, and which vary with z^2, must each be small, and s must have a value close to unity. However, taking s as unity represents only a small part of the simplification which is proposed in the following.

Comparison of Exact and Approximate Strains for Case of Axially Loaded Thin Cylinder*—Exact Strains

To demonstrate the applicability of the exact strain expressions (6.8) used with (6.2), (6.4), (6.5), (6.6), and to compare them with commonly used approximate expressions, the case was selected of axial compression of a perfect right circular cylinder of a material for which Hooke's Law and conventional energy methods apply. This is an important problem on which a large literature exists, and it is known from many experiments that the buckling deflection pattern is a simple repetitive one and that the exact satisfaction of end boundary conditions is unimportant for all except very short cylinders. This is because for longer cylinders there are many buckling waves in the length direction, and the waves in the middle of the length, where the conditions have been critical in tests, are little affected by constraints on the end waves. This case will be discussed in more detail later in Sec. 7.2. This case is very representative since all six membrane and flexural stresses have the same order of magnitude, and the deflection components which we will use to study it represent a wide range of deflection types.

* L. H. Donnell, 'General shell displacement–strain relations', *Archiwum Mechaniki Stosowanej*, vol. 17, 1965, p. 277.

For a cylinder of wall thickness $h = 2c$ and radius R, the usual cylindrical coordinate lines are the lines of curvature. We take $\alpha = x$ as the axial coordinate, $\beta = y$ as the circumferential coordinate, $A, B = 1$, $\mathbf{a}, \mathbf{c}, \mathbf{d} = 0$, $\mathbf{b} = 1/R$, as in Fig. 6.7(a) and Table 6.2. Using these in Eqs. (6.8) and taking $(1 - az/A)$, etc., equal to unity, the strains can be written in the form (the dots represent terms involving higher powers of z):

$$\epsilon_x = \epsilon_{xm} + \epsilon_{x1}z + \epsilon_{x2}z^2/2 + \cdots, \qquad \epsilon_y = \epsilon_{ym} + \epsilon_{y1}z + \epsilon_{y2}z^2/2 + \cdots,$$
$$\epsilon_{xy} = \epsilon_{xym} + \epsilon_{xy1}z + \epsilon_{xy2}z^2 + \cdots, \qquad \text{where:}$$

$$\epsilon_{xm} = f + (f^2 + g^2 + h^2)/2, \quad \epsilon_{ym} = j + (i^2 + j^2 + k^2)/2, \quad \epsilon_{xym} = g + i + fi + gj + hk$$
$$\epsilon_{x1} = f' + ff' + gg' + hh', \qquad \epsilon_{x2} = f'^2 + g'^2 + h'^2$$
$$\epsilon_{y1} = j' + ii' + jj' + kk' + (1 + 2j + i^2 + j^2 + k^2)/R$$
$$\epsilon_{y2} = i'^2 + j'^2 + k'^2 + (j' + ii' + jj' + kk')/R + 3\epsilon_{y1}/R \tag{6.11}$$
$$\epsilon_{xy1} = g' + i' + if' + fi' + gj' + jg' + hk' + kh' + (g + i + fi + gj + hk)/R$$
$$\epsilon_{xy2} = f'i' + g'j' + h'k' + \epsilon_{xy1}/R, \qquad \text{and:}$$

$$f = \frac{\partial u}{\partial x}, \quad g = \frac{\partial v}{\partial x}, \quad h = \frac{\partial w}{\partial x}, \quad i = \frac{\partial u}{\partial y}, \quad j = \frac{\partial v}{\partial y} - \frac{w}{R}, \quad k = \frac{\partial w}{\partial y} + \frac{v}{R}$$

$$f' = \frac{\partial(ms)}{\partial x}, \quad g' = \frac{\partial(ns)}{\partial x}, \quad h' = \frac{\partial(s + ls)}{\partial x},$$

$$i' = \frac{\partial(ms)}{\partial y}, \quad j' = \frac{\partial(ns)}{\partial y} - \frac{s + ls}{R}, \quad k' = \frac{\partial(s + ls)}{\partial y} + \frac{ns}{R}$$

and m, n, l, and s are defined by Eqs. (6.5).

As mentioned above it is not necessary to satisfy exact end boundary conditions. Then from the symmetry of each buckling wave, indicated in all experiments, the displacements can be completely described by the expressions:

$$w = h \sum_p \sum_q W_{pq} \cos \frac{p\lambda n' x}{R} \cos \frac{qn' y}{R}$$
$$v = h \sum_p \sum_q V_{pq} \cos \frac{p\lambda n' x}{R} \sin \frac{qn' y}{R} \tag{6.12}$$
$$u = -\epsilon x + h \sum_p \sum_q U_{pq} \sin \frac{p\lambda n' x}{R} \cos \frac{qn' y}{R}$$

where h is the thickness and ϵ is the average unit shortening of the cylinder and $(p + q)$ must be even, that is p, q have only the values:

$$\begin{array}{l} pq = 00 \quad 20 \diagup 40 \quad 60 \quad \cdots \\ 11 \diagup 31 \quad \cdots \\ 02 \diagup 22 \quad \cdots \\ \diagup 13 \quad 33 \quad \cdots \\ \cdots \end{array} \tag{6.13}$$

The number of circumferential waves n, the ratio between the circumferential and longitudinal wave length λ of the primary term for $pq = 11$, and W_{pq}, V_{pq}, U_{pq} are unknown coefficients to be determined. The formulas (6.12) conform to the pattern predicted in the earlier discussion of Fig. 6.1(d).

If expressions (6.12) are substituted in Eqs. (6.11) and trigonometric transformations are used for the squares and products of the trigonometric functions involved, as given in Table 6.3, the expression for each of the quantities defined involves only trigonometric terms of the same type, whose coefficients can then be combined to form one series. Thus f, j, l, s, f', j', ϵ_x and ϵ_y become cosine-cosine, k, n and k' cosine-sine, h, m, h' sine-cosine, and g, i, g', i', ϵ_{xy} sine-sine series in x and y respectively. The strain energy:

$$\mathscr{E} = \int dx \int dy \int_{-c}^{c} dz \left(\epsilon_x^2 + 2\nu\epsilon_x\epsilon_y + \epsilon_y^2 + \frac{1-\nu}{2}\epsilon_{xy}^2\right) \tag{6.14}$$

is then found as a function of the unknown coefficients λ, n', W_{pq}, V_{pq}, U_{pq}, which can be determined in the usual manner by the law of virtual work. This of course involves the solution of as many nonlinear algebraic equations as the number of these coefficients.

Approximate Expressions for Strains

When n' is large compared to one, as in the case being considered, the approximate expressions commonly used for the strains in a thin cylindrical shell subjected to large displacements are:

$$\epsilon_x = \frac{\partial u}{\partial x} + \frac{1}{2}\left(\frac{\partial w}{\partial x}\right)^2 - \frac{\partial^2 w}{\partial x^2} z, \qquad \epsilon_y = \frac{\partial v}{\partial y} - \frac{w}{R} + \frac{1}{2}\left(\frac{\partial w}{\partial y}\right)^2 - \frac{\partial^2 w}{\partial y^2} z$$
$$\epsilon_{xy} = \frac{\partial v}{\partial x} + \frac{\partial u}{\partial y} + \frac{\partial w}{\partial x}\frac{\partial w}{\partial y} - 2\frac{\partial^2 w}{\partial x\, \partial y} z \tag{6.15}$$

When n' is small—wz/R^2 must be added to ϵg, and other flexural terms may become of minor importance, as is indicated in Eqs. (6.20) derived later. If this approximation is used directly for calculating the strain energy, the algebraic equations which must be satisfied to minimize the energy are linear in V_{pq} and U_{pq}, which greatly simplifies a solution.

The work can be further simplified by using an Airy stress function ϕ for the membrane stresses, Eqs. (3.16a). This satisfies the same equations of equilibrium in the x and y directions as in two-dimensional elasticity, and therefore ignores the effects on equilibrium in the x, y directions of initial curvature and finite displacements. Equating the membrane strains (the terms independent of z) in expression (6.15) to the membrane strains expressed in terms of ϕ by using Hooke's Law, Eqs. (3.11a) we obtain:

$$\epsilon_{xm} = \frac{\partial u}{\partial x} + \frac{1}{2}\left(\frac{\partial w}{\partial x}\right)^2 = \frac{1}{E}\left(\frac{\partial^2 \phi}{\partial y^2} - \nu\frac{\partial^2 \phi}{\partial x^2}\right), \qquad \epsilon_{ym} = \frac{\partial v}{\partial y} - \frac{w}{R} + \frac{1}{2}\left(\frac{\partial w}{\partial y}\right)^2 = \frac{1}{E}\left(\frac{\partial^2 \phi}{\partial x^2} - \nu\frac{\partial^2 \phi}{\partial y^2}\right)$$
$$\epsilon_{xym} = \frac{\partial v}{\partial x} + \frac{\partial u}{\partial y} + \frac{\partial w}{\partial x}\frac{\partial w}{\partial y} = -\frac{2(1+\nu)}{E}\frac{\partial^2 \phi}{\partial x\, \partial y} \tag{6.16}$$

Eliminating u and v by applying the operator $\partial^2/\partial y^2$ to the first equation of (6.16), the operator $\partial^2/\partial x^2$ to the second, and $-\partial^2/\partial x\,\partial y$ to the third, and adding the three equations, we obtain, instead of Eq. (3.16c), the relation*:

$$\nabla^4\phi = E\left[\left(\frac{\partial^2 w}{\partial x\,\partial y}\right)^2 - \frac{\partial^2 w}{\partial x^2}\frac{\partial^2 w}{\partial y^2} - \frac{1}{R}\frac{\partial^2 w}{\partial x^2}\right] \tag{6.17}$$

With this relation, it is only necessary to assume an expression for w (satisfying the boundary conditions when necessary) and find ϕ by integrating Eq. (6.17), using trigonometric transformations for the squares and products of terms (see Table 6.3). The law of virtual work can then be applied, using expressions (4.70), (4.71) for the flexural and membrane strain energies respectively. These expressions were developed for plates, but we are using the same flexural strains in Eq. (6.15), $-(\partial^2 w/\partial x^2)z$, etc., as for plates, and the effect of curvature on the membrane energy is given by the $1/R\,\partial^2 w/\partial x^2$ term in expression (6.17). A solution then requires the satisfaction of only as many equations as λ, n' and the number of W_{pq} used. In applying Eq. (6.17) to cases where boundary conditions are important, the previous warning should be remembered that solutions obtained by raising the order of the differential equation by application of the operators $\partial^2/\partial x^2$, etc., are not meaningful for satisfying such conditions.

Comparison of Approximate and Exact Strains

In a 1963 paper, Almroth** used this approximate method and successively increased the number of terms in the w series until the results shown in Fig. 6.10(a), which gives average stress σ plotted against unit shortening ϵ, were not appreciable changed. He found that it is necessary to use at least seven terms to adequately study this case in the postbuckling range. To obtain a comparable solution using the exact relation (6.11) by conventional energy techniques would require the solution of seventeen nonlinear simultaneous equations, which may be impractical even with the help of large computers. However, a complete solution by the exact theory was not necessary for the present purposes; the approximate theory can be compared with the exact theory by simply making a parallel calculation of the strains and strain energy, using the same values of λ, n', W_{pq}, V_{pq}, and U_{pq} in Eqs. (6.11) and (6.15) respectively.

In any computation with series, the series must be cut off somewhere. If we start with a certain number of terms (that is values of p, q) many more terms are added each time the series are multiplied together, but for the desired purposes it was considered sufficient to cut off all the series uniformly with ten terms, using

* L. H. Donnell, 'A new theory for the buckling of thin cylinders under axial compression and bending', *Trans. ASME*, vol. 56, 1934, p. 795. For $1/R = 0$ (flat plate) this is the same as Kármán's finite deflection plate theory: *Encyklopädie der Math. Wiss.*, vol. 4, 1910, p. 349. See also Eq. (6.31j) developed later.

** B. O. Almroth, 'Postbuckling behavior of axially compressed circular cylinders', *A.I.A.A. Journal*, vol. 1, 1963, p. 630.

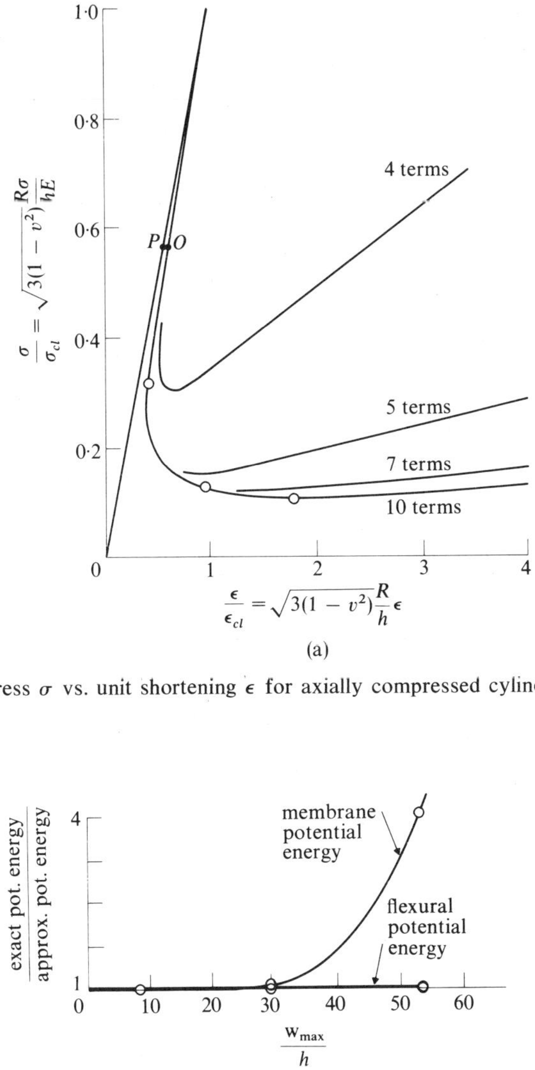

(a)

(a) Average stress σ vs. unit shortening ϵ for axially compressed cylinder (Almroth)

(b)

FIGURE 6.10

the values of pq shown in (6.13). These favor the higher harmonics in x (consistent with the known fact that the term 20 is far more important than the term 02). These were the terms used in his final solution by Almroth, who kindly provided the values of λ, n' and W_{pq} which he found in his solution. The corresponding values of V_{pq} and U_{pq} were found from Eqs. (6.16) by solving them for the derivatives of u and v and integrating the first and second equations and determining the functions of integration from the third equation.

It is well known that the relations of the approximate theory can be written in terms of parameters which encompass all the elastic and geometric properties of the cylinder. Thus the calculations result in a single relation which applies to all cylinders, such as that graphed in Fig. 6.10(a). In the case of the exact theory the radius thickness ratio R/h, cannot be so incorporated in the theory. The results obtained may thus be somewhat different for different R/h ratios, but this variation must be very slight so long as the exact and approximate theories give essentially the same results. For computational purposes $R/h = 1000$ was assumed (it would not be realistic to use a much smaller value and assume, as we do, that the material remains elastic under the very large deflections which we will consider), but the conclusions reached should not be limited to this ratio and are believed to be typical for thin shells in general. For this value of the radius–thickness ratio, the terms in the denominators of Eqs. (6.8) would not affect the numerical results obtained to the accuracy with which the calculations were carried out.

The results of applying the trigonometric transformations such as $\cos x \cos y = [\cos(x-y) + \cos(x+y)]/2$ and $\sin x \cos y = [\sin(x-y) + \sin(x+y)]/2$ to the multiplication or squaring of trigonometric series can be tabulated as illustrated in Table 6.3.

Table 6.3

(a) Coefficients of series representing product of two series

pq	00	20	11	02	40	31	22	13	60	33
$\Sigma\Sigma$ sin sin	—	—	c	—	—	f	g	h	—	j
$\times\Sigma\Sigma$ sin sin	—	—	c'	—	—	f'	g'	h'	—	j'
$=\frac{1}{4}$ $\Sigma\Sigma$ cos cos	cc'	$-cc'$	cg'	$-cc'$	$-cf'$	$-cg'$	cc'	$-cg'$	$-ff'$	cg'
	ff'	cf'	gc'	ch'	$-fc'$	$-gc'$	$-cf'$	$-gc'$	$-jj'$	gc'
	gg'	fc'	fg'	hc'	$-gg'$	$-gh'$	$-fc'$	$-fg'$		
	hh'	$-hh'$	gf'	$-ff'$	$-hj'$	$-hg'$	$-ch'$	$-gf'$		
	jj'	hj'	gh'	fj'	$-jh'$		$-hc'$			
		jh'	hg'	jf'			cj'			
			gj'				jc'			
			jg'				fh'			
							hf'			

(b) Coefficient of series representing square of series

pq	00	20	11	02	40	31	22	13	60	33
$(\Sigma\Sigma \sin\sin)^2$	—	—	c	—	—	f	g	h	—	j
$=\frac{1}{4}$ $\Sigma\Sigma$ cos cos	c^2	$-c^2$	$2cg$	$-c^2$	$-2cf$	$-2cg$	c^2	$-2cg$	$-f^2$	$2cg$
	f^2	$2cf$	$2fg$	$2ch$	$-g^2$	$-2gh$	$-2cf$	$-2fg$	$-j^2$	
	g^2	$-h^2$	$2gh$	$-f^2$	$-2hj$		$-2ch$			
	h^2	$2hj$	$2gj$	$2fj$			$2cj$			
	j^2						$2fh$			

pq	00	20	11	02	40	31	22	13	60	33
$(\Sigma\Sigma \cos\cos)^2$	a	b	c	d	e	f	g	h	i	j
$=\frac{1}{4}$ $\Sigma\Sigma \cos\cos$	$4a^2$ $2b^2$ c^2 $2d^2$ $2e^2$ f^2 g^2 h^2 $2i^2$ j^2	$8ab$ $4be$ c^2 $2cf$ $4dg$ $4ei$ h^2 $2hj$	$8ac$ $4bc$ $4bf$ $4cd$ $2cg$ $4dh$ $4ef$ $2fg$ $2gh$ $2gj$	$8ad$ $4bg$ c^2 $2ch$ f^2 $2fj$	$8ae$ $2b^2$ $4bi$ $2cf$ g^2 $2hj$	$8af$ $4bc$ $4ce$ $2cg$ $4df$ $4dj$ $4fi$ $2gh$	$8ag$ $8bd$ c^2 $2cf$ $2ch$ $2cj$ $4eg$ $2fh$	$8ah$ $4bh$ $4bj$ $4cd$ $2cg$ $4ej$ $2fg$	$8ai$ $4be$ f^2 j^2	$8aj$ $4bh$ $2cg$ $4df$ $4eh$ $4ij$
$(\Sigma\Sigma \sin\cos)^2$	—	b	c	—	e	f	g	h	i	j
$=\frac{1}{4}$ $\Sigma\Sigma \cos\cos$	$2b^2$ c^2 $2e^2$ f^2 g^2 h^2 $2i^2$ j^2	$4be$ $-c^2$ $2cf$ $4ei$ $-h^2$ $2hj$	$4bc$ $4bf$ $2cg$ $4ef$ $2fg$ $2gh$ $2gj$	$4bg$ c^2 $2ch$ f^2 $2fj$	$-2b^2$ $4bi$ $-2cf$ $-g^2$ $-2hj$	$-4bc$ $4ce$ $-2cg$ $4fi$ $-2gh$	$-c^2$ $2cf$ $-2ch$ $2cj$ $4eg$ $2fh$	$4bh$ $4bj$ $2cg$ $4ej$ $2fg$	$-4be$ $-f^2$ $-j^2$	$-4bh$ $-2cg$ $4eh$ $4ij$
$(\Sigma\Sigma \cos\sin)^2$	—	—	c	d	—	f	g	h	—	j
$=\frac{1}{4}$ $\Sigma\Sigma \cos\cos$	c^2 $2d^2$ f^2 g^2 h^2 j^2	c^2 $2cf$ $4dg$ h^2 $2hj$	$4cd$ $2cg$ $4dh$ $2fg$ $2gh$ $2gj$	$-c^2$ $2ch$ $-f^2$ $2fj$	$2cf$ g^2 $2hj$	$2cg$ $4df$ $4dj$ $2gh$	$-c^2$ $-2cf$ $2ch$ $2cj$ $2fh$	$-4cd$ $-2cg$ $-2fg$	f^2 j^2	$-2cg$ $-4df$

Results of Comparison of Approximate and Exact Strains

The calculations described in the foregoing paragraphs were carried out for the three values of the unit shortening $\epsilon = 10^{-4} \times 2{\cdot}51$, $5{\cdot}70$, $10{\cdot}63$ (or $\epsilon/\epsilon_{cl} = 0{\cdot}415$, $0{\cdot}95$, $1{\cdot}75$) indicated by circles in Fig. 6.10(a). These correspond to maximum deflections of $w_{max} = 8{\cdot}3h$, $29{\cdot}6h$, and $53{\cdot}6h$ (h = thickness) thus extending far into the 'large deflection' range. The final details of the strain calculations are shown in Tables 6.4, which shows the amplitude found for each harmonically varying strain component. The maximum flexural strains (at the surface $z = c$) are given. The blank spaces indicate values which are zero or below the minimum to which the calculations were carried.

In the results obtained for the exact strains from expressions (6.11), the parts of the strains $\epsilon_{x2}z^2/2$, $\epsilon_{y2}z^2/2$, $\epsilon_{xy2}z^2$ and parts containing still higher powers of z are not shown in Table 6.4 because all the values found for these terms turned out to be

Table 6.4

(a) Strain calculations for $\epsilon/\epsilon_{cl} = 1{\cdot}75$, $w_{max} = 53{\cdot}6h$

	λ n / pq	00	20	11	02	40	31	22	13	60	33
0·252 7·25	$10^3\,W_{pq}$	9440	10230	36000	−1460	1600	1980	−5280	−270	54	1370
	$10^3\,V_{pq}$			6143	501		1011	176	−378		−172
	$10^3\,U_{pq}$		43	−639		77	101	120	−27	18	−80
	$10^3\,\phi_{pq}$		80·0	46·4	−28·6	−11·1	−20·6	−13·4	14·8	−32·7	7·3
membrane strains × 10^6	$\partial u/\partial x$	−1063	157	−1164		561	553	437	−49	197	−437
	$+(\partial w/\partial x)^2/2$	997	−146	1080	208	−564	−522	−342	−193	−238	320
	$= \epsilon_{xm}$ (approx.)	−66	11	−84	208	−3	31	95	−242	−41	−117
	$+(\partial v/\partial x)^2/2$	20		2	−23	−16	−2	−2	−1	−4	2
	$+(\partial u/\partial x)^2/2$	1		1		−1	−1				
	$= \epsilon_{xm}$ (exact)	−45	11	−81	185	−20	28	93	−243	−45	−115
	$\partial v/\partial y - w/R$	−9441	−10202	8406	8703	−1603	5330	7820	−7932	−54	−5103
	$+(\partial w/\partial y)^2/2$	9460	10165	−8386	−8765	1623	−5320	−7843	8003	184	5131
	$= \epsilon_{ym}$ (approx.)	19	−37	20	−62	20	10	−23	71	130	28
	$+(\partial w/\partial y)v/R + v^2/2R^2$	−384	−454	108	422	−81	95	479	−125	−9	−116
	$+(\partial v/\partial y - w/R)^2/2$	121	177	−152	−133	71	−98	−187	208	15	135
	$+(\partial u/\partial y)^2/2$	4	−3	−3	−2		2	4	2		−2
	$= \epsilon_{ym}$ (exact)	−240	−317	−27	225	10	9	273	156	136	45
	$\partial v/\partial x + \partial u/\partial y$			−6570			−6260	−2374	1274		2672
	$+(\partial w/\partial x)(\partial w/\partial y)$			6515			6334	2438	−1327		−2750
	$= \epsilon_{xym}$ (approx.)			−55			74	64	−53		−78
	$+(\partial w/\partial x)v/R$			−139			−185	−99	18		35
	$+(\partial v/\partial x)(\partial v/\partial y - w/R)$			112			148	−32	−45		−42
	$+(\partial u/\partial x)(\partial u/\partial y)$			−3							−1
	$= \epsilon_{xym}$ (exact)			−85			37	−67	−80		−86
flexural strain × 10^6	$-c\,\partial^2 w/\partial x^2$		68	60		43	30	−35	−1	3	21
	ϵ_{x1}		68	59		43	30	−36	−1	3	20
	$-c\,\partial^2 w/\partial y^2$			942	−153		52	−553	−64		323
	ϵ_{y1}			926	−168	−1	52	−559	−68	−1	318
	$-2c\,\partial^2 w/\partial x\,\partial y$			−474			−78	278	11		−163
	ϵ_{xy1}			−471			−78	284	12		−161

below the minimum to which the calculations were carried out. These terms of course represent symmetrical strain distributions and so should be combined with the membrane strains ϵ_{xm}, ϵ_{ym}, ϵ_{xym}. In order to do this their average value over the thickness was calculated, that is $1/(2c)\int_{-c}^{c} dz(\epsilon_{x2}z^2/2) = \epsilon_{x2}c^2/6$, etc. The largest numerical value found for such average values of these parts of the strains was 0·2, in the scale of the numbers listed in Table 6.4, which are given only to the nearest unit (this maximum value of one-fifth of a unit was found for the circumferential membrane strain $\epsilon_y = \epsilon_{my}$, for $w_{max} = 53{\cdot}6h$ and $pq = 22$; an equal negative maximum value was found for $pq = 11$).

The negligible value of these strain components proportional to z^2 and higher powers of z also facilitated the calculation of the strain energy from Eq. (6.14) as two separate parts, membrane and flexural strain energy. The results of these calculations are listed in Table 6.5 and plotted in Fig. 6.10(b). It can be seen

Table 6.4 *(continued)*

(b) Strain Calculations for $\epsilon/\epsilon_{cl} = 0{\cdot}95$, $w_{max} = 29{\cdot}6\,h$

λ	n	pq	00	20	11	02	40	31	22	13	60	33
0·257	9·33	$10^3\,W_{pq}$	4700	5200	20300	−310	880	1780	−1950	−460	−20	330
		$10^3\,V_{pq}$			2428	253		372	188	−102		−51
		$10^3\,U_{p1}$		−3	−308		33	44	56	−6	7	−26
		$10^3\,\phi_{pq}$		−264	27·2	−20·3	−17·2	−13·5	−13·6	6·2	−11·3	3·1
membrane strains × 10^6		$\partial u/\partial x$	−574	−14	−740		315	318	269	−14	103	−187
		$+(\partial w/\partial x)^2/2$	501	−37	676	195	−327	−292	−142	−119	−126	123
		$= \epsilon_{xm}$ (approx.)	−73	−51	−64	195	−12	26	127	−133	−23	−64
		$+(\partial v/\partial x)^2/2$	5		2	−6	−4	−1		−2	−1	1
		$+(\partial u/\partial x)^2/2$										
		$= \epsilon_{xm}$ (exact)	−68	−51	−62	189	−16	25	127	−135	−24	−63
		$\partial v/\partial y - w/R$	−4697	−5213	2354	5030	−880	1687	5444	−2386	20	−1756
		$+(\partial w/\partial y)^2/2$	4717	5400	−2344	−5079	922	−1678	−5465	2418	47	1769
		$= \epsilon_{ym}$ (approx.)	20	187	10	−49	42	9	−21	32	67	13
		$+(\partial w/\partial y)v/R + v^2/2R^2$	−114	−137	3	119	−26	13	138		−2	−14
		$+(\partial v/\partial y - w/R)^2/2$	30	44	−22	−39	17	−15	−56	34	3	23
		$+(\partial u/\partial y)^2/2$	1	−1	−1	−1		1	2	1		−1
		$= \epsilon_{ym}$ (exact)	−63	93	−10	30	33	8	63	67	68	21
		$\partial v/\partial x + \partial u/\partial y$			−2936			−3006	−1958	409		1089
		$+(\partial w/\partial x)(\partial w/\partial y)$			2834			3051	2045	−429		−1123
		$= \epsilon_{xym}$ (approx.)			−102			45	87	−20		−34
		$+(\partial w/\partial x)v/R$			−37			−50	−40	−1		
		$+(\partial v/\partial x)(\partial v/\partial y - w/R)$			29			40	−1	−11		−18
		$+(\partial u/\partial x)(\partial u/\partial y)$			−1							
		$= \epsilon_{xym}$ (exact)			−111			35	46	−32		−52
flex. strain × 10^6		$-c\,\partial^2 w/\partial x^2$		60	59		41	46	−23	−2	−2	9
		$-c\,\partial^2 w/\partial y^2$			885	−54		78	−339	−180		129
		$-2c\,\partial^2 w/\partial x\,\partial y$			−454			−119	174	31		−67

from Eq. (6.14) that 'cross products' of the membrane parts of the strains (terms independent of z) in the strain energy cancel between the top and bottom of the shell wall. On the other hand, cross products between both the membrane and flexural parts and the strain components proportional to z^2 do not cancel, but because of the negligible magnitudes of the latter components these cross product parts of the strain energy are also negligible even for the relatively enormous deflection of more than fifty times the shell wall thickness. This finding is in conformity with the conclusions of Koiter* and others.

Conclusions

Among the elastic strain energies shown in Table 6.5, which were calculated to the scale of areas under the curves in Fig. 6.10(a), those calculated from the

* *Loc. cit.* Page 308.

Table 6.4 *(continued)*

(c) *Strain Calculations for* $\epsilon/\epsilon_{cl} = 0{\cdot}415$, $w_{max} = 8{\cdot}3h$

λ n	pq	00	20	11	02	40	31	22	13	60	33
0·257 13·23	$10^3 W_{pq}$	555	900	5150	310	210	770	420	23		−20
	$10^3 V_{pq}$			350	33		41	43	13		5
	$10^3 U_{pq}$		4	−71		4	1	5	1	−1	2
	$10^3 \phi_{pq}$		127·5	38·3	−4·0	6·4	12·0	−2·6	−0·7	−1·6	−0·2
membrane strains × 10^6	$\partial u/\partial x$	−251	27	−242		52	8	34	4	−13	18
	$+(\partial w/\partial x)^2/2$	58	7	114	54	−45	−42	1	17	9	−12
	$= \epsilon_{xm}$ (approx.)	−192	34	−128	54	7	−34	35	21	−4	6
	$+(\partial v/\partial x)^2/2$										
	$+(\partial u/\partial x)^2/2$										
	$= \epsilon_{xm}$ (exact)	−192	34	−128	54	7	−34	35	21	−4	6
	$\partial v/\partial y - w/R$	−555	−900	−475	563	−210	−227	714	482		218
	$+(\partial w/\partial y)^2/2$	613	786	506	−579	187	215	−722	−489	13	−219
	$= \epsilon_{ym}$ (approx.)	58	−114	31	−16	−23	−12	−8	−7	13	−1
	$+(\partial w/\partial y)v/R + v^2/2R^2$	−6	−8	−5	6	−2	−2	7	5		2
	$+(\partial v/\partial y - w/R)^2/2$	1	1	1	−1	1		−1	−1		−1
	$+(\partial u/\partial y)^2/2$										
	$= \epsilon_{ym}$ (exact)	53	−121	27	−11	−24	−14	−2	−3	13	
	$\partial v/\partial x + \partial u/\partial y$		−250				−429	−424	−91		−122
	$+(\partial w/\partial x)(\partial w/\partial y)$		204				386	436	94		124
	$= \epsilon_{xym}$ (approx.)		−46				−43	12	3		2
	$+(\partial w/\partial x)v/R$		−1				−2	−2			−1
	$+(\partial v/\partial x)(\partial v/\partial y - w/R)$						1				
	$+(\partial u/\partial x)(\partial u/\partial y)$										
	$= \epsilon_{xym}$ (exact)		−47				−43	11	2		1
flex. strain × 10^6	$-c\,\partial^2 w/\partial x^2$		42	60		39	80	19			−2
	$-c\,\partial^2 w/\partial y^2$			903	217		135	293	36		−32
	$-2c\,\partial^2 w/\partial x\,\partial y$			−464			−208	−151	−6		16

Table 6.5 STRAIN ENERGIES CALCULATED BY APPROXIMATE AND EXACT THEORIES

ϵ/ϵ_{cl}		1·75	0·95	0·42
w_{max}		53·6h	29·6h	8·3h
membrane strain energy	approx. theory	0·085	0·074	0·070
	exact theory	0·353	0·075	0·070
flexural strain energy	approx. theory	0·189	0·137	0·036
	exact theory	0·190	0·137	0·036

approproximate strain expressions (6.15) are practically identical with those calculated from the exact expressions (6.11) except for the membrane strain energies at the very large deflections of 53·6h, more than fifty times the wall thickness. For this deflection the approximate value is less than a quarter of the exact value. At the next smaller deflection the approximate value of the membrane strain energy is slightly less than the exact value, but close enough for any ordinary purpose. This result seems to indicate that the approximate expressions for membrane strains take care of the effects of curvature adequately, and that their provisions for large deflection effects, the $(\partial w/\partial x)^2/2$ term in ϵ_x, etc., may be adequate for deflections up to possibly as much as thirty times the thickness, but not beyond that. This limitation is not at all serious because the exact expressions for membrane strains are relatively simple and easy to apply, so we will use these exact expressions in our most general thin shell theory.

It is very significant that the *flexural* strain energies calculated from the approximate strain expressions check closely those calculated from the exact expressions, even at the highest deflection, a deflection larger than is likely to occur in practical problems. This suggests the possibility for a drastic simplification of the general thin shell theory, since most of its complications are in the flexural strains and in the strain terms proportional to higher powers than one of z.

However, before drawing final conclusions, it should be remembered that the calculation of strain energies represents an averaging process which may not reveal wide divergence in parts of the strains, which might cancel each other in the strain energy calculations. Table 6.4 shows not only the values of individual parts of the strain expressions, but also the strains for a wide variety of deflection shape components.

Consider first the comparison of flexural strains at the outer surface $z = c$, in the first section of Table 6.4 for $w_{max} = 53{\cdot}6h$. The greatest discrepancy is in the circumferential strains, where the approximate strain is simply $-c\,\partial^2 w/\partial y^2$, and the exact strain is designated as ϵ_{y1}, defined in Eqs. (6.11). But even here the largest discrepancies are about 9 percent in the values for $pq = 02$, and 6 percent for $pq = 13$, except for the very small strains where the maximum discrepancy is one unit. The average discrepancy in all the flexural strains is 2 percent, or one unit for the small strains. These errors for the case of such extremely large deflections would certainly not justify retaining the extremely complex exact expressions for flexural strains in the general thin shell theory. The exact flexural strains were not calculated for the smaller deflections of $w_{max} = 29{\cdot}6h$ and $8{\cdot}3h$, because the calculations are very lengthy and seemed unnecessary, since it can be assumed that the discrepancies for such deflections would be even smaller.

It is interesting to note that the approximate expressions for flexural strains used in this comparison are the same as they would be for flat plates, with no allowances for either large deflections or initial curvature. As has been noted before, for small values of the number of circumferential waves the effects of initial curvature should be considered. The complications in the exact expressions are chiefly those for large deflection effects and we will discard these, but we will retain all the allowances for initial curvature, which are relatively simple.

Comparison of the exact and approximate values of membrane strains shows again great discrepancies for the maximum deflections. There are also quite large discrepancies for the medium deflections $w_{max} = 29{\cdot}6h$, with a fairly good check at the smallest deflections. This indicates that the approximate expressions (6.15) should not be used for deflections greater than eight or ten times the thickness, at least when local conditions are being investigated.

It is important to note in Table 6.4 that the difference between the approximate and exact strains is *not* due to the additional terms, appearing in the exact theory only, being of the same order of magnitude as the terms which are common to the two theories. As a matter of fact the additional terms involving v average only about 3 percent of the common terms in absolute magnitude, and the additional terms involving u are far smaller. A mere comparison of the orders of magnitude of the various terms would thus give the entirely misleading impression that the additional terms can be neglected. They *are* important because the common terms nearly cancel each other, and the additional terms involving v are of the same order of magnitude as their difference. The additional terms involving u are small even compared to this difference, and in the present case could be neglected without great error. However, the distinction between u and v in general shell theory is not usually obvious, and the retention of terms involving both does not greatly add to the difficulties involved in retaining one.

The Simplified Strain Expressions

As indicated in the discussion above, we will simplify the exact strain expressions (6.8) by combining the fairly simple exact expressions (6.8a) for the membrane strains with the equally simple expressions for flexural strains obtained by linearizing the parts of the exact theory involving z, and discarding terms involving higher power of z. For most purposes we will find that it is sufficient to retain the denominators of the strain expressions (6.8) by multiplying the numerators by $1+2\mathbf{a}z/A+\cdots$, $1+2\mathbf{b}z/B+\cdots$, $1+(\mathbf{a}/A+\mathbf{b}/B)z+\cdots$ respectively, and then discarding terms involving z which are nonlinear in the displacements or their derivatives, and all terms involving higher powers of z. The justification for this is not only the small size found above for such terms, but in the complete study of the effects of such minor terms made later in the development of Eq. (6.36).

Since m, n, l, s are always multiplied by z, Eqs. (6.4), we can take $s = 1/(1+l) \approx 1-l$, $U = u - hz$, $V = v - kz$, $W = w + z$, and so on. We thus obtain the following much simplified strain expressions:

$$\epsilon_\alpha = f + (f^2+g^2+h^2)/2 - h_\alpha z, \qquad \epsilon_\beta = j + (i^2+j^2+k^2)/2 - k_\beta z, \qquad \epsilon_{\alpha\beta} = g + i + fi + gj + hk - (h_\beta + k_\alpha)z, \quad \text{where:} \tag{6.18}$$

$$f = (\partial u/\partial\alpha - \mathbf{a}w - \mathbf{c}v)/A, \qquad g = (\partial v/\partial\alpha + \mathbf{c}u)/A, \qquad h = (\partial w/\partial\alpha + \mathbf{a}u)/A,$$
$$i = (\partial u/\partial\beta + \mathbf{d}v)/B, \qquad j = (\partial v/\partial\beta - \mathbf{b}w - \mathbf{d}u)/B, \qquad k = (\partial w/\partial\beta + \mathbf{b}v)/B. \tag{6.19}$$

$$h_\alpha = (\partial h/\partial\alpha - \mathbf{c}k - \mathbf{a}f)/A, \qquad h_\beta = (\partial h/\partial\beta + \mathbf{d}k - \mathbf{b}i)/B,$$
$$k_\alpha = (\partial k/\partial\alpha + \mathbf{c}h - \mathbf{a}g)/A, \qquad k_\beta = (\partial k/\partial\beta - \mathbf{d}h - \mathbf{b}j)/B.$$

The physical meanings of the two terms h_β and k_α in the flexural shear strain are illustrated later in Fig. 6.13(d).

These relations between strains and displacements in general shells should be valid for any practical problem involving thin shells made of ordinary engineering materials in their elastic range. It may be noted that in Almroth's example, upon which the above conclusion is based, such materials were assumed, for which the maximum unit strains are of the order of magnitude of 0·001 (mild steel) to 0·01 (high strength steels, hard plastics, etc.). Expressions (6.8) are not limited to such small strains, and because of this may be useful in studies of rubber-like materials or of plastic flow, where the strains may be of the order of unity.

However, in the pages which follow we will assume that strains are small compared to unity like those of the materials mentioned above, but we will be careful not to ignore strains compared to unity at one stage in a development when the terms corresponding to unity cancel at a later stage, making the strains compared possibly important. More general studies of structures in which strains of the order of unity occur would involve both great theoretical complications and the almost certain necessity of restating or replacing Hooke's law (even if the material is 'elastic', and whether definitions (6.7) or (6.9) are used to define strains). Such considerations are beyond the scope of the present treatment.

These general shell strain expressions (6.18), in terms of the displacements u, v, w and the scale and curvature functions A, B, $\mathbf{a}$, $\mathbf{b}$, $\mathbf{c}$, $\mathbf{d}$ defined in Fig. 6.3, can easily be applied to any type of shell by substitution of the proper values of α, β, A, B, $\mathbf{a}$, $\mathbf{b}$, $\mathbf{c}$, $\mathbf{d}$ such as those given in Table 6.2. For the cylinder, for example, using the circumferential coordinate $\beta = y$, we obtain:

$$
\begin{aligned}
\epsilon_x &= \frac{\partial u}{\partial x} + \left[\left(\frac{\partial w}{\partial x}\right)^2 + \left(\frac{\partial v}{\partial x}\right)^2 + \left(\frac{\partial u}{\partial x}\right)^2\right]\Big/2 - \frac{\partial^2 w}{\partial x^2} z \\
\epsilon_y &= \frac{\partial v}{\partial y} - \frac{w}{R} + \left[\left(\frac{\partial w}{\partial y} + \frac{v}{R}\right)^2 + \left(\frac{\partial v}{\partial y} - \frac{w}{R}\right)^2 + \left(\frac{\partial u}{\partial y}\right)^2\right]\Big/2 - \left(\frac{\partial^2 w}{\partial y^2} + \frac{w}{R^2}\right) z \qquad (6.20) \\
\epsilon_{xy} &= \frac{\partial v}{\partial x} + \frac{\partial u}{\partial y} + \frac{\partial w}{\partial x}\left(\frac{\partial w}{\partial y} + \frac{v}{R}\right) + \frac{\partial v}{\partial x}\left(\frac{\partial v}{\partial y} - \frac{w}{R}\right) + \frac{\partial u}{\partial x}\frac{\partial u}{\partial y} - \left(2\frac{\partial^2 w}{\partial x\,\partial y} + \frac{1}{R}\frac{\partial v}{\partial x} - \frac{1}{R}\frac{\partial u}{\partial y}\right) z
\end{aligned}
$$

An idea of the relative importance of the terms in these expressions in certain cases can be gained from the results given in Table 6.4. Further simplifications which are valid in special cases will be discussed later in Secs. 6.5, and 6.6.

6.4 GENERAL THIN SHELL EQUILIBRIUM CONDITIONS

For the elastic case in which the material follows Hooke's law, complete solutions can of course be obtained by replacing equilibrium equations by the law of virtual work, using expression (6.14) for the elastic strain energy, and expression (6.18) for the strains. However, energy methods have many disadvantages, such as the fact that only series solutions can be obtained with them, and the series converge poorly for calculating local conditions, as has previously been pointed out. Hence, in this section we will develop general thin shell equilibrium conditions. To match the generality of our strain–displacement relations, we will try to do this initially without using any more approximations than those inherent in the basic Love–Kirchhoff assumption and the simplification which we have accepted as valid of ignoring large deflection effects in the flexural strains, with accompanying limitations to thin shells and small strains.

We could theoretically develop the equilibrium equations from the strain energy, using the calculus of variations, but as discussed in the first chapter, this would not be easy with the indirect strain expressions (6.8) or (6.18), and the physical meaning would be less clear than deriving them directly, as we will do, from the simple concept of equilibrium.

Effects of Curvatures and Deformation

Since we have defined strains and stresses in terms of *initial* lengths and areas, we need only take into consideration the initial geometry of an element of the shell wall in determining the magnitude of the forces which act upon its sides. But in setting up the conditions of equilibrium of those forces we must consider their moment arms, directions and lines of action, and these depend upon the *final* geometry of the element.

Stress–Strain Relations

We need relations between the strains for which we have expressions and the stresses, which we need for setting up the equilibrium conditions. For the elastic case which we are considering, these relations are of course given by Hooke's law. Since the α and β directions are perpendicular to each other, like the x and y directions used in Eqs. (3.5) and (3.11a, b, c, d), we can use these statements of Hooke's law simply by substituting α, β for x, y in the subscripts.

But a question then arises which is somewhat perplexing. Shall we assume a condition of plane stress in the shell wall, that is assume that $\sigma_z = 0$, as we did in setting up plate theory and as is also customary in shell theory? Or, should we use the plane strain relations between the strains and stresses, that is assume that $\epsilon_z = 0$, because we actually made that assumption in using the Love–Kirchhoff approximation that transverse lines are unchanged in length during the deformation?

Actually, of course, there *are* some σ_z stresses (and they may be somewhat more important than in plate theory, because such stresses are evidently produced by transverse components of the σ_α or σ_β stresses due to the angles which initial curvature causes these stresses to have relative to each other at the ends of elements). There are also actually some ϵ_z strains, due principally to the 'Poisson's ratio effect' of the large σ_α and σ_β stresses. But it would introduce far too great complications to take these actual stresses and strains into consideration (we will have to face these complications in the next chapter in studying 'thick shell' problems, for which they are too important to ignore, but we will restrict that study to small deflections).

It might be argued that it would be more 'consistent' to use plane strain theory, but consistency is meaningless in comparing one approximation (ignoring small transverse strains as a means of simplifying the displacement–strain relations) with another completely unrelated approximation (ignoring the unknown transverse strains or stresses in order to simplify the relation between strains and stresses in the α and β directions).

The only thing which we really know is that the σ_z stresses will be small compared to the σ_α, σ_β stresses for thin shells. Consider for example the effect of the thickness–radius ratio on the relation between these stresses. The simplest test of this is given by elementary 'boiler' theory, that is the case of a thin cylinder of radius R subjected to a uniform internal pressure p; in this case the average circumferential stress is $(R/h)p$, while the transverse normal stress varies between zero and p. The effect of the ratio between the thickness and the half wavelength l of the loading distribution is exemplified in a simply supported beam of length l with a uniform loading p; in this case the maximum bending stresses are more than $(l/h)^2p$, while the transverse normal stress again varies between zero and p.

Thus (except for local conditions such as those in the neighborhood of a concentrated load, which no classical theory can consider, but which can be corrected for as in Figs. 3.15 and 5.12) it is clear that σ_z stresses approach zero compared to σ_α, σ_β stresses as the thickness approaches zero compared to the other dimensions. On the other hand if we made an assumption that transverse normal *strains* are zero, we are really assuming the presence of the σ_z stresses which would be required to make these strains zero, although the actual σ_z stresses are determined by completely different considerations, and are almost certain to be far smaller and as likely as not opposite in direction to those required to make the strains zero.

Thus the best approximation which we can make, and one which can be considered to be an excellent approximation for thin shells, is to assume that the shell wall is in a condition of plane stress, for which from Eqs. (3.11b), changing x, y to α, β:

$$\sigma_\alpha = \frac{E}{1-\nu^2}(\epsilon_\alpha + \nu\epsilon_\beta), \qquad \sigma_\beta = \frac{E}{1-\nu^2}(\epsilon_\beta + \nu\epsilon_\alpha), \qquad \sigma_{\alpha\beta} = \frac{E}{2(1+\nu)}\epsilon_{\alpha\beta} \tag{6.21}$$

Forces and Moments on Sides of Elements

Figure 6.11 shows the middle surface and the two rear sides of an element of the shell wall before it is deformed. On the side normal to the α direction the

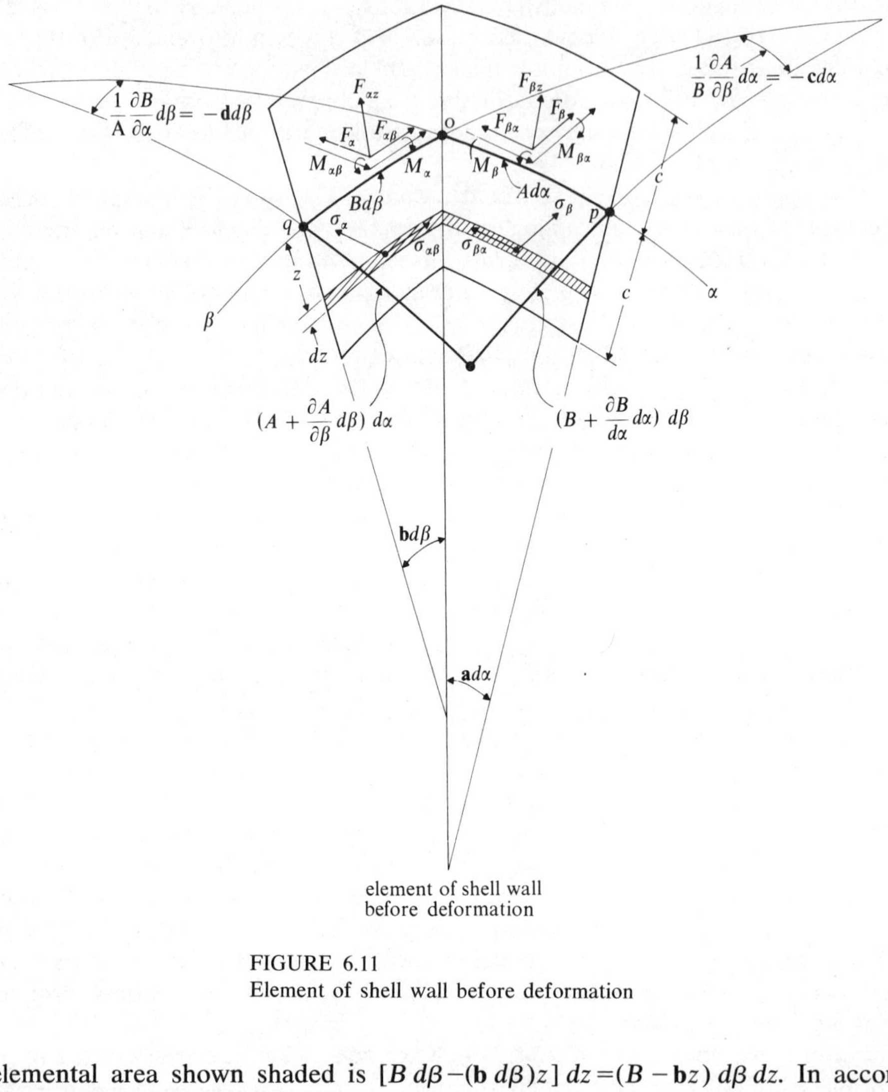

FIGURE 6.11
Element of shell wall before deformation

elemental area shown shaded is $[B\,d\beta-(\mathbf{b}\,d\beta)z]\,dz=(B-\mathbf{b}z)\,d\beta\,dz$. In accordance with usual practice (which we followed in setting up plate theory), we define the F's and M's shown as forces and moments per unit length of section at the middle surface. Then the total normal force on this side of the element is

$$F_\alpha B\,d\beta = \int_{-c}^{c} \sigma_\alpha (B-\mathbf{b}z)\,d\beta\,dz,$$

and similarly for the other F's and M's, from which:

$$F_\alpha = \int_{-c}^{c} \left(1-\frac{\mathbf{b}}{B}z\right)\sigma_\alpha\,dz, \qquad F_{\alpha\beta} = \int_{-c}^{c} \left(1-\frac{\mathbf{b}}{B}z\right)\sigma_{\alpha\beta}\,dz$$

$$M_\alpha = \int_{-c}^{c} \left(1 - \frac{\mathbf{b}}{B} z\right) z\sigma_\alpha \, dz, \qquad M_{\alpha\beta} = \int_{-c}^{c} \left(1 - \frac{\mathbf{b}}{B} z\right) z\sigma_{\alpha\beta} \, dz \tag{6.22}$$

The expressions F_β, $F_{\beta\alpha}$, M_β, $M_{\beta\alpha}$ can be found by interchanging α, $\mathbf{a}$, A with β, $\mathbf{b}$, B respectively.

The subscripts used for the F's and M's (as we have used them in the past and as is customary) are the same as those used for the stresses which compose them, the first subscript corresponding to the direction of the normal to the surface on which the stresses act and the second to the direction of the stress, using only one subscript when these directions are the same. (An argument might be made for having the second subscript correspond to the directions of the axes in the case of moments, especially when they are represented by moment vectors as in Fig. 6.12, but this would be confusing in view of custom and the practice followed for the other quantities.)

For the thin shell problems it might seem that it would be legitimate to simplify expressions (6.22) by ignoring the $\mathbf{b}z/B$ or $\mathbf{a}z/A$ in the parentheses in comparison to unity, in which case we would have $F_{\beta\alpha} = F_{\alpha\beta}$ and $M_{\beta\alpha} = M_{\alpha\beta}$. For the shells listed in Table 6.2 $\mathbf{b}z/B$ and $\mathbf{a}z/a$ are either zero or the ratio between z (which cannot exceed half the thickness) and a radius of curvature. If they are retained the expressions for the F's will contain terms involving $\mathbf{b}/B$ or $\mathbf{a}/A$ times the flexural strains, and the expressions for the M's will contain similar terms involving the membrane strains.

There may be a question as to whether such terms can become important when one of these strains is much larger than the other, for example if the number of circumferential waves in the deformation of a cylinder is very small or very large. But in the elementary 'boiler' case, where the number of these waves is zero, these terms involve only an approximation to the variation of the circumferential stresses across the thickness, which *is* important for a thick cylinder, but negligible for thin cylinders. And if the number of circumferential waves is large enough for such terms to be important, the wavelength of the deformation is no longer large compared to the thickness, which is one of our requirements for the use of classical shell theory.

For these reasons we *will* make these approximations and ignore the $\mathbf{b}z/B$ and $\mathbf{a}z/A$ terms in the expressions (6.22) for the F's and the M's. However, in recognition of the fact that minor terms may possibly prove of some significance in later stages of a study, we will further justify this decision by a complete analysis of the effects of these terms in the derivation of Eq. (6.36). In view of the finding there that these terms *are* negligible, we will also ignore the certainly less important effect of transverse strains upon the moment arms in the expressions for the moments.

If we make these simplifications and use expressions (6.21) for the stresses and (6.18) for the strains and carry out the integrations, we obtain the following expressions for the forces and moments per unit length of section in terms of the strains and hence the displacements:

$$\begin{aligned} F_\alpha &= C(\epsilon_{\alpha m} + \nu\epsilon_{\beta m}), \qquad F_\beta = C(\epsilon_{\beta m} + \nu\epsilon_{\alpha m}) \\ F_{\alpha\beta} &= F_{\beta\alpha} = \frac{1-\nu}{2} C\epsilon_{\alpha\beta m} \\ M_\alpha &= -D(h_\alpha + \nu k_\beta), \qquad M_\beta = -D(k_\beta + \nu h_\alpha) \\ M_{\alpha\beta} &= M_{\beta\alpha} = -\frac{1-\nu}{2} D(h_\beta + k_\alpha) \end{aligned} \tag{6.23}$$

where $\epsilon_{\alpha m}$, $\epsilon_{\beta m}$, $\epsilon_{\alpha\beta m}$ can be taken the same as ϵ_α, ϵ_β, $\epsilon_{\alpha\beta}$ at $z = 0$ given by Eqs. (6.8a), h_α, h_β, k_α, k_β are defined in Eqs. (6.19), and C and D are respectively the 'membrane' and 'flexural' stiffnesses of the shell wall, defined as:

$$C = \frac{2Ec}{1-\nu^2}, \qquad D = \frac{2Ec^3}{3(1-\nu^2)} \tag{6.23a}$$

If we use the values of A, B, **a**, **b**, **c**, **d** in the first line of Table 6.2 these expressions reduce to the corresponding expressions for plates, Eqs. (4.14), (4.16).

Multiplying these forces and moments per unit length of section by the middle surface length of the sides of the element on which they act, we obtain the total forces and moments acting on the element which are shown in Figs. 6.12(a), (b). The force in the α direction on the side oq is $BF_\alpha\, d\beta$. On the side opposite oq, which is the same except for a change $d\alpha$ in α, it is then $\{BF_\alpha + [\partial(BF_\alpha)/\partial\alpha]\, d\alpha\}\, d\beta$, and similarly for the other forces and moments on the sides opposite oq and op. Because of our use of the Love–Kirchhoff assumption, we cannot obtain an expression for the transverse shear forces per unit length of section $F_{\alpha z}$, $F_{\beta z}$, but they exist and must be considered in the equilibrium equations (from which we can determine their magnitudes) although they should be small compared to F_α, F_β, $F_{\alpha\beta}$ in thin shell problems for which the Love–Kirchhoff assumption is a good approximation.

These figures also show the total forces and moments caused by external forces and moments per unit area p, f_α, etc. The moments might be produced by tangential forces acting in opposite directions on the top and bottom surfaces, while the forces might be the difference between such tangential forces, or, in the case of p, normal forces on the top or bottom surfaces (classical shell theory cannot distinguish between these), or body forces such as gravity or 'inertia' forces.

Relative Displacements and Rotations of Sides of Element After Deformation

Figures 6.12(c), (d) show the displacements and rotations respectively of the side of an element of the shell wall on which they are shown relative to the opposite side. As was discussed earlier, for use in setting up the equilibrium conditions these must include both the initial displacements and rotations and those caused by the deformation. In the following paragraphs we will discuss how these results were obtained.

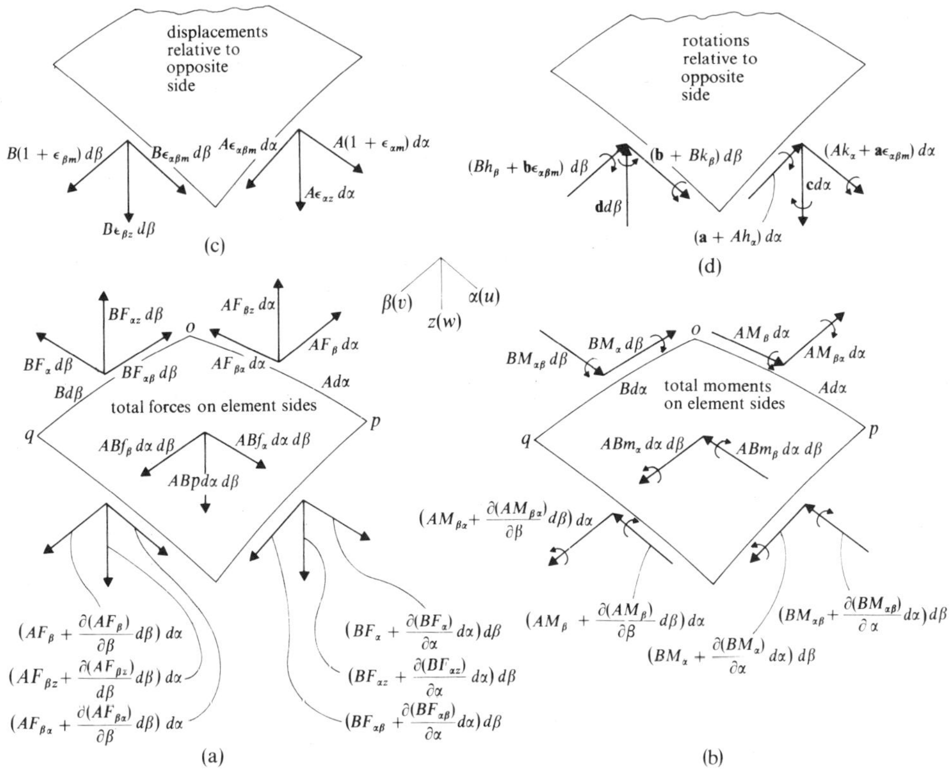

FIGURE 6.12

Referring to Fig. 4.8, which is similar to Fig. 6.12 for the special case of flat plates, it can be seen that all the displacements and rotations contain one increment of the coordinates dx, or dy, which correspond to $Ad\alpha$ and $Bd\beta$ in the general case. The expressions for displacements and rotations in the general case must contain the same terms as for flat plates, plus possible additional terms. Any additional terms containing more than one increment $d\alpha$ or $d\beta$ can therefore be ignored as a small quantity of higher order, which disappears in the limit as the increment approaches zero. The sines or tangents and the cosines of the angles of rotation can therefore be taken as the angles themselves and unity respectively. The parts of the displacements involving strains already include one increment, and any change in them due to curvature can therefore be ignored, so these parts are the same as for flat plates.

Displacements enter the equilibrium conditions only as moment arms of couples formed by forces on opposite sides of an element. Figure 6.13(a) shows a cross section of an element and pertinent forces. It can be seen that any moment arm for the F_α forces, and any difference between $Ad\alpha$ and the moment arm of the $F_{\alpha z}$ forces would involve more than one $d\alpha$, and hence these can also be ignored. Thus, the whole of the displacement expressions are the same as for flat

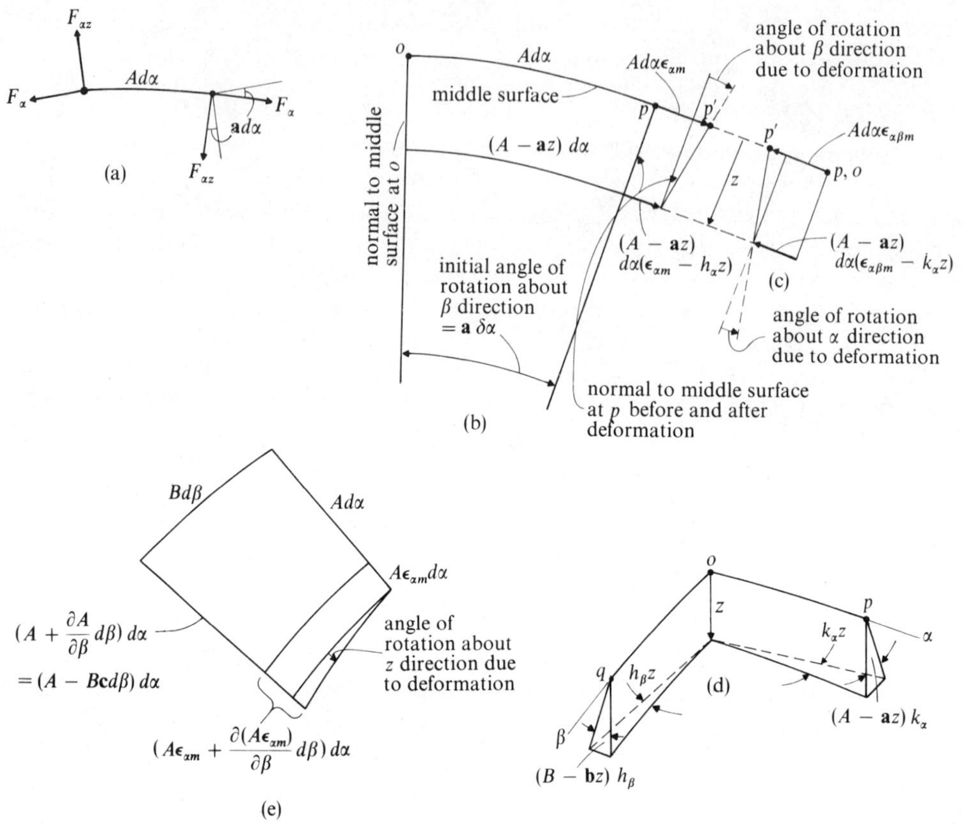

FIGURE 6.13

plates, Fig. 4.8(a), with the substitution of $Ad\alpha$, $Bd\beta$ for dx, dy, and of α, β for x, y in the subscripts. We retain for the time being the $(1+\epsilon_{\alpha m})$, $(1+\epsilon_{\beta m})$ in the displacements in the α, β directions because later, in the equilibrium equation of moments about the z direction, the main terms cancel and the terms containing the strains remain.

As in the case of flat plates, we have no values for the transverse shear strains $\epsilon_{\alpha z}$ and $\epsilon_{\beta z}$, which we have neglected so far in accordance with the Love–Kirchhoff assumption. Their average values can be determined as the average transverse shear stresses $F_{\alpha z}/h$, $F_{\beta z}/h$ divided by the shear modulus $G = E/2(1+\nu)$. Using Eqs. (6.23a), this gives:

$$\epsilon_{\alpha z} = \frac{2}{1-\nu}\frac{F_{\alpha z}}{C}, \qquad \epsilon_{\beta z} = \frac{2}{1-\nu}\frac{F_{\beta z}}{C} \tag{6.23b}$$

However, any terms containing these transverse shear strains are obviously likely to be unimportant for any case for which the Love–Kirchhoff assumption is a good approximation, with the possible exception of buckling problems in which these strains are multiplied by the finite forces producing the buckling.

The rotation about the β direction of the side opposite oq relative to oq is illustrated in Fig. 6.13(b). This figure shows a true view of the side of the element containing op, including the middle surface line op itself, and normals to the middle surface at o and p (which are corners of the element). Of course the side of an element is not in general a plane surface either before or after deformation, but deviations from a plane surface will be proportional to $d\alpha$, and their effects on the following analysis will be small quantities of higher order.

The normals to the middle surface at o and p are initially at the angle $\mathbf{a}d\alpha$ to each other. The additional angle of rotation due to deformation is caused by the unit strain $\epsilon_\alpha = \epsilon_{\alpha m} - h_\alpha z$, which produces different total strains at different distances z from the middle surface. It can be seen from the figure that this angle is, using Eqs. (6.19) and (6.8a): $[A d\alpha\, \epsilon_{\alpha m} - (A - \mathbf{a}z)\, d\alpha (\epsilon_{\alpha m} - h_\alpha z)]/z$. Adding to this the initial rotation $\mathbf{a}d\alpha$, and ignoring $\mathbf{a}z/A$ and $\epsilon_{\alpha m}$ compared to unity for thin shells and small strains, the total relative rotation becomes $(\mathbf{a} + Ah_\alpha)d\alpha$.

Any initial rotation about the α direction of the side opposite oq relative to oq would represent a twist, which is zero here because oq is a line of curvature. The corresponding rotation due to deformation is a little more difficult to analyze because it is affected by conditions in three dimensions, instead of everything being essentially two-dimensional as in the preceding case. However, we can show the final results of the deformation in the same manner as in the preceding case.

Figure 6.13(c) shows a true view of the displaced side of the element containing oq (again we can disregard the fact that this side is not exactly a plane surface). Because of the zero initial twist of the element, at the beginning of the displacement (during which we will think of this side as stationary) the projection of the normal to the middle surface through p on this view will coincide with the normal through o (the effect of the curvature described by $\mathbf{c}$ is a higher order effect involving the square of $d\alpha$). As displacement proceeds, due to the membrane shear strain $\epsilon_{\alpha\beta m}$ the projection of the normal through p will move relative to that through o a distance $A d\alpha\, \epsilon_{\alpha\beta m}$, independent of z.

There will also be a movement proportional to z and to the part $-k_\alpha z$ of the flexural shear strain; the other part $-h_\beta z$ of the flexural strain does not affect this motion. Figure 6.13(d) shows why this is so. The flexural shear strain causes a rotation of $(A - \mathbf{a}z)k_\alpha$ of the side of the element containing op about op in the direction shown, causing the movement described above, and a rotation $(B - \mathbf{b}z)h_\beta$ of the side of the element containing oq about oq which does not affect the movement being studied because its motion is perpendicular to the view of Fig. 6.13(c).

How the flexural shear strain divides up in this way can be seen by reference to the definitions of k_α and h_β in Eqs. (6.19), which show that the principal term of k_α is $(1/A)\partial[(1/B)\partial w/\partial\beta]/\partial\alpha$; this represents physically the rate of change in the α direction of the slope of the middle surface in the β direction, and produces the kind of rotation of the side of the element containing op shown in the figure. On the other hand, the principal term in h_β is $(1/B)\partial[(1/A)\partial w/\partial\alpha]/\partial\beta$, which represents the rate of change in the β direction of the slope in the α direction and produces the rotation of the side oq shown, but does not contribute to the rotation of the side op. It is also interesting to note that k_α comes from the same term, G in Eqs. (6.8) as the $\partial v/\partial\alpha$ term in the membrane shear strain, which suggests that G really describes the motion of side op due to shear, while the term I in Eqs. (6.8) similarly describes the motion of oq. If k_α is expanded into functions of the displacements it is found to contain no function of u (which can cause angular movement of side oq but not of op), but only functions of w and v (which can cause angular movements of op). Similarly, h_β contains only w and u.

Then from Fig. 6.13(c), which shows movements resulting from the actions discussed above, it can be seen that the rotation about the α direction of the normal through p (and hence of the side opposite oq) about the normal through o (and hence of the side oq) is: $[Ad\alpha\,\epsilon_{\alpha\beta m} - (A - \mathbf{a}z)d\alpha(\epsilon_{\alpha\beta m} - k_\alpha z)]/z$. Expanding, and ignoring $\mathbf{a}z/A$ compared to unity for thin shells, this rotation becomes $(Ak_\alpha + \mathbf{a}\epsilon_{\alpha\beta m})d\alpha$.

Finally the initial angle of rotation about the z direction of the side opposite oq relative to oq is, from Fig. 6.11, $(\partial A/\partial\beta)d\alpha/B = -\mathbf{c}d\alpha$. The additional rotation due to deformation is shown in Fig. 6.13(e). It can be seen that it is $[\partial(A\epsilon_{\alpha m})/\partial\beta]d\alpha/B$. Adding the initial rotation we have $\{\partial[A(1+\epsilon_{\alpha m})]/\partial\beta\}d\alpha/B$, which reduces to the initial rotation if we ignore unit strains compared to unity. The expressions for the rotations of the side opposite op relative to op are found in similar manner.

Equations of Equilibrium

The information given in Fig. 6.12 can now be used to write down the six equations of equilibrium, stating that the total forces acting on an element of the shell wall in the α, β, z directions, and the moments of forces about the β, α, z directions respectively add up to zero. Table 6.6 gives these six equations in the order mentioned above. Each term obtained from Fig. 6.12 contains the factor $d\alpha d\beta$ (terms containing more increments of course being ignored as small quantities of higher order) and the equations have been simplified by dividing through by this factor, and in the case of the fifth equation by -1 and in the case of the sixth equation by AB. As far as possible the terms which seem most likely to be important (for example terms which reduce to those given in Eqs. (4.8) for plates) are given first, and those likely to be less important last.

The terms are of the following four types, the same as those listed in deriving Eqs. (4.8):

(1) External forces and moments. After dividing by $d\alpha d\beta$ these are, in the order given above: ABf_α, ABf_β, ABp, ABm_α, ABm_β, 0. Of these the force ABp in the z direction normal to the shell's surfaces, which could be a pressure acting on the upper surface, is the one most commonly occurring in practical problems. Other

external forces or moments can of course be applied to the boundary edges, but these are not directly applied to the general wall element which we are considering, and merely form one of the F's or M's on an element at the edge.

(2) Changes in forces or moments over the width of the element. Consider, say, the force $BF_\alpha d\beta$ in the α direction on the side oq and the same force plus $[\partial(BF_\alpha)/\partial\alpha]d\alpha d\beta$ on the opposite side. The main parts $BF_\alpha d\beta$ (which contain only one coordinate increment, $d\beta$) cancel out between the two sides, except for small resultant forces in the β and z directions equal to their magnitude times the small angles $\mathbf{c}d\alpha$ and $(\mathbf{a}+Ah_\alpha)d\alpha$ between the sides on which they act and hence between them; these are considered below as type (3). The change in force, $[\partial(BF_\alpha)/\partial\alpha]d\alpha d\beta$, is in the α direction except for these small angles; since it

Table 6.6 GENERAL SHELL EQUILIBRIUM EQUATIONS

	(1) External forces and mom.	(2) Changes in forces or moments over width of element	(3) Resultants of forces or moments on opposite sides, due to angles between sides of elements	(4) Couples due to forces on opposite sides
1	ABf_α	$+\dfrac{\partial(BF_\alpha)}{\partial\alpha}$ $+\dfrac{\partial(AF_{\beta\alpha})}{\partial\beta}$	$+AF_\beta\mathbf{d}-BF_{\alpha\beta}\mathbf{c}-BF_{\alpha z}(\mathbf{a}+Ah_\alpha)$ $-AF_{\beta z}(Bh_\beta+\mathbf{b}\epsilon_{\alpha\beta m})=0$	
2	ABf_β	$+\dfrac{\partial(AF_\beta)}{\partial\beta}$ $+\dfrac{\partial(BF_{\alpha\beta})}{\partial\alpha}$	$+BF_\alpha\mathbf{c}-AF_{\beta\alpha}\mathbf{d}-AF_{\beta z}(\mathbf{b}+Bk_\beta)$ $-BF_{\alpha z}(Ak_\alpha+\mathbf{a}\epsilon_{\alpha\beta m})=0$	
3	ABp	$+\dfrac{\partial(BF_{\alpha z})}{\partial\alpha}$ $+\dfrac{\partial(AF_{\beta z})}{\partial\beta}$	$+BF_\alpha(\mathbf{a}+Ah_\alpha)+AF_\beta(\mathbf{b}+Bk_\beta)$ $+ABF_{\alpha\beta}\left(k_\alpha+\epsilon_{\alpha\beta m}\dfrac{\mathbf{a}}{A}\right)+ABF_{\beta\alpha}\left(h_\beta+\epsilon_{\alpha\beta m}\dfrac{\mathbf{b}}{B}\right)=0$	
4	ABm_α	$+\dfrac{\partial(BM_\alpha)}{\partial\alpha}$ $+\dfrac{\partial(AM_{\beta\alpha})}{\partial\beta}$	$+AM_\beta\mathbf{d}-BM_{\alpha\beta}\mathbf{c}$	$-AB(F_{\alpha z}+F_{\beta z}\epsilon_{\alpha\beta m}$ $-F_\alpha\epsilon_{\alpha z}-F_{\beta\alpha}\epsilon_{\beta z})=0$
5	ABm_β	$+\dfrac{\partial(AM_\beta)}{\partial\beta}$ $+\dfrac{\partial(BM_{\alpha\beta})}{\partial\alpha}$	$+BM_\alpha\mathbf{c}-AM_{\beta\alpha}\mathbf{d}$	$-AB(F_{\beta z}+F_{\alpha z}\epsilon_{\alpha\beta m}$ $-F_\beta\epsilon_{\beta z}-F_{\alpha\beta}\epsilon_{\alpha z})=0$
6			$+M_\beta\left(h_\beta+\epsilon_{\alpha\beta m}\dfrac{\mathbf{b}}{B}\right)-M_\alpha\left(k_\alpha+\epsilon_{\alpha\beta m}\dfrac{\mathbf{a}}{A}\right)$ $+M_{\alpha\beta}\left(h_\alpha+\dfrac{\mathbf{a}}{A}\right)-M_{\beta\alpha}\left(k_\beta+\dfrac{\mathbf{b}}{B}\right)$	$+F_{\beta\alpha}(1+\epsilon_{\beta m})-F_{\alpha\beta}(1+\epsilon_{\alpha m})$ $-\epsilon_{\alpha\beta m}(F_\beta-F_\alpha)=0$

already contains two increments, $d\alpha$ and $d\beta$, the effects of these angles, which each contain another increment, can be ignored as small quantities of higher order. This leaves the force $\partial(BF_\alpha)/\partial\alpha$ in the α direction, after dividing by $d\alpha d\beta$. The forces and moments of this type, listed in the second column of Table 6.6 are all important in any shell theory.

(3) Resultants of forces or moments on opposite sides, due to angles between sides of elements. Two forces of this type, listed in the third column of Table 6.6 for the second and third equations, were calculated in the last paragraph. Some of the forces or moments of this type are important in all shell theories and some are negligible in all theories, depending upon the sizes of the forces and of the angles between them.

As we have pointed out before, the transverse shear forces $F_{\alpha z}$, $F_{\beta z}$ should be small compared to F_α, F_β, $F_{\alpha\beta}$ for our thin-shell case. In stability problems one or more of the latter forces produce the buckling and are finite in magnitude while the strains and forces and moments produced by the buckling displacement are still infinitesimal. In the expressions for the angles between the sides, Ah_α, Bk_β, etc., are in general small compared to **a**, **b**, **c**, **d** in 'small-deflection' problems, but can grow and become as or more important in 'large-deflection' problems. Terms involving products of strains by curvatures are probably never important.

(4) Couples due to forces on opposite sides, with the displacements of the sides on which they act as moment arms. For example, the main parts of the forces $AF_{\beta z}d\alpha$ are separated by the displacement in the α direction $B\epsilon_{\alpha\beta m}$ of the sides on which they act, giving a moment about the β direction $ABF_{\beta z}\epsilon_{\alpha\beta m}$. As before, the effects of the changes in the forces and angles the sides make with each other prove to be higher order quantities. Again some of the terms of this type given in Table 6.6 are important and some unimportant, using the criteria discussed in the preceding paragraphs.

6.5 SIMPLIFICATIONS AND SOLUTIONS OF GENERAL THIN SHELL THEORY

In this section we will consider possibilities of simplifying the general thin shell relations which we have derived, and the requirements for solving them.

General Methods and Requirements for Obtaining Solutions

Equations (6.18) relate the membrane and flexural strains to the three displacements u, v, w. Equations (6.23) give the relations between these strains and the forces F_α, F_β, $F_{\alpha\beta}$ and moments M_α, M_β, $M_{\alpha\beta}$ on cross sections. And finally Table 6.6 gives the six equilibrium equations relating these forces and moments to each other, and to the unknown transverse shear forces $F_{\alpha z}$, $F_{\beta z}$, as well as to known external forces and moments p, f_α, etc., acting on the shell wall. The material properties, wall thickness and the geometric functions A, B, **a**, **b**, **c**, **d** will also be known.

If the F's and M's in Table 6.6 are written in terms of u, v, w by using relations (6.18) and (6.23), we obtain six differential equations involving the five unknown functions u, v, w, $F_{\alpha z}$, $F_{\beta z}$. These five unknowns can theoretically be determined from the first five of the equilibrium equations. A solution of the thin shell problem obtained in this way satisfies not only the requirement of equilibrium of forces in the α, β, z directions, and of moments about the α, β directions, but also, by expressing F_α, F_β, $F_{\alpha\beta}$, M_α, M_β, $M_{\alpha\beta}$ in terms of continuous displacement functions u, v, w, it satisfies the equally important requirement of continuity in the α, β directions (which is as much as can be satisfied while using the Love–Kirchhoff approximation).

We will disregard the last equation of equilibrium, of moments about the z axis. The principal terms of this equation, $ABF_{\beta\alpha}$ and $ABF_{\alpha\beta}$, are not shown because they cancel each other identically as in the last of Eqs. (4.8) for plates. The remaining terms in this equation shown in Table 6.6 correspond to the unimportant terms which were ignored in writing the equation for plates; any lack of equilibrium which there may be between these remaining unimportant terms represents unavoidable errors which we are making.

Simplification of General Equations

We can now solve for $F_{\alpha z}$ and $F_{\beta z}$ from the fourth and fifth equations, which represent equilibrium of moments about the β and α directions. This is greatly facilitated by the fact that the parentheses at the ends of these equations containing type (4) couples reduce to $F_{\alpha z}$ and $F_{\beta z}$ respectively for small strains. Thus, in the fourth equation, using Eqs. (6.23b) for $\epsilon_{\alpha z}$ and $\epsilon_{\beta z}$ and the third of Eqs. (6.23) for $\epsilon_{\alpha\beta m}$, the fourth term in the parenthesis, in which the $\epsilon_{\beta z}$ should obviously have its average value over the cross section, the same as in Eqs. (6.23b), becomes $-F_{\beta z}[2/(1-\nu)]F_{\alpha\beta}/C = -F_{\beta z}\epsilon_{\alpha\beta m}$, cancelling the second term in the parenthesis. Similarly the first and third terms become $F_{\alpha z}\{1-[2/(1-\nu)]F_\alpha/C\}$, which can be reduced to $F_{\alpha z}$ for small strains because, from Eq. (6.23), F_α/C represents a unit strain.

The fourth and fifth terms, containing the curvature functions **c** and **d**, of the first, second, fourth, and fifth equations of Table 6.6 are related to the second and third terms of the same equations. Expanding the latter terms and replacing **c**, **d** by using Eqs. (6.10), these terms in the fourth equation, for example, become:

$$\begin{aligned} B(\partial M_\alpha/\partial\alpha)&+M_\alpha(\partial B/\partial\alpha)+A(\partial M_{\beta\alpha}/\partial\beta)+M_{\beta\alpha}(\partial A/\partial\beta)-M_\beta(\partial B/\partial\alpha)+M_{\alpha\beta}(\partial A/\partial\beta)\\ &=B(\partial M_\alpha/\partial\alpha)+A(\partial M_{\beta\alpha}/\partial\beta)+(M_\alpha-M_\beta)\partial B/\partial\alpha+(M_{\alpha\beta}+M_{\beta\alpha})(\partial A/\partial\beta)\end{aligned}$$

a form which involves simpler terms, although it is no shorter than the original form.

An obvious simplification can be made by neglecting the terms $-AF_{\beta z}\mathbf{b}\epsilon_{\alpha\beta m}$ and $-BF_{\beta z}\mathbf{a}\epsilon_{\alpha\beta m}$ in the first two equations, and $ABF_{\alpha\beta}\epsilon_{\alpha\beta m}(\mathbf{a}/A+\mathbf{b}/B)$ in the third equation. For example, multiplying the parenthesis containing $\epsilon_{\alpha\beta m}$ in the first equation by z/B, the first term, $h_\beta z$, represents a unit strain, while the second term, $z(\mathbf{b}/B)\epsilon_{\alpha\beta m}$ represents a unit strain multiplied by z over a radius of curvature, a small quantity for thin shells; moreover, even the first term is not very important, as is discussed in the next paragraph. While such an argument may not be rigorously conclusive it is hard to imagine any practical problems involving thin shells and small strains for which the above terms could be important.

Terms such as $BF_\alpha Ah_\alpha$, $AF_\beta Bk_\beta$, etc., in the third equation are important for buckling problems where F_α, F_β or $F_{\alpha\beta}$ may represent the finite forces producing the buckling which are present at the beginning of the buckling. However, the nonlinear terms involving the transverse shear forces in the first two equations, such as $-BF_{\alpha z}Ah_\alpha$ and $-AF_{\beta z}Bh_\beta$ in the first equation, are another matter, since the transverse shear forces $F_{\alpha z}$, $F_{\beta z}$ are small compared even to ordinary F_α, etc., forces produced by the deformation. We ignored the corresponding terms in setting up Eqs. (4.8) for plates, and they undoubtedly are negligible in problems involving the moderately large deflections (of the order of magnitude of the thickness) which were being considered then. But we will retain these terms for our most general theory to be sure of its applicability with very large deflections, up to fifty times the thickness, which we considered in Sec. 6.3.

With these modifications the equations of equilibrium become, using the symbols $F'_{\alpha z} = BF_{\alpha z}$, $F'_{\beta z} = AF_{\beta z}$:

$$
\begin{aligned}
&ABf_\alpha + B\frac{\partial F_\alpha}{\partial \alpha} + A\frac{\partial F_{\beta\alpha}}{\partial \beta} + \frac{\partial B}{\partial \alpha}(F_\alpha - F_\beta) + \frac{\partial A}{\partial \beta}(F_{\alpha\beta} + F_{\beta\alpha}) \\
&\qquad\qquad - F'_{\alpha z}(\mathbf{a} + Ah_\alpha) - F'_{\beta z}Bh_\beta = 0 \\
&ABf_\beta + A\frac{\partial F_\beta}{\partial \beta} + B\frac{\partial F_{\alpha\beta}}{\partial \alpha} + \frac{\partial A}{\partial \beta}(F_\beta - F_\alpha) + \frac{\partial B}{\partial \alpha}(F_{\alpha\beta} + F_{\beta\alpha}) \\
&\qquad\qquad - F'_{\beta z}(\mathbf{b} + Bk_\beta) - F'_{\alpha z}Ak_\alpha = 0 \\
&ABp + \frac{\partial F'_{\alpha z}}{\partial \alpha} + \frac{\partial F'_{\beta z}}{\partial \beta} + BF_\alpha(\mathbf{a} + Ah_\alpha) + AF_\beta(\mathbf{b} + Bk_\beta) + ABF_{\alpha\beta}(h_\beta + k_\alpha) = 0
\end{aligned}
\tag{6.24}
$$

where:

$$
\begin{aligned}
F'_{\alpha z} &= Bm_\alpha + \frac{1}{A}\left[B\frac{\partial M_\alpha}{\partial \alpha} + A\frac{\partial M_{\beta\alpha}}{\partial \beta} + \frac{\partial B}{\partial \alpha}(M_\alpha - M_\beta) + \frac{\partial A}{\partial \beta}(M_{\alpha\beta} + M_{\beta\alpha})\right] \\
F'_{\beta z} &= Am_\beta + \frac{1}{B}\left[A\frac{\partial M_\beta}{\partial \beta} + B\frac{\partial M_{\alpha\beta}}{\partial \alpha} + \frac{\partial A}{\partial \beta}(M_\beta - M_\alpha) + \frac{\partial B}{\partial \alpha}(M_{\alpha\beta} + M_{\beta\alpha})\right]
\end{aligned}
\tag{6.25}
$$

The differences between $F_{\alpha\beta}$ and $F_{\beta\alpha}$ or between $M_{\alpha\beta}$ and $M_{\beta\alpha}$ are minor quantities for thin shells and can be ignored for most purposes, as is shown later in the discussion of Eq. (6.36). Equations (6.24) and (6.25) together with Eqs. (6.18), (6.23) form the most general thin shell theory. They, and modifications of them developed in the following for particular ranges, can be converted to theories for shells of specific geometric shapes merely by substituting in them the corresponding values of the geometric functions, such as those given in Table 6.2.

Initial Deviations From Perfect Shape

Let $w_0(\alpha, \beta)$ represent an initial lateral deviation (that is in the z direction) of the unloaded shell from its assumed theoretical shape, and $w(\alpha, \beta)$ its lateral movement under load. The net strains occurring under load can then be calculated as those which would be produced by a lateral deflection of $(w_0 + w)$ minus those which would be produced by a lateral deflection of w_0 alone. All the terms in the expressions (6.18) for strains which are linear in w will remain unchanged; thus the $\partial w/\partial\beta$ in k will become $\partial(w_0 + w)/\partial\beta - \partial w_0/\partial\beta = \partial w/\partial\beta$. All the flexural strains will be unchanged, since they are linear in w. But the nonlinear parts of the membrane strains, Eq. (6.18), will be changed. For instance the $h^2/2$ term in $\epsilon_\alpha = \epsilon_{\alpha m}$ would become $\{[\partial(w_0 + w)/\partial\alpha + \mathbf{a}u]^2 - (\partial w_0/\partial\alpha)^2\}/2A^2$. Such expressions can be considerably simplified in specific cases, as will be illustrated later in Eqs. (6.29f).

Solutions and Boundary Conditions

As outlined above we can use Eqs. (6.25) to eliminate $F'_{\alpha z}$, $F'_{\beta z}$ from the three equations of (6.24), which, after using Eqs. (6.23) and (6.18) will involve only known functions and the three unknown displacements u, v, w. We must then solve these three differential equations for the three unknown functions, preferably in an 'uncoupled' form in which one equation is obtained which involves only w, with two more equations relating u to w and v to w respectively (the limitations of such solutions will be discussed on p. 373. To carry out such operations for the general theory is impractical, but we will find that it can be done for special cases.

The three equations (6.24) can theoretically not only be solved for the displacements, but should also contain enough functions of integration to satisfy at least the most important boundary conditions. The first two equations of (6.24) will be of the second order, sufficient together to satisfy two 'membrane' conditions, on the u, v displacements or on the membrane forces F_α, F_β, $F_{\alpha\beta}$, along four edges of a curved panel. The third equation will be of fourth order and, like Eq. (4.18) or (4.19) for flat plates, can satisfy two 'flexural' conditions, on the w displacement or slopes $F_{\alpha z}$, $F_{\beta z}$, or on M_α, M_β, $M_{\alpha\beta}$ along four edges of a curved panel.

It should be mentioned here that what has just been said does not rule out the possibility that in some cases more boundary conditions than those stated may be satisfied without having to use supplementary solutions. Even if we paid no attention to boundary conditions, whatever solution we get will involve *some* kind of stresses and displacements at every point of the boundaries, and it might happen by chance that these are the stresses or displacements which the problem being studied requires. The number of conditions stated above are the number which we can be *sure* of satisfying with the equations considered.

Most shells of course are not in the form of a curved panel, and some or all of the edge conditions are replaced by continuity conditions across sections which for other problems could be edges of panels. Thus in a complete cylinder there are two edges at the cylinder ends, while the other two edge conditions are replaced by the condition that displacements, etc., must be cyclical functions of the circumference, to insure continuity across axially directed cross sections. In the case of a complete sphere there are no edges but continuity conditions in both latitude and longitude directions.

In many important cases the satisfaction of two membrane and two flexural conditions will be sufficient for practical purposes, for example for hinged or fixed edges, while in other cases, for example for free edges, Kirchhoff's device for combining transverse shear forces and twisting moments can be used for satisfying at least gross boundary conditions, with much the same limitations and approximations as for the case of flat plates which were discussed in Secs. 4.5, and 5.5. The satisfaction of gross boundary conditions, that is conditions on resultant forces or moments or the displacements of one surface such as the middle surface, should be sufficient for problems involving what we have defined as 'thin' shells, but if it seems necessary to satisfy more complete conditions at every point of cross sections, many of the supplementary methods and solutions discussed for flat plates can be applied with reasonable accuracy to most shell problems, as was discussed when these methods were presented. More specific discussion and illustration of these matters for cylindrical shells will be given in Chap. 7.

Another method of solution of the thin shell problem was illustrated for the case of a flat plate by Eq. (4.13) and for circular cylindrical shells by Eq. (6.17). In this method of solution the membrane stresses (or forces) are represented by a stress function $\phi(\alpha, \beta)$, which satisfies the first two equations of (6.24), and when substituted into the third equation reduces the number of unknown functions to two, ϕ and w. This satisfaction of the equilibrium equations must be supplemented by satisfaction of continuity in the α, β directions, which can be accomplished by equating the expressions for the three membrane strains in terms of ϕ to their expressions in terms of continuous displacements u, v, w. The resulting three equations are reduced to one by eliminating u and v between them; this gives the second of the two equations involving only ϕ and w which are required to solve for these two functions. These two equations will be of fourth order, like Eqs. (4.13) and (4.18) for flat plates, and theoretically contain the same number of functions for satisfying edge conditions as the three equations involving u, v, w discussed earlier.

While this method may simplify the solution, it is evident that the first two equations of (6.24) for the general thin shell can not be satisfied by an ordinary Airy stress function. Even if the terms containing $F'_{\alpha z}$ and $F'_{\beta z}$ were ignored, it seems difficult if not impossible to satisfy these equations with a more complex type of stress function. And even if this hurdle were overcome, the task of eliminating u and v from the equations obtained from the expressions for the membrane strains, for the general case when the geometric functions A, B, $\mathbf{a}$, $\mathbf{b}$, $\mathbf{c}$, $\mathbf{d}$ vary with α and β, appears to be hopeless. However, this method of solution is feasible for certain shell geometries, as will be demonstrated in Sec. 6.7.

6.6 GENERAL THIN SHELL THEORIES FOR PARTICULAR RANGES

We can particularize the theory, while still keeping it general enough to apply to various shell types such as those listed in Table 6.2, in various ways. Two ways which have been found very useful are: (A) on the basis of the ratio of the maximum deflection to the thickness; (B) on the basis of the relation between the shortest half wavelength of the deformation and the radius (or thickness); this relation seems to determine the relative importance of the membrane and flexural parts of the straining. There are also some geometric restrictions which do not entirely limit the type of shell to which they apply, for example the axisymmetric case, which we will consider.

A. Ranges of Deflection–Thickness Ratio

We will distinguish three ranges: (1) 'Very large deflections', to which the general theory which we have presented applies, consisting of equations (6.18), (6.23), (6.24), (6.25). (2) 'Large deflections', of the order of magnitude of the thickness. (3) 'Small deflections', that is deflections so small that changes caused by them in the geometry of the shell can be ignored, resulting in a theory which is linear in the deflections and their derivatives.

Large Deflections

If we can consider the case studied in Sec. 6.3, some results of which were shown in Fig. 6.10(b), as typical, the type of simplification used in expression (6.15) for strains gives a good approximation for w_{max}/h ratios up to 10 or more. It would probably be over optimistic to draw such a conclusion for the general case on the basis of a single example, however representative, but it seems safe to use this type of simplification whenever the deflections are of the order of magnitude of the thickness, say for w_{max}/h ratios from 1/5 to 5, which covers most practical problems. For the case studied in Sec. 6.3 the simplification consisted of ignoring the squares or products of derivatives of v and u in comparison to those of w; this is consistent with the conclusion, made at the beginning of this chapter under the title 'General Effects of Curvature', that the displacements v and u are generally small compared to w (however, in such an extreme case as that noted, of pure bending of a cylindrical tube, the squares of at least the derivatives of v must obviously be included).

For the general case, the principal terms of the functions f, g, i, j are derivatives of u or v, while those of h and k are derivatives of w. Two other terms contained in f and j are $(\mathbf{a}/A)w$ and $(\mathbf{b}/B)w$; these will be of the order of the thickness–radius ratio and should not be of a greater order of magnitude than the principal terms. The other terms in f, g, i, j contain $\mathbf{c}/A$ or $\mathbf{d}/B$, which in some cases may be larger than $\mathbf{a}/A$ and $\mathbf{b}/B$, but since they are multiplied by the smaller displacements u or v they should also be of no greater order of magnitude than the principal terms. Hence for the general case it seems safe to ignore squares and products of f, g, i, j compared to those of h or k.

Equations (6.18) for the membrane strains can then be simplified for the moderately large deflection range to:

$$\epsilon_{\alpha m} = f + h^2/2, \qquad \epsilon_{\beta m} = j + k^2/2, \qquad \epsilon_{\alpha\beta m} = g + i + hk \tag{6.26a}$$

Some similar simplifications should perhaps be made in the expressions for h_α, h_β, k_α, k_β for the flexural strains, but since these strains are linear in the displacements this is not very important for computational purposes. Hence we will keep Eqs. (6.23) unchanged for this range, and also Eqs. (6.25) and the third equation of (6.24). In the first two equations of (6.24) it should be safe, for this case, to neglect the nonlinear terms containing the transverse shear strains, which reduces these equations to:

$$\begin{aligned} &ABf_\alpha + B\frac{\partial F_\alpha}{\partial \alpha} + A\frac{\partial F_{\alpha\beta}}{\partial \beta} + \frac{\partial B}{\partial \alpha}(F_\alpha - F_\beta) + 2\frac{\partial A}{\partial \beta}F_{\alpha\beta} - F'_{\alpha z}\mathbf{a} = 0 \\ &ABf_\beta + A\frac{\partial F_\beta}{\partial \beta} + B\frac{\partial F_{\alpha\beta}}{\partial \alpha} + \frac{\partial A}{\partial \beta}(F_\beta - F_\alpha) + 2\frac{\partial B}{\partial \alpha}F_{\alpha\beta} - F'_{\beta z}\mathbf{b} = 0 \end{aligned} \tag{6.26b}$$

General Small Deflection (Linear) Shell Theory for Lateral or Other Ordinary Loading

For this range, of deflections which are small compared to the thickness, say $w_{max}/h < 1/5$, we have merely to delete all terms which are nonlinear in the displacements, that is containing squares or products of the displacements or their derivatives. The membrane strains then become:

$$\epsilon_{\alpha m} = f, \qquad \epsilon_{\beta m} = j, \qquad \epsilon_{\alpha\beta m} = g + i \tag{6.27a}$$

Equations (6.23) remain unchanged, with f, h_α, etc., defined in Eq. (6.19). Of the equilibrium equations, (6.26b) and (6.25) remain unchanged for the first and second and fourth and fifth equations. In the third equation of equilibrium (the most important of them all, since it represents equilibrium of forces tending to deform the shell in its weakest direction, the direction of the small thickness) the nonlinear terms $BF_\alpha Ah_\alpha$, etc., must be eliminated for the case of ordinary lateral loading, giving:

$$ABp + \frac{\partial F'_{\alpha z}}{\partial \alpha} + \frac{\partial F'_{\beta z}}{\partial \beta} + BF_\alpha\mathbf{a} + AF_\beta\mathbf{b} = 0 \tag{6.27b}$$

General Small Deflection Theory for Classical Stability Studies

In studies of the classical stability limit of shells which can have such a limit (for example perfect cylindrical or conical shells under uniform axial loading or twist, and these as well as spherical shells under uniform external pressure) the deformation can be separated into two phases, the prebuckling phase, during which the forces F_α, F_β, or $F_{\alpha\beta}$ are brought up to the limit at which the shell becomes unstable, and the buckling phase during which such forces remain substantially constant.

It should be pointed out here that in this discussion it will be assumed that the end conditions permit the ends of shells such as cylinders or conical shells to expand or contract freely in the same manner as the middle parts of these shells during the prebuckling phase, so that generators remain straight—otherwise there will be a local deformation at the ends of the shells which acts during the buckling like an initial defect or deviation from perfect geometrical shape. [The whole concept of 'stability' is actually an academic one, since real shells always have defects, but it is nevertheless a useful concept even in cases where, as we will find, it does not give a good approximation to the real buckling load. Very large or large deflection theory must in general be used to study the effects of initial defects, since these deviations from perfect shape or equivalent defects are usually already of the order of magnitude of the thickness before loading starts, and they become still larger before the maximum resistance is reached. These questions will be discussed more fully and illustrated with several examples in the next chapter.]

During the buckling phase, for shells of the type considered, the forces F_α, F_β or $F_{\alpha\beta}$ remain constant, that is they are not functions of the buckling displacements, and hence the terms such as $BF_\alpha Ah_\alpha$ are linear in these displacements (there will of course also be additional F_α, F_β, $F_{\alpha\beta}$ forces produced by and growing with these displacements, but these can be ignored since it is sufficient for determining the stability limits to take the buckling displacements as infinitesimal).

During the prebuckling phase, for shells of the type considered, there will be negligible flexure and hence negligible transverse shear forces $F'_{\alpha z}$ and $F'_{\beta z}$ and negligible flexural strains h_α, etc. Hence the third equation of equilibrium can be reduced to good approximation to the same form, Eq. (6.28), as in the membrane theory discussed below:

$$ABp + BF_\alpha \mathbf{a} + AF_\beta \mathbf{b} = 0 \tag{6.27c}$$

which gives a relation between any external pressure p and the resulting F_α or F_β forces.

If now we subtract this Eq. (6.27c) from the third Eq. (6.24) written for the buckling phase, we obtain the following equation of equilibrium of forces in the z direction for studying stability:

$$\frac{\partial F'_{\alpha z}}{\partial \alpha} + \frac{\partial F'_{\beta z}}{\partial \beta} + AB[\mathbf{F}_\alpha h_\alpha + \mathbf{F}_\beta k_\beta + \mathbf{F}_{\alpha\beta}(k_\alpha + h_\beta)] = 0 \tag{6.27d}$$

where the symbols $\mathbf{F}_\alpha$, $\mathbf{F}_\beta$, $\mathbf{F}_{\alpha\beta}$ will be used to designate membrane forces which are present when buckling starts and are assumed to remain constant during buckling, the terms in which they appear being consequently linear in the displacements. This equation can be used with Eqs. (6.25) and (6.26b) as discussed above.

B. Ranges of the Relation Between the Half Wavelength of the Deflection and the Radius

In this classification we can distinguish three or four different ranges of practical significance. As intimated in the discussion at the beginning of this section, the ratio between the important half wavelength of the deformation (say the distance between nodes) in each direction and the radius of curvature of the shell in that direction is significant because of its effect upon the relative importance of the two types of straining in the shell, membrane and flexural. Hence the complete span of applications is from cases where the smallest of these ratios is very large and hence flexure is so unimportant that it can be ignored ('membrane theory') to the opposite extreme where the largest ratio is very small and at least linear membrane straining can be ignored and we can apply flat plate theory to the shell problem.

Figure 6.14 shows diagrammatically this complete span as it applies to the particular case of circular cylindrical shells having a deformation involving n whole sine or cosine waves in the circumference; a further discussion of this question will be given in Sec. 7.1 and Fig. 7.2. There is no important bending in the circumferential direction for the cases of $n = 0$ or 1, and if the same is true in the longitudinal direction these cases can be studied by membrane theory. If the longitudinal half wavelength is large compared to the diameter then the first of these cases can also be studied by elementary 'boiler theory', and the second by elementary bending theory by treating the whole cylinder as a beam (since circumferential deformation in one wave produces a lateral translation of the cross section without distorting it), with corrections for transverse shear effects, as discussed in Sec. 3.5, unless the longitudinal half wavelength is very long.

If the longitudinal half wavelength is only large compared to the thickness, the case of $n = 0$ can be studied by applying elementary beam theory to longitudinal strips of the wall, which are under the condition of an 'elastic

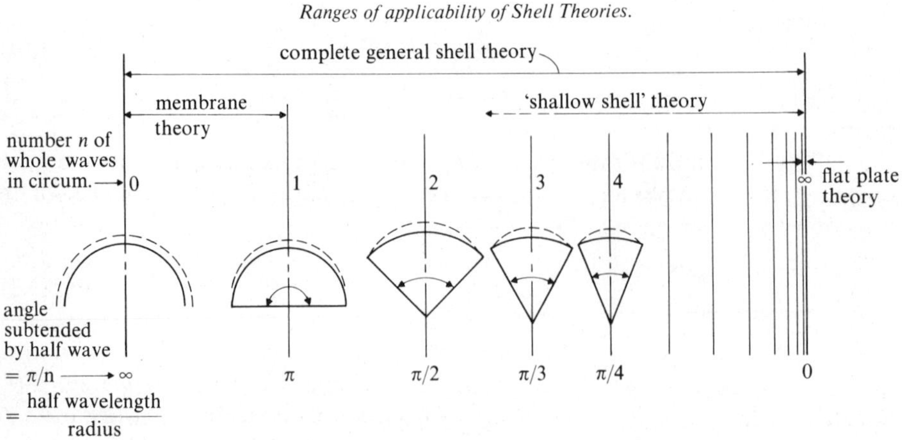

FIGURE 6.14

foundation' due to the radial components of the circumferential stresses, which resist the deflection in the same manner as for the strut discussed in Sec. 2.5, Eq. (2.3.1). These cases are further discussed in Sec. 7.1.

At the other extreme, when n is very large so that the half wavelength of the deformation is small compared to the radius (but large compared to the thickness if classical theory is to apply), the flat plate theory developed in Chap. 4 should givc a fair approximation. However this is not likely to be very useful, since the range of its applicability would be narrow and hard to predict.

The complete general shell theory is of course applicable over the whole span, but a modification of it which has come to be known as 'shallow shell theory' has been found to be very useful, giving good approximations in the upper range of values of n, say for values down to three or four (and a not very bad approximation for values as low as two); this is the same as half wavelength–radius ratios from zero to unity or a little greater.

The name 'shallow shell theory' is rather misleading, as it seems to imply that such a theory is applicable only to a shallow shell such as a spherical segment or 'cap' whose height is rather small compared to the diameter of its base; actually, as we have seen, it can apply to shells which are not 'shallow' at all, provided that they are deformed in a number of waves. But the name is well established, and is satisfactory if one interprets it as we will here, as applying only to the part of a shcll bctween nodes of the deformation or loading.

Membrane Theory

If we ignore flexural strains, taking h_α, h_β, k_α, $k_\beta = 0$ and hence M_α, M_β, $M_{\alpha\beta}$, $F'_{\alpha z}$, $F'_{\beta z} = 0$, we leave unchanged Eqs. (6.8a) and the expressions for the F's in Eqs. (6.23). We have only the first three equilibrium equations to satisfy, and these become:

$$
\begin{aligned}
&ABf_\alpha + B\frac{\partial F_\alpha}{\partial \alpha} + A\frac{\partial F_{\alpha\beta}}{\partial \beta} + \frac{\partial B}{\partial \alpha}(F_\alpha - F_\beta) + 2\frac{\partial A}{\partial \beta}F_{\alpha\beta} = 0 \\
&ABf_\beta + A\frac{\partial F_\beta}{\partial \beta} + B\frac{\partial F_{\alpha\beta}}{\partial \alpha} + \frac{\partial A}{\partial \beta}(F_\beta - F_\alpha) + 2\frac{\partial B}{\partial \alpha}F_{\alpha\beta} = 0 \qquad (6.28) \\
&ABp + BF_\alpha \mathbf{a} + AF_\beta \mathbf{b} = 0
\end{aligned}
$$

The last equation is the same as Eq. (6.27c), but F_α and F_β must be either tensions or, if they are compressions, they must be below the stability limits, because buckling always involves flexure (the 'easiest' buckling shape can be regarded as the one whose potential energy involves the best compromise between membrane and flexural energy).

The above discussion is for very large deflections. For moderately large or small deflections the membrane theory would be the same except for the substitution of Eqs. (6.26a) or (6.27a) respectively for Eq. (6.8a), as the expressions for the membrane strains. The type of problem to which membrane theory is applicable was considered on p. 38, in the discussion of Eq. (2.4).

Shallow Shell Theory

For the large range of shell problems to which this type of shell theory has been found to apply, we retain all the forces and moments but ignore terms in the expressions for them and in the equilibrium equations which would be small compared to the other terms when the largest half wavelength–radius of curvature ratio is less than unity. While the possible neglections on this basis are not very many, they result in very important simplification, and many practical problems fall into this class.

To compare terms in an expression on this basis we can take $w = hW \cos(A\pi\alpha/l_\alpha)\cos(B\pi\beta/l_\beta)$, $v = hV \cos(\,)\sin(\,)$, $u = hU \sin(\,)\cos(\,)$. We assume that $l_\alpha \mathbf{a}/A \leq 1$, $l_\beta \mathbf{b}/B \leq 1$, and that W is much larger than V or U and can be large compared to unity for very large deflections, of the order of unity for moderately large deflections, and small compared to unity for small deflections. At the same time l_α/h, $l_\beta/h \geq 8$ for classical thin shell theory to apply.

Using these criteria we find for example that the second term in the function k, Eqs. (6.19), is $(1/\pi)(l_\beta \mathbf{b}/B)V/W$ times the first, and so can be ignored without great error, and similarly for the third term in k_α, etc. We thus find that for shallow shells we can take:

$$\begin{aligned}
f &= \left(\frac{\partial u}{\partial \alpha} - \mathbf{a}w - \mathbf{c}v\right)/A, & g &= \left(\frac{\partial v}{\partial \alpha} + \mathbf{c}u\right)/A, & h &= \frac{\partial w}{\partial \alpha}/A \\
i &= \left(\frac{\partial u}{\partial \beta} + \mathbf{d}v\right)/B, & j &= \left(\frac{\partial v}{\partial \beta} - \mathbf{b}w - \mathbf{d}u\right)/B, & k &= \frac{\partial w}{\partial \beta}/B \\
h_\alpha &= \left(\frac{\partial h}{\partial \alpha} - \mathbf{c}k\right)/A, & h_\beta &= \left(\frac{\partial h}{\partial \beta} + \mathbf{d}k\right)/B & & \\
k_\alpha &= \left(\frac{\partial k}{\partial \alpha} + \mathbf{c}h\right)/A, & k_\beta &= \left(\frac{\partial k}{\partial \beta} - \mathbf{d}h\right)/B & &
\end{aligned} \tag{6.29a}$$

For the three ranges of deflections, respectively:

$$\begin{aligned}
\epsilon_{\alpha m} &= f + (f^2 + g^2 + h^2)/2, & & f + h^2/2, & & f \\
\epsilon_{\beta m} &= j + (i^2 + j^2 + k^2)/2, & & j + k^2/2, & & j \\
\epsilon_{\alpha\beta m} &= g + i + fi + gj + hk, & & g + i + hk, & & g + i
\end{aligned} \tag{6.29b}$$

The expressions for forces and moments (6.23) and the fourth and fifth equilibrium equations (6.25) are unchanged. The only change in the other equilibrium equations is that the terms $F'_{\alpha z}\mathbf{a}$ in the first and $F'_{\beta z}\mathbf{b}$ in the second equation are small compared to the second and third terms in the same equations. The first equation is then, for the very large deflection case:

$$ABf_\alpha + B\frac{\partial F_\alpha}{\partial \alpha} + A\frac{\partial F_{\alpha\beta}}{\partial \beta} + \frac{\partial B}{\partial \alpha}(F_\alpha - F_\beta) + 2\frac{\partial A}{\partial \beta}F_{\alpha\beta} - AF'_{\alpha z}h_\alpha - BF'_{\beta z}h_\beta = 0 \tag{6.29c}$$

while for moderate and small deflections it is:

$$ABf_\alpha + B\frac{\partial F_\alpha}{\partial \alpha} + A\frac{\partial F_{\alpha\beta}}{\partial \beta} + \frac{\partial B}{\partial \alpha}(F_\alpha - F_\beta) + 2\frac{\partial A}{\partial \beta}F_{\alpha\beta} = 0 \tag{6.29d}$$

The second equation is the same as these except with $\alpha\beta$, AB, hk interchanged. The third equation of equilibrium is then the same as in (6.24) for very large and moderately large deflections, and the same as (6.27b) (for lateral loading) or (6.27d) (for calculating stability limits) for small deflections.

Initial Deviations of Shallow Shells

As was pointed out in the discussion of small deflection theory, the effect of initial defects may be very important in the buckling of shells or other structures, and these effects must be studied by large deflection theory since the initial deviations are usually already of the order of magnitude of the thickness. Since experiments show that thin shells buckle in many waves around any circumference, shallow shell theory can be used.

It is also well known that, although initial deviations from perfect theoretical form are in general of random shape, they can be separated into components and only the components which are of the same or nearly the same shape as that of the buckling deflection are of importance, since they grow rapidly into the large buckling deflection while the other components remain nearly their original size, as was demonstrated for struts at the beginning of Sec. 2.5.

It is a reasonable and simplifying approximation, therefore, to ignore the other components and consider only the component of the initial deviation w_0 which has the same shape as the buckling deflection w. It is convenient to assume that:

$$w_0 = \frac{K-1}{2}w, \qquad \text{or} \qquad K = 1 + 2\frac{w_0}{w} \tag{6.29e}$$

where K is constant with respect to the coordinates α, β for the particular shell studied. For a perfect specimen its value would be $K = 1$. For actual imperfect specimens its value will be different for each specimen, but should be characteristic of the proportions, material and method of manufacture of each type of shell. Using Eqs. (6.29e) in (6.29b) for the case of moderately large deflections of a shallow shell, we find the strain $\epsilon_{\alpha m} = f + [\partial(w_0 + w)/\partial\alpha]^2/2A^2 - [\partial w_0/\partial\alpha]^2/2A^2 = f + K(\partial w/\partial\alpha)^2/2A^2$, and similarly for $\epsilon_{\beta m}$ and $\epsilon_{\alpha\beta m}$:

$$\epsilon_{\alpha m} = f + \frac{K}{2A^2}\left(\frac{\partial w}{\partial \alpha}\right)^2, \quad \epsilon_{\beta m} = j + \frac{K}{B^2}\left(\frac{\partial w}{\partial \beta}\right)^2, \quad \epsilon_{\alpha\beta m} = g + i + \frac{K}{AB}\frac{\partial w}{\partial \alpha}\cdot\frac{\partial w}{\partial \beta} \tag{6.29f}$$

General Thin Shell Theory for Axisymmetric Case

For axisymmetric loads and structures we will take β as the circumferential direction about the axis of symmetry, in which case, v, **c**, f_β, $m_\beta = 0$ and the other quantities in our equations do not vary with β. Hence, in Eqs. (6.18) g, i, k, h_β, k_α,

$\epsilon_{\alpha\beta n} = 0$, and:

$$
\begin{aligned}
&f = (\partial u/\partial\alpha - \mathbf{a}w)/A, \qquad h = (\partial w/\partial\alpha + \mathbf{a}u)/A, \qquad j = (-\mathbf{b}w - d\mathbf{u})/B \\
&h_\alpha = (\partial h/\partial\alpha - \mathbf{a}f)/A, \qquad k_\beta = (-\mathbf{d}h - \mathbf{bj})/B \\
&\epsilon_\alpha = f + (f^2 + h^2)/2 - h_\alpha z, \qquad \epsilon_\beta = j + j^2/2 - k_\beta z \quad \text{(very large deflections)} \\
&\epsilon_\alpha = f + h^2/2 - h_\alpha z, \qquad \epsilon = j - k_\beta z \quad \text{(large deflections)} \\
&\epsilon_\alpha = f - h_\alpha z, \qquad \epsilon_\beta = j - k_\beta z \quad \text{(small deflections)}
\end{aligned}
\tag{6.30a}
$$

In Eqs. (6.23) $F_{\alpha\beta}$, $M_{\alpha\beta} = 0$, while Eqs. (6.25) become:

$$
F'_{\alpha z} = Bm_\alpha + \frac{B}{A}\frac{\partial M_\alpha}{\partial\alpha} + \frac{1}{A}\frac{\partial B}{\partial\alpha}(M_\alpha - M_\beta), \qquad F'_{\beta z} = 0 \tag{6.30b}
$$

And finally, Eqs. (6.24) reduce to the two equations:

$$
\begin{aligned}
&ABf_\alpha + B\frac{\partial F_\alpha}{\partial\alpha} + \frac{\partial B}{\partial\alpha}(F_\alpha - F_\beta) + F'_{\alpha z}(\mathbf{a} + Ah_\alpha) = 0 \\
&ABp + \frac{\partial F'_{\alpha z}}{\partial\alpha} + BF_\alpha(\mathbf{a} + Ah_\alpha) + AF_\beta(\mathbf{b} + Bk_\beta) = 0
\end{aligned}
\tag{6.30c}
$$

for very large deflections. For large deflections these become:

$$
\begin{aligned}
&ABf_\alpha + B\frac{\partial F_\alpha}{\partial\alpha} + \frac{\partial B}{\partial\alpha}(F_\alpha - F_\beta) + F'_{\alpha z}\mathbf{a} = 0 \\
&ABp + \frac{\partial F'_{\alpha z}}{\partial\alpha} + BF_\alpha(\mathbf{a} + Ah_\alpha) + AF_\beta(\mathbf{b} + Bk_\beta) = 0
\end{aligned}
\tag{6.30d}
$$

For small deflections these two equilibrium equations are

$$
\begin{aligned}
&ABf_\alpha + B\frac{\partial F_\alpha}{\partial\alpha} + \frac{\partial B}{\partial\alpha}(F_\alpha - F_\beta) + F'_{\alpha z}\mathbf{a} = 0 \\
&ABp + \frac{\partial F'_{\alpha z}}{\partial\alpha} + BF_\alpha\mathbf{a} + AF_\beta\mathbf{b} = 0 \quad \text{(for lateral loading)} \\
&\frac{\partial F'_{\alpha z}}{\partial\alpha} + AB(\mathbf{F}_\alpha h_\alpha + \mathbf{F}_\beta k_\beta) = 0 \quad \text{(for stability studies)}
\end{aligned}
\tag{6.30e}
$$

Shallow Shell Theory for Axisymmetric Case

For this case Eqs. (6.30a) apply except that:

$$
h = (\partial w/\partial\alpha)/A, \qquad h_\alpha = (\partial h/\partial\alpha)/A, \qquad k_\beta = -\mathbf{d}h/B \tag{6.30f}
$$

while the equations of equilibrium are the same except that the $F'_{\alpha z}\mathbf{a}$ can be dropped from the first of the two equations (6.30c,d,e).

6.7 SOME PARTICULAR THIN SHELL SOLUTIONS

In Chap. 7 a number of solutions of specific problems will be given. In the present section, we will confine the discussion to certain interesting and important types of solutions.

General Cylindrical Shells

The principal difficulty in trying to obtain more specific solutions in general form for the cases discussed in the last section, as was intimated there, is the fact that the geometrical parameters A, B, **a**, **b**, **c**, **d** must be considered to be functions of the coordinates α, β; this makes carrying out solutions in a general form impossible or impractically complex. We will therefore consider here shells having constant scale factors A, B, for which these difficulties are greatly reduced.

This class of shells is of course relatively restricted, but includes some of the most important types of shells involved in practical problems. From Eqs. (6.10), if A, B are constants, then $\mathbf{c}, \mathbf{d} = 0$, and the middle surface can be developed into a flat surface with the coordinate lines in it forming a rectangular grid (the class of course includes flat plates). In order for the coordinate lines to remain lines of curvature, this developed middle surface can only be rolled into a curved shell in one way, so that one set of coordinates, say the α lines, remain straight, forming the generators of a general cylindrical surface, the other β coordinate lines forming right sections of this surface.

It can be seen that the curvature of the surface in the α direction measured by **a** will be zero, while $\mathbf{b} = \mathbf{b}(\beta)$ is a function of β only. This function determines the type of cylindrical shell; for instance if **b** is a constant we have a right circular cylindrical shell, which for simplicity we will usually call simply a 'cylindrical shell', as distinguished from a 'general cylindrical' shell when $\mathbf{b} = \mathbf{b}(\beta)$.

Furthermore, if the coordinate $\alpha = x$ is the actual distance along the generators, and $\beta = y$ the actual distance measured along the middle surface at right angles to x (that is along the circumference of a right section of the middle surface of the cylindrical shell) then the scale factors A, B will have values of unity, $A, B = 1$, and **b** will then represent the actual curvature; for example for **b** constant, $\mathbf{b} = 1/R$, where R is the radius of a circular shell. Also $F'_{xz} = F_{xz}$ and $F'_{yz} = F_{yz}$. Summing up, for the general cylindrical type of shell considered:

$$\alpha = x, \qquad \beta = y, \qquad A, B = 1, \qquad \mathbf{a}, \mathbf{c}, \mathbf{d} = 0, \qquad \mathbf{b} = \mathbf{b}(y) \qquad (6.31a)$$

As an illustration of the wide range of cross sectional shapes of potential practical importance which are included in this type of shells, consider those which can be described by a simple function $\mathbf{b} = b_1 + b_2 \cos(ny/P)$; for a closed tube P would be the perimeter of the middle surface cross section with a value of a little over $2\pi/b_1$. If $b_1 = 0$ we obtain the corrugated sheet with cross section shown at Fig. 6.15(a), while if b_1 is small compared to b_2 we have a curved corrugated sheet as shown at (b). If $n = 0$ or $b_2 = 0$, we obtain the right circular

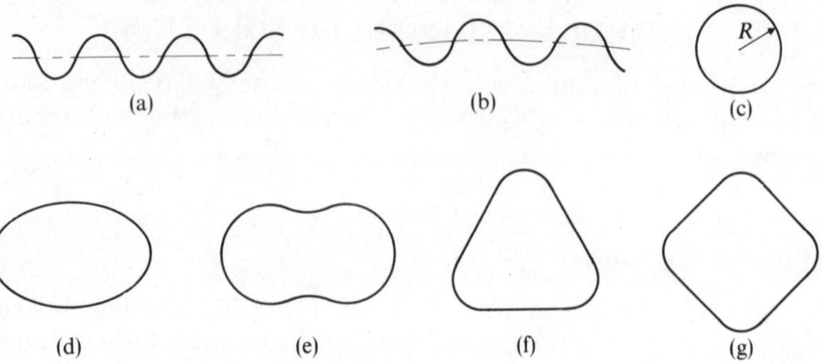

FIGURE 6.15

cylinder of radius $R = 1/(b_1 + b_2)$ or $R = 1/b_1$ respectively. If $n = 2$ we can obtain the ellipse-like cross section shown at (d), or, if b_2 is a little larger compared to b_1, the dumbell shaped section at (e). If $n = 3$ or 4 we can obtain three- or four-lobed sections approximating a triangle or square with rounded corners, etc. With more complex functions we can of course obtain any shape of cross section desired, consistent with the small ratio of thickness to minimum radius of curvature required in using thin shell theory.

Large Deflections of General Cylindrical Shells Using Shallow Shell Theory

In order to obtain a solution of the type of Eqs. (4.13) and (4.18) for flat plates, in which the membrane stresses are represented by a stress function, we must confine ourselves to problems which can be studied by shallow shell theory, that is in which at each point the wave length of the deformation is not much greater than the radius of curvature ($1/\mathbf{b}$) at that point. We can consider moderately large deflections, of the order of magnitude of the thickness. We can also consider deviations from perfect form, w_0, having the same shape as the deflection w, so that w_0/w and $K = 1 + 2w_0/w$ are constant with respect to x and y.

Using Eqs. (6.29f) and (6.31a) in (6.29a) and (6.29b), for the large deflection range, we obtain:

$$f = \frac{\partial u}{\partial x}, \quad g = \frac{\partial v}{\partial x}, \quad h = \frac{\partial w}{\partial x}, \quad i = \frac{\partial u}{\partial y}, \quad j = \frac{\partial v}{\partial y} - \mathbf{b}w, \quad k = \frac{\partial w}{\partial y}$$
$$h_x = \frac{\partial^2 w}{\partial x^2}, \quad h_y = k_x = \frac{\partial^2 w}{\partial x \partial y}, \quad k_y = \frac{\partial^2 w}{\partial y^2} \tag{6.31b}$$

$$\epsilon_{xm} = \frac{\partial u}{\partial x} + \frac{K}{2}\left(\frac{\partial w}{\partial x}\right)^2, \quad \epsilon_{ym} = \frac{\partial v}{\partial y} - \mathbf{b}w + \frac{K}{2}\left(\frac{\partial w}{\partial y}\right)^2$$
$$\epsilon_{xym} = \frac{\partial v}{\partial x} + \frac{\partial u}{\partial y} + K\frac{\partial w}{\partial x}\cdot\frac{\partial w}{\partial y} \tag{6.31c}$$

From Eq. (6.29d) the equations for the equilibrium of forces in the α, β directions are

$$f_x + \frac{\partial F_x}{\partial x} + \frac{\partial F_{xy}}{\partial y} = 0, \qquad f_y + \frac{\partial F_y}{\partial y} + \frac{\partial F_{xy}}{\partial x} = 0 \tag{6.31d}$$

For the common case where f_x, f_y are body forces which have a potential we can take:

$$f_x = h\frac{\partial \rho}{\partial x}, \qquad f_y = h\frac{\partial \rho}{\partial y} \tag{6.31e}$$

and Eqs. (6.31d) are then identically satisfied by using a stress function $\phi(x, y)$ defined by:

$$F_x = h\left(\frac{\partial^2 \phi}{\partial y^2} - \rho\right), \qquad F_y = h\left(\frac{\partial^2 \phi}{\partial x^2} - \rho\right), \qquad F_{xy} = -h\frac{\partial^2 \phi}{\partial x \partial y} \tag{6.31f}$$

where h is the thickness. For $\rho = 0$, ϕ is the Airy stress function for the membrane stresses, defined by Eqs. (3.16a).

Using expressions (6.31b,c) in (6.23), (6.25) we find:

$$F_{xz} = m_x - D\frac{\partial}{\partial x}\nabla^2 w, \qquad F_{yz} = m_y - D\frac{\partial}{\partial y}\nabla^2 w \tag{6.31g}$$

while from Eqs. (6.24), using (6.31f,g), the important third equation of equilibrium of forces in the transverse direction can be written in the form:

$$D\nabla^4 w = p + \frac{\partial m_x}{\partial x} + \frac{\partial m_y}{\partial y} + h\left[\left(\frac{\partial^2 \phi}{\partial y^2} - \rho\right)\frac{\partial^2 w}{\partial x^2} + \left(\frac{\partial^2 \phi}{\partial x^2} - \rho\right)\left(\mathbf{b} + \frac{\partial^2 w}{\partial y^2}\right)\right.$$
$$\left. - 2\frac{\partial^2 \phi}{\partial x \partial y} \cdot \frac{\partial^2 w}{\partial x \partial y}\right] \tag{6.31h}$$

A second equation involving only the functions w and ϕ can be obtained from the continuity condition in the α, β directions, which requires that:

$$\left[\epsilon_{xm} = \frac{F_x - \nu F_y}{Eh} = \right]\frac{1}{E}\left[\frac{\partial^2 \phi}{\partial y^2} - \nu\frac{\partial^2 \phi}{\partial x^2} - (1-\nu)\rho\right] = \frac{\partial u}{\partial x} + \frac{K}{2}\left(\frac{\partial w}{\partial x}\right)^2$$
$$\left[\epsilon_{ym} = \frac{F_y - \nu F_x}{Eh} = \right]\frac{1}{E}\left[\frac{\partial^2 \phi}{\partial x^2} - \nu\frac{\partial^2 \phi}{\partial y^2} - (1-\nu)\rho\right] = \frac{\partial v}{\partial y} - \mathbf{b}w + \frac{K}{2}\left(\frac{\partial w}{\partial y}\right)^2 \tag{6.31i}$$
$$\left[\epsilon_{xym} = \frac{2(1+\nu)F_{xy}}{Eh} = \right] - \frac{2(1+\nu)}{E}\frac{\partial^2 \phi}{\partial x \partial y} = \frac{\partial v}{\partial x} + \frac{\partial u}{\partial y} + K\frac{\partial w}{\partial x} \cdot \frac{\partial w}{\partial y}$$

We can eliminate u and v from these equations by applying the operator $\partial^2/\partial y^2$ to the first, $\partial^2/\partial x^2$ to the second (remembering that $\mathbf{b}$ is a function of y only) and $-\partial^2/\partial x \partial y$ to the third, and adding. The resulting equation can be written in the form:

$$\nabla^4 \phi = EK\left[\left(\frac{\partial^2 w}{\partial x \partial y}\right)^2 - \frac{\partial^2 w}{\partial x^2} \cdot \frac{\partial^2 w}{\partial y^2}\right] - E\mathbf{b}\frac{\partial^2 w}{\partial x^2} + (1-\nu)\left(\frac{\partial f_x}{\partial x} + \frac{\partial f_y}{\partial y}\right) \tag{6.31j}$$

Except for the K and the external body forces f_x, f_y and the fact that $\mathbf{b}$ can be a function of y, this is the same as Eq. (6.17), which was published (with K) by the author in 1934, and which is an extension of von Kármán's equation for perfect flat plates published in 1910.* The two nonlinear terms on the right hand side of Eq. (6.31j) determine the nonlinear membrane stresses caused by large deflections, while the $E\mathbf{b}\partial^2 w/\partial x^2$ term determines the linear membrane stresses caused by the initial curvature.

For zero deflection and body forces, Eq. (6.31j) becomes $\nabla^4\phi = 0$, the same as Eq. (3.16c) of plane stress elasticity theory. Any solutions of this equation (such as the power or trigonometric-hyperbolic functions discussed in Sec. 3.3) can evidently be added to solutions of Eq. (6.31j), and used to satisfy two boundary conditions on the membrane forces or u, v displacements of the middle surface on all four sides of a curved panel or equivalent continuity conditions. Similar solutions of $\nabla^4 w = 0$ can be added to solutions of Eq. (6.31h) and used to satisfy the same number of boundary conditions on the transverse shear forces or moments or lateral deflections or slope of the middle surface.

Solutions of nonlinear differential equations such as Eqs. (6.31h, j) are difficult to obtain, and problems are sometimes solved by using conventional energy methods, that is assuming suitable expressions, such as Eqs. (6.12), for u, v, w, with unknown coefficients such as U_{pq}, V_{pq}, W_{pq}. These expressions can be substituted into the formulas for strains in (6.31c), and the strains into Eqs. (4.69) for the strain energy, and then applying the law of virtual work to obtain as many equations as unknown coefficients.

This requires solution of a great many simultaneous equations, and their number can almost be divided by three by using a very convenient combination of equilibrium and energy methods which was outlined by the author in 1934 in the paper in which Eq. (6.17) was introduced.* By this method only the expression for w is assumed, the choice of which can be guided by experiments in which the shape of the lateral displacement w can readily be measured, or even estimated fairly accurately by eye. Substituting this expression (and the known values of f_x and f_y, if any) into Eq. (6.31j) we obtain an expression for $\nabla^4\phi$ from which it is usually not difficult to calculate ϕ, after which Eqs. (4.70), (4.71) for the strain energy and the law of virtual work can be used as above, but with only the coefficients of the w expression to be determined. As was stated in the discussion of Eq. (6.17), the flexural strain energy is exactly the same as for plates because the expressions we are using for flexural strains, in Eqs. (6.31b), are the same as for plates, and the effect of curvature on the membrane strain energy, Eq. (4.71) is given by the $E\mathbf{b}\,\partial^2 w/\partial x^2$ term in expression (6.31j) for ϕ (and the effect of loads f_x, f_y if they are present is given by the $\partial f_x/\partial x$, $\partial f_y/\partial y$ terms).

Solutions for Circular Cylindrical Shells

As was discussed at the end of Sec. 6.5 the most convenient form for a shell theory is probably the 'uncoupled' form, in which we obtain one equation containing only the lateral deflection w and the loading, with four other equations

* Loc. cit. Eq. (6.17).

giving u, v, $F_{\alpha z}$, $F_{\beta z}$ respectively in terms of w. The single equation in w, obtained from the most important equilibrium condition of forces in the weak lateral direction, is sufficient to get a reasonably accurate solution in many practical problems, although the other equations, obtained respectively from equilibrium conditions of forces in the α, β directions and moments about these directions, are needed for satisfying boundary conditions unless these are of the type, such as fixed or hinged, which involve chiefly w and its derivatives.

Unfortunately obtaining such a solution in general terms is impossible or impractical except in the case of circular cylindrical shells because of the complication caused by a variable curvature. However this is probably the most important type of shell in practical problems. Its characteristics are given by line 3, Table 6.2. Using Eqs. (6.2):

$$\alpha = x, \quad \beta = y, \quad A, B = 1, \quad \mathbf{a}, \mathbf{c}, \mathbf{d} = 0, \quad \mathbf{b} = 1/R$$

$$f = \frac{\partial u}{\partial x}, \quad g = \frac{\partial v}{\partial x}, \quad h = \frac{\partial w}{\partial x}, \quad i = \frac{\partial u}{\partial y}, \quad j = \frac{\partial v}{\partial y} - \frac{w}{R}, \quad k = \frac{\partial w}{\partial y} + \frac{v}{R} \tag{6.32}$$

We will now develop some useful solutions for such shells, many of which will be applied in Chap. 7.

Large Deflections of Shallow Circular Cylindrical Shells

This case is of course included in the solution just found in terms of the deflection w and a stress function ϕ of the membrane stresses, for large deflections of general cylindrical shells using shallow shell theory, and that solution is applied to cylindrical shells simply by taking $\mathbf{b} = 1/R$, where R is the constant radius of a circular cylindrical shell.

We can also obtain at least a partial solution in the 'uncoupled' form for this case, which may be useful in some cases. To do this we can use Eqs. (6.31b, c, d) as they are, remembering that $\mathbf{b}$ is now a constant. Instead of satisfying Eqs. (6.31d) by means of a stress function we write them out in terms of the displacements, using Eqs. (6.23) and (6.31b, c). The results can be written in the form:

$$Zv + Xu + \frac{\partial w}{\partial x}\left(KXw - \frac{v}{R}\right) + \frac{\partial w}{\partial y} KZw + \frac{f_x}{C} = 0$$

$$Zu + Yv + \frac{\partial w}{\partial y}\left(KYw - \frac{1}{R}\right) + \frac{\partial w}{\partial x} KZw + \frac{f_y}{C} = 0 \tag{6.32a}$$

where C is defined by Eq. (6.23a) and X, Y, Z represent the operators:

$$X = \frac{\partial^2}{\partial x^2} + \frac{1-\nu}{2}\frac{\partial^2}{\partial y^2}, \quad Y = \frac{\partial^2}{\partial y^2} + \frac{1-\nu}{2}\frac{\partial^2}{\partial x^2}, \quad Z = \frac{1+\nu}{2}\frac{\partial^2}{\partial x \partial y} \tag{6.32b}$$

We then eliminate v by applying the operator Y to the first and Z to the second and subtracting the resulting equations. Similarly we eliminate u by applying Z to the first and X to the second and subtracting. The results can be written in the following form:

$$\nabla^4 u = \frac{1}{R}\left[\nu\frac{\partial^3 w}{\partial x^3} - \frac{\partial^3 w}{\partial x\,\partial y^2}\right] + \frac{2}{1-\nu}\frac{Zf_y - Yf_x}{C}$$
$$+\frac{2}{1-\nu}K\left[Z\left(\frac{\partial w}{\partial x}\cdot Zw + \frac{\partial w}{\partial y}\cdot Yw\right) - Y\left(\frac{\partial w}{\partial x}\cdot Xw + \frac{\partial w}{\partial y}\cdot Zw\right)\right] \quad (6.32c)$$

$$\nabla^4 v = \frac{1}{R}\left[(2+\nu)\frac{\partial^3 w}{\partial x^2\,\partial y} + \frac{\partial^3 w}{\partial y^3}\right] + \frac{2}{1-\nu}\frac{Zf_x - Xf_y}{C}$$
$$+\frac{2}{1-\nu}K\left[Z\left(\frac{\partial w}{\partial x}\cdot Xw + \frac{\partial w}{\partial y}\cdot Zw\right) - X\left(\frac{\partial w}{\partial x}\cdot Zw + \frac{\partial w}{\partial y}\cdot Yw\right)\right]$$

The next step in getting an uncoupled solution would be to use Eqs. (6.32c) to eliminate u and v from the third equation of equilibrium; the latter is the same as in (6.24), but expressed in terms of the displacements by using Eqs. (6.31b, c, g) and (6.23). The only way to do this for this nonlinear large deflection case is by rewriting Eqs. (6.32c) as $u = \nabla^{-4}P$, where P is the right-hand side of the first equation, and similarly for v. The symbol ∇^{-4} is an 'integral operator' signifying that $u(x, y)$ is the function which, if ∇^4 is applied to it, gives P. The integro-differential equation which would result from this procedure would technically involve w only, and could theoretically be solved for w, but its practical usefulness would be questionable.

As a matter of fact, as we have pointed out before, the usefulness of even such conventional pairs of differential equations as Eqs. (6.31h, j), or (4.13), (4.18) for flat plates, is somewhat questionable, as such simultaneous nonlinear differential equations can seldom be solved directly, and the series solutions which have to be used may or may not have any advantage over a straightforward energy method approach. But the combination equilibrium–energy method described above for the use of Eq. (6.31j) has very definite advantages, and Eqs. (6.32c) can be used in the same way, again assuming only an expression for w with unknown coefficients, finding corresponding expressions for u and v from Eqs. (6.32c), and again completing the problem by using energy methods. Using Eq. (6.31j) would certainly be preferable in cases where boundary conditions involving the membrane forces must be satisfied, but Eqs. (6.32c) may be useful when boundary conditions involve u and v.

Small Deflections of Shallow Circular Cylindrical Shells

Any 'large deflection' theory can of course be used also for problems involving only small deflections by discarding nonlinear terms. For example for small deflections and loading p only, Eqs. (6.31h, j) become:

$$D\nabla^4 w = p + h\mathbf{b}\frac{\partial^2\phi}{\partial x^2}, \qquad \nabla^4\phi = -E\mathbf{b}\frac{\partial^2 w}{\partial x^2} \quad (6.32d)$$

To obtain an uncoupled form of solution for this case, we can use Eqs. (6.31b) with $\mathbf{b} = 1/R$, and (6.31d, g). The membrane strains are:

$$\epsilon_{xm} = \frac{\partial u}{\partial x}, \quad \epsilon_{ym} = \frac{\partial v}{\partial y} - \frac{w}{R}, \quad \epsilon_{xym} = \frac{\partial v}{\partial x} + \frac{\partial u}{\partial y} \tag{6.33a}$$

Equations (6.32a, c) become:

$$Zv + Xu - \frac{\nu}{R}\frac{\partial w}{\partial x} + \frac{f_x}{C}, \quad Zu + Yv - \frac{1}{R}\frac{\partial w}{\partial y} + \frac{f_y}{C} \tag{6.33b}$$

$$\begin{aligned} \nabla^4 u &= \frac{1}{R}\left[\nu\frac{\partial^3 w}{\partial x^3} - \frac{\partial^3 w}{\partial x\,\partial y^2}\right] + \frac{1+\nu}{Ec}(Zf_y - Yf_x) \\ \nabla^4 v &= \frac{1}{R}\left[(2+\nu)\frac{\partial^3 w}{\partial x^2\,\partial y} + \frac{\partial^3 w}{\partial y^3}\right] + \frac{1+\nu}{Ec}(Zf_x - Xf_y) \end{aligned} \tag{6.33c}$$

where X, Y, Z are defined by Eqs. (6.32b).

To save space we will combine Eq. (6.27b) designed for ordinary loading with Eq. (6.27d) designed for studying stability problems (it being assumed that p is omitted in stability problems, being cancelled by an $\mathbf{F}_y/R$ term which is not shown). This equation is then:

$$p + \frac{\partial F_{xz}}{\partial x} + \frac{\partial F_{yz}}{\partial y} + \frac{F_y}{R} + \mathbf{F}_x h_x + \mathbf{F}_y k_y + \mathbf{F}_{xy}(k_x + h_y) = 0 \tag{6.33d}$$

which becomes, after using Eqs. (6.31g) for F_{xz}, F_{yz} and (6.23), (6.31b, c) for F_y:

$$\begin{aligned} p + \frac{\partial m_x}{\partial x} + \frac{\partial m_y}{\partial y} - D\nabla^4 w + \frac{C}{R}\left(\frac{\partial v}{\partial y} - \frac{w}{R} + \nu\frac{\partial u}{\partial x}\right) \\ + \mathbf{F}_x\frac{\partial^2 w}{\partial x^2} + \mathbf{F}_y\frac{\partial^2 w}{\partial y^2} + 2\mathbf{F}_{xy}\frac{\partial^2 w}{\partial x \partial y}\Big) = 0 \end{aligned} \tag{6.33e}$$

We can now eliminate u and v from this equation by applying the operator ∇^4 to it, and subtracting the two equations (6.33c) to which $(\nu C/R)\partial/\partial x$ and $(C/R)\partial/\partial y$ respectively have been applied. The resulting equation contains only w and known loads, and is:*

$$\begin{aligned} \frac{Eh^3}{12(1-\nu^2)}\nabla^8 w + \frac{Eh}{R^2}\frac{\partial^4 w}{\partial x^4} &= \nabla^4\left(p + \frac{\partial m_x}{\partial x} + \frac{\partial m_y}{\partial y} + \mathbf{F}_x\frac{\partial^2 w}{\partial x^2} + \mathbf{F}_y\frac{\partial^2 w}{\partial y^2} + 2\mathbf{F}_{xy}\frac{\partial^2 w}{\partial x\partial y}\right) \\ &\quad - \frac{1}{R}\left[(2+\nu)\frac{\partial^3 f_y}{\partial x^2\partial y} + \frac{\partial^3 f_y}{\partial y^3} + \nu\frac{\partial^3 f_x}{\partial x^3} - \frac{\partial^3 f_x}{\partial x\partial y^2}\right] \end{aligned} \tag{6.34}$$

The first term on the left-hand side represents the flexural resistance, and the second term the membrane resistance (due to the curvature) to the loading which is given by the right-hand side of this equation.

As in Eq. (6.33d) from which it is derived, $\mathbf{F}_x$, $\mathbf{F}_y$, $\mathbf{F}_{xy}$ are to be omitted when Eq. (6.34) is applied to problems involving ordinary loading, while in applying Eq. (6.34) to classical stability studies it is assumed that p is to be omitted, being cancelled by an $\mathbf{F}_y/R$ term not shown. Small deflection equations such as (6.34) are

* L. H. Donnell, 'Stability of thin-walled tubes under torsion', *NACA Report* No. 479, 1933, p. 107.

not applicable to problems involving both ordinary loading and buckling forces; as discussed previously, large deflection theory should be used for such problems. All these remarks apply also to Eq. (6.36) derived below.

Equations (6.33c), (6.34), first published (except for the effects of external loads m_x, m_y, f_x, f_y) in 1933, have become known as the 'Donnell equations', and represent probably the first published shallow shell theory, as well as the first 'uncoupled' equations for shells. They have proven to be very useful, especially the principal equation of transverse equilibrium (6.34), which for cylinders with hinged or fixed end edges is sufficient for a complete solution if the relatively unimportant conditions on u and v are ignored. Equations (6.33c) as well as (6.31g) are needed to satisfy other types of boundary conditions. The range of applicability of these equations will be considered in more detail in Sec. 7.1, Fig. 7.2.

Small Deflections of General Circular Cylindrical Shells

It is also feasible to obtain an uncoupled solution for small deflections of circular cylindrical shells which are not limited to the shallow shell condition and so apply to the entire range illustrated in Fig. 6.14. This case also presents us with our first opportunity to test the importance of numerous minor terms such as the $\mathbf{b}z/B$ terms in Eqs. (6.22). We have so far neglected these, but they are potentially significant and are a common cause of differences between the results obtained by different authors, and the controversy which this can cause. Such terms are even smaller than the thickness–radius ratio and so for thin shells are small compared to terms with which they are associated. However, we have found that apparently unimportant terms may turn out to be significant in later stages of a development.

This proves to be a possibility here, for the following reason. The principal internal forces in the basic equation of equilibrium of transverse forces are of two types: (1) Changes in transverse shear forces (flexural resistance to deflection); and (2) transverse components of membrane forces due to curvature or twist. In the transverse equilibrium equation some of the terms of type (2) contain a factor of R^2/c^2 which is not present in the other terms. As a consequence terms which are of the order of c^2/R^2 times other terms may, when inserted in the transverse equilibrium equation, be of the same order of magnitude as other important terms.

It is impossible to tell how important these terms may be until their effects can be compared on a common basis, and this is only possible in the final transverse equilibrium condition of uncoupled equations, because there is no way to compare the relative importance of terms involving displacements in the plane of the middle surface with terms involving transverse displacements. Since terms often partially or entirely cancel each other the results might be misleading if only part of the minor terms are considered, so we will consider them all.

Terms containing factors of c^2/R^2 enter the expressions for membrane forces, and terms of similar importance containing the membrane strains enter the expressions for the flexural moments, in three ways: (1) from the $\mathbf{b}z/B$ and $\mathbf{a}z/A$

terms in Eqs. (6.22), which describe the variation in width of an element due to curvature; (2) from the similar terms in the denominators of the strain expressions (6.8b); and (3) from the transverse normal stress σ_z caused by the effect of curvature on the bending stresses, which affects the strains in the middle surface through Poisson's ratio. (The effect of the transverse strain on the moment arms of the moment expression is nonlinear and so has no place in small-deflection theory.)

Transverse Normal Stress σ_z Due to Bending and Curvature

Due to curvature the flexural stresses cause σ_z stresses in the middle of the shell wall in the same manner as the stresses in a necking tensile specimen, Fig. 1.4(a), except that in the latter case the curvatures are in opposite directions above and below the middle while the longitudinal stresses are in the same direction, whereas in the former case the curvatures are in the same direction and the longitudinal stresses in opposite directions. Since σ_z represents a small corrective quantity in thin-shell analysis we can use first approximation expressions for the flexural stresses $\sigma_{\alpha f}$, $\sigma_{\beta f}$ in calculating it. Linearizing Eqs. (6.18) and using the two-dimensional stress–strain relations (3.11b), these expressions are:

$$\sigma_{\alpha f} = -\frac{E}{1-\nu^2}(h_\alpha + \nu k_\beta)z, \qquad \sigma_{\beta f} = -\frac{E}{1-\nu^2}(k_\beta + \nu h_\alpha)z \tag{6.35a}$$

where h_α, k_β are defined in Eqs. (6.19).

The pertinent transverse forces acting on a slice of thickness dz, two edges of which are shown shaded in Fig. 6.11 are:

$$\frac{\partial \sigma_z}{\partial z}\,dz(A\,d\alpha\,B\,d\beta) + (\sigma_{\alpha f}\,dz\,B\,d\beta)\mathbf{a}\,d\alpha + (\sigma_{\beta f}\,dz\,A\,d\alpha)\mathbf{b}\,d\beta \tag{6.35b}$$

The twisting stresses $\sigma_{\alpha\beta f}$ would also enter this expression except that there is no twist along α, β lines. There are of course other forces on this element, due to the membrane stresses $\sigma_{\alpha m}$, $\sigma_{\beta m}$, the transverse shear stresses $\sigma_{\alpha z}$, $\sigma_{\beta z}$, and the external transverse loading p (that is the forces included in the transverse equilibrium equation); these transverse forces all act in the same direction over the whole element of thickness $2c$, and whatever they contribute to σ_z cannot be considered in thin shell theory because this depends upon how p is applied. The effect of the bending stresses on σ_z is obtained by equating expression (6.35b) to zero; using expressions (6.35a) and dividing through by $A\,d\alpha\,B\,d\beta\,dz$, this gives:

$$\frac{\partial \sigma_z}{\partial z} = \frac{E}{1-\nu^2}\left[(h_\alpha + \nu k_\beta)\frac{\mathbf{a}}{A} + (k_\beta + \nu h_\alpha)\frac{\mathbf{b}}{B}\right]z \tag{6.35c}$$

which is satisfied together with the boundary conditions $z = \pm c$: $\sigma_z = 0$ if:*

$$\sigma_z = \frac{E}{1-\nu^2}\left[(h_\alpha + \nu k_\beta)\frac{\mathbf{a}}{A} + (k_\beta + \nu h_\alpha)\frac{\mathbf{b}}{B}\right]\frac{z^2 - c^2}{2} \tag{6.35d}$$

* A. E. H. Love, *The Mathematical Theory of Elasticity*, 4th edn, p. 533, Eq. (43).

Solving the first two of the three-dimensional stress–strain relations (3.5) (with x, y replaced by α, β) for σ_α, σ_β the second approximation expressions for σ_α, σ_β, $\sigma_{\alpha\beta}$ are then:

$$\sigma_\alpha = \frac{E}{1-\nu^2}(\epsilon_\alpha + \nu\epsilon_\beta) + \frac{\nu}{1-\nu}\sigma_z, \quad \sigma_\beta = \frac{E}{1-\nu^2}(\epsilon_\beta + \nu\epsilon_\alpha) + \frac{\nu}{1-\nu}\sigma_z,$$

$$\sigma_{\alpha\beta} = \frac{E}{2(1+\nu)}\epsilon_{\alpha\beta} \tag{6.35e}$$

where we will use the exact expressions (6.8b) for ϵ_α, ϵ_β, $\epsilon_{\alpha\beta}$ and (6.35d) for σ_z. This is a consistent second approximation because expression (6.35d) is of the same order of magnitude as the minor terms in Eqs. (6.8b).

Corrected Force and Moment Expressions for Cylinders

We restrict the following to cylindrical shells because the general expressions are too long. For this case, using Eqs. (6.32), Eqs. (6.35d, e), (6.8b) and (6.22) become:

$$\sigma_z = \frac{E}{1-\nu^2}\left(\frac{\partial^2 w}{\partial y^2} + \frac{w}{R^2} + \nu\frac{\partial^2 w}{\partial x^2}\right)\frac{z^2 - c^2}{2R} \tag{6.35f}$$

$$\epsilon_x = \frac{\partial u}{\partial x} - \frac{\partial^2 w}{\partial x^2}z, \quad \epsilon_y = \left[\left(\frac{\partial v}{\partial y} - \frac{w}{R}\right) - \left(\frac{\partial^2 w}{\partial y^2} + \frac{1}{R}\frac{\partial v}{\partial y}\right)z\right]\Big/\left(1 - \frac{z}{R}\right) \tag{6.35g}$$

$$F_x = \int_{-c}^{c}\left(1-\frac{z}{R}\right)\sigma_x dz, \quad F_{xy} = \int_{-c}^{c}\left(1-\frac{z}{R}\right)\sigma_{xy} dz. \quad F_y = \int_{-c}^{c}\sigma_y dz, \quad F_{yx} = \int_{-c}^{c}\sigma_{xy} dz \tag{6.35h}$$

while the moment expressions are the same as the force expressions having the same subscripts except for an added factor z. We next use Eqs. (6.35e) with α,β replaced by x, y and Eqs. (6.35f, g) for σ_z, ϵ_x, ϵ_y in the above expressions for forces and moments. When these expressions involve $1/(1-z/R)$ we use as many terms of the expansion $1 + z/R + z^2/R^2 + \cdots$ as required to include all the terms involving c^2 in the expressions. We thus find:

$$F_x = C\left\{\frac{\partial u}{\partial x} + \nu\frac{\partial v}{\partial y} - \nu\frac{w}{R} + \frac{c^2}{3R}\left[\frac{1-\nu-\nu^2}{1-\nu}\frac{\partial^2 w}{\partial x^2} - \frac{\nu}{1-\nu}\left(\frac{\partial^2 w}{\partial y^2} + \frac{w}{R^2}\right)\right]\right\}$$

$$F_y = C\left\{\nu\frac{\partial u}{\partial x} + \frac{\partial v}{\partial y} - \frac{w}{R} - \frac{c^2}{3R}\left[\frac{\nu^2}{1-\nu}\frac{\partial^2 w}{\partial x^2} + \frac{1}{1-\nu}\left(\frac{\partial^2 w}{\partial y^2} + \frac{w}{R^2}\right)\right]\right\}$$

$$F_{xy} = \frac{1-\nu}{2}C\left[\frac{\partial v}{\partial x} + \frac{\partial u}{\partial y} + \frac{c^2}{3R}\left(\frac{\partial^2 w}{\partial x \partial y} + \frac{1}{R}\frac{\partial v}{\partial x}\right)\right],$$

$$F_{yx} = \frac{1-\nu}{2}C\left[\frac{\partial v}{\partial x} + \frac{\partial u}{\partial y} - \frac{c^2}{3R}\left(\frac{\partial^2 w}{\partial x \partial y} - \frac{1}{R}\frac{\partial u}{\partial y}\right)\right] \tag{6.35i}$$

$$M_x = -D\left(\frac{\partial^2 w}{\partial x^2} + \nu\frac{\partial^2 w}{\partial y^2} + \frac{1}{R}\frac{\partial u}{\partial x} + \frac{\nu}{R}\frac{\partial v}{\partial y}\right), \quad M_y = -D\left(\nu\frac{\partial^2 w}{\partial x^2} + \frac{\partial^2 w}{\partial y^2} + \frac{w}{R^2}\right)$$

$$M_{xy} = -\frac{1-\nu}{2} D\left(2\frac{\partial^2 w}{\partial x \partial y} + \frac{2}{R}\frac{\partial v}{\partial x}\right), \quad M_{yx} = -\frac{1-\nu}{2} D\left(2\frac{\partial^2 w}{\partial x \partial y} + \frac{1}{R}\frac{\partial v}{\partial x} - \frac{1}{R}\frac{\partial u}{\partial y}\right)$$

The equations of equilibrium of moments about the x, y directions (6.25) become for this case:

$$F_{xz} = m_x + \frac{\partial M_x}{\partial x} + \frac{\partial M_{yx}}{\partial y}, \quad F_{yz} = m_y + \frac{\partial M_y}{\partial y} + \frac{\partial M_{xy}}{\partial x} \tag{6.35j}$$

from which, using Eqs. (6.35i):

$$F_{xz} = m_x - D\left[\frac{\partial}{\partial x}\nabla^2 w + \frac{1}{R}\left(\frac{\partial^2 u}{\partial x^2} - \frac{1-\nu}{2}\frac{\partial^2 u}{\partial y^2} + \frac{1+\nu}{2}\frac{\partial^2 v}{\partial x\,\partial y}\right)\right]$$

$$F_{yz} = m_y - D\left[\frac{\partial}{\partial y}\left(\nabla^2 w + \frac{w}{R^2}\right) + \frac{1-\nu}{R}\frac{\partial^2 v}{\partial x^2}\right] \tag{6.35k}$$

The equations of equilibrium of forces in the x, y directions in Eqs. (6.24) are now:

$$f_x + \frac{\partial F_x}{\partial x} + \frac{\partial F_{yx}}{\partial y} = 0, \qquad f_y + \frac{\partial F_y}{\partial y} + \frac{\partial F_{xy}}{\partial x} - \frac{F_{yz}}{R} = 0 \tag{6.35l}$$

Substituting the values of the F's from Eqs. (6.35i, k) into these equations, the result can be put in the form:

$$X'u + Zv = W, \qquad Zu + Y'v = W' \tag{6.35m}$$

where:

$$X' = \frac{\partial^2}{\partial x^2} + \left(1 + \frac{c^2}{3R^2}\right)\frac{1-\nu}{2}\frac{\partial^2}{\partial y^2}, \quad Y' = \left(1 + \frac{c^2}{R^2}\right)\frac{1-\nu}{2}\frac{\partial^2}{\partial x^2} + \frac{\partial^2}{\partial y^2}, \quad Z = \frac{1+\nu}{2}\frac{\partial^2}{\partial x\,\partial y}$$

$$W = \left(\nu + \frac{c^2}{3R^2}\frac{\nu}{1-\nu}\right)\frac{1}{R}\frac{\partial w}{\partial x} + \frac{c^2}{3R}\left(\frac{-1+\nu+\nu^2}{1-\nu}\frac{\partial^3 w}{\partial x^3} + \frac{1+\nu^2}{2(1-\nu)}\frac{\partial^3 w}{\partial x\,\partial y^2}\right) + \frac{f_x}{C} \tag{6.35n}$$

$$W' = \left(1 + \frac{c^2}{3R^2}\frac{\nu}{1-\nu}\right)\frac{1}{R}\frac{\partial w}{\partial y} + \frac{c^2}{3R}\left(\frac{-3+4\nu+\nu^2}{2(1-\nu)}\frac{\partial^3 w}{\partial x^2\,\partial y} + \frac{\nu}{1-\nu}\frac{\partial^3 w}{\partial y^3}\right) + \frac{f_y - m_y/R}{C}$$

As before, we now eliminate v from Eqs. (6.35m) by applying the operator Y' to the first equation and Z to the second and subtracting the resulting equations; similarly we eliminate u by applying Z to the first and X' to the second equation and subtracting, from which:

$$(X'Y' - ZZ)u = (Y'W - ZW'), \quad (X'Y' - ZZ)v = (X'W' - ZW) \tag{6.35o}$$

For the third equation of equilibrium of transverse forces we again combine Eq. (6.27b) for ordinary loading with Eq. (6.27d) for studying stability problems, to obtain Eq. (6.33d), making the same proviso as previously regarding the omission of p in applications to problems involving stability under external pressure. While this is the same equation which we used for the shallow shell case we now use in it the more complete expressions (6.35i, k) for the F's and (6.19) for h_x, k_y, k_x, h_y in the expressions for buckling forces. Equation (6.33d) then becomes:

$$D\left[\nabla^4 w+\frac{1}{R}\left(\frac{\partial^3 u}{\partial x^3}-\frac{1-\nu}{2}\frac{\partial^3 u}{\partial x\,\partial y^2}+\frac{3-\nu}{2}\frac{\partial^3 v}{\partial x^2\,\partial y}\right)\right.$$
$$\left.+\frac{1}{R^2}\left(\frac{\nu^2}{1-\nu}\frac{\partial^2 w}{\partial x^2}+\frac{2-\nu}{1-\nu}\frac{\partial^2 w}{\partial y^2}+\frac{1}{1-\nu}\frac{w}{R^2}\right)+\frac{C}{R}\left(\frac{w}{R}-\nu\frac{\partial u}{\partial x}-\frac{\partial v}{\partial y}\right)\right]$$
$$=p+\frac{\partial m_x}{\partial x}+\frac{\partial m_y}{y}+\mathbf{F}_x h_x+\mathbf{F}_y k_y+\mathbf{F}_{xy}(k_x+h_y) \qquad (6.35\text{p})$$

We next eliminate u and v from this equation by applying the operator $(X'Y'-ZZ)$ to the whole equation and then applying Eqs. (6.35o) to the terms involving u and v. This is the same thing as applying $(X'Y'-ZZ)$ to the terms involving w or the external loading, and then the operators involving u (such as $\partial^3/\partial x^3$) to $(Y'W-ZW')$ and the operators involving v to $(X'W'-ZW)$. This is legitimate only because the order of application is immaterial for the operators involved.

Before stating the final form of the resulting equation let us consider how it differs from the simplified shallow shell equation (6.34). The difference is entirely in certain additional terms in the new equation, the most important of which are added to the $\nabla^8 w$ term on the left-hand side, and which (as for all other solutions for this case) can be expressed in the form:

$$\frac{D}{R^4}\left[R^2\nabla^2\left(c_1\frac{\partial^4 w}{\partial x^4}+c_2\frac{\partial^4 w}{\partial x^2\partial y^2}+c_3\frac{\partial^4 w}{\partial y^4}\right)+c_4\frac{\partial^4 w}{\partial x^4}+c_5\frac{\partial^4 w}{\partial x^2\,\partial y^2}+c_6\frac{\partial^4 w}{\partial y^4}\right] \qquad (6.35\text{q})$$

Line (1) in Table 6.7 gives the values of the coefficients $c_1, c_2, \ldots$ which are obtained by the above development in which all the minor terms are included. Line (2) gives the values of the same coefficients which are found when the development is carried out in exactly the same way except that σ_z is ignored. Line (3) gives the values found if we ignore σ_z and all the other minor terms, as we have done in previous developments in this chapter, the strains and force and moment expressions being taken as given by Eqs. (6.20) and (6.23).

The remaining lines represent the other published solutions for cylinders under general loading which have been put in uncoupled form, including Flügge's well known solution which has been used as a standard of comparison for many years. Wu and Lee's solution, which is discussed in detail later in Sec. 7.5, is a most interesting one. It was not intended as a solution for the thin-walled cylinder, but was obtained as a by-product of the development of a series solution in loading functions for thick-walled cylinders, its terms being determined step by step, with no preconceived idea of the final result, by starting with the simple Eq. (6.34) and satisfying three-dimensional elasticity theory at each step.

Morley's solution was derived as a modification of Eq. (6.34), empirically chosen to make it check closely with Flügge's solution in loading ranges where Eq. (6.34) is well known to be inaccurate. It has the advantage that the terms shown can be written in the compact form $2\nabla^6 w+\nabla^4 w$, or combined with the $\nabla^8 w$ term to give the even more compact form $\nabla^4(\nabla^2+1)^2 w$; this somewhat simplifies its application to cases in which p and w are trigonometric functions of x and y.

Table 6.7 COEFFICIENTS OF TERMS WHICH ARE IN EQ. (6.36) BUT NOT IN EQ. (6.34)

		c_1 $\nabla^2 \dfrac{\partial^4 w}{\partial x^4}$	c_2 $\nabla^2 \dfrac{\partial^4 w}{\partial x^2 \partial y^2}$	c_3 $\nabla^2 \dfrac{\partial^4 w}{\partial y^4}$	c_4 $\dfrac{\partial^4 w}{\partial x^4}$	c_5 $\dfrac{\partial^4 w}{\partial x^2 \partial y^2}$	c_6 $\dfrac{\partial^4 w}{\partial y^4}$
1	Considering all minor terms	$2\nu + \nu^2$	$6 - \nu$	2	$4 + \nu - 3\nu^2$	$4 - \nu$	1
2	Considering all minor terms but σ_z	2ν	$6 - 2\nu$	2	$4 - 3\nu^2$	$4 - 2\nu$	1
3	Neglecting all minor terms	ν	5	2	$1 - \nu^2$	3	1
4	Wu and Lee (see Sec. 7.5)	$-\dfrac{12(1-\nu^2)}{5}$	6	2	$\dfrac{12(1-\nu^2)}{5}$	4	1
5	Flügge*	2ν	$6 - 2\nu$	2	0	$4 - 2\nu$	1
6	Morley,** or Simmonds† with $\lambda, \lambda_* = 1$	2	4	2	1	2	1
7	Simmonds† with $\lambda, \lambda_* = 0$	0	2	2	0	0	1

* *Stresses in Shells*, Springer, 1960, Chapter 5.
** 'An improvement on Donnell's approximation . . . ,' *Quart. J. Mech. Appl. Math.*, 1959, vol. XII, part 1, p. 89.
† 'A set of simple, accurate equations for circ. cyl. elast. shells,' *Int. J. Solids Structures*, 1966, vol. 2, p. 537.

However, the choice of coefficients in Morley's solution (and partially in Simmond's solution) merely to give the desired results in certain particular applications, rather than deriving them from basic principles as was done in Flügge's and our own solutions, makes their accuracy somewhat questionable in applications to unchecked problems, even when these problems are in the same general range of wavelengths as those for which the coefficients were chosen.

Comparison of the different solutions in Table 6.7 reveals that all, in spite of wide differences in their method of development, have the same values $c_3 = 2$ and $c_6 = 1$. This fact indicates that these terms are very important, and it can be shown that these values for c_6 and the $\partial^6 w/\partial y^6$ part of c_3 are necessary for the solutions to reduce to ring theory for the case $\partial w/\partial x = 0$ (and thus to give, for example, the correct uniform buckling pressure for long tubes which buckle to an oval shape, as demonstrated later in Sec. 7.3). In a previous early study,* all the essential elements of the above development (including most of the minor terms) were carried through to the final equation in w, where these two terms were correctly

* L. H. Donnell, 'A discussion of thin shell theory', *Proc. 5th Int. Congr. Appl. Mech.*, 1938, p. 70.

determined to be important and retained; however, all the rest were incorrectly (though somewhat tentatively) discarded as unimportant.

As a matter of fact, probably only one other condition for these coefficients *is* very important, and that is the condition that $c_2 = 2 + c_5$. The importance of this condition is again indicated by the fact that all these diverse solutions satisfy it. It is shown later in Sec. 7.1, case $n = 1$, that this is the condition that the solutions reduce to the elementary but exact solution for the case of pure longitudinal bending of a cylindrical tube as a beam. The absolute value of c_5 (and hence of c_2 to satisfy the above condition) is probably much less important. The fact that its value can vary from 4 down to zero in solutions advanced as having broad application is one indication of this, as is perhaps the fact that apparently no known elementary solution like those discussed above exists, by which its magnitude can be checked.

It can be seen that the importance of terms seems to diminish as we decrease the order of differentiation with respect to y and increase the order with respect to x, and this would indicate that the terms with coefficients c_1 and c_4 are probably unimportant. This unimportance is indicated by the diverse magnitudes given for these coefficients by the different theories, just as having the same magnitudes for c_3, c_6 and $c_2 - c_5$ was an indication of their importance. There is an elementary case, that for axisymmetric deformation and loading ($\partial w/\partial y = 0$) which indicates that taking these coefficients as zero gives an excellent approximation. If this is done, the shell solutions reduce to the elementary case of a beam (represented by a longitudinal element of the shell wall) on an elastic foundation produced by the circumferential curvature, as is demonstrated later in Sec. 7.1, case $n = 0$. Since this elementary solution for the shell is known to be an excellent approximation for thin shells, we will ignore these terms (the negative value for c_1 and equal positive value for c_4 given by Wu and Lee's solution is apparently needed, however, for thick shells).

Comparison of solutions (1), (2), (3) indicates that the minor terms have an unimportant effect upon the final results, as we had assumed earlier. We will therefore take the value of $c_5 = 3$ given in line (3), which was obtained by ignoring the minor terms. We do this both for simplicity and, more important, for consistency with the other shell theories which were discussed in the first part of this chapter—since it now seems obvious that it would not pay to consider the effects of the minor terms for them. An argument might be made for using a value of $c_5 = 4$, as given by Wu and Lee's calculations, as this would be just as simple and might be more accurate; however, the choice is certainly not critical.

We thus obtain, for general small deflections of a thin circular cylindrical shell:

$$\frac{Eh^3}{12(1-\nu^2)}\left[\nabla^8 w + \frac{1}{R^2}\nabla^2\left(5\frac{\partial^4 w}{\partial x^2\,\partial y^2} + 2\frac{\partial^4 w}{\partial y^4}\right) + \frac{1}{R^4}\left(3\frac{\partial^4 w}{\partial x^2\,\partial y^2} + \frac{\partial^4 w}{\partial y^4}\right)\right] + \frac{Eh}{R^2}\frac{\partial^4 w}{\partial x^4}$$
$$= \nabla^4\left[p + \frac{\partial m_x}{\partial x} + \frac{\partial m_y}{\partial y} + \mathbf{F}_x\frac{\partial^2 w}{\partial x^2} + \mathbf{F}_y\left(\frac{\partial^2 w}{\partial y^2} + \frac{w}{R^2}\right) + 2\mathbf{F}_{xy}\frac{\partial^2}{\partial x\,\partial y}\left(w + \frac{1}{R^2}\nabla^{-2}w\right)\right.$$
$$\left. - \frac{1}{R}\left\{\left[(2+\nu)\frac{\partial^3}{\partial x^2\,\partial y} + \frac{\partial^3}{\partial y^3}\right]\left(f_y - \frac{m_y}{R}\right) + \left(\nu\frac{\partial^3}{\partial x^3} - \frac{\partial^3}{\partial x\,\partial y^2}\right)f_x\right\} \qquad (6.36)$$

Equations (6.35o) for u and v, and Eqs. (6.35k) for F_{xz}, F_{yz} can now be expanded and then simplified like Eq. (6.36). When this is done it is found that the values of u, v, F_{xz}, F_{yz} given by Eqs. (6.33c) and (6.31g) for shallow circular cylindrical shells and their use in satisfying boundary conditions, which was discussed under Eq. (6.34), are usually also sufficient for problems involving more general circular cylindrical shells. The relative magnitude of the terms contained in Eq. (6.36) but not in (6.34) compared to terms common to the two equations is further discussed in Sec. 7.1 and plotted in Fig. 7.2.

Limitations of 'Uncoupled' Solutions

Equations (6.36), and the simpler shallow shell Eqs. (6.33c), (6.34) are obviously 'escalated' in order, and the warning made about such equations in the discussion of Eq. (3.8) should be heeded. Except for this mathematical escalation, these equations are physically the same as, and part of, the basic thin shell equations (6.18), (6.23), (6.24), (6.25). Users must not expect to be able to satisfy more boundary conditions than was discussed for these basic equations (namely two flexural plus two membrane conditions on each of four sides of a panel, or their equivalent) unless additional independent solutions are superposed. This will be further discussed in connection with Eq. (7.3e).

Another disadvantage of the escalation required to obtain uncoupled solutions is that the differentiation involved can obviously eliminate some simpler functions and thus invalidate certain types of solutions. For example, the original equations of equilibrium of transverse forces such as Eq. (6.33d) can only be satisfied for a constant circumferential force F_y by a corresponding constant lateral unit force p. But uncoupled cylinder solutions such as those shown in Table 6.7, together with end boundary and continuity conditions, can be satisfied if $w = w_0$, $u = u_l x$, where w_o, u_l are constants, and all the external forces are zero; this gives a constant $F_y = C(\nu u_l + w_o/R)$ together with zero p, an obviously impossible condition.

It might be at least psychologically helpful to apply the integral operator ∇^{-4} to the equation for w and ∇^{-2} to the equations for u and v, reducing these to equations of the type: $D\nabla^4 w + \nabla^{-4}[\cdots] = p + \cdots$ and $\nabla^2 u = \nabla^{-2}[\cdots]$, etc. This would bring the equations back to the order of the original physically-based equations, and might give the user a more realistic idea of the number and kind of functions of integration which are available to him. However, this would be in the direction of complicating a technique ('uncoupling') the principal value of which has been in the simplification which it affords. Fortunately, anomalous 'solutions' of the type cited above, which are made possible by losses due to escalation, are probably relatively simple cases which can easily be ruled out by comparison with much more easily obtained elementary solutions.

Energy Solutions—Buckling of Thin Spherical Shells Under External Pressure

In the foregoing discussions of shells the emphasis has been upon solutions of equilibrium equations, with only a few mentions of the alternative of energy solutions—and even these have been confined to combinations of energy and equilibrium solutions. A number of energy solutions will be given in the next chapter but these will be confined to cylinder problems.

Actually energy solutions of shell problems are very useful when the desired result depends upon overall rather than local conditions, as for instance in buckling or vibration problems, or general magnitudes of deflections under transverse loads. Let us consider the problem of the buckling of a thin spherical shell under uniform external pressure. While the final collapse of such a spherical shell is likely to be asymmetrical and complex in shape, tests indicate that buckling generally starts as a small circular dimple; the remainder of this article will be devoted to studying the requirements for producing such a dimple and its characteristics.

The tests indicate that there is little disturbance of the shell wall even a short distance beyond the buckled region; thus even if the shell is a spherical segment or cap rather than a complete sphere, the boundary conditions at the edge of the cap should have little effect upon the buckling if the diameter of the cap is fairly large compared to the diameter of the dimple. The buckling will of course tend to occur where the wall is weakest, due to slightly smaller thickness or initial deviation from spherical shape, but it will also tend to occur at sufficient distance from the edge to avoid the stiffening effect of the edge constraints, whatever these may be. Hence if we ignore edge constraints as we will, the results should apply not merely to complete spheres but with good approximation to caps with diameters several times that of the dimple, which is approximately indicated by l in Fig. 6.16(a).

To study this phenomenon we can evidently use axisymmetric shell theory by taking the axis of symmetry through the center of the dimple. We can also use shallow shell theory because calculations based on such an assumption yield a dimple diameter small enough compared to the radius to justify this assumption. Because all actual shells have imperfections which may be of the order of magnitude of the thickness we will use large deflection theory.

From Eqs. (6.30a, f), for large deflections of axisymmetric shallow shells, the strains are given by:

$$f = \left(\frac{\partial u}{\partial \alpha} - \mathbf{a}w\right)/A, \qquad h = \frac{\partial w}{\partial \alpha}/A, \qquad j = -(\mathbf{b}w + \mathbf{d}u)/B \tag{6.37a}$$

$$h_\alpha = \frac{\partial}{\partial \alpha}\left(\frac{1}{A}\frac{\partial w}{\partial \alpha}\right)/A, \qquad k_\beta = -\mathbf{d}\frac{\partial w}{\partial \alpha}/AB$$

$$\epsilon_\alpha = f + \frac{1}{2}h^2 - h_\alpha z, \qquad \epsilon_\beta = j - k_\beta z, \qquad \epsilon_{\alpha\beta} = 0 \tag{6.37b}$$

Using line 6 of Table 6.2, this becomes, for the case of the spherical shell:

$$\begin{aligned} \epsilon_\varphi &= \frac{1}{R}\left[\frac{\partial u}{\partial \varphi} - w + \frac{1}{2R}\left(\frac{\partial w}{\partial \varphi}\right)^2 - \frac{\partial^2 w}{\partial \varphi^2}\frac{z}{R}\right], \\ \epsilon_\theta &= \frac{1}{R}\left[\cot\varphi\, u - w - \cot\varphi \frac{\partial w}{\partial \varphi}\frac{z}{R}\right] \end{aligned} \tag{6.37c}$$

In the calculations for this case which were made in 1959,* the results of which will be presented here, the terms $(w/R^2)z$, which we have ignored in shallow shell theory, were included in these expressions. Since these terms are unimportant when shallow shell conditions prevail, this should have negligible effect upon the results which were obtained.

Suitable expressions for the displacements w and u, which can give distributions of these displacements like those pictured by the dotted lines in Fig. 6.16(a), are:

$$w = h\,e^{-a\varphi^2}\sum_m W_m\varphi^m, \qquad u = h\,e^{-a\varphi^2}\sum_m U_m\varphi^{m+1} \quad (m = 0, 2, 4, \ldots) \tag{6.37d}$$

where the coefficient a determines the size of the dimple and the coefficients W_m, U_m its depth and shape, and all must be determined by use of the law of virtual work. Using Eq. (4.69) the strain energy $\mathscr{E}_s$ and the work $\mathscr{E}_e$ done by the uniform external pressure p are respectively:

$$\mathscr{E}_s = \frac{\pi R^2 E}{1-\nu^2}\int_0^\pi d\varphi\,\sin\varphi\int_{-c}^{c} dz\,(\epsilon_\varphi^2 + 2\nu\epsilon_\varphi\epsilon_\theta + \epsilon_\theta^2)$$

$$\mathscr{E}_e = 2\pi R^2 p\int_0^\pi d\varphi\,\sin\varphi\,w \tag{6.37e}$$

For convenience the integration limits are given for the whole surface of a sphere, but it would make little difference if only a part of a sphere is involved since the integrals will be practically zero outside the neighborhood of the buckle. It is also convenient in this case to express the law of virtual work in the form:

$$\frac{\partial\mathscr{E}}{\partial a}, \frac{\partial\mathscr{E}}{\partial W_m}, \frac{\partial\mathscr{E}}{\partial U_m} = 0, \qquad \text{where} \qquad \mathscr{E} = \mathscr{E}_s - \mathscr{E}_e \tag{6.37f}$$

* L. H. Donnell, 'Shell theory', *Proc. of 4th Midwestern Conference on Solid Mechanics*, published by Univ. of Texas, 1959, p. 1.

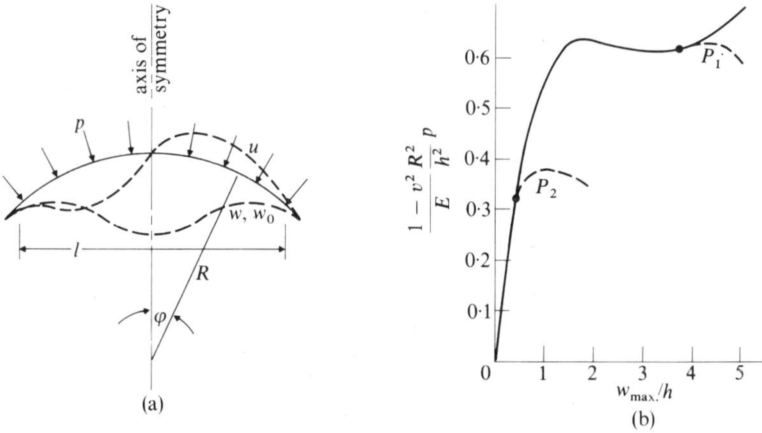

FIGURE 6.16 (from L. H. Donnell, 'Shell theory', *Proc. of 4th Midwestern Conference on Solid Mechanics*, p. 1, 1959, courtesy of University of Texas at Austin)

In the 1959 paper the calculations outlined above were carried out using only the first term in the w series and the first two terms in the u series, giving the results shown by the solid line in Fig. 6.16(b) for the relation between the loading and the deflection. The diameter of the dimple l, Fig. 6.16(a) is found to be:

$$l \approx 7\sqrt{Rh} \tag{6.37g}$$

Since the loading–deflection curve apparently continues to rise indefinitely for elastic conditions, the ultimate pressure is reached at about the point when yielding first occurs, after which the resistance can be expected to fall off rapidly. Both the membrane and flexural stresses increase as we go along the curve, the membrane stresses chiefly with the vertical and the flexural stresses with the horizontal distance. If yielding occurs at the point P_l after a large movement in both directions, the ultimate pressure would be about:

$$p = \frac{0{\cdot}63}{1-\nu^2}\frac{Eh^2}{R^2} \tag{6.37h}$$

In a test reported by Kármán and Tsien* the size of the dimple was about that given by Eq. (6.37g), while the buckling pressure was about half of that given by Eq. (6.37h), indicating perhaps that yielding started at a point such as P_2, Fig. 6.16(b).

These results at least establish a characteristic size of the buckling dimple and show that it is small enough to justify the use of shallow shell theory, since, from Eq. (6.37g):

$$\frac{l}{R} \approx 7\sqrt{\frac{h}{R}} \tag{6.37i}$$

which should be less than unity for any truly thin shell.

* 'The buckling of thin cylindrical shells under axial compression', *J. Aero. Sci.*, 1941, vol. 8, p. 303.

7

CIRCULAR CYLINDRICAL SHELLS

Introduction

It seems desirable and logical to thoroughly explore in this last chapter the applications of shell principles to one particular type of shell, as examples of their application to shells in general, since it is impractical to cover the whole range of shells in a complete way. The choice of the circular cylindrical as the type of shell to treat in this manner is rather obvious, since it is at the same time the simplest type which embodies most of the features of shells in general, more is known about it than about any other type, and it is the most important type for practical applications.

7.1 SMALL DEFLECTION APPLICATIONS

As a simple but important application of Eqs. (6.34) or (6.36), consider the cylinder shown in Fig. 7.1, of length L and mean radius R, with edges hinged at the ends,

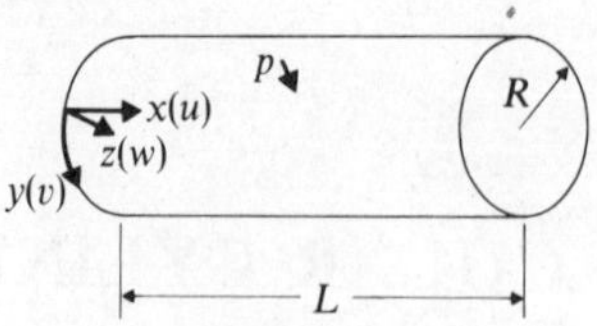

FIGURE 7.1

under a harmonically distributed radial loading p. Equations (6.34) or (6.36), as well as continuity around the circumference, and zero edge radial displacements and resistance M_x to edge rotations, can be satisfied if:

$$w = \sum_m \sum_n W_{mn} \sin \frac{m\pi x}{L} \sin \frac{ny}{R}, \qquad p = \sum_m \sum_n P_{mn} \sin \frac{m\pi x}{L} \sin \frac{ny}{R} \tag{7.1a}$$

where m, n are integers. This results in a relation between W_{mn} and P_{mn} for each set of values of m, n. If we take u and v as similar cosine-sine and sine-cosine functions respectively, we could similarly satisfy also Eqs. (6.33c), which would make v but not u zero at the ends if the physical problem requires it. However, as was shown in the discussion of Fig. 6.1, u should be smaller than v and much smaller than w, so this condition would be relatively unimportant even if physical constraints actually prevent u displacements at the ends.

Comparative Importance of Terms in Eq. (6.36) and Not in (6.34)

The half wavelengths of the deflection (7.1a) in the x and y directions are:

$$l_x = L/m, \qquad l_y = \pi R/n, \qquad \text{or} \qquad L = ml_x, \qquad R = nl_y/\pi \tag{7.1b}$$

Substituting expressions (7.1a) into Eq. (6.36), using (7.1b) to replace L and R, and multiplying by $12(1-\nu^2)l_y^8/(\pi^4 Eh^3 w)$ the left-hand side of this equation can be written in the form, for $\nu = 0{\cdot}3$:

$$\pi^4\left(\frac{l_y^2}{l_x^2}+1\right)^4 + \left[-\pi^2\frac{l_y^2}{R^2}\left(\frac{l_y^2}{l_x^2}+1\right)\left(5\frac{l_y^2}{l_x^2}+2\right)+\frac{l_y^4}{R^4}\left(3\frac{l_y^2}{l_x^2}+1\right)\right] + 10{\cdot}9\frac{l_y^2}{h^2}\frac{l_y^2}{R^2}\frac{l_y^4}{l_x^4} \tag{7.1c}$$

The terms in the brackets are the four terms contained in Eq. (6.36) which are not in Eq. (6.34). Taking $l_y/h = 10$ in the last term as about the minimum value for which the Love–Kirchhoff approximation applies, the ratio of the absolute magnitudes of each of these four terms to the largest of the remaining two terms (that is, the terms common to the two equations) can readily be calculated for all values of l_y/R and l_y/l_x from zero to infinity.

Figure 7.2 shows the results of these calculations. The shaded areas in the plots show the values of l_y/R and l_y/l_x for which the simpler equation (6.34) can certainly be used, since the terms it neglects can never have values greater than 1 percent of the terms retained. But we can get a fairly good approximation by using Eq. (6.34) even if the terms neglected are merely less than 10 percent of those

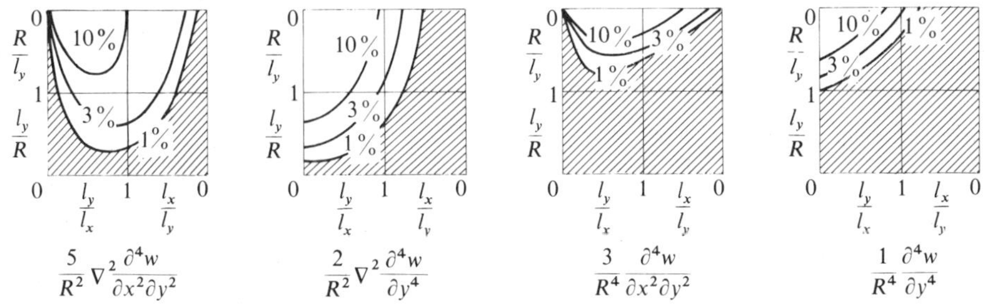

FIGURE 7.2
Ratios of terms neglected in Eq. (6.34) to those retained

retained. For such an approximation the simpler equation might be used for roughly the whole region shown in the plots except for the upper left quadrant. This would mean that the simpler equation (6.34) should give reasonably accurate results if *either* the half wavelength in the circumferential or the axial direction is smaller than the radius, but that it should not be used if *both* these half wavelengths are larger than the radius. These findings do not necessarily have any significance for shallow shell theories in general, since the particular case of the circular cylindrical shell is exceptional in having curvature in one direction only; each case could presumably be studied in a similar manner.

Laterally Loaded Cylindrical Shell

Expressions (7.1a) can evidently be used for any distribution of lateral loading (including concentrated loads) by applying double harmonic analysis in the same manner as plates. The coefficients P_{mn} can be determined by multiplying both sides of the second Eq. (7.1a) by $\sin(m\pi x/L)\sin(ny/R)$ and integrating over the middle surface, giving:

$$\int_0^L dx \int_0^{2\pi R} dy\, p \sin\frac{m\pi x}{L}\sin\frac{ny}{R} = \frac{\pi RL}{2} P_{mn} \tag{7.2a}$$

Any distributions of the other types of loads considered in Eqs. (6.34) or (6.36) can be similarly studied by taking m_x, f_x as comparable cosine-sine functions or m_y, f_y, as sine-cosine functions.

According to the above discussion, in calculating the displacement w and quantities depending upon it, Eq. (6.36) should be used for calculating the components having long wavelengths while the simpler Eq. (6.34) can be used for the shorter wavelength components. Actually this is probably not very important unless the load distribution itself has a long wavelength in both directions. For other loads (a concentrated load for instance) the components having long wavelengths contribute only a small part to the total deflection, so the error made in calculating everything by Eq. (6.34) is not very great.

Free Vibrations of Cylindrical Shell

The above operations should involve no difficulties, so let us consider a slightly different type of problem for a worked-out example, the free vibration of a thin cylindrical shell. As in our similar studies of bars, Eqs. (2.18)–(2.20), and plates, Eqs. (4.30)–(4.32), we take the deflection as a function of time as well as of the coordinates x, y:

$$w = W_{mn} \sin \frac{m\pi x}{L} \sin \frac{ny}{R} \sin 2\pi N_{mn} t, \qquad p = -\rho h \frac{\partial^2 w}{\partial t^2} \tag{7.2b}$$

where ρ is the mass per unit volume of the material, W_{mn} is the amplitude of vibration, and N_{mn} is the frequency of vibration in cycles per unit of time, for the vibration mode of m half waves in the axial and n full waves in the circumferential direction.

Substituting expressions (7.2b) into Eq. (6.36), and then dividing through by the expression (7.2b) for w, we obtain:

$$\frac{Eh^3}{12(1-\nu^2)}\left[\left(\frac{m^2\pi^2}{L^2}+\frac{n^2}{R^2}\right)^4 - \frac{1}{R^2}\left(5\frac{m^2n^2\pi^2}{L^2R^2}+2\frac{n^4}{R^4}\right)\left(\frac{m^2\pi^2}{L^2}+\frac{n^2}{R^2}\right) + \frac{1}{R^4}\left(3\frac{m^2n^2\pi^2}{L^2R^2}+\frac{n^4}{R^4}\right)\right] + \frac{Eh}{R^2}\frac{m^4\pi^4}{L^4} = 4\pi^2\rho h N_{mn}^2\left(\frac{m^2\pi^2}{L^2}+\frac{n^2}{R^2}\right)^2 \tag{7.2c}$$

from which the frequency can readily be found; this frequency, as in other linear (elastic, small deflections) vibration problems, is independent of the amplitude.

Since the more general Eq. (6.36) was used in deriving Eq. (7.2c), the results should apply to vibrations of any wavelength down to wavelengths which are no longer large compared to the thickness, when both transverse strains and rotary inertia would need to be considered, as in Eq. (5.87d) for plates.

Local Stress Fields for Satisfying Boundary Conditions

As was indicated in Secs. 5.3, 5.4, 5.5, 6.5, many of the local stress fields which are useful for satisfying edge boundary conditions in plates can also be used to sufficient approximation for thin shells. Such local stress fields are also contained in Eqs. (6.36) or (6.34), and (6.33c) as the solutions of the homogeneous equations obtained by setting all the external loads p, f_x, . . . equal to zero, namely:

$$\nabla^8 w + \left[\frac{1}{R^2}\nabla^2\left(5\frac{\partial^4 w}{\partial x^2\,\partial y^2}+2\frac{\partial^4 w}{\partial y^4}\right)+\frac{1}{R^4}\left(3\frac{\partial^4 w}{\partial x^2\,\partial y^2}+\frac{\partial^4 w}{\partial y^4}\right)\right]+\frac{12(1-\nu^2)}{R^2h^2}\frac{\partial^4 w}{\partial x^4}=0 \tag{7.3a}$$

$$R\nabla^4 u = \nu\frac{\partial^3 w}{\partial x^3}-\frac{\partial^3 w}{\partial x\,\partial y^2}, \qquad R\nabla^4 v = (2+\nu)\frac{\partial^3 w}{\partial x^2\,\partial y}+\frac{\partial^3 w}{\partial y^3}, \tag{7.3b}$$

where the terms enclosed in brackets are present in Eq. (6.36) but absent in Eq. (6.34).

We will confine the discussion chiefly to obtaining the thin cylinder analogs of the most important of the local stress fields developed for plates, Eqs.

(5.78a, b); this should also given an idea of the conditions under which local fields developed for plates or beams can be applied with satisfactory approximation to shells in general.

Harmonically Distributed Edge Loading, Membrane Case

The simplest solution of Eqs. (7.3a, b) is for w equal to zero or a constant, in which case the first equation is satisfied identically and the other two reduce to $\nabla^4 u = 0$, $\nabla^4 v = 0$. Solution (5.78a), with $u_x = u$, $u_y = v$, $u_z = w$, satisfies all these conditions, from which we can conclude that this solution for harmonically distributed membrane edge loads on plates applies equally, within the approximations of thin shell theory, to membrane forces on the ends of a cylinder, distributed harmonically along the circumference. Thus the solution of Eq. (5.78a) for $a = 2b/(1+\nu)$, $b = E\pi u_{x0}/[3(1-\nu)l]$, taking $l = \pi R/n$, could be superposed on solution (7.1a) to make u zero at the ends (or, for a very short cylinder, a similar correction obtained by using $\sinh x'$ and $\cosh x'$ instead of $e^{-x'}$ to satisfy the conditions at both ends simultaneously).

Harmonically Distributed End Loading, Flexural Case

This case is more complicated than the membrane case because the w displacements which must accompany it are resisted by the radial components of the circumferential stresses and other effects produced by the curvature, as well as by flexural resistances like those in the flat plate case. Hence we will need a more general solution of Eqs. (7.3a, b).

For a complete cylinder these equations and the condition of continuity in the circumferential direction can be satisfied if:

$$w = \cos\frac{ny}{R}X_1(x), \qquad u = \cos\frac{ny}{R}X_2(x), \qquad v = \sin\frac{ny}{R}X_3(x) \tag{7.3c}$$

where n is an integer and the X's can be power functions of x, or exponential functions or functions expressed in terms of exponential functions such as hyperbolic and trigonometric functions.

A solution could equally well be obtained with $\cos(ny/R)$ and $\sin(ny/R)$ interchanged; this would not represent an independent solution, since it would really be the same solution with the coordinate axes shifted circumferentially; however it may be convenient in a particular problem to use either form, or to combine them. If the problem involves an incomplete cylinder such as a curved panel with sides parallel to the coordinate axes, n need not be an integer and can be chosen to help satisfy the boundary conditions on the edges parallel to the axis, or other types of functions of y might be used. We will confine the discussion here to complete cylinders and solutions of the type of Eqs. (7.3c).

The Case $n = 1$

This is a special case for which we can get solutions most easily by taking the X's as power functions of x rather than the exponential functions required for $n = 0, 2, 3, \ldots$, so we will deal with it first. For the case $n = 1$ displacements w proportional to $\cos(y/R)$, together with suitable v displacements proportional to $\sin(y/R)$, produce a lateral movement of the whole cross section of the cylinder without changing its circular shape, and thus evidently can describe the bending of a long cylindrical tube. For the case of pure bending of a tube under couples M_0 on its ends, for instance, we can take:

$$w = C\cos\frac{y}{R}(x^2), \qquad u = C\cos\frac{y}{R}(2Rx), \qquad v = C\sin\frac{y}{R}(x^2 - 2\nu R^2) \tag{7.3d}$$

Substituting these expressions into the complete equations (7.3a, b) (the simpler equation without the terms in the brackets gives incorrect results for this case, as might be expected from using the curves given in Fig. 7.2), we find that they are satisfied identically; the coefficients of the terms in Eqs. (7.3d) were of course chosen to satisfy these and the following conditions. If the first of Eqs. (7.3d) is substituted into Eq. (7.3a) in which the coefficient 5 of the second term is replaced by c_2 and the coefficient 3 of the fourth term by c_5, as in Table 6.7 of Sec. 6.7, it will be found that the condition for the satisfaction of Eq. (7.3a) is that $c_2 - c_5 = 2$, as was stated in the discussion of this table.

Using Eqs. (6.20) and (6.23) we find that F_y, $F_{xy} = 0$, $F_x = C\cos(y/R)2EhR$, and $M_0 = \int_0^{2\pi R} dy\, F_x R\cos(y/R) = 2\pi CEhR^3$, or $C = M_0/(2\pi EhR^3)$. It can readily be checked that this solution gives the same lateral displacements and relative rotations of cross sections as well as stresses as is given by elementary classical beam theory, considering the cylindrical tube as a simple beam.

Other cases of bending of cylindrical tubes as beams, such as a cantilever loaded by a force on the end (produced by transverse shear forces F_{xz}, F_{yz}, given by Eqs. (6.25)), or uniformly loaded beams using loadings p, f_x, or f_y, can be studied in the same manner, using higher powers of x in the X functions, Eqs. (7.3c). Such solutions would be more accurate than those given by elementary beam theory, since they would consider transverse shear strains (not present in the above case of pure bending) in a nearly exact manner; however they should check reasonably well with the corrected beam theory developed in Sec. 3.5.

Complete Solution of Simplified Homogeneous Equations

A general solution of Eqs. (7.3a, b) without the terms in the brackets, that is of the homogeneous equations obtained by eliminating the loading terms from the simplified cylinder equations (6.33c), (6.34), has been presented by Hoff,* using exponential type functions of x in the X's of Eqs. (7.3c). Since we are principally

* 'Boundary-value problems of the thin-walled circular cylinder', *Proc. ASME*, 1954, Paper No. 54. APM-4.

interested in deflections, slopes, and bending moments, which can all be expressed in terms of the radial displacement only, it will be sufficient here to consider the general solution for w, which we will express here in the form:

$$w = \cos\frac{ny}{R}\, e^{px/R}\left(a\cos\frac{qx}{R} + b\sin\frac{qx}{R}\right), \quad \text{where:} \tag{7.3e}$$

$$p = \pm\frac{1}{2}\left(\sqrt[4]{\mu} + \frac{n^2}{\gamma} + \gamma\right), \qquad q = \left(\sqrt[4]{\mu} + \frac{n^2}{\gamma} - \gamma\right),$$
$$\gamma = \left\{\sqrt{\frac{\mu}{4} + n^4} - \sqrt{\frac{\mu}{4}}\right\}^{1/2}, \qquad \mu = 3(1-\nu^2)\frac{R^2}{h^2} \tag{7.3f}$$

and all roots are to be taken as positive, in which case p and q are real numbers. Considering the arbitrary coefficients a and b and the plus or minus sign in the expression for p, expressions (7.3e, f) represent four independent solutions, which correspond to the fourth order of the original equation of equilibrium of forces in the transverse direction from which Eqs. (6.34), (6.36), and Eq. (7.3a) were derived by escalation.

Hoff's 'general solution', on the other hand, included eight independent solutions, which can be represented by adding a $\pm$ sign in front of the term $\sqrt[4]{\mu}$ in the expressions (7.3f) for p and q; these eight solutions correspond to the eighth order of Eq. (7.3a). However, we shall see that the solutions obtained from Eqs. (7.3e, f) by using a positive value for $\sqrt[4]{\mu}$, are consistent with the physical problem, whereas corresponding solutions obtained with a negative value for $\sqrt[4]{\mu}$ are inconsistent with the physical problem. We therefore conclude that the four solutions obtainable with a negative $\sqrt[4]{\mu}$, while mathematically solutions of Eq. (7.3a), are related only to the mathematical operations of escalation and not to the original physical problem, and we therefore discard them.

The positive and negative values of p in Eq. (7.3e) correspond to positive and negative exponential functions (or the combinations of them which we call hyperbolic functions); they can be used for satisfying conditions at both ends of a very short cylinder. For simplicity we will confine ourselves to the negative values of p and hence negative exponential functions, as in Eqs. (5.78a, b), which give solutions which die out so rapidly that they can be used without serious error for all but very short cylinders.

The Case $n = 0$

This is the axisymmetric case where the displacements are independent of y. Equations (7.3f) give indeterminate values for p and q, since for this case $n^2/\gamma = 0/0$. We can get the correct values of p and q most easily by going back to Eq. (7.3a). In this case w is independent of y so all the derivatives with respect to y drop out; this also means that all the terms in the brackets drop out, and the simpler equation (6.34) gives the same results in this case as the complete equation (6.36). All that remains is: $\partial^8 w/\partial x^8 + (4\mu/R^4)\,\partial^4 w/\partial x^4 = 0$ (note that this relation depends upon taking $\epsilon_1, \epsilon_4 = 0$ in Eq. (6.35q) and Table 6.7). Substituting expression

(7.3e) into this relation and multiplying by $R^8/e^{px/R}$, we obtain:

$$(Ab + Ba)\sin(qx/R) + (Aa - Bb)\cos(qx/R) = 0, \quad \text{where:} \tag{7.3g}$$

$$A = (p^8 - 28p^6q^2 + 70p^4q^4 - 28p^2q^6 + q^8) + 4\mu(p^4 - 6p^2q^2 + q^4)$$
$$B = 8pq[(p^6 - 7p^4q^2 + 7p^2q^4 - q^6) + 2\mu(p^2 - q^2) \tag{7.3h}$$

Equation (7.3g) can be satisfied for all values of x only if $Ab + Ba = 0$ and $Aa - Bb = 0$. Eliminating a and b between these relations (if a, b are zero there is no solution) we obtain $A^2 + B^2 = 0$, which, if A and B are to be real, can only be satisfied if $A = 0$ and $B = 0$. Setting the value of B from (Eq. 7.3h) equal to zero we find $p = q$. Substituting this into $A = 0$ we obtain $p^8 - p^4\mu = 0$, which requires either that $p = q = 0$ or $p = q = \sqrt[4]{\mu} = \sqrt[4]{3(1-\nu^2)}\sqrt{R/h}$. The first solution reduces to $w = b$, which gives the membrane case which was treated above, while the second is the same as the solution for a beam on an elastic foundation.* Here the beam is a longitudinal strip of the cylinder wall with width dy and depth h, having the bending stiffness:

$$E\,dy\,h^3/[12(1-\nu^2)] \tag{7.3i}$$

We must use the plane strain modulus $E/(1-\nu)^2$ instead of E because the beams fibers are not free to expand and contract in the lateral or y direction. This beam is on an elastic foundation with a 'modulus':

$$\beta = Eh\,dy/R^2, \tag{7.3j}$$

that is the foundation exerts a resisting force β per unit length of beam per unit deflection. This value of β can also be derived by noting that the radial displacement w produces a unit circumferential strain w/R, which must be multiplied by E to get the circumferential stress and by h to get the circumferential force per unit length, and by the angle dy/R to get the resisting radial resultant of the circumferential forces on each side of the element, which are at this angle to each other.

The Cases $n > 1$

Equations (7.3f) give the values of p and q for $n > 1$ for any value of μ. As has been discussed previously, in order for the Love–Kirchhoff assumption to be applicable $l_y = \pi R/n > 10h$, from which μ must have a value of more than about $28n^2$. For example, for $\mu = 36n^2$, Eqs. (7.3f) give, for $n = 2, 3, 10, 100$: $p = \pm 4{\cdot}10$, $\pm 5{\cdot}42$, $\pm 14{\cdot}0$, $\pm 112{\cdot}3$; $q = 2{\cdot}99$, $3{\cdot}48$, $5{\cdot}36$, $13{\cdot}75$, respectively.

From Eq. (7.3e), using Eqs. (6.20), (6.23):

$$\partial w/\partial x = \frac{1}{R}\cos\frac{ny}{R}\,e^{px/R}\left[(pa + qb)\cos\frac{qx}{R} + (pb - qa)\sin\frac{qx}{R}\right]$$

* See, for example, Timoshenko, *Strength of Materials*, part II, second edition, p. 2. The buckling of such a beam under axial load was discussed in Sec. 2.5, Eq. (2.30).

$$M_x = \frac{D}{R^2} \cos \frac{ny}{R}\, e^{px/R} \Big\{ [(p^2 - q^2 - \nu n^2 + \nu)a + 2pqb] \cos \frac{qx}{R} \qquad \text{(7.3k)}$$

$$+ [-2pqa + (p^2 - q^2 - \nu n^2 + \nu)b] \sin \frac{qx}{R} \Big\}$$

For the case in which $x = 0$: $w = 0$, $\partial w/\partial x = (\partial w/\partial x)_0 \cos(ny/R)$: $a = 0$, $b = (R/q)(\partial w/\partial x)_0$ and:

$$\partial w/\partial x = (\partial w/\partial x)_0 \cos \frac{ny}{R}\, e^{px/R} \left(\cos \frac{qx}{R} + \frac{p}{q} \sin \frac{qx}{R} \right)$$

$$M_x = \frac{D}{R} (\partial w/\partial x)_0 \cos \frac{ny}{R}\, e^{px/R} \left(2p \cos \frac{qx}{R} + \frac{p^2 - q^2 + \nu(1 - n^2)}{q} \sin \frac{qx}{R} \right) \qquad \text{(7.3 l)}$$

Figure 7.3 shows in dotted lines nw plotted against nx/R for this case of $a = 0$, $b = (R/q)(\partial w/\partial x)_0$, for $\mu = 36n^2$, $n = 2$, 3, 10, 100, using the negative values of p given above. As can be seen such solutions die out to negligible values at $x = R$ for $n = 2$, to $x \approx 8R/n$ for very large values of n, and thus could be used for satisfying boundary conditions at one end of a cylinder of this length without appreciably affecting the conditions at the other end.

This solution could be superposed on solutions for hinged ends such as Eqs. (7.1a) to reduce the slope at each end to zero and thus convert the solution to one for fixed ends. For very short cylinders a similar solution using both the positive and negative values of p could be used for the same purpose.

For comparison, a similar solution obtained from Eqs. (5.78b) for plates is plotted in full lines. It can be seen that for the flexural case the plate solution could not be used for cylinders, even for quite large values of n. As would be expected, the deflections die out much faster for small values of n than for plates because of the added resistance to deflection produced by curvature, and again, as expected, this effect of curvature decreases as n becomes larger.

The nearly vertical dash-dot line shows the results obtained for $\mu = 36n^2$, $n = 2$ by using a negative value for the term $\sqrt[4]{\mu}$ in Eqs. (7.3f); it can be seen that this result would require an impossible negative effect of curvature so that such solutions should be discarded, as was discussed above. For comparison also, the dashed line shows the above case for $\mu = 144n^2$, $n = 2$; comparing this with the case $\mu = 36n^2$, $n = 2$ we see that, again as might be expected, the results are not very sensitive to the radius–thickness ratio.

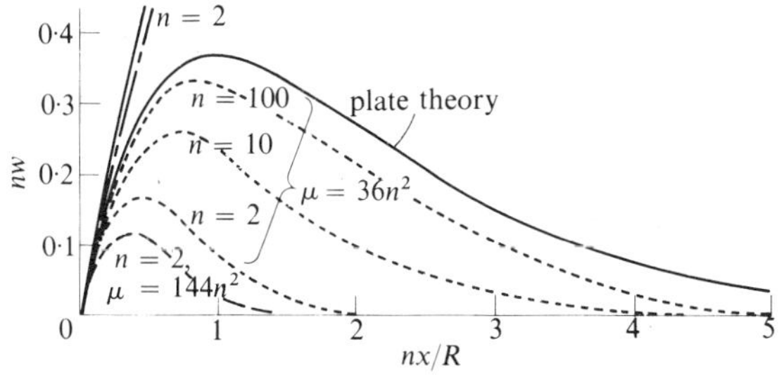

FIGURE 7.3

Equations (7.3f) and the above results for $n > 1$ were obtained with the simplified cylinder equation (6.34). No general solution of the homogeneous equation derived from the more exact solution (6.36) is available. If we take $l_y = \pi R/2$ and $l_x = R$ for $n = 2$, we have $l_x/l_y = R/l_y \approx 0{\cdot}6$, which indicates in Fig. 7.2 that the first term in the brackets in Eq. (7.3a) has about 4 percent and the other terms less than 1 percent of the values of the terms common to the two cylinder theories. For a large n we can take $l_y = \pi R/n$, $l_x = 8R/n$, from which $l_y/l_x \approx 0{\cdot}4$, $l_y/R = \pi/n$, which indicates even less error for the simpler theory.

As a further check, a solution was obtained for the most unfavorable case of $n = 2$ by starting with the values of p and q given by Eq. (7.3f) and trying small positive and negative corrections, and interpolating graphically so as to satisfy the full Eq. (7.3a). In this way it was found that for $n = 2$, $\mu = 36n^2 = 144$, the more exact cylinder equations are satisfied by $p = \pm 4{\cdot}08$, $q = 2{\cdot}96$ as compared to $p = \pm 4{\cdot}10$, $q = 2{\cdot}99$ for the simpler equations. These values give curves in Fig. 7.3 which can hardly be distinguished from those shown. It is therefore concluded that Eqs. (7.3f), based on the simpler theory, is adequate for calculating the effects of harmonically distributed flexural end loads on thin cylinders.

7.2 BUCKLING OF THIN CYLINDERS UNDER AXIAL COMPRESSION

Introduction

There are three basic types of buckling of thin cylinders, corresponding to the three types of membrane stresses: axial and circumferential normal stresses, and shear stresses on axial and circumferential surfaces. To cause buckling, the axial and circumferential stresses must of course be compressive. The most common loading conditions causing buckling are thus axial compression producing uniformly distributed axial compressive stress, pure bending producing axial compressive stress uniform in the axial direction but varying over half the circumference, uniform circumferential compressive stress produced by uniform external pressure or internal vacuum, and uniform shear stress produced by simple torsion.

Certain combinations also frequently occur, such as of axial and circumferential compressive stress produced by external pressure or internal vacuum acting on the end closures as well as on the surface of cylinders, and combinations of axial and shear stresses produced by bending accompanied by transverse shear, or by torsion combined with axial compression or tension (in the latter case of course the torsion must be great enough to overcome the tendency of the tension to prevent buckling).

What we have called the 'classical stability' study of a cylinder of perfect shape and elastic properties involves two stages. The first, the study of the distribution of stresses during the prebuckling period until they reach the critical values, is the simplest because for thin cylinders the elementary theories of bending of tubular beams, of 'boiler' theory, and of compression and torsion of thin tubes will be sufficient for our purposes.

The second or buckling stage involves studying the conditions under which the loaded cylinder can be in equilibrium in a deflected state. Usually it can be in such a state of equilibrium in many different deflection shapes, and then we must find the shape which requires the smallest loads, as it can be assumed that buckling will occur at that loading. As was discussed in Sec. 2.5 for the simpler case of struts, theoretically a truly perfect and undisturbed specimen would be in equilibrium and could remain unbuckled even when the loads are at or above the critical values at which the specimen could also be in equilibrium in a buckled state; however, we must be realistic enough to assume the presence of small disturbances which will initiate buckling even though their magnitudes are too small to appreciably affect the buckling load (see the discussion of 'perfect struts' in Sec. 2.5).

The study of the buckling of real cylinders, on the other hand, must take into account the fact that all constructions, no matter how carefully fabricated, do have defects sufficiently large to not only initiate buckling but to affect the buckling load and process; we will find that for some types of loading this effect is large, while for others it is quite small.

Under the definition of 'defects' we include initial deviations from the assumed perfect geometric shape, or equivalent deviations from uniform, isotropic elastic behavior (due for instance to the multicrystalline composition of engineering metals, or to inclusions, cavities, etc.), as well as deviations produced during the loading process, or disturbances such as those produced by incidental lateral loads such as weight. Transitory disturbances, such as soundwaves, vibrations transmitted through foundations, accidental jarring or air currents, etc., may be sufficient to trigger buckling of the theoretical 'perfect' specimen, but experience indicates that such passing disturbances can be ignored in practical problems compared to the more permanent defects listed above. This is fortunate since the latter are generally properties of the specimen or known conditions under which it is used in a practical application, and there is thus some hope of predicting at least some systematic part of their effects, and thereby, by taking this part into consideration, of eliminating some of the variation which would otherwise have to be regarded as unpredictable 'scatter'.

When the defects of real specimens are considered, the two 'stages', prebuckling and buckling, of the classical stability study merge into one, because, as was illustrated in the simpler case of struts in Sec. 2.5, the buckling deformation starts from the very beginning of loading. Since the defects of shape or equivalent defects of elastic behavior are likely to be of the order of magnitude of the thickness, it is necessary to use large deflection theory for this type of study of shells. In the next three sections we will discuss all of the types of problems described above.

Stability of Perfect Thin Cylinder Under Uniform Axial Compression

Consider a perfect hinged-edge cylinder of mean radius R, length L and thickness h under a compressive axial load P. In the prebuckling stage of the loading we assume that this produces a uniform axial compressive stress $\sigma = -\sigma_x = P/(2\pi Rh)$, the other membrane stresses and the flexural stresses being zero. When the classical stability limit is reached $\sigma = \sigma_{cl} = -\mathbf{F}_x/h$, where σ_{cl} is the axial compressive stress at the classical stability limit, and $\mathbf{F}_x$ was defined under Eq. (6.27d). We also assume that the cylinder is long enough so that end conditions are not very critical (tests indicate that this is so if $L > 1{\cdot}5R$) and short enough (say $L < 30R$) so that the possibility of its buckling as a tubular strut can be ignored.

Due to Poisson's ratio effects the axial compression tends to produce a circumferential and consequent radial strain of $\nu\sigma/E$, and hence an increase in radius of $w = \nu R\sigma/E$. If this radial movement is prevented at the ends by the boundary restraint during the loading process, then a specimen with an initially true cylindrical shape would no longer have such a shape when the critical stress is reached, and would no longer be capable of the kind of true instability which we are now trying to study. We will therefore assume for the present that the end restraints permit free radial movement, leaving discussion of the effects of initial deviations from cylindrical shape until later.

For the buckling stage, substituting $\mathbf{F}_x = -h\sigma_{cl}$ into Eq. (6.34) and taking the other loads on the right-hand side as zero, we see that, since all the terms involve even derivatives of w, this equation and the boundary condition that w, $M_x = 0$ at $x = 0, L$ can be satisfied if the deflection has the form given by Eq. (7.1a), $w = W \sin(m\pi x/L)\sin(ny/R)$, and σ_{cl} is a constant; that is the cylinder can be in equilibrium in this shape, provided that σ_{cl} has a certain value which we must find. It should be noted that due to the condition that σ_{cl} and hence $\mathbf{F}_x$ is constant, equations such as (6.34) used to study classical stability problems are *not* nonlinear, as they are sometimes erroneously stated to be.

The value of σ_{cl} for which such equilibrium in a deflected position can exist is found by substituting the above expression for w into Eq. (6.34) and solving for σ_{cl}. The result can be written in the form:

$$\sigma_{cl} = \frac{Eh}{\sqrt{12(1-\nu^2)}R}\left(A + \frac{1}{A}\right), \quad \text{where} \quad A = \frac{Rh}{\sqrt{12(1-\nu^2)}} \frac{\left(\dfrac{m^2\pi^2}{L^2} + \dfrac{n^2}{R^2}\right)^2}{\dfrac{m^2\pi^2}{L^2}} \tag{7.4a}$$

The fact that W and the functions of x and y cancel out shows that σ_{cl} is constant not only with respect to the coordinates but also with respect to the amplitude W of the deflection.

Equation (7.4a) shows that there are theoretically an infinite number of shapes, characterized by different values of m and n, in which the cylinder can be in equilibrium in a deflected state. Each of these shapes requires a different value of the uniform compressive stress σ_{cl}, and, as has been discussed, buckling will actually occur when the lowest of these required buckling stresses is reached.

Since m and n must be integers to satisfy end conditions and circumferential continuity, it is not very simple in any given instance to pick out the values of m and n which give the smallest σ_{cl}. However if m and n are fairly large (as turns out to be true except for the value of m in very short cylinders) we can get a good approximation to their values by ignoring the fact that they must be integers; after we have found their values which minimize σ_{cl} we could finish the problem by trying the few combinations of integers closest to these values and picking the combination giving the lowest value for σ_{cl}.

Treating σ_{cl} as a continuous function of m, n, and hence of A, its minimum value will occur when $d\sigma_{cl}/dA = 0$, which gives $A = 1$ (thus $0{\cdot}9 + 1/0{\cdot}9 \approx 2{\cdot}01$, $1 + 1/1 = 2$, $1{\cdot}1 + 1/1{\cdot}1 \approx 2{\cdot}01$). Then the lowest value which σ_{cl} can have is:*

$$\sigma_{cl} = \frac{E}{\sqrt{3(1-\nu^2)}}\frac{h}{R} \approx 0{\cdot}605E\frac{h}{R}, \qquad \frac{\left(\dfrac{m^2\pi^2}{L^2} + \dfrac{n^2}{R^2}\right)^2}{\dfrac{m^2\pi^2}{L^2}} = \frac{2\sqrt{3(1-\nu^2)}}{Rh} \tag{7.4b}$$

This result determines only a relation between m and n but not their values. According to this theory, then, the same critical buckling stress could produce an axisymmetric deflection in which the cylinder would collapse like a 'Sylphon' bellows, with $n = 0$, $l_x = L/m = \pi\{Rh/[2\sqrt{3(1-\nu^2)}]\}^{1/2}$, or in a checkerboard of waves as shown in Figs. 6.1 and 7.5, in which the ratio between the circumferential and axial wavelengths $\lambda = l_y/l_x = \pi mR/(nL)$ and $n = \sqrt[4]{12(1-\nu^2)}\sqrt{R/h}/(\lambda + 1/\lambda)$. The axisymmetric case coincides with the case of the buckling of struts consisting of longitudinal strips of the cylinder wall, with bending stiffnesses (7.3i), each on an elastic foundation with modulus β given by (7.3j) (due to the interaction between adjacent strips) the buckling load for which was given in Sec. 2.5 by Eq. (2.30).

Comparison of Results of Classical Stability Study with Tests

A great many tests have been made of thin cylinders under axial compression. These tests all show that, although thick walled cylinders buckling in the plastic range buckle in the axisymmetric, bellows-like shape, thin cylinders always buckle in small waves in which the ratio λ of wavelength in the circumferential direction to that in the axial direction is close to unity, with an average value of about $\lambda = 0{\cdot}75$; for this wavelength ratio our theory would give the number of whole waves in the circumference as $n \approx 0{\cdot}87\sqrt{R/h}$.

Unfortunately this theoretical value for the number of waves, and the value for the buckling stress given by Eq. (7.4b) check poorly with the values found in the tests, as shown in Fig. 7.4, which gives the locus of the hundreds of test points which have been published. As can be seen, the discrepancies increase with the thinness except for some test series of fairly thick walled cylinders whose failure, as we shall see, was probably precipitated by yielding. In some tests the buckling

* This result was probably first published by R. Lorenz in 1911, *Phys. Zeit.*, vol. 13, p. 241.

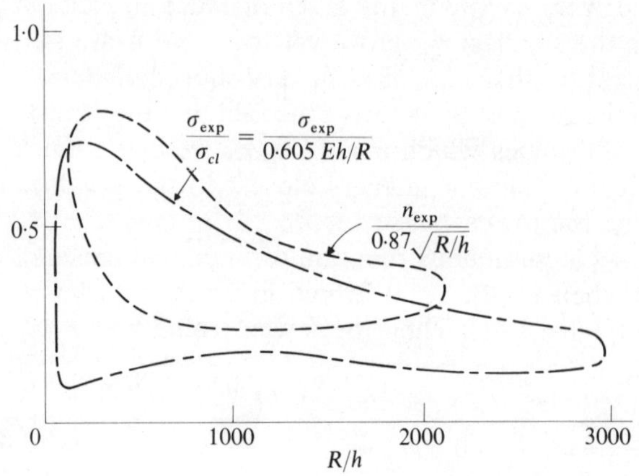

FIGURE 7.4

strength was only 10 percent or 15 percent of that predicted by the classical stability theory which we have described.

A voluminous literature has grown up attempting to explain these discrepancies. The first explanation which comes to mind is that the theory is not accurate enough. To be sure, the simplified equation (6.34) used above has limitations which we have discussed, but the small-wave type of deformation which both theory and experiment indicate for this case are in the range where this equation has proven to be very accurate. If we add the terms included in Eq. (6.36) or in any of the other theories listed in Table 6.7, we obtain results which are practically indistinguishable from Eqs. (7.4b). The extra terms added do indeed eliminate the indeterminancy of the ratio between the circumferential and axial wavelengths, but when we attempt to solve for this ratio these extra terms become isolated and yield physically impossible results. The truth is that these terms are very small in this case, and whatever tendencies they measure are of no importance compared to the true cause of the discrepancies. This cause undoubtedly lies in the great sensitivity discussed below of the many-circumferential-wave type of buckling to defects and large-deflection effects. As was shown above, the axisymmetric or zero-circumferential-wave type of buckling is essentially strut-type buckling which has none of this sensitivity; cylinders thus have a greatly lowered resistance to many-circumferential-wave buckling.

The neglection of the condition that m and n should be integers certainly cannot provide the explanation; in fact, in most of the tests buckling occurred over only a portion of the length and circumference, in which case m and n need not be integers. Some investigators have attempted to explain the smallness of the experimental strengths by showing that the end conditions which are unsatisfied in analyses like that given above result in end regions which are much weaker than the middle part of the shell, to which our analysis applies with accuracy. Such an end weakness is a possibility, but even if it exists it cannot in any way explain the discrepancies which have been noted. This is because practically all the tests (and

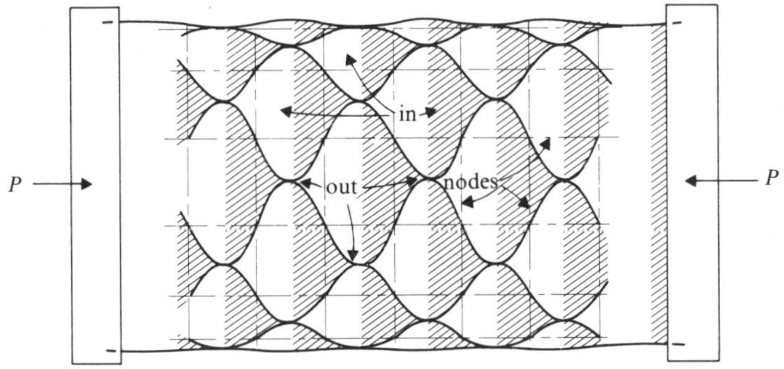

FIGURE 7.5

most practical applications) involve fixed edges and, far from buckling first in the end regions, their end regions were obviously stronger than the middle part of the cylinders, because buckling deformations remained entirely absent near the ends, as shown by the typical side view of a buckled cylinder in Fig. 7.5; the effect of the fixed edges on the resistance of the middle part is probably slight and whatever effect there is should *increase* the resistance.

Application of Large-Deflection Theory to Buckling of Axially Compressed Cylinders

If the large discrepancies indicated by Fig. 7.4 cannot be explained by shortcomings of the classical stability study, then it seems to follow that they must be due to the factors which a classical stability study does not consider, namely initial defects and the use of large deflection theory. The author's paper of 1934* was perhaps the first to consider these factors in this case. However, aside from the development of a practical large deflection theory for thin cylinders which has been presented in Eqs. (6.17) and (6.31j), and which has been used by practically all subsequent investigators of the problem, this study was rather unsatisfactory because of excessive simplifications which were made in the application of the theory to facilitate a solution, and because of its limitation to failures initiated by yielding. Kármán and Tsien** were probably the first to demonstrate the extraordinary effect of using large-deflection instead of linear theory in analyzing the buckling of perfect cylinders, as exemplified by the curves shown in Fig. 6.10(a).

To study this problem with large deflection theory as well as a consideration of initial defects we will use Eq. (6.31j), with $\mathbf{b} = 1/R$ and f_x, $f_y = 0$. Instead of using its companion equation (6.31h) for w, which would require the solution of two nonlinear simultaneous partial differential equations, we will use the combination of equilibrium and energy methods which was discussed in Sec. 6.7 in

* Loc. cit., Eq. (6.17).
** Loc. cit., Eq. (6.37h).

connection with these two equations. In this method we assume an expression for w with unknown coefficients, determine ϕ by integration of Eq. (6.31j), and finish the solution by using the law of virtual work, calculating the strain energy from Eqs. (4.70) and (4.71). By this method the number of nonlinear algebraic equations which have to be solved simultaneously is limited to the number of unknown coefficients in the expression for w plus the wavelengths of the primary term.

Let us consider first the application of large deflection theory to perfect cylinders, in which case the constant K in Eqs. (6.31j) equals unity and, for $f_x, f_y = 0$, this equation reduces to Eq. (6.17). In the discussion of this latter equation we considered this problem, but incompletely and for an entirely different purpose; we will now complete that discussion.

The hinged-edge condition that $w, M_x = 0$ at the ends, can readily be satisfied with the expression (6.12) for w, by proper choice of origin of coordinates and λ; as discussed in the small-deflection study above, the large number of waves makes it unimportant to limit the choice of λ because of end boundary conditions except for very short cylinders, which we will not consider here. We will also assume that the conditions are the same at the ends as in the middle part of the cylinder; this means that any radial movement at the nodes must be permitted by the end constraints (or, more practically, that we can ignore such conditions at the ends because the end regions remained unbuckled in the tests which have to be explained). It should be mentioned in this connection that besides the outward radial movement $\nu R\sigma/E$ due to the Poisson's ratio effect of the average axial compression σ, there is now an added tendency for *inward* radial movement because the circumferential waving would otherwise produce an average circumferential tension; this is represented by the $pq = 00$ term below.

Kármán and Tsien used the method of studying the problem described above, with the four terms to the left of the dashed line in (6.13), $pq = 00$, 20, 11, 02; the $pq = 11$ term is the primary term describing the checkerboard buckling wave form, while the other terms merely modify its shape to decrease the potential energy. In practice the term $pq = 00$ drops out in the differentiations required in using Eqs. (6.17) and (4.70) and need not be further considered, although its value can be found if desired from the condition of zero average circumferential stress. This leaves the parameters W_{02}, W_{11}, W_{20}, λ, n which must be determined from energy considerations or assumed. For convenience we will call the amplitude of the primary term $W_{11} = a$, $W_{20}/W_{11} = b$, $W_{02}/W_{11} = c$, $W_{00}/W_{11} = d$ giving us the expression for w:

$$w = ah\left(\cos\frac{\lambda nx}{R}\cos\frac{ny}{R} + b\cos\frac{2\lambda nx}{R} + c\cos\frac{2ny}{R} + d\right) \tag{7.4c}$$

Kármán and Tsien assumed b to be equal to c, used energy consideration to determine a and b (they also determined d) for various assumed values of λ and n, and obtained relations somewhat similar in shape to the curve labelled '4 terms' in Fig. 6.10(a). The peak points in all the curves were at the height 1·0 given by the classical theory. Leggett and Jones* refined these studies by determining λ

* Aero. Res. Council Rep. and Mem. No. 2190, 1942.

and n also from energy considerations, and obtained one curve, essentially the curve labelled '4 terms' in Fig. 6.10(a). Almroth further extended these studies by using more and more of the terms in (6.13), and obtained all the curves in Fig. 6.10(a).

The behavior of cylinders buckling under axial compression indicated by these curves is in great contrast to our experience with the elastic buckling of struts, for which elastic postbuckling resistance remains constant until the buckling deflection becomes large, and then rises somewhat. The catastrophic drop off in postbuckling resistance of compressed cylinders indicates first that the failure should be explosive in nature, since the buckling, once started, will incur little resistance and will be carried on by momentum; the tests on thin cylinders verify this conclusion. Secondly and even more importantly, the sharp narrow razor-edge-like shape of the resistance curve indicates tremendous sensitivity to defects in this type of buckling, because a real imperfect specimen should give a continuous rounded resistance curve which should lie below and inside that for the perfect specimen, and the narrower the latter is the more the rounded peak of the former should be depressed, as illustrated later in Fig. 7.8.

In spite of the logic of this argument, other interpretations of these results have been advanced. When only the curve labelled '4 terms' was available it was proposed that the low point or 'valley' of this curve, which was at about one-third of the classical stability resistance, obviously represents a very stable condition and should be used as a safe and conservative value for design. Concerning this it could first be said that the height of this valley has no demonstrable relation to the peak resistance which must actually be overcome before this valley can be reached, while under some conditions, especially for large R/h ratios, this value is not at all conservative, and under other conditions it would represent wasteful design, since we can actually rely on considerably higher resistances. In any case Almroth's work shows conclusively that no such valley exists; if we could use an indefinite number of terms, as well as the more exact large deflection theory which the studies of Sec. 6.3 show is needed, the resistance curve would probably be found to approach zero asymptotically.

Another proposed interpretation involved the concept of the specimen 'jumping' from one equilibrium configuration to another, specifically from a point such as P, Fig. 6.10(a), to Q, the stress at which this jump is postulated being taken as that for which the potential energy at Q equals that at P. It was assumed that when these energies are equal, any accidental disturbance could cause the jump from one configuration to the other. The acceptance of this concept was perhaps aided by (or perhaps caused) the labelling of the explosive type of failure which characterizes this phenomenon as a 'snap-through' or 'oil-canning' type of failure.

Actually the idea that any system requires only a triggering minor disturbance to change from one state to a state involving the same or even a much smaller potential energy is a minor scientific heresy. If it were true, then we could hardly count on anything in the world 'staying put',—a breath of wind could cause the water behind the dam to transfer itself to the rocks below the dam where its potential energy is smaller, and so on. All we can say definitely is that if a continuous 'path' exists from the one state to the other, at every point of which the energy has a downward slope, transfer will occur spontaneously, as water finds its way from the continental divide to the sea. To be sure, a sufficiently large impulse can carry it over an obstacle, but no one has shown that any 'impulse' less than the axial load required to raise it over the peak can get a cylinder from states such as 'P' to 'Q'.

The reason such 'explanations' gained wide acceptance is probably to a great extent their convenience. They gave designers definite values for the strengths of their shells, strengths they 'ought to' have, backed by the prestige of a plausible 'theory', even if these values had no relation to the real factors determining the strength, even if they had no real justification other than that they fell in the range between 0·15 and 0·6 of the classical stability limit shown by most tests, and even if these values were demonstrably too conservative in some cases and unsafe in others. Without such values engineers had to face the awful truth that each of their constructions had different defects which significantly affected their strengths and which only individual nondestructive (and nonexistent) tests could determine. But engineers must live and work in the real world, and, as we shall see, a considerably part of this variation can be predicted and reliably allowed for.

The Equivalent Magnitude of Actual Defects

It was shown in Sec. 2.5, for the simpler case of buckling of imperfect struts, that the component of the initial equivalent deviation from perfect shape which has the same form as the buckling deflection is by far the most important during the critical part of the loading. We will therefore ignore all other components and take the initial equivalent deviation as:

$$w_0 = a_0 h \left(\cos \frac{\lambda n x}{R} \cos \frac{n y}{R} + b \cos \frac{2\lambda n x}{R} + c \cos \frac{2 n y}{R} + d \right) \tag{7.5a}$$

The principal amplitude $a_0 h$ of this initial deviation is analogous to the amplitude w_{01} of the important component of the deviation of a strut, which we gave reason for defining by Eq. (2.33) as $w_{01} = (U/\pi^2) l^2/h$; in this formula l is the half wavelength of the deformation (equal to the length, in the case of a hinged-end strut), h is the thickness, and U/π^2 is a dimensionless number, an 'unevenness' factor, whose average value should depend chiefly upon the manufacturing process; the π^2 was introduced to simplify the resulting expressions.

Since the manufacturing processes for columns and cylinders are not too dissimilar—in fact the same sheet could be used for forming both—it seems reasonable to assume a formula similar to that of Eq. (2.33) for the amplitudes of the initial deviations in cylinders. If there were no preferred directions in the cylinder wall it would seem reasonable to replace the l^2 in the formula for the one-directional column by the product $l_x l_y$ for the cylinder, where $l_x = \pi R/\lambda n$, $l_y = \pi R/n$ are the half wavelengths of the primary term of expressions (7.4c), (7.5a) for w and w_0.

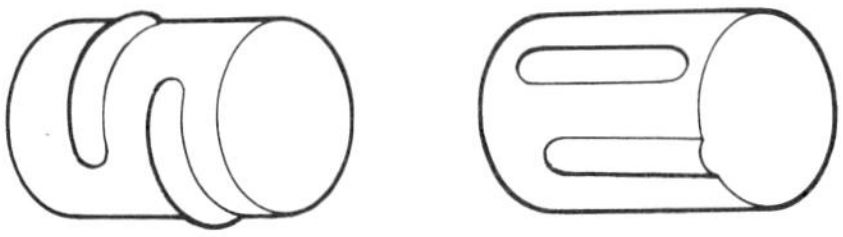

(a) (b)

FIGURE 7.6 (from L. H. Donnell and C. C. Wan, 'Effect of imperfections on buckling of thin cyl. and columns under axial compression, *Trans. ASME*, vol. 72, p. 73, 1950, courtesy ASME)

Actually there generally *is* a preferred direction, because thin-walled cylinders in most tests and in practically all applications are made by bending flat sheets, and this bending into a cylinder tends to flatten out waves which are long in the circumferential direction, as shown at (a) in Fig. 7.6, much more than those which are long in the axial direction as at (b). To allow for this effect and still keep the unevenness factor U nondimensional, we can replace the l^2 of the column theory by $l_x^{1+\alpha} l_y^{1-\alpha}$, thereby augmenting the effect on the expression for initial deviation of components such as those at (b) which are less reduced by the forming process. Since the half wavelength l_y in the circumferential direction should evidently not have a negative exponent, the number α should have a value between zero and unity, and in want of definite data we will use the intermediate value of 0·5. We therefore take:

$$a_0 h = (U/\pi^2) l_x^{1.5} l_y^{0.5}/h, \qquad \text{from which:} \tag{7.5b}$$

$$a_0 = UR^2/(\lambda^{1.5} n^2 h^2), \quad K = 1 + 2w_0/w = 1 + 2a_0/a$$

As to the value of the factor U, which describes the magnitudes of the initial defects, it is reassuring that similar values of this factor can explain test results in not only the cases of struts and of cylinders under compression, but also in the case of cylinders under circumferential compression due to external pressure, considered in the next section; the case of cylinders under torsion has not been studied by exactly the same criteria, but studies which have been made at least partially confirm their application to this case also. These findings are all the more remarkable in that this initial defect formula, similarly used in each case, explains discrepancies between classical stability theories and test results which differ widely in the different cases—from fairly small discrepancies in the case of struts, medium discrepancies in the case of cylinders under external pressure, to very large discrepancies in the present case of cylinders under axial compression.

While the magnitudes of the equivalent initial deviations from perfect shape were calculated in all these studies as the magnitude required to explain test results, it is important to point out that in the case of struts actual magnitudes can be and have been* checked independently and found to agree reasonably with the calculated values. It should also be explained that by 'equivalent' deviations is meant geometric deviations of the shape assumed, which are equivalent in their effects upon the buckling to the combinations of components of this and neighboring shapes of the geometric and elastic defects, and incidental lateral

* Loc. cit., Eq. (2.34).

loadings due to weight of the specimen, etc., all of which may actually be present in typical specimens.

The fact that the defects needed to explain tests for struts and for cylinders are comparable may seem remarkable in view of the circumstance considered above that the act of forming a sheet into a cylinder tends to markedly decrease the geometric unevennesses—a fact which anyone who has ever rolled up a badly wrinkled sheet of paper or other material into a small tube can readily corroborate. As at least partial explanation of this it may be pointed out that there is an important offsetting factor which increases the effective magnitude of defects in cylinders but not in struts; this is the fact that in cylinder buckling tests only a small portion of the wall usually buckles (at least in the first stages of the buckling), so that the controlling factor must be the *worst* portion of the wall area, large enough only to contain a few waves, for which the effective U may be much larger than the average for the entire cylinder. Evidence of this is shown in the rather remarkable plot given in Fig. 7.7 which shows comparably obtained experimental buckling stresses plotted as a function of the ratio between the area of one wave and the total area of the cylinder wall.

Calculation of Strength of Cylinders with Defects

For this calculation, which was carried out by the author and C. C. Wan* in 1948, we use Eq. (6.31j) with $\mathbf{b} = 1/R$, and $f_x, f_y = 0$. We will use expression (7.4c) for w (although, as discussed previously, the term d does not enter into the following

* L. H. Donnell and C. C. Wan, 'Effect of imperfections on the buckling of thin cylinders and columns under axial compression', 1950, *J. Appl. Mech.*, vol. 72, p. 73. Presented at Seventh Int. Cong. of Appl. Mech., London, 1948.

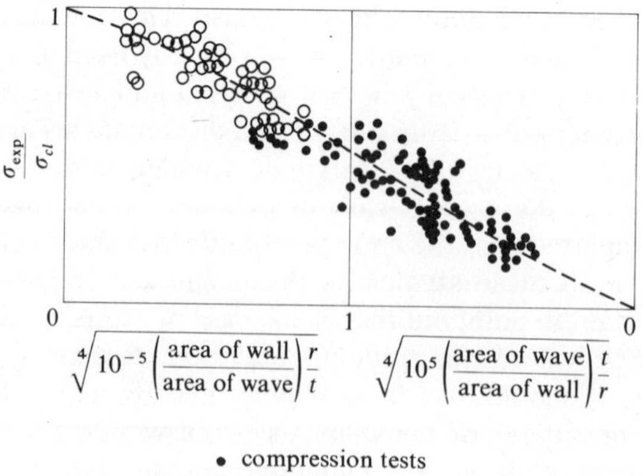

FIGURE 7.7 (from L. H. Donnell and C. C. Wan, 'Effect of imperfections on buckling of thin cyl. and columns under axial compression, *Trans. ASME*, vol. 72, p. 73, 1950, courtesy ASME)

calculations). As stated earlier, instead of attempting to solve Eq. (6.31j) simultaneously with (6.31h) for w and ϕ, we will use the combination equilibrium and energy method in which (6.31j) is used to find ϕ in terms of the five unknowns a, b, c, λ, n in the expression for w, after which these unknowns are determined by using the law of virtual work. For convenience in calculations we will use the following symbols α and k, which can replace a and n as unknowns along with b, c, and λ. Using Eqs. (7.5b):

$$\alpha = \frac{\sqrt{3(1-\nu^2)}}{2}\, a, \qquad k = \frac{n^2 ah}{2R} K = \frac{n^2 ah}{2R}\left(1 + 2\frac{a_0}{a}\right) = \frac{n^2 ah}{2R} + \frac{1}{\lambda^{1\cdot 5}}\frac{UR}{n} \tag{7.6a}$$

Substituting expression (6.12) into (6.31j), and using trigonometric transformations for the resulting squares and products of trigonometric functions, as tabulated in Table 6.3, we obtain:

$$\begin{aligned}\nabla^4\phi = \frac{E\lambda^2 n^2 ah}{R^3}\Big\{&\cos\frac{\lambda nx}{R}\cos\frac{ny}{R} + 4b\frac{2\lambda nx}{R} \\ &- k\Big[4(b+c)\cos\frac{\lambda nx}{R}\cos\frac{ny}{R} + 32bc\cos\frac{2\lambda nx}{R}\cos\frac{2ny}{R} \\ &+ 4b\cos\frac{3\lambda nx}{R}\cos\frac{ny}{R} + 4c\cos\frac{\lambda nx}{R}\cos\frac{3ny}{R} + \cos\frac{2\lambda nx}{R} + \cos\frac{2ny}{R}\Big]\Big\}\end{aligned} \tag{7.6b}$$

Solving this equation for the Airy stress function of the membrane stresses ϕ, it can readily be checked that:

$$\begin{aligned}\phi = \frac{E\lambda^2 ahR}{n^2}\Big[&\frac{1-4k(b+c)}{(\lambda^2+1)^2}\cos\frac{\lambda nx}{R}\cos\frac{ny}{R} - \frac{2kbc}{(\lambda^2+1)^2}\cos\frac{2\lambda nx}{R}\cos\frac{2ny}{R} \\ &- \frac{4kb}{(9\lambda^2+1)^2}\cos\frac{3\lambda nx}{R}\cos\frac{ny}{R} - \frac{4kc}{(\lambda^2+9)^2}\cos\frac{\lambda nx}{R}\cos\frac{3ny}{R} \\ &+ \frac{4b-k}{16\lambda^4}\cos\frac{2\lambda nx}{R} - \frac{k}{16}\cos\frac{2ny}{R}\Big] - \frac{\sigma y^2}{2}\end{aligned} \tag{7.6c}$$

The term $-\sigma y^2/2$ is the only solution which we will use from the homogeneous equation $\nabla^4\phi = 0$; besides such power function terms we could also include trigonometric–hyperbolic solutions of this equation and use them to satisfy more complete boundary conditions, but this would complicate rather unnecessarily an already complicated problem. From Eq. (6.31f) the axial membrane stress $\sigma_x = F_x/h = \partial^2\phi/\partial y^2$. The harmonic terms in Eq. (7.6c) give values of σ_x which are either zero or have alternating positive and negative values around the circumference, with zero resultants over the whole circumference. Hence the average value of $\sigma_x = \partial^2(-\sigma y^2/2)/\partial y^2 = -\sigma$, where σ is the average axial compressive stress in the cylinder.

The difference between the values at the two ends of the cylinder of the displacement in the axial direction u represents the total shortening of the cylinder, and this divided by the length is the average unit shortening, which we will designate by ϵ. To find ϵ it is not necessary to go through the usual rather laborious process of finding u by integrating the first two of Eqs. (6.31i) for u and

v and determining the functions of integration from the third equation. It is sufficient to note that the unit shortening at every point is, from the first of Eqs. (6.31i):

$$-\frac{\partial u}{\partial x}=-\frac{1}{E}\left(\frac{\partial^2\phi}{\partial y^2}-\nu\frac{\partial^2\phi}{\partial x^2}\right)+\frac{K}{2}\left(\frac{\partial w}{\partial x}\right)^2 \tag{7.6d}$$

If we substitute expressions (7.6c) for ϕ and (7.4c) for w into the right-hand side of this equation and use trigonometric transformations for the resulting squares and products of trigonometric functions, we obtain an expression with many periodic terms which do not contribute to the average shortening of the cylinder in the axial direction, plus the non-periodic terms $\sigma/E+k\lambda^2ha(1+8b^2)/(4R)$. Hence the average unit shortening of the cylinder in the axial direction can be written as:

$$\epsilon=\frac{\sigma}{E}+\frac{k\lambda^2h}{4R}a(1+8b^2) \tag{7.6e}$$

This can be changed to the following nondimensional form by dividing through by $\epsilon_{cl}=\sigma_{cl}/E=h/(\sqrt{3(1-\nu^2)}R)$ from Eq. (7.4b). We thus obtain:

$$\frac{\sigma}{\sigma_{cl}}=\frac{\epsilon}{\epsilon_{cl}}-\frac{k\lambda^2\alpha(1+8b^2)}{2} \tag{7.6f}$$

Substituting expressions (7.4c) for w and (7.6c) for ϕ into the expressions (4.70), (4.71) for strain energy (as explained before, in the present approximation we use the same flexural strains as in plate theory, so these expressions apply here) and integrating over the circumference and the length L, and using expressions (7.5b) for K and (7.6f) for σ, we obtain the following expression proportional to the total strain energy $\mathscr{E}$:

$$\frac{3(1-\nu^2)R}{\pi ELh^3}\mathscr{E}=\alpha^2(k^2A-kB+C)+D\left(k-\frac{1}{\lambda^{1\cdot 5}}\frac{UR}{h}\right)-\alpha kF\frac{\epsilon}{\epsilon_{cl}}+\left(\frac{\epsilon}{\epsilon_{cl}}\right)^2 \tag{7.6g}$$

where A, B, C, D, F are functions of b, c, and λ only:

$$\begin{aligned}
A&=\left(\frac{4(b+c)\lambda^2}{\lambda^2+1}\right)^2+\left(\frac{8bc\lambda^2}{\lambda^2+1}\right)^2+\left(\frac{4b\lambda^2}{9\lambda^2+1}\right)^2+\left(\frac{4c\lambda^2}{\lambda^2+9}\right)^2+\frac{\lambda^4+1}{8}+\frac{\lambda^4}{4}(1+8b^2)^2\\
B&=\frac{8(b+c)\lambda^4}{(\lambda^2+1)^2}+b,\qquad C=\left(\frac{\lambda^2}{\lambda^2+1}\right)^2+2b^2\\
D&=\left(\frac{\lambda^2+1}{2}\right)^2+8(b^2\lambda^4+c^2),\qquad F=\lambda^2(1+8b^2)
\end{aligned} \tag{7.6h}$$

Expression (7.6g) for the strain energy involves only UR/h, ϵ/ϵ_{cl} and the five unknowns α, b, c, λ, and k. The simplest way to use the law of virtual work to find these five unknowns is to treat ϵ/ϵ_{cl} as a constant, as it would be if the cylinder were subjected to a compressive load in a rigid testing machine. Then for a given cylinder both ϵ/ϵ_{cl} and UR/h will be given and fixed, and, since there will be no change in length, the external axial compression will do no work during virtual displacements such as provided by small changes in the five unknowns. By the law of virtual work, therefore, the partial derivatives of the strain energy, and hence of the expression on the right-hand side of Eq. (7.6g), with respect to each of the unknowns α, b, c, λ, and k can be set equal to zero, giving us five equations to be solved simultaneously for the five unknowns (this is of course equivalent to choosing their values to minimize the strain energy).

The resulting equations are nonlinear in the unknowns, and complex. At the time when this work was done, 1948, computers were not available. The work of solution had to be done by cut-and-try methods, assuming values for b, c, and λ and using cross plotting to determine values of the remaining unknowns, after which σ/σ_{cl}, ϵ/ϵ_{cl}, and UR/h can be plotted as in Fig. 7.8. The work was very difficult because the values chosen for b, c, λ had to be correct within a narrow range to even obtain real values of the remaining quantities. However, solutions were obtained for all five unknowns for the upper quarter of the values of σ/σ_{cl}.

When efforts were made to extend the plot downward the difficulties increased greatly, but at the same time the values of b, c, and λ which had to be assumed stabilized around the values: $b = 0{\cdot}18$, $c = 0{\cdot}03$, and $\lambda = 0{\cdot}728$. The calculations were therefore continued using a simpler type of solution, taking b, c, λ as constants with the values given above, with only the two unknowns α and k.

It may be remarked that the value $\lambda = 0{\cdot}728$ is consistent with test results, and that b is shown to be six times larger than c, so the assumption of Kármán and Tsien and of Leggett and Jones that $c = b$ was not a very good guess. As a matter of fact, elementary physical reasoning would indicate that b is much more

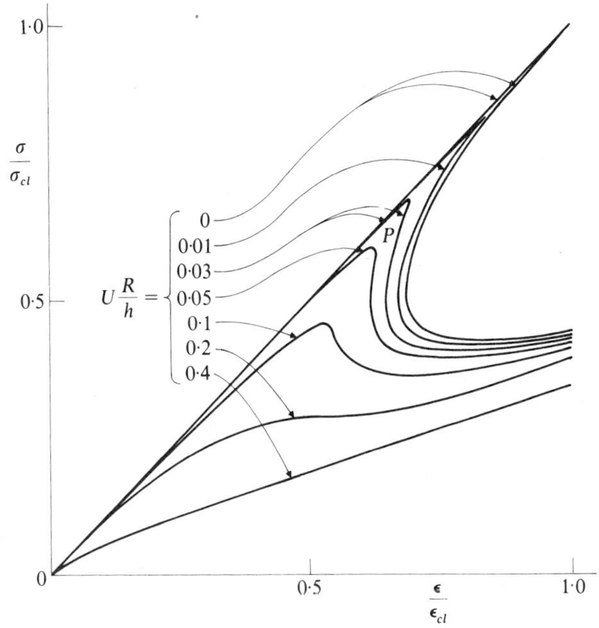

FIGURE 7.8 (from L. H. Donnell and C. C. Wan, 'Effect of imperfections on buckling of thin cyl. and columns under axial compression, *Trans. ASME*, vol. 72, p. 73, 1950, courtesy ASME)

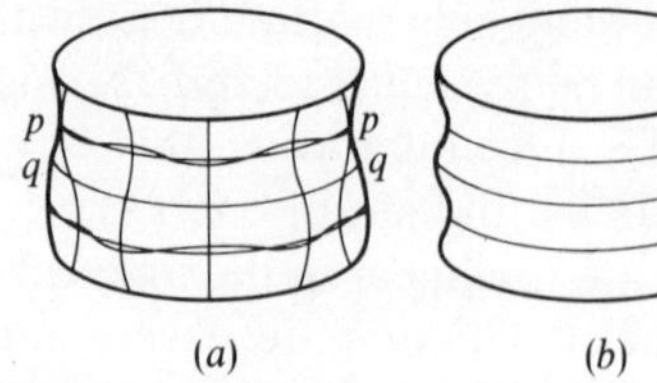

FIGURE 7.9 (from L. H. Donnell, 'A new theory for buckling of thin cyl. under axial comp. and bending', *Trans. ASME*, vol. 56, p. 795, 1934, courtesy of ASME)

important than c. In the author's paper of 1934* only the terms involving a and b were used, Figs. 7.9(a) and (b) were shown, and the following argument was presented: 'The primary (e.g. a) term is shown at (a). It is evident that at section pp the material is on the average circumferentially stretched, while the material at qq is on the average circumferentially compressed to have equilibrium in the circumferential direction. By superposing the symmetrical deformation shown at (b) (e.g. the b term) we can annul this average circumferential stretching and compressing, and although we introduce a certain amount of bending in the longitudinal direction, the total internal energy is considerably reduced'. On the other hand no such obvious need can be cited for the c term.

Substituting the above values for b, c, λ in Eqs. (7.6f) for σ, and into expression (7.6g) for the strain energy, setting the derivatives of the latter with respect to α and k equal to zero and simplifying the resulting equations, they can be put in the form:

$$\frac{\sigma}{\sigma_{cl}} = \frac{\epsilon}{\epsilon_{cl}} - 0{\cdot}333\alpha k \tag{7.6i}$$

$$\alpha = \frac{0{\cdot}922k\epsilon/\epsilon_{cl}}{k^2 - 1{\cdot}057k + 0{\cdot}513} = \left[\frac{k(k - 1{\cdot}61UR/h)}{0{\cdot}278 - 0{\cdot}287k}\right]^{1/2} \tag{7.6j}$$

Then by assuming values of UR/h and k (values of k between 0·5 and $2UR/h$ are found to be suitable) we can solve for α and ϵ/ϵ_{cl}, and σ/σ_{cl}. Corresponding values of σ/σ_{cl}, ϵ/ϵ_{cl}, and UR/h thus obtained are plotted in Fig. 7.8, along with those found previously for the upper part of this diagram, with which values found from Eqs. (7.6i,j) check closely. The elastic buckling curve in Fig. 7.10 is plotted from the peak points of Fig. 7.8 (such as the point p for the $UR/h = 0{\cdot}05$ curve), each peak giving a set of values of σ/σ_{cl} and UR/h.

The portion of this curve for values of UR/h from 0 to 0·1 should be quite accurate, because the peak points of the σ/σ_{cl} vs. ϵ/ϵ_{cl} curves, Fig. 7.8, on which it is based are close to the straight line which represents elastic loading of a perfect cylinder with no deflections. For points close to this line the deflections are small, so the moderately large deflection theory used should be very adequate, and, as can be seen from Fig. 6.10(a), the four terms used in getting a solution to this theory should be sufficient to give reasonably accurate results. The right half

* Loc. cit., Eq. (6.17).

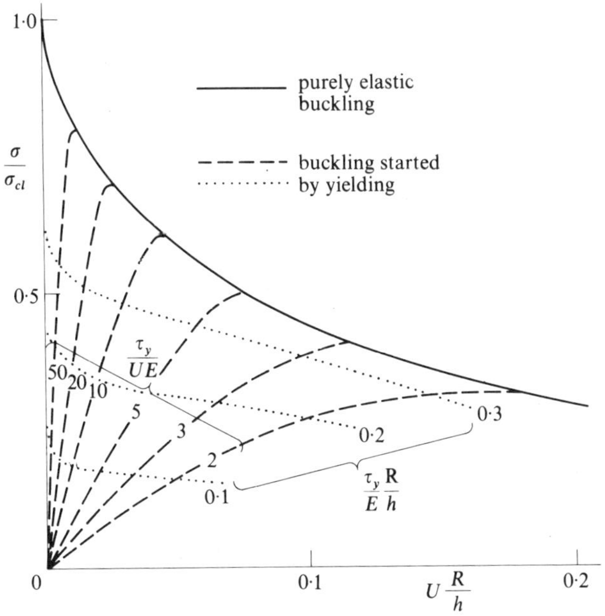

FIGURE 7.10 (from L. H. Donnell and C. C. Wan, 'Effect of imperfections on buckling of thin cyl. and columns under axial compression, *Trans. ASME*, vol. 72, p. 73, 1950, courtesy ASME)

of the elastic buckling curve, for $UR/h > 0{\cdot}1$ is somewhat speculative but the general trend of the curve is certainly correct.

Failure Initiated by Yielding

All the results given above are based on the assumption that the material remains elastic. If yielding occurs after peak points such as *p*, Fig. 7.8, are reached, as of course it must eventually, the resistance will drop off even more rapidly than is indicated by the figure and the failure will be even more explosive in character. The elasticity of the testing machine, or of the adjacent structure in a practical application, will also greatly increase the explosiveness of the failure by providing more stored up energy to be released. But the cylinder should take load smoothly and gradually up to the peak point, with no appreciable dynamic effects, and what happens after failure is usually of little practical importance.

However, the stresses, both membrane and flexural increase as the loading increases from *O* to the peak point *p*. For some materials or geometry yielding may occur at a certain point in the cylinder at a loading between *O* and *p*, and experience indicates that complete failure will then occur at a loading not much greater than this. Two kinds of buckling failure are thus possible in this case. The first kind is precipitated under purely elastic conditions, and occurs in cylinders which, because they are very thin walled or made of high yield material, do not incur yielding until after peak points such as *p* are reached. The second kind, occurring to thicker-walled cylinders or those made of lower-yield material, is precipitated by yielding before such a peak is reached.

We will use the shear energy criteria for yielding, Eq. (1.2):

$$(\sigma_1-\sigma_2)^2+(\sigma_1-\sigma_3)^2+(\sigma_2-\sigma_3)^2=2\tau_y^2 \tag{7.7a}$$

where σ_1, σ_2, σ_3 are the principal stresses at the most highly stressed point and τ_y is the yield point stress of the material. By Eq. (1.1):

$$\sigma_1, \sigma_2=\frac{\sigma_x+\sigma_y}{2}\pm\left[\frac{(\sigma_x-\sigma_y)^2}{4}+\sigma_{xy}^2\right]^{1/2}, \qquad \sigma_3=0 \tag{7.7b}$$

Substituting these into Eq. (7.7a) the condition for yielding becomes:

$$\sigma_x^2-\sigma_x\sigma_y+\sigma_y^2+3\sigma_{xy}^2=\tau_y^2 \tag{7.7c}$$

The stresses are the sum of the membrane and flexural stresses, or:

$$\begin{aligned}
\sigma_x &= \frac{\partial^2\phi}{\partial y^2}-\frac{Ez}{1-\nu^2}\left(\frac{\partial^2 w}{\partial x^2}+\nu\frac{\partial^2 w}{\partial y^2}\right)\\
\sigma_y &= \frac{\partial^2\phi}{\partial x^2}-\frac{Ez}{1-\nu^2}\left(\frac{\partial^2 w}{\partial y^2}+\nu\frac{\partial^2 w}{\partial x^2}\right)\\
\sigma_{xy} &= -\frac{\partial^2\phi}{\partial x\partial y}-\frac{Ez}{1+\nu}\frac{\partial^2 w}{\partial x\partial y}
\end{aligned} \tag{7.7d}$$

Trial indicates that the stress condition is probably most unfavorable at the nodes of the primary deformation ($pq=11$) on the inner surface, that is for:

$$x=\frac{\pi R}{2\lambda n}, \qquad y=\frac{\pi R}{2n}, \qquad z=h/2 \tag{7.7e}$$

Using values (7.4c) and (7.6c) for w and ϕ, and the values assumed for b, c, λ, the yield condition (7.7c) becomes:

$$\left(\frac{\tau_y}{UE}\right)^2\left(\frac{UR}{h}\right)^2=X^2-XY+Y^2+3Z^2, \qquad \text{where} \tag{7.7f}$$

$$X=0{\cdot}149\alpha k+0{\cdot}605\frac{\sigma}{\sigma_{cl}}+0{\cdot}457\left(k-1{\cdot}61\frac{UR}{h}\right)$$

$$Y=0{\cdot}296\alpha k-0{\cdot}218\alpha+0{\cdot}258\left(k-1{\cdot}61\frac{UR}{h}\right)$$

$$Z=-0{\cdot}135\alpha k+0{\cdot}200\alpha+0{\cdot}560\left(k-1{\cdot}61\frac{UR}{h}\right)$$

Using sets of values of α, k, UR/h, and σ/σ_{cl} found from Eqs. (7.6i,j) for points up to the peaks of Fig. 7.8, we can find $\tau_y/(UE)$ from Eq. (7.7f). By cross plotting, curves of σ/σ_{cl} vs. UR/h for constant values of $\tau_y/(UE)$ were found and plotted in Fig. 7.10 in dashed lines. An alternative way of representing the conditions for failure initiated by yielding is by plotting constant $\tau_y R/Eh = (\tau_y/UE)(UR/h)$ lines, which are shown by dotted lines in Fig. 7.10.

Comparison with Test Results

Figure 7.11(a) shows the elastic buckling line, and some of the pertinent lines for buckling started by yielding from Fig. 7.10, superposed upon the envelope of experimental points, and it can be seen that the theoretical elastic buckling line predicts roughly the same dependence of the buckling strength upon the thinness, or R/h ratio, as the experimental results.

With the value shown of the uneveness factor U the theoretical curve is approximately in the middle of the experimental points. The assumption of higher values of U would of course raise the theoretical curve and smaller values lower it, and values of U from about 0·0001 to 0·0004 would be required to completely

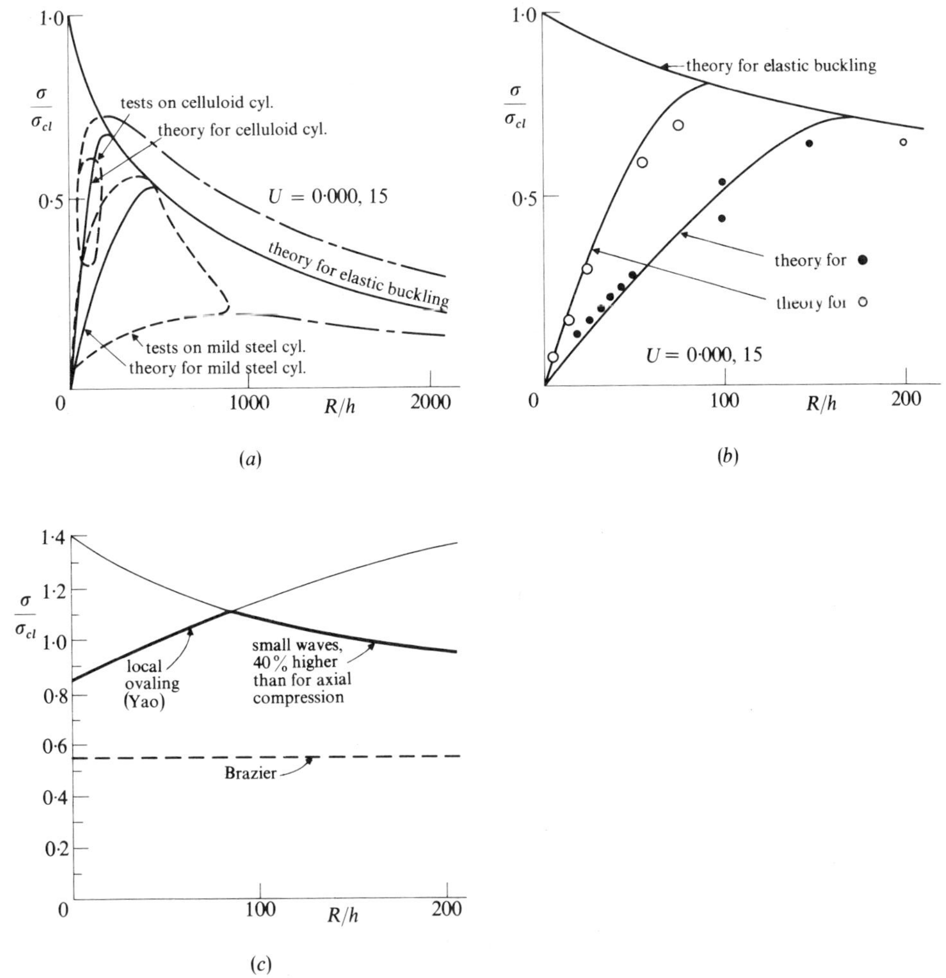

FIGURE 7.11 ((a) is taken from L. H. Donnell, 'A new theory for buckling of thin cyl. under axial comp. and bending', *Trans. ASME*, vol. 56, p. 795, 1934, courtesy of ASME)

enclose the experimental points which represent elastic failures uninfluenced by yielding.

This range of values required for U is not surprising considering that the tests included specimens in which the greatest care was taken to reduce defects to a minimum, to specimens which were merely rolled up from sheets of 'shim stock', which does not need to be flat for the purposes for which it is ordinarily used and in which the deviations from perfect cylindrical shape were sometimes obvious to the unaided eye. For comparison, the values of U required to duplicate commonly used empirical formulas for columns, Fig. 2.10 (in which E/τ_y can be taken as about 1000, since the empirical design formulas were designed for columns of structural steel) ranged from 0·0002 to 0·0010.

Figure 7.11(a) also shows two curves for buckling initiated by yielding for the values of $\tau_y/(UE)$ (taking $U = 0{\cdot}000\,15$) corresponding to two groups of tests, one of celluloid cylinders* and the other of mild steel specimens, at least part of which obviously underwent failure initiated by yielding.** An even more surprising check between the theory for failure initiated by yielding and actual test results, is shown by Fig. 7.11(b) which shows the results of testing two groups of fairly thick cylinders of two kinds of aluminum alloy.†

In applying the above results to design problems, if there is any question about whether failure will be initiated elastically or by yielding, theoretical strengths should of course be calculated for both types of failure and the smaller value accepted.

The values of n predicted by the above solution considering large deflections and defects are smaller than those predicted by the classical stability theory, but not as small as those found in tests (Fig. 7.4), being roughly half way between these extremes.

Buckling of Thin Cylinders Under Pure Bending

Since even the smallest value of n given by test results is around ten, the angle subtended by each half wave is 18° or less, and the compressive stress on each half wave in the middle of the compressed side in a cylinder undergoing bending is nearly constant; even in a group of four half waves the average stress is only about 10 percent less than the maximum bending stress. This, and the fact that, even under uniform compression, failure usually occurs over only a part of the circumference, indicates that thin cylinders should buckle in bending when the maximum bending stress reaches a value of β times the stress required in axial compression, where β is a number close to unity. Tests on similar specimens tested in axial compression and in bending indicate that $\beta = 1{\cdot}4$ is a conservative

* W. Flügge, 'Die Stabilität der Kreiszylinderschale', *Ing. Arch.*, 1932, vol. 3, p. 463.

** W. M. Wilson and N. M. Newmark, 'The strength of thin cylindrical shells', *Bulletin* No. 255, Univ. of Illinois, 1933.

† Discussion by M. Holt and J. W. Clark of paper by Donnell and Wan, loc. cit., p. 396.

value.* Only classical stability studies of this case have been made; they indicate about the same value for β.

Figure 7.11(c) shows such a curve for the small wave buckling of thin cylinders under pure bending (taken 40 percent higher than that in Fig. 7.10 for cylinders under axial compression) compared to the curve obtained by Yao** for local buckling under bending to an oval shape (two waves in the circumferential direction, with the one buckle in the longitudinal direction dying out exponentially away from the center of the buckle). Yao in his analysis used the large deflection terms which we have found to be important, and this type of buckling is commonly observed in the buckling under bending of thick walled tubes such as rubber hoses and thick metal tubes buckling nonelastically.

The lowest parts of these two curves are shown by the heavy line, which indicates the moments at which tubes of various radius-thickness ratio should first buckle in some way. It can be seen that according to these analyses tubes with $(R/h) < 80$ should fail first by ovaling, while thinner tubes should fail first by small wave buckling; this result seems reasonably consistent with common experience.

Brazier† has presented a very interesting analysis for a cylindrical tube under pure bending, which he assumes to oval uniformly throughout its length. As the bending proceeds, the resulting elastic curvature of the tube's axis acts to reduce the tube's resistance in the same way as the initial curvature of a curved tube under a bending moment which acts in the direction tending to increase the curvature. The latter case was studied by von Kármán,†† who showed that due to the curvature both the compressive stresses on the concave side and the tensile stresses on the convex side on the ends of an elemental length have resultants which act toward the middle of the tube and thus tend to squeeze the originally circular cross section into an oval shape (this deformation would be negligible in a solid curved bar but can evidently be more important for hollow bars). The effect upon the moment of inertia of the cross section is not important, but the movement of the shorter fibers on the concave side and the longer fibers on the convex side toward the middle produces rotations of the ends which add to those which we would have in a straight tube, thus reducing the tube's effective bending stiffness.

Brazier plotted the bending moment as a function of the curvature of the initially straight tube and obtained a curve whose slope starts the same as in elementary bending theory but then decreases until the curve becomes horizontal, the bending moment at the peak representing the maximum moment which the tube can resist. Unfortunately the ultimate moment which he thus obtains, which is independent of R/h and is shown by the dotted line in Fig. 7.11(c), is

* See author's 1934 paper, loc. cit., Eq. (6.17).

** J. C. Yao, 'Large deflection analysis of buckling of a cylinder under bending', *J. Appl. Mech.*, paper No. 62-WA-46, 1961.

† L. G. Brazier, 'Of the flexure of thin cylindrical shells and other thin sections', *Proc. Roy. Soc. Lond.*, ser. A, vol. 116, 1927, p. 104.

†† *Ver. Deutsch. Ing.*, vol. 55, 1911, p. 1889. See also Timoshenko, *Strength of Materials*, part II, 2nd edn, p. 107.

unrealistically low, and his analysis seems to be deficient in its consideration of large deflection effects which would increase the resistance.

An independent solution of Brazier's case can easily be obtained from Kármán's solution by merely using the curvature at any instant instead of the initial curvature which Kármán considered. Kármán obtained the approximate formula:

$$\theta = \frac{Ml}{\left(1 - \frac{9}{10 + 12(hR'/R^2)^2}\right) EI}$$

where θ is the relative bending rotation of the ends of a length l of the tube and R' is the initial radius of curvature of the centerline of the curved tube. If we take R' as infinite for an initially straight tube this formula reduces to the $\theta = Ml/EI$ of elementary beam theory. Instead we will take $R' = l/\theta$, that is the instantaneous radius of curvature. Making this substitution, Kármán's formula can be written:

$$\frac{\sigma}{\sigma_{cl}} = \frac{M}{M_{cl}} = \sqrt{3(1-\nu^2)}\lambda\left(1 - \frac{9}{10 + 12\lambda^2}\right) \tag{7.7g}$$

where $\lambda = R^2\theta/(hl)$ and $M_{cl} = \pi ERh^2/\sqrt{3(1-\nu^2)}$ is the moment corresponding to the maximum stress σ_{cl}. Plotting M/M_{cl} against λ we obtain a curve whose slope, like Brazier's, starts at the value given by elementary bending theory and then decreases. The slope never quite reaches zero but becomes very small at an M/M_{cl} of around 1·0, which might be taken as an effective limit for M/M_{cl}, since yielding or other type of local failure would prevent an indefinite increase of λ and hence of θ. This is a far more realistic value than Brazier's 0·55, and may be considered to represent a third possible type of failure, which, however, requires somewhat special end conditions, as the moment must be applied while permitting ovaling at the ends.

7.3 BUCKLING OF THIN CYLINDERS UNDER CIRCUMFERENTIAL COMPRESSION

Introduction

The cases of buckling of cylinders under uniform circumferential compression caused by external pressure (or internal vacuum) and under uniform shear stress caused by torsion differ in one important respect from the case of axial compression. In both of the former cases the buckling waves extend in the axial direction from one end of the cylinder to the other, whereas in the latter case of course there are many waves in the axial direction unless the cylinder is extremely short. As a consequence of this the end edge boundary conditions cannot be entirely ignored in the former cases. There is an analogy here to the case of buckling of an ordinary strut and that of a strut on an elastic foundation which buckles in many waves; however, because of the rapidity with which cyclically varying moments (such as would be required to fix edges, for instance) applied to cylinder edges die out (as shown in Fig. 7.3), the effect of end edge conditions in cylinders is a relatively minor one.

There are also two different practical types of loading of cylinders under external pressure: (1) the case when the external pressure acts not only on the outside of the cylinder wall but also on the end closures of the cylinder, as for example in a vacuum tank or submarine shell; and (2) the case when it acts only on the outside of the cylinder, as it may for example in boiler tubes. Case (1) produces of course a combination of circumferential compression and axial compression; however the axial compression is too small to affect the character of the buckling, and this combination is so common in practical applications that it represents a basic case which should be considered along with simple circumferential compression, as we will do below.

Stability of Perfect Thin Cylinders Under External Pressure

In the prebuckling stage of the loading we assume that for thin cylinders we can use elementary 'boiler' theory, from which an external pressure or internal suction p applied to the cylinder wall produces a uniform circumferential stress $\sigma_y = -Rp/h$, and if applied to the end closures produces a uniform axial stress $\sigma_x = -Rp/(2h)$, with other stresses zero. When the stability limit is reached $\sigma_y = \mathbf{F}_y/h$, $\sigma_x = \mathbf{F}_x/h$ or $\mathbf{F}_y = -Rp$, $\mathbf{F}_x = -Rp/2$, where $\mathbf{F}_y$, $\mathbf{F}_x$ were defined under Eq. (6.27d).

For a classical stability study of this case we can use most conveniently either Eqs. (6.34) or (6.36). From experience and trial we know that the number of waves around the circumference decreases as the length increases, and takes its minimum value of two only for very long tubes, for which case we will have to use the complete equation (6.36). However if we try to cover the entire range of geometries with Eq. (6.36), while this offers no theoretical difficulty it results in a relation between three quantities such as p/E, R/h, and R/L, and the algebra is complex and requires laborious cross plotting to get numerical results. On the other hand Eq. (6.34) gives results which can be expressed immediately in terms of only two quantities which can be plotted as a single curve, the numerical calculations are simple, and reference to Fig. 7.2 shows that it is amply accurate for the range of short and medium length cylinders of greatest practical interest.

Either of these equations, as well as the most important hinged-edge conditions at the ends w, v, $M_x = 0$, can be satisfied if we take the origin at the middle of the tube, and:

$$w = W \cos \frac{\pi x}{L} \cos \frac{ny}{R} \tag{7.8a}$$

We substitute this into Eq. (6.34), with $\mathbf{F}_y = -Rp$, $\mathbf{F}_x = -\alpha Rp/2$, where $\alpha = 0$ when only the cylinder wall is subjected to the external pressure, while $\alpha = 1$ if the cylinder end closures are both subjected to this pressure. Simplifying the result by multiplying through by $L^8/(\pi^4 Eh^3 w)$, we obtain the relation:

$$\frac{\pi^4}{12(1-\nu^2)}(1+N)^4+\gamma^4=\pi^2 P\gamma(1+N)^2\left(N+\frac{\alpha}{2}\right) \tag{7.8b}$$

where:

$$P=\frac{R^{3/2}L}{h^{5/2}}\frac{p}{E}, \qquad \gamma=\frac{L}{\sqrt{Rh}}, \qquad N=\frac{n^2L^2}{\pi^2R^2} \tag{7.8c}$$

Although n should theoretically be an integer to satisfy continuity in the circumferential direction, this is unimportant for the range of short and moderately long cylinders we are now studying, where, as we will find, n is large. Treating n and hence N as a number which can have any value, the smallest pressure p for which there can be equilibrium in the deformed shape (7.8a) is given when $\partial P/\partial N=0$, which yields, after simplification:

$$\frac{\pi^4}{3(1-\nu^2)}(1+N)^2=\pi^2 P\gamma(1+\alpha+3N) \tag{7.8d}$$

The quantity N can easily be eliminated between Eqs. (7.8b) and (7.8d), and P obtained as a function of γ for either of the values of α. The results of these calculations can not be expressed as a simple formula like Eq. (7.4b) for the axial compression case, but are most conveniently shown graphically in Fig. 7.12 in dashed lines for $\alpha=0$ and full lines for $\alpha=1$. From the values of $\sqrt{h/R}\,n=\pi\sqrt{N}/\gamma$ it can be seen that, in the range shown, n is more than 6 or 7 even for an R/h as small as 100 and is larger for thinner cylinders.

Also shown are all the test data* which were available at the time when the large deflection calculations given later were made. All the specimens buckled in one wave in the lengthwise direction as we have assumed, and the number of waves around the circumference checks quite well with the theoretical values. The tests were made with pressure on the end closures, $\alpha=1$, but the buckling strengths, measured by P, averaged 20 or 30 percent lower than the theoretical value, a discrepancy which must be attributed largely to initial defects.

Two test points are *above* the theoretical curve, which can mean either that the theory is inaccurate or that the test boundary conditions were more restrictive than the theory assumed. The theory is certainly an excellent approximation, but the tests were all made with nominally fixed end edges. A solution for the latter boundary condition would undoubtedly give theoretical strengths higher than all the experimental strengths. However, a solution for this case is much more difficult than the simple solution which we were able to obtain for the hinged-edge condition above. It could be found in the same way as for the torsion buckling case, details of which will be given in Sec. 7.4, but we will not consider this case here, since it would obviously only increase the discrepancy between classical stability strengths and most of the experimental strengths which must be

* *David Taylor Model Basin Reports* No. 997 and 1062, by A. F. Kirstein, R. C. Slankard, and E. Wenk, Jr., 1956, and paper by D. F. Windenburg and C. Trilling, *Trans. ASME*, vol. 56, 1934, p. 819.

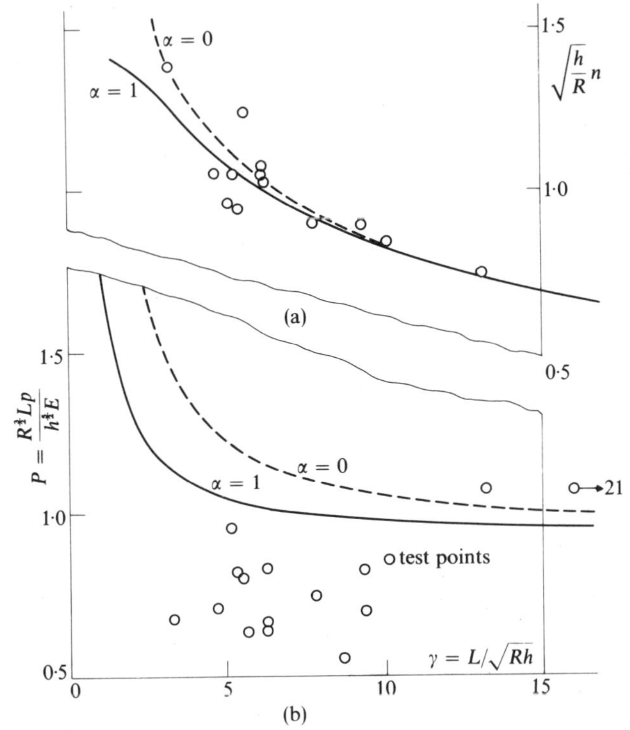

FIGURE 7.12
Classical stability curves for hinged-edge cylinders under external pressure

explained. In the discussion of the effects of initial defects below both hinged- and fixed-edge conditions will be considered.

As mentioned above, the approximate Eq. (6.34) is not a good approximation for the extreme case of a very long tube, for which the number of circumferential waves has its minimum value of two. For this case, at least in the central part of the tube, we can ignore variation of the deflection w in the axial direction and take:

$$w = W \cos \frac{2y}{R} \tag{7.8e}$$

Substituting this into Eq. (6.36), taking $\mathbf{F}_y = -Rp$, and multiplying through by R^5, we obtain:

$$\frac{Eh^3}{12(1-\nu^2)R^3}(2^8 - 2\cdot 2^6 + 2^4) = p\,2^4(2^2-1), \qquad p = \frac{Eh^3}{4(1-\nu^2)R^3} \tag{7.8f}$$

If we had used Eq. (6.34) in the same way, we would have obtained

$$\{Eh^3/[12(1-\nu^2)R^3]\}(2^8) = p\,2^4(2^2), \qquad p = Eh^3/[3(1-\nu^2)R^3],$$

or 4/3 of the correct value. The error should decrease very rapidly as n increases, and in the range shown in Fig. 7.12 should be negligible.

Substantially the same results as those found above for the stability of perfect hinged edge cylinder under external pressure were first obtained, in less simple form, by von Mises.*

Calculation of Strength of Cylinders with Defects Under External Pressure

The following calculations, made in 1956 and 1958,** consider only the case $\alpha = 1$, since the available test results and most applications are for this case; calculations for the case $\alpha = 0$ would require rather obvious modifications. Since the method used was the same as for the corresponding case of buckling under axial compression, the calculations for which were described in great detail in Sec. 7.2, Eqs. (7.5a, b), (7.6a, . . . , j), (7.7a, . . . , f), it will not be necessary to go into all the particulars here.

There are of course many important differences between the buckling failures induced by external pressure and by axial compression. Thus, the wavelength of the buckling deflection in the axial direction will no longer be arbitrary but will be fixed by the length of the cylinder. We will also find that the shape of the load–deformation curve given by large-deflection theory involves no sudden drop-off of resistance; in fact in most cases the resistance rises after buckling starts, and consequently the ultimate failure load is usually determined by yielding.

As before, we use Eq. (6.31j) with $\mathbf{b} = 1/R$ and $f_x, f_y = 0$, and use the combination of equilibrium and energy methods for solution. For the hinged-edge case, taking the origin in the middle of the cylinder, we can take the deflection w in the form:

$$w = ah\left[\cos\frac{\lambda nx}{R}\cos\frac{ny}{R} + b\left(\cos\frac{\lambda nx}{R} + 0{\cdot}3\cos\frac{3\lambda nx}{R}\right)\right] \tag{7.9a}$$

while for the fixed-edge case we take:

$$w = ah\left(1 + \cos\frac{2\lambda nx}{R}\right)\left(\cos\frac{ny}{R} + b\right) \tag{7.9b}$$

where $\lambda = \pi R/(nL)$ and a, b, and n are parameters which must be determined from energy considerations.

The primary displacements (the parts of the above expressions not involving b) satisfy the most important boundary conditions for the two cases: $x = \pm L/2$: $w, M_x = 0$ and $w, \partial w/\partial x = 0$, respectively. The secondary displacements (the parts

* 'The critical external pressure of cylindrical tubes under uniform radial and axial load'. *Stodola's Festschrift*, Zurich, Switzerland, 1929, p. 418. Von Mises also published the solution for the case $\alpha = 0$ (apparently with some inadvertent errors which were pointed out by D. F. Windenburg, *US Exp. Model Basin Report* 309) *in* 1914, *Z. Ver. deut. Ing.*, vol. 58, p. 750.

** L. H. Donnell, 'Effect of imperfections on buckling of thin cylinders under external pressure', *J. Appl. MEch.*, 1956, p. 569; L. H. Donnell, 'Effect of imperfections on buckling of thin cylinders with fixed edges under external pressure', *Proc. 3rd US Nat. Congr. of Appl. Mech., ASME*, 1958, p. 305.

of the above expressions proportional to b) can represent both the inward deflection of the cylinder due to the pressure (which we ignored in our classical stability study) as well as the more important axisymmetric inward deflection which is essential in all large-deflection theories to partially cancel the nonlinear circumferential strains due to circumferential waves with linear curvature-induced circumferential strains.

In the fixed-end case the secondary displacement also satisfies the fixed-end condition, as it obviously should. In the hinged-end case this condition is more complex. The second term in the parenthesis, Eq. (7.9a), should be zero in order for the secondary displacement to satisfy hinged-end conditions, as it should for example for a single cylinder hinged at the end edges. On the other hand, for the probably more important case (for submarine skins, for example) of a bay which forms one of many bays, with a continuous skin but separated by bulkhead rings of open section (so that they are rigid radially, but have little resistance to twist), the primary buckling wave would be inward in one bay and outward in the adjoining bays with nodes at the rings; on the other hand the axisymmetric secondary deflection would have to be inward in all bays, with a horizontal tangent at the rings, which would require a coefficient of 1/3 for the second term in the parenthesis, Eq. (7.9a). The coefficient of 0·3 used is a compromise between these extremes; its exact value probably does not have a major effect upon the results.

Again, as in the axial compression case, we consider only the component w_0 of the equivalent initial defects which has the same expression as w, but with a replaced by $a_0 = [(K-1)/2]a$, where $a_0 = (U/\pi^2)l_x^{1\cdot5}l_y^{0\cdot5}/h^2$ as in Eq. (7.5b). Taking l_x, l_y as the distances between the nodes of the primary deformation, we have $l_y = \pi R/n$, and $l_x = \pi R/(\lambda n)$ for the hinged-edge case while $l_x = \pi R/(2\lambda n)$ for the fixed-edge case. Using $\lambda = \pi R/(nL)$, $n = \pi R/(\lambda L)$, we find:

$$K = 1 + \frac{2UL^2\lambda^{1/2}}{a\pi^2h^2} \quad \text{(hinged edges)}$$
$$K = 1 + \frac{UL^2\lambda^{1/2}}{a\pi^2h^2\sqrt{2}} \quad \text{(fixed edges)} \qquad (7.9c)$$

As before we will use the symbol $k = Kahn^2/(2R)$, from which:

$$k = \frac{ahn^2}{2R} + \frac{1}{\lambda^{1\cdot5}}\frac{UR}{h} \quad \text{(hinged edges)}$$
$$k = \frac{ahn^2}{2R} + \frac{1}{(2\lambda)^{1\cdot5}}\frac{UR}{h} \quad \text{(fixed edges)} \qquad (7.9d)$$

Substituting the expression for w into Eq. (6.31j), using trigonometric transformations for the resulting squares and products of trigonometric functions, as tabulated in Table 6.3 and as used in developing Eq. (7.6b) for the axial compression case, and integrating the resulting expression for $\nabla^4\phi$ to obtain ϕ, as in Eq. (7.6c), we find for the hinged-edge case:

$$\phi = \frac{E\lambda^4 hL^2}{\pi^2 R} a\left[\frac{1}{(\lambda^2+1)^2}\cos\frac{\lambda nx}{R}\cos\frac{ny}{R} - k\left(\frac{1}{16\lambda^4}\cos\frac{2\lambda nx}{R} + \frac{1}{16}\cos\frac{2ny}{R}\right)\right.$$
$$- kb\cos\frac{ny}{R}\left(1 + \frac{3\cdot 7}{(4\lambda^2+1)^2}\cos\frac{2\lambda nx}{R} + \frac{2\cdot 7}{(16\lambda^2+1)^2}\cos\frac{4\lambda nx}{R}\right) \tag{7.9e}$$
$$\left. + \frac{b}{\lambda^4}\left(\cos\frac{\lambda nx}{R} + \frac{1}{30}\cos\frac{3\lambda nx}{R}\right)\right] + C_1x^2 + C_2y^2$$

For the fixed-edge case we obtain:

$$\phi = \frac{E\lambda^4 hL^2}{\pi^2 R} a\left[\frac{4}{(4\lambda^2+1)^2}\cos\frac{2\lambda nx}{R}\cos\frac{ny}{R}\right.$$
$$- k\left(\frac{1}{4\lambda^4}\cos\frac{2\lambda nx}{R} + \frac{1}{4}\cos\frac{2ny}{R} + \frac{1}{64\lambda^4}\cos\frac{4\lambda nx}{R} + \frac{1}{4(\lambda^2+1)^2}\cos\frac{2\lambda nx}{R}\cos\frac{2ny}{R}\right)$$
$$\left. + \frac{b}{4\lambda^4}\cos\frac{2\lambda nx}{R} - 4kb\cos\frac{ny}{R}\left(1 + \frac{2}{(4\lambda^2+1)^2}\cos\frac{2\lambda nx}{R} + \frac{1}{(16\lambda^2+1)^2}\cos\frac{4\lambda nx}{R}\right)\right]$$
$$+ C_1x^2 + C_2y^2 \tag{7.9f}$$

From the condition of equilibrium in the axial direction of one end of the cylinder, we have for both the hinged- and fixed-edge cases, using Eqs. (3.16a) and the above expressions for ϕ:

$$-\pi R^2 p = h\int_0^{2\pi R} dy\,\sigma_{xm} = h\int_0^{2\pi R} dy\frac{\partial^2\phi}{\partial y^2} = 4\pi RhC_2, \quad C_2 = -\frac{pR}{4h} \tag{7.9g}$$

The simplest way to determine C_1 and the volume change of the middle surface of the shell due to its deformation (which we will need in applying the law of virtual work) is probably to find expressions for the parts of the displacements u and v which are non-periodic in y. Calling these parts u' and v', using the above expressions for w and ϕ in Eqs. (6.31i) (with $\rho = 0$), integrating the first two equations with respect to x and y respectively and determining the functions of integration from the third equation, we find for the hinged-edge case:

$$u' = -\frac{h\lambda ka}{n}\left[\left(\frac{\lambda n}{4R}x + \frac{\nu-\lambda^2}{8\lambda^2}\sin\frac{2\lambda nx}{R}\right) - \frac{\nu b}{\lambda^2 k}\left(\sin\frac{\lambda nx}{R} + \frac{1}{10}\sin\frac{3\lambda nx}{R}\right)\right.$$
$$\left. + \frac{b^2}{4}\left(\frac{3\cdot 62\lambda n}{R}x + 0\cdot 8\sin\frac{2\lambda nx}{R} - 0\cdot 9\sin\frac{4\lambda nx}{R} - 0\cdot 27\sin\frac{6\lambda nx}{R}\right)\right] + \frac{2}{E}(C_2 - \nu C_1)x$$

$$v' = \left[-\frac{ahk}{4R} + \frac{2}{E}(C_1 - \nu C_2)\right]y \tag{7.9h}$$

For the fixed-edge case:

$$u' = \frac{hLka}{16\pi R}\left[\frac{8\nu(b-k)}{k}\sin\frac{2\lambda nx}{R} + (4\lambda^2 - \nu + 8\lambda^2 b^2)\sin\frac{4\lambda nx}{R}\right]$$
$$+ \left[\frac{2}{E}(C_2 - \nu C_1) - \frac{h\lambda^2 ka}{R}(1 + 2b^2)\right]x \tag{7.9i}$$
$$v' = \left[\frac{ha}{4R}(4b - 3k) + \frac{2}{E}(C_1 - \nu C_2)\right]y$$

For continuity in the circumferential direction v' must be zero (the parts of v which are periodic in y, which are omitted in v, obviously satisfy continuity). Using the above expressions and (7.9g) and solving for C_1, we find:

$$C_1 = \frac{Ehka}{8R} - \frac{\nu Rp}{4h} \qquad \text{(hinged edges)}$$

$$C_1 = \frac{Eha}{8R}(3k - 4b) - \frac{\nu Rp}{4h} \qquad \text{(fixed edges)} \tag{7.9j}$$

The decrease ΔV in volume of the cylinder is:

$$\Delta V = \pi R^2(u'_{(x=-L/2)} - u'_{(x=L/2)}) + \int_{-L/2}^{L/2} dx \int_0^{2\pi R} w\, dy \tag{7.9k}$$

Using the symbols:

$$P = \frac{R^{3/2}Lp}{h^{5/2}E}, \qquad \gamma = \frac{L}{\sqrt{Rh}}, \qquad V = \frac{\Delta V}{(Rh)^{3/2}} \tag{7.9l}$$

and expressions (7.9a, b, d, h, i), Eq. (7.9k) can be put in the form:

$$P = \frac{2V}{\pi(1-\nu^2)} - \frac{\gamma^3}{\pi^2(1-\nu^2)}\left(\lambda^2 k - \sqrt{\lambda}\frac{UR}{h}\right)\left[\frac{7{\cdot}2(2-\nu)b}{\pi} + k(\lambda^2 + \nu + 3{\cdot}62\lambda^2 b^2)\right]$$

$$\text{(hinged edges)} \tag{7.9m}$$

$$P = \frac{2V}{\pi(1-\nu^2)} - \frac{4\gamma^3}{\pi^2(1-\nu^2)}\left(\lambda^2 k - \sqrt{\frac{\lambda}{8}}\frac{UR}{h}\right)\left[(2-\nu)b + k\left(\lambda^2 + \frac{3\nu}{4} + 2\lambda^2 b^2\right)\right]$$

$$\text{(fixed edges)}$$

We now can use expressions (7.9a, b, e, f) for w and ϕ in Eqs. (4.70), (4.71) for the strain energy $\mathscr{E}$. Using expressions (7.9g, j) for C_1, C_2, the resulting expression will involve the pressure p, which we replace by P and in turn by the expressions (7.9m). Using $\nu = 0{\cdot}3$, we thus obtain for the hinged-edge case:

$$\frac{\pi^3 R^3}{EhL^5}\mathscr{E} = \frac{\pi^2}{1-\nu^2}\left(\frac{V}{\gamma^3}\right)^2 - \frac{V}{\gamma^3}\left(\lambda^2 k - \sqrt{\lambda}\frac{UR}{h}\right)[13{\cdot}5b + k(\lambda_1 + \lambda_2 b^2)]$$
$$+ \left(\lambda^2 k - \sqrt{\lambda}\frac{UR}{h}\right)^2\left[\frac{8{\cdot}92}{\gamma^4}\left(\frac{1}{\lambda_3} + 16{\cdot}6b^2\right) + \lambda_3 + 6{\cdot}35b^2\right.$$
$$\left. + kb(\lambda_4 + \lambda_5 b^2) + k^2(\lambda_6 + \lambda_7 b^2 + \lambda_8 b^4)\right] \tag{7.9n}$$

For the fixed case we have:

$$\frac{\pi^3 R^3}{EhL^5}\mathscr{E} = \frac{\pi^2}{1-\nu^2}\left(\frac{V}{\gamma^3}\right)^2 - \frac{V}{\gamma^3}\left(\lambda^2 k - \sqrt{\frac{\lambda}{8}}\frac{UR}{h}\right)[23{\cdot}5b + k(\lambda_9 + \lambda_{10} b^2)]$$
$$+ \left(\lambda^2 k - \sqrt{\frac{\lambda}{8}}\frac{UR}{h}\right)^2\left[\frac{8{\cdot}92}{\gamma^4}(\lambda_{11} + 32b^2) + \lambda_{12} + 18{\cdot}7b^2\right.$$
$$\left. + kb(\lambda_{13} + \lambda_{14} b^2) + k^2(\lambda_{15} + \lambda_{16} b^2 + \lambda_{17} b^4)\right] \tag{7.9o}$$

where λ_1, λ_2, etc., are functions of λ only.

Expressions (7.9n, o) for the strain energy involve only UR/h and γ (which will be known for any given cylinder), V and λ, k, b. The unknown $\lambda = \pi R/(nL)$ contains n, while k, given by Eqs. (7.9d) contains a, so λ, k, b are independent unknowns which can take the place of the original three unknowns n, a, b. As in the axial compression case, where we treated the axial shortening as constant in applying the law of virtual work so as to eliminate the work done by external forces, we will treat V and hence ΔV as constant (as it could theoretically be if the cylinder were immersed in an incompressible fluid in a perfectly rigid container) during virtual changes in λ, k, b. Hence by the law of virtual work the external pressure p will do no work during such changes, and $\partial\mathscr{E}/\partial\lambda$, $\partial\mathscr{E}/\partial k$, $\partial\mathscr{E}/\partial b = 0$. These three equations can then be solved for λ, k, b in terms of UR/h, γ, and V, after which a, n, P and the expressions for w and ϕ can be evaluated.

While these three equations are nonlinear in the unknowns it proved possible to solve them by assuming values for some of the quantities and solving for others, cross plotting, etc. In the fixed edge case more difficulty was encountered and an approximate method of determining λ was used, which trial showed to be reliable. Further details of these calculations are given in the 1956 and 1958 papers cited at the beginning of this discussion of cylinders with defects under external pressure.

Figure 7.13 shows some results of these calculations. Figure 7.13(a) shows typical load–deformation curves for medium length cylinders as defined by the geometric characteristic $\gamma = L/\sqrt{Rh} = 9{\cdot}7$. Comparison of this with the comparable figure for axial compression, Fig. 7.8, shows a great contrast in postbuckling behavior. Figure 7.13(a) shows that for fixed edges there is practically no drop-off

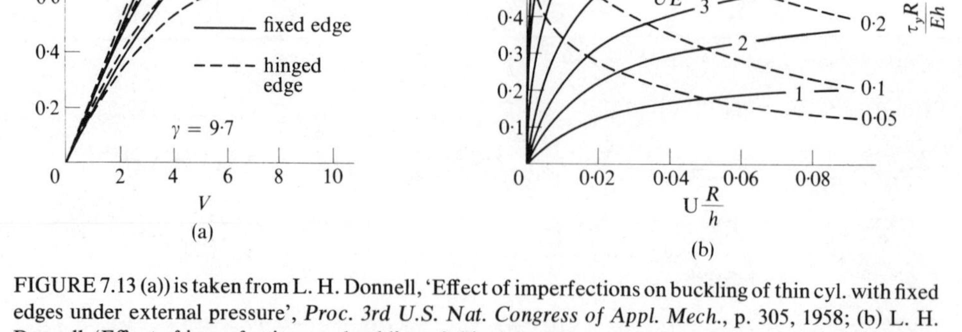

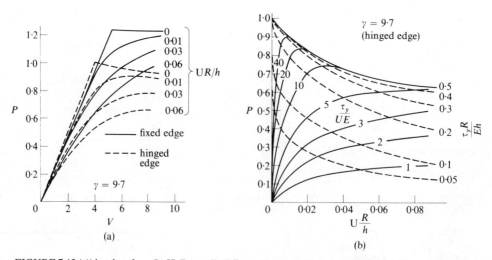

FIGURE 7.13 (a)) is taken from L. H. Donnell, 'Effect of imperfections on buckling of thin cyl. with fixed edges under external pressure', *Proc. 3rd U.S. Nat. Congress of Appl. Mech.*, p. 305, 1958; (b) L. H. Donnell, 'Effect of imperfections on buckling of thin cyl. under external pressure', *Trans. ASME*, vol. 78, p. 569, 1956, both courtesy of ASME)

in resistance after buckling, while for hinged edges the drop-off is slight. For short cylinders there is no drop-off for either case, and the load curves continue to rise indefinitely if elastic conditions are assumed.

As a consequence, as noted at the beginning of this discussion, most failure of cylinders under external pressure must be initiated by yielding at the most highly stressed point. Trials and test indications showed that yielding should occur first at:

$$\begin{aligned} &x = \pi R/(2\lambda n), \qquad y = \pi R/(2n), \qquad z = h/2 \qquad \text{(hinged edges)} \\ &x = 0, \qquad y = \pi R/n, \qquad \text{or} \quad x, y = 0, \qquad z = \pm h/2 \qquad \text{(fixed edges)} \end{aligned} \tag{7.9p}$$

Using Eqs. (7.7d) in (7.7c) and expressions (7.9a, b, e, f) for w, ϕ we can obtain relations between P, UR/h and τ_y/UE or $\tau_y R/Eh$ as we did in Eq. (7.7f) for the axial compression case. Figure 7.13(b) shows these relations for hinged edges for the same case $\gamma = 9{\cdot}7$ shown in Fig. 7.13(a). It can be seen that these curves are very similar to Fig. 7.10 for cylinders under axial compression.

However, curves such as those of Fig. 7.13(b) are inconvenient to use because we must have different sets of curves for different values of γ, and interpolation between such sets of curves would be difficult. Fortunately it was found that, at least in the range studied, plots of P against corresponding values of $(\tau_y/E)\sqrt{R/(Uh)} = [\tau_y R/(Eh)]/\sqrt{UR/h}$ fall very nearly upon one line for all values of γ and of $\tau_y R/(Eh)$. These lines are shown in Fig. 7.14(a). Even the lines for elastic buckling can be included, by substituting for them the lines $\tau_y R/(Eh) = 1/2$, which practically coincide with the elastic buckling lines, as can be seen in Fig. 7.13(b) for $\gamma = 9{\cdot}7$. The P vs. $(\tau_y/E)\sqrt{R/(Uh)}$ points do not fall as closely on a single line for the hinged-edge case as they do for the fixed-edge case, but they

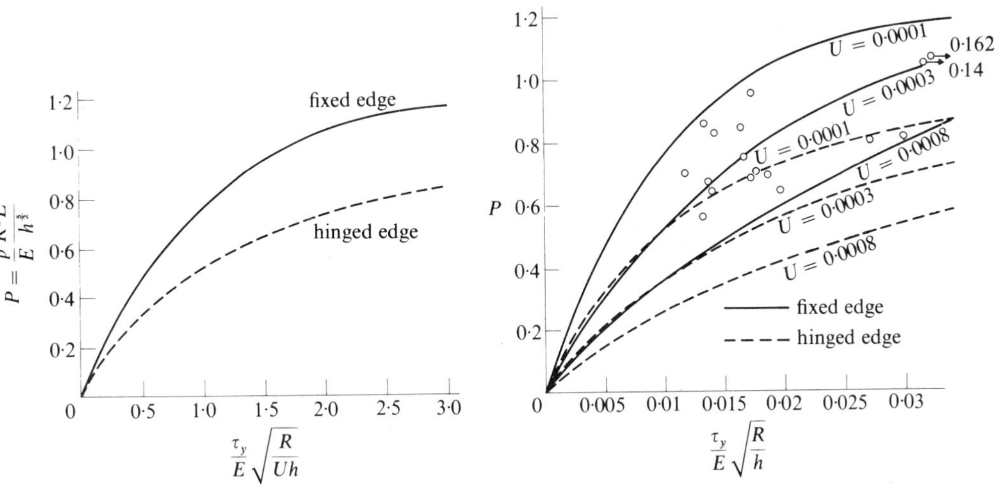

FIGURE 7.14 (from L. H. Donnell, 'Effect of imperfections on buckling of thin cyl. with fixed edges under external pressure', *Proc. 3rd U.S. Nat. Congress of Appl. Mech.*, p. 305, 1958, courtesy of ASME)

cluster about it with maximum deviations of about ±10 percent in P, and it is felt that both lines in Fig. 7.14(a) give as good a picture of the resistance of thin cylinders to external pressure for both edge conditions as can be obtained, considering the uncertainties inherent in actual cylinders with defects.

Figure 7.14(b) shows the same information as in Fig. 7.14(a) but with P plotted against $(\tau_y/E)\sqrt{R/h}$ for different values of U, and with points for the same tests as were plotted on Fig. 7.12. It will be seen that it requires values for U between 0·0001 and 0·0008 to explain the test results, which is consistent with the finding of similar studies of struts and of cylinders under axial compression.

7.4 BUCKLING OF THIN CYLINDERS UNDER TORSION

Introduction

As mentioned at the beginning of the last section, this case is similar to that of cylinders under circumferential compression in that in both cases the buckling waves extend from one end of the cylinder to the other and end-edge boundary conditions cannot be ignored. Also in both cases the number n of waves around the circumference is large for short and medium length cylinders, decreasing as the length increases and taking its minimum value of two only for very long tubes. As in the previous case, reference to Fig. 7.2 shows that the approximate Eq. (6.34) (which was originally developed for this torsion buckling case) gives an excellent approximation except for the limiting case of $n = 2$, where it overestimates the buckling stress by about 13 percent.

Because the waves spiral around the cylinder it is not possible to satisfy edge boundary conditions in such simple ways as we could in obtaining classical stability solution (7.4b) for cylinders under axial compression, and (7.8b, d), Fig. 7.12, for cylinders under circumferential compression. In the following we will primarily follow the author's 1933 study,* which was the first satisfactory general solution of this problem, and which was based on Eqs. (6.34), (6.33c); for the case $n = 2$ we will improve on that study by using Eq. (6.36).

Stability of Perfect Thin Cylinders Under Torsion

In the prebuckling stage of the loading we assume that for thin cylinders the shearing stress is uniformly distributed over the cylinder cross sections, being $\sigma_{xy} = T/(2\pi R^2 h)$ where T is the torque applied to the cylinder. When the classical stability limit is reached $\sigma_{xy} = S = -\mathbf{F}_{xy}/h$ where $\mathbf{F}_{xy}$ was defined under Eq. (6.27d) (the negative sign has no particular significance in this case, being needed because of a difference between our present sign conventions and those used in the 1933 study).

* Loc. cit. Eq. (6.34), p. 365.

For the buckling stage of loading we will use the symbols:

$$A = (1-\nu^2)\frac{S}{E}\frac{L^2}{h^2}, \qquad B = \sqrt{1-\nu^2}\,\frac{S}{E}\frac{L}{h}, \qquad H = \frac{\sqrt{1-\nu^2}}{2}\frac{L^2}{Rh},$$

$$J = \frac{1}{8\sqrt{1-\nu^2}}\frac{L^2h}{R^3}, \qquad k = \frac{nL}{2R} \tag{7.10a}$$

Taking the origin of coordinates in the middle of the cylinder as before, we take the displacements in the form:

$$w = \sum_m W_m \cos\frac{n}{R}\left(\frac{\lambda_m}{k}x + y\right)$$

$$v = \sum_m V_m \sin\frac{n}{R}\left(\frac{\lambda_m}{k}x + y\right) \tag{7.10b}$$

$$u = \sum_m U_m \sin\frac{n}{R}\left(\frac{\lambda_m}{k}x + y\right)$$

where n is an integer and λ_m is a number which may be complex. However, trial shows that the important λ_m's are real and λ_m/k evidently represents the tangent of the spiral angle (or roughly the angle itself) which the corresponding component of the buckling deflection makes with the axial direction.

From tests and an exact solution for long flat strips under shear,* which is the limiting case for cylinders under torsion as their length approaches zero, we know that the angle of twist of the whole deflection has a maximum value of about 45° for very short tubes and rapidly decreases as the length increases. Tests as well as the results obtained in the following solution (see Fig. 7.17(b)) indicate that all the values of λ_m/k are small enough so that their squares can be neglected compared to unity without serious error for all except very short cylinders. This fact forms the basis for simplifications which greatly reduced the work of calculation; they would otherwise have been impractically laborious, since at the time when they were made computer help was unavailable.

Substituting expressions (7.10b) into Eqs. (6.34) and (6.33c) we find that in order for these to be satisfied for all values of x, y the following relations must be satisfied for each value of m:

$$k^8(1+\lambda_m^2/k^2)^2 + 3\left(\frac{H}{1+\lambda_m^2/k^2}\right)^2\lambda_m^4 = 6Ak^5\lambda_m \tag{7.10c}$$

$$U_m = \frac{W_m\lambda_m}{nk}\,\frac{1-\nu\lambda_m^2/k^2}{(1+\lambda_m^2/k^2)^2}$$

$$V_m = -\frac{W_m}{n}\,\frac{1+(2+\nu)\lambda_m^2/k^2}{(1+\lambda_m^2/k^2)^2} \tag{7.10d}$$

* R. V. Southwell and Sylvia W. Skan, 'On the stability under shearing forces of a flat elastic strip', *Proc. Roy. Soc. London*, vol. 105A, 1924, p. 582.

If we neglect λ_m^2/k^2 compared to one in Eqs. (7.10d) we obtain the simplified expressions for U_m, V_m:

$$U_m = \frac{W_m \lambda_m}{nk}, \qquad V_m = -\frac{W_m}{n} \tag{7.10e}$$

The error introduced by this approximation is zero when $L/R = 0$, because at this extreme n approaches infinity and so U_m and V_m are zero; we would reach the same conclusion by noting that at this extreme the cylinder approaches a flat strip. The error is also negligible at the opposite extreme when $L/R \to \infty$ because λ_m/k then becomes very small. The error is also small for any intermediate case because when λ_m/k is not small compared to one, n is large and U_m and V_m are of little importance. For example when $L/R = 2$ the error in V_m is about 3 percent and in U_m about 14 percent, but investigation of the final results shows that V_m is of little importance here (becoming important only when L/R is large) while U_m is never important. Hence we can use Eqs. (7.10e) for U_m, V_m for the entire range without serious error.

Simplification of Eq. (7.10c) is a little more complicated, but it can be done in several ways. The simplest way is of course merely to neglect λ_m^2/k^2 compared to one; the result of doing this, after multiplying by $n^4h^2/[4(1-\nu^2)R^2]$, can be written in the form:

$$\lambda_m^4 - 2n^5 BJ\lambda_m + n^8J^2/3 = 0 \tag{7.10f}$$

This would give a very poor approximation for very short tubes where λ_m^2/k^2 is not very small compared to one, but it is an excellent approximation for medium length and long tubes, and we can obtain most of our results in the range of practical interest from it.

For the range of shorter cylinders we can obtain a worthwhile approximation by taking:

$$H/(1+\lambda_m^2/k^2) = H' \tag{7.10g}$$

where H' is taken as independent of λ_m until we obtain a solution. This gives:

$$(k^4 + 3H'^2)\lambda_m^4 + 2k^6\lambda_m^2 - 6Ak^5\lambda_m + k^8 = 0 \tag{7.10h}$$

The error introduced by this is zero for $l/R = 0$ since then $H = 0$ (and the error is also very small for $l/R \to \infty$ where λ_m/k is very small). For intermediate cases (7.10h) gives a fairly good first approximation because when the error in neglecting λ_m/k is fairly great H is small and not very important. However, we can get a much better second approximation, where it is needed, by taking $H = H'(1+\lambda_m^2/k^2)$ where λ_m^2/k^2 is taken as a weighted average of the values of λ_m^2/k^2 found by using the first approximation. In applying this, it is found that the value of A (and therefore of S) found for a given H by the second approximation in the range where the two approximations differ the most (for $H \approx 10$) is about 20 percent less than that given by the first approximation; hence if general experience is a safe guide the error in the second approximation should not exceed a few percent, and this is borne out by the tests.

Edge Boundary Conditions

Both equations (7.10f) and (7.10h) are of fourth degree in λ_m, so there are four roots of the equation for any given set of values of H or J (which will be known in any application) and n or k, A or B (which will be unknown). The four values of λ_m and the corresponding values of W_m, U_m, V_m must be used to satisfy Eqs. (7.10e), (7.10f) or (7.10h) and the boundary conditions, to give us the desired relation between n or k, A or B and H or J.

As we have discussed previously, we should be able to satisfy two flexural and two membrane conditions on all boundaries. The conditions on longitudinal sections are replaced by continuity conditions which are automatically satisfied since we have taken our displacements as cyclical about the circumference. We take the end-edge boundary conditions as $x = \pm L/2$:

$$w = 0, \qquad \frac{\partial w}{\partial x} = 0 \quad \text{(fixed)}, \qquad \frac{\partial^2 w}{\partial x^2} + \nu \frac{\partial^2 w}{\partial y^2} = 0 \quad \text{(hinged)} \tag{7.10i}$$
$$v = 0, \qquad u = 0$$

Substituting Eqs. (7.10b) for u, v, w, and (7.10e) for U_m, V_m into these conditions, simplifying them as much as possible by dividing through by common factors and combining them, and using trigonometric formulas for the sines and cosines of the sum of two numbers, we find that all but the unimportant condition $u = 0$ for hinged edges can be satisfied for all values of y if:

$$\sum_m W_m \sin \lambda_m = 0, \qquad \sum_m W_m \cos \lambda_m = 0$$
$$\sum_m W_m\lambda_m \sin \lambda_m = 0, \qquad \sum_m W_m\lambda_m \cos \lambda_m = 0 \quad \text{(fixed edges)} \tag{7.10j}$$
$$\sum_m W_m\lambda_m^2 \sin \lambda_m = 0, \qquad \sum_m W_m\lambda_m^2 \cos \lambda_m = 0 \quad \text{(hinged edges)}$$

Southwell and Skan* obtained the same conditions for the limiting case of a flat strip under shear, and showed that they could be satisfied if the four λ_m s are related as follows:

$$\begin{aligned}(\lambda_1 - \lambda_2)(\lambda_3 - \lambda_4) \sin (\lambda_1 - \lambda_3) \sin (\lambda_2 - \lambda_4) \\ = (\lambda_1 - \lambda_3)(\lambda_2 - \lambda_4) \sin (\lambda_1 - \lambda_2) \sin (\lambda_3 - \lambda_4)\end{aligned} \quad \text{(fixed edges)}$$
$$\begin{aligned}(\lambda_1^2 - \lambda_2^2)(\lambda_3^2 - \lambda_4^2) \sin (\lambda_1 - \lambda_3) \sin (\lambda_2 - \lambda_4) \\ = (\lambda_1^2 - \lambda_3^2)(\lambda_2^2 - \lambda_4^2) \sin (\lambda_1 - \lambda_2) \sin (\lambda_3 - \lambda_4)\end{aligned} \quad \text{(hinged edges)} \tag{7.10k}$$

Trial shows that, as in the flat strip case, the four λ_m s can be written as follows:

$$\lambda_1 = \mathbf{a} + \mathbf{b}, \qquad \lambda_2 = \mathbf{a} - \mathbf{b}, \qquad \lambda_3 = -\mathbf{a} + i\mathbf{c}, \qquad \lambda_4 = -\mathbf{a} - i\mathbf{c} \tag{7.10l}$$

where **a**, **b**, **c** are positive real numbers and $i = \sqrt{-1}$. Substituting expressions

* Loc. cit. Eq. (7.10b).

(7.10 l) into the boundary conditions (7.10k) and using well known relations between trigonometric functions of imaginary numbers and hyperbolic functions of real numbers, these relations can be put in the form:

$$4\mathbf{a}^2 = \mathbf{b}^2 - \mathbf{c}^2 + \frac{2\mathbf{bc}}{N \tan 2\mathbf{b}} \quad \text{(fixed edges)} \tag{7.10m}$$

$$4\mathbf{a}^2 = \frac{(\mathbf{b}^2 + \mathbf{c}^2)^2}{\mathbf{b}^2 - \mathbf{c}^2 - \dfrac{2\mathbf{bc}}{N \tan 2\mathbf{b}}} \quad \text{(hinged edges)} \tag{7.10n}$$

where $N = \tanh 2\mathbf{c}/[1 - \cos 4\mathbf{a}/(\cos 2\mathbf{b} \cosh 2\mathbf{c})]$ can be taken as unity for the values of **a**, **b**, **c** which trial shows we need to consider.

The general equation of which expressions (7.10 l) are the roots is:

$$[\lambda_m - (\mathbf{a}+\mathbf{b})][\lambda_m - (\mathbf{a}-\mathbf{b})][\lambda_m - (-\mathbf{a}+i\mathbf{c})][\lambda_m - (-\mathbf{a}-i\mathbf{c})] = 0, \quad \text{or:}$$

$$\lambda_m^4 - (2\mathbf{a}^2 + \mathbf{b}^2 - \mathbf{c}^2)\lambda_m^2 - 2\mathbf{a}(\mathbf{b}^2 + \mathbf{c}^2)\lambda_m + (\mathbf{a}^2 - \mathbf{b}^2)(\mathbf{a}^2 + \mathbf{c}^2) = 0 \tag{7.10o}$$

Equating the coefficients in this equation to those of equilibrium condition (7.10f), which we found to be a good approximation for long and medium length cylinders, we find that:

$$2\mathbf{a} + \mathbf{b}^2 - \mathbf{c}^2 = 0, \qquad \mathbf{a}(\mathbf{b}^2 + \mathbf{c}^2) = n^5 BJ$$
$$3(\mathbf{a}^2 - \mathbf{b}^2)(\mathbf{a}^2 + \mathbf{c}^2) = n^8 J^2 \tag{7.10p}$$

We can now readily eliminate **a**, **b**, **c** between Eqs. (7.10p) and either (7.10m) or (7.10n) by algebraic and graphical means and obtain sets of corresponding values of n^8J^2 and n^5BJ, examples of which are shown in the upper part of Table 7.1. From these, for any value of $n = 2, 3, 4, \ldots$, we can calculate corresponding values of J and B.

Figure 7.15 shows these relations between B (which defines the buckling stress S) and J plotted on a logarithmic scale for several values of n. Only the portion of each curve which is below the other curves, that is the portion between intersections with the adjacent curves, has practical significance and is shown, as buckling will occur at the lowest stress at which equilibrium in a buckled state can exist. In the range covered by each segment buckling should theoretically occur with the number of circumferential values n which corresponds to that segment.

The curves thus obtained for $n = 3, 4, 5, 6$ are shown by full lines and the curves for $n = 2$ by dashed lines. As discussed at the beginning of this article, Eqs. (7.10p), are based on Eq. (6.34) and are not very accurate for $n = 2$. If we go through the same development by using Eq. (6.36) instead of Eq. (6.34) we obtain the same equations as (7.10p) except in the second equation n^5BJ is changed to $(n^5 - n^3)BJ$, while in the third equation n^8J^2 is changed to $(n^8 - 2n^6 + n^4)J^2$. With these modifications we obtain the full line curves for $n = 2$ which, as was mentioned at the beginning, gives values of B and hence S about 13 percent lower than the values found by using the unmodified Eq. (7.10p). The modified equations could of course also be used for the other values of n, but the difference is minor for them, even for $n = 3$.

Table 7.1

a	b	c	n^8J^2	n^5BJ	$B/J^{1/4}$
			Fixed edges		
2·75	1·728	4·25	348	57·7	1·49
4·10	1·669	6·03	2236	160·6	1·293
5·62	1·641	8·13	8442	387	1·36
7·36	1·623	10·54	25620	839	1·47
			Hinged edges		
2·16	1·342	3·33	135	27·8	1·29
3·21	1·391	4·74	822	78·3	1·182
3·81	1·414	5·57	1707	125·7	1·20
5·17	1·449	7·44	6046	297	1·29

a	b	c	k	H'	H	A
			Fixed edges			
1·977	1·804	4·334	1·96		0	7·39
2·03	1·796	4·52	2·18	0·67	1·69	7·73
2·14	1·781	4·81	2·71	2·58	5·30	9·47
2·47	1·751	5·22	4·10	13·0	20·1	17·04
2·91	1·721	5·59	6·21	53·6	69·5	36·3
3·53	1·691	5·95	10·92	377	426	128
3·86	1·678	6·06	17·60	2180	2300	440
			Hinged edges			
1·445	1·383	2·977	1·18		0	4·40
1·53	1·390	3·16	1·73	1·52	3·69	6·22
1·70	1·395	3·43	2·61	6·07	10·38	10·06
2·15	1·404	4·06	4·63	33·1	43·6	23·3
2·56	1·404	4·46	7·11	131	153	55·3
2·94	1·410	4·72	11·60	726	783	180

As can be seen from the figure, for J very large, say $J > 20$, the curves for the two edge conditions merge together and approach asymptotically a straight line. The equation for this line can be obtained in various ways, for example from the observation that as J increases, **a** and **c** increase without limit while **b** approaches $\pi/2$. Hence, from the first of Eqs. (7.10p), in the limit $\mathbf{c}^2 = 2\mathbf{a}^2$. Using this in the modified second and third equations and ignoring **b** compared to **a**, they become:

$$2\mathbf{a}^3 = (n^5 - n^3)BJ, \qquad 9\mathbf{a}^4 = (n^8 - 2n^6 + n^4)J^2$$

Raising the third equation to the 3/4 power and dividing by the second to eliminate **a**, and setting $n = 2$, the resulting relation can be expressed as:

$$S = 0{\cdot}253E\left(\frac{h}{R}\right)^{3/2} \qquad \text{(for } J > 20\text{)} \qquad (7.10q)$$

which is almost the same as the formula which was found for this case by Schwerin* in 1924, shown on Fig. 7.15 by the diagonal dotted line.

* *Proc. of First Int. Congr. for Appl. Mech.*, Delft, p. 255.

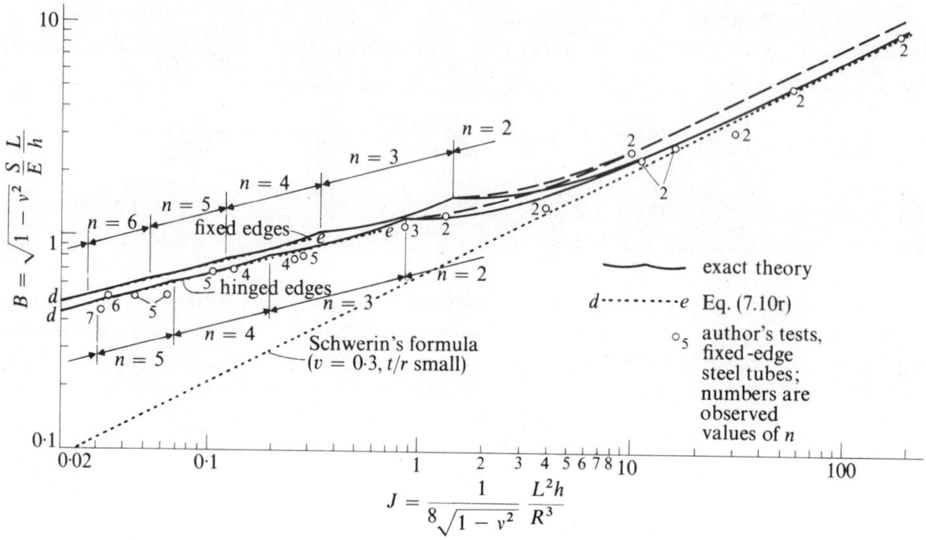

FIGURE 7.15
Critical torsion stress for slender tubes (from L. H. Donnell, NACA Report No. 479, 1933, courtesy of NASA)

Shown also on Fig. 7.15 are a dozen or so experimental points. These were made with fixed edges, and the observed value of n is shown for each. It can be seen that the observed buckling strengths are 10 percent or so less than those predicted by theory, which represents a rather moderate effect of initial defects compared to that for the cases of axial and circumferential compression. The number n of circumferential waves also checks closely with that predicted by the theory.

At the left end of Fig. 7.15, as J decreases and n increases, the jagged lines made up of the lower portions of the curves for different values of n approach closely the dotted straight lines *de*, which represent the envelopes of the jagged lines. These envelopes can be used without serious error for obtaining the buckling strengths for $J < 1/2$ or $n > 3$. The equations of these envelopes can be found by choosing n to minimize S or B while ignoring the fact that n should theoretically be an integer.

These equations can easily be found by calculating from our sets of values in Table 7.1 values for $n^5BJ/(n^8J^2)^{5/8} = B/J^{1/4}$. When this is plotted against n^8J^2, we find that the minimum value of $B/J^{1/4}$, and hence of B or S for any value of J, occurs when $n^8J^2 = 2236$ (fixed edges) or 822 (hinged edges) and has values of 1·293 and 1·182 respectively. From this the smallest critical loading and the number of circumferential waves n at which it occurs is given by:

$$B = 1{\cdot}293J^{1/4}, \qquad n = 2236^{1/8}/J^{1/4} = 2{\cdot}62/J^{1/4} \quad \text{(fixed edges)}$$
$$B = 1{\cdot}182J^{1/4}, \qquad n = 822^{1/8}/J^{1/4} = 2{\cdot}31/J^{1/4} \quad \text{(hinged edges)} \qquad (7.10r)$$

In order to show the complete range of short and medium length cylinders, it is necessary to use different variables. By multiplying the above expressions for B by $\sqrt{1-\nu^2}L/h$ and for n by $L/2R$, Eqs. (7.10r) become:

$$\begin{aligned} A &= 1{\cdot}293H^{3/4}, \qquad k = 2{\cdot}62H^{1/4} \quad \text{(fixed edges)} \\ A &= 1{\cdot}182H^{3/4}, \qquad k = 2{\cdot}31H^{1/4} \quad \text{(hinged edges)} \end{aligned} \tag{7.10s}$$

These relations were used to plot the right hand side of Figs. 7.17(a),(c) given later. To fill in the rest of these figures for shorter cylinders it is necessary to use Eqs. (7.10g,h) instead of (7.10f).

Working in the same way as above, if we equate the coefficients of Eq. (7.10o) to those of Eq. (7.10h) divided through by $(k^4+3H'^2)$, we obtain the following relations for determining **a**, **b**, **c**:

$$\begin{gathered} 2\mathbf{a}^2+\mathbf{b}^2-\mathbf{c}^2 = -\frac{2k^6}{k^4+3H'^2}, \qquad \mathbf{a}(\mathbf{b}^2+\mathbf{c}^2) = \frac{3Ak^5}{k^4+3H'^2} \\ (\mathbf{a}^2-\mathbf{b}^2)(\mathbf{a}^2+\mathbf{c}^2) = \frac{k^8}{k^4+3H'^2} \end{gathered} \tag{7.10t}$$

We must now eliminate **a**, **b**, **c** between these equations and the boundary conditions (7.10m) or (7.10n) and obtain the required relations between k, H', and A. This is more difficult than the previous calculations using Eq. (7.10f). Particular sets of values were found as follows.

For a first approximation, values of **b** and **c** are assumed, and the value of **a** found from (7.10m) or (7.10n). The value of k is then found from an equation obtained by dividing the third by the first equation of (7.10t), after which H' can be found from the first and A from the second of these equations. To get a second approximation the value of H is computed from Eq. (7.10g), using a weighted average of the values of λ_m^2/k^2 found from the values of **a**, **b**, **c** used in the first approximation. Trial shows that the terms containing λ_3 and λ_4 are very small compared to those containing λ_1 and λ_2, so we will make an error of only a percent or so if we ignore them and take

$$H = H'\left(1+\frac{\lambda_1^2+\lambda_2^2}{2k^2}\right) = H'\left(1+\frac{\mathbf{a}^2+\mathbf{b}^2}{k^2}\right) \tag{7.10u}$$

We now have corresponding values of A and H satisfying equilibrium and the boundary conditions and thus representing possible buckling configurations. But the original choice of **b** and **c** was guesswork, and with different values we may obtain lower values of A and therefore of S for the same value of H. We must therefore find the value of **b** and **c** which give the lowest value of A for each value of H. If we had to try values of **b** and **c** blindly the task would be very difficult as only a small range of values even result in real values of **a**, k, A, and H.

Fortunately we know the values of **b** and **c** for the extreme cases when $H=0$, given by Southwell and Skan's solution for the flat strip under shear, and for $H\to\infty$, given by our previous solution given above. These sets of values are represented by the points p and q, Fig. 7.16. The desired values for **b** and **c** for intermediate values of H are evidently given by points on lines connecting p and

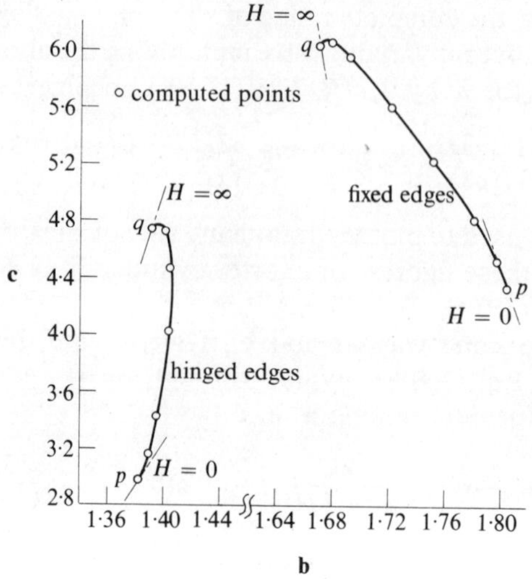

FIGURE 7.16
Values found for **b** and **c** from $H = 0$ to $H = \infty$ (from L. H. Donnell, NACA Report No. 479, 1933, courtesy of NASA)

q. By trying a number of points distributed over the area between p and q, plotting the results on Fig. 7.17(a) and making use of cross plotting, we locate the lines shown in Fig. 7.16, points on either side of which give points above the full- and dashed-line curves in Fig. 7.17(a).

Sets of values of **a**, **b**, **c**, k, H', H, and A found in this way are given in the lower part of Table 7.1 and plotted in Fig. 7.17. The approximate values of the angle θ which the spiral buckling waves make with the axial direction are found from:

$$\theta \approx \tan^{-1}\frac{\lambda_1 + \lambda_2}{2k} = \tan^{-1}\frac{\mathbf{a}}{k} \tag{7.10v}$$

The results of all the tests on metal cylinders which were available when this work was done are also plotted; all were with fixed edges. It can be seen that the classical stability theory predicts the shape of the buckling deflections and the general trend of the buckling strengths remarkably well considering the enormous range of sizes, proportions, and materials which these results represent, but the experimental strengths are consistently lower than those predicted by classical stability theory, ranging from a minimum of about 60 percent up to close to 100 percent of the theoretical value. To explain this discrepancy we must consider the initial defects.

Comparison of Classical Stability Strengths Under Different Loadings

Before doing that it is of interest to note that the classical stability limits for the three basic cases of axial and circumferential compression and shear can be

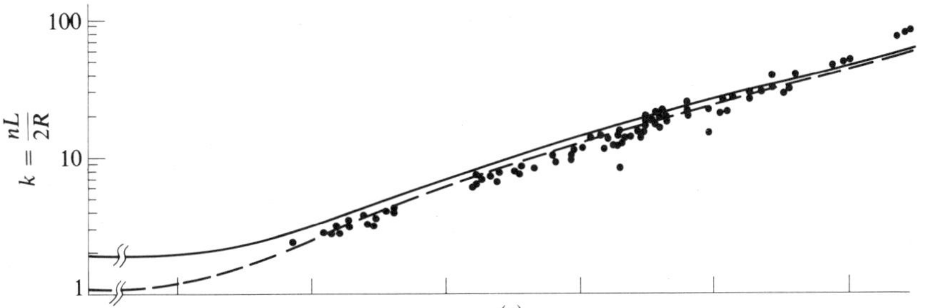

(c)

(c) Number of circumferential waves (from L. H. Donnell, NACA Report No. 479, 1933, courtesy of NASA)

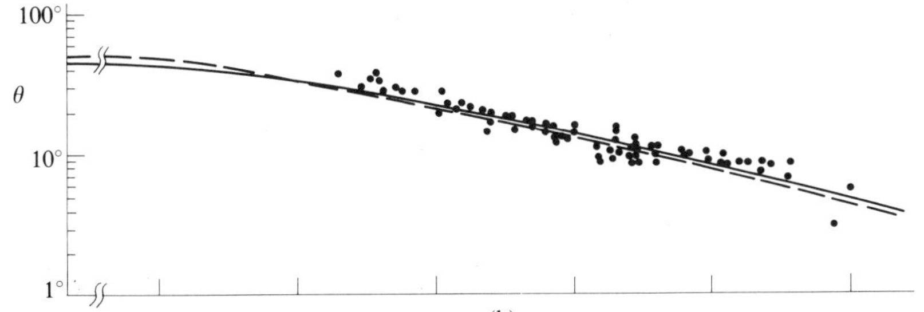

(b)

(b) Spiral angle of waves

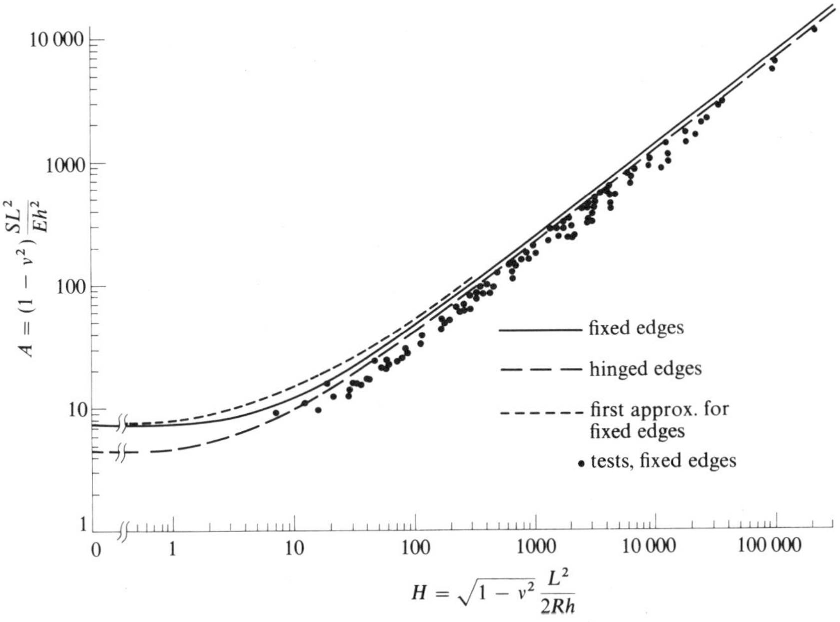

(a)

(a) Stability limit of cylinders under torsion

FIGURE 7.17

directly compared by plotting $\sigma R/(Eh)$ against $\gamma = L/\sqrt{Rh}$, where σ is the buckling stress in each case, as shown in Fig. 7.18. The differences which are evident in this comparison can be simply and plausibly explained by the different effects on the three cases of two major stabilizing factors: (1) curvature, (2) the end constraint against w displacements and hence against waving. For long cylinders the effect of end constraints is unimportant and curvature is the dominant factor. The axial compression case utilizes the full stabilizing effect of the curvature of the cylinder wall, as can easily be experienced by comparing the axial crushing resistance of a rolled up sheet of paper to its resistance when flat. Curvature has little or no particular stabilizing effect against circumferential compression

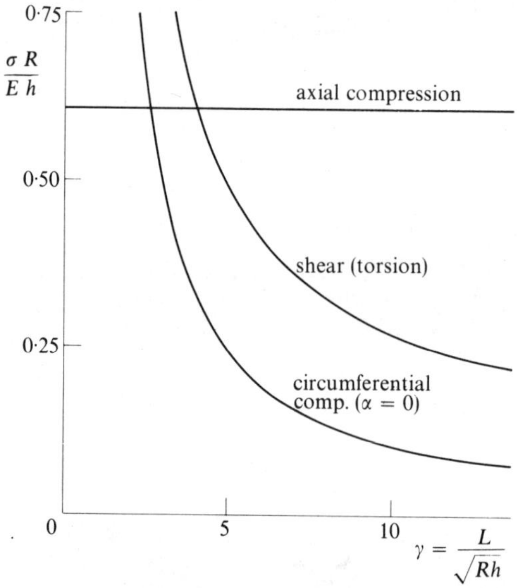

FIGURE 7.18

buckling—a long rolled up sheet of paper shows no more resistance to circumferential waving than before rolling. Shear buckling is an intermediate case; the waves are at an angle and so utilize curvature only partially.

For short cylinders the effect of end constraint is very important for the two cases of circumferential and shear buckling, because their waves go from end to end and so are greatly restrained by constraints against waving at the ends. On the other hand, as we have discussed before, end constraint has little effect on axial compression buckling (unless the cylinder is extremely short, a case which we did not consider and is not included here); along the length waving is zero anyway at every node.

Calculation of Strength of Cylinders with Defects Under Torsion

This problem has been studied by T. T. Loo,* using somewhat similar methods to those which have been described in the previous two articles for the axial and circumferential compression cases. In this study the deflection was assumed to have the following shape, using the origin of coordinates and symbols similar to those used previously:

$$w = ah\left[\cos\frac{n}{R}\left(\frac{\lambda_0}{k}x + y\right)\cos\frac{\pi x}{L} + b\cos\frac{2\pi x}{L}\right] \tag{7.11a}$$

The initial deviation was taken to have a value $w_0 = (a_0/a)w$. A constant term was also used in Eq. (7.11a), like the d in Eq. (7.4c), but, as in that case, this term drops out in the differentiations and does not affect the following results. By taking $\lambda_0 = 0$ this expression could also be applied to the axial compression case, as it resembles Eq. (7.4c) except for the latter's relatively unimportant third term proportional to c.

As in our previous studies of cylinders with defects using large-deflection theory, expression (7.11a) was substituted into Eq. (6.31j) with $\mathbf{b} = 1/R$, which was integrated to find the stress function of the membrane stresses ϕ, using the solution of the homogeneous equation $-Sxy$ (or $-\sigma y^2/2$ for application to the axial compression case). The resulting expression for ϕ and expression (7.11a) for w were then used in Eqs. (4.70), (4.71) for the strain energy, after which the unknowns a, n, λ_0, b could be found as in the previous cases by use of the law of virtual work.

However, to reduce the great mathematical difficulties of having to solve the resulting four nonlinear equations, the approximate assumption was made that λ_0/k (which evidently represents the tangent of the spiral angle θ of the deformation, and hence approximately the angle itself) and the number of circumferential waves n, have the same values as had been found in the classical stability study. These values, for long and medium length cylinders such as were used in most of the tests, are given by the right-hand ends of the curves in Figs. 7.17(b) and (c). Using the values for fixed-end cylinders (which conforms to the tests, although the expression (7.11a) used for w satisfies only the most important condition $w = 0$ at the ends), these values are, using Eq. (7.10s):

$$\begin{aligned} n &= (5{\cdot}24R/L)H^{1/4} \\ \lambda_0/k \approx \theta &= 1{\cdot}44/H^{1/4} \end{aligned} \tag{7.11b}$$

In applying the law of virtual work to determine the remaining unknowns, a and b, the work of the external forces is eliminated, as in the two previous cases, by treating the twist or relative rotation of the ends of the cylinder as constant with respect to these unknowns. The angle of twist is $(\partial v/\partial x + \partial u/\partial y)L/R$. From

* 'Effects of large deflections and imperfections on the elastic buckling of cylinders under torsion and axial compression', *Proc. Second US Nat. Congr. of Appl. Mech.*, ASME, 1955, p. 345. Also submitted as a doctoral dissertation at Illinois Inst. of Technology, 1952.

the third of Eqs. (6.31i),

$$\frac{\partial v}{\partial x}+\frac{\partial u}{\partial y}=-\frac{2(1+\nu)}{E}\frac{\partial^2\phi}{\partial x\partial y}-K\frac{\partial w}{\partial x}\cdot\frac{\partial w}{\partial y} \tag{7.11c}$$

Substituting expression (7.11a) for w and the expression found for ϕ into this expression, using trigonometric transformations for the resulting products of trigonometric functions and dropping the periodic terms which do not effect the average value of $(\partial u/\partial y + \partial v/\partial x)$, which is designated by γ_{xy}, relation (7.11c) can be written, using $E/2(1+\nu) = G$, in the form:

$$S = G\left(\gamma_{xy} + \frac{h^2a^2n\lambda_0}{2RL}\right) \tag{7.11d}$$

Substituting this expression for S and expressions (7.11b) for n and λ_0 and $K = 1 + 2a_0/a$ into the strain energy $\mathscr{E}$, and setting $\partial\mathscr{E}/\partial a = 0$, $\partial\mathscr{E}/\partial b = 0$ (treating γ_{xy} and therefore the twist as constant) a and b can be eliminated from the resulting two equations. In this way corresponding values of $(S/G)(L^2/h^2)$, $\gamma_{xy}(L^2/h^2)$, and a_0 were found for several values of L^2/Rh.

Figures 7.19(a) shows these results for $L^2/Rh = 20$. It will be seen that the resistance increases continuously and so there is no purely elastic buckling. There can of course be buckling initiated by yielding for both this and other values of L^2/Rh, but Loo did not consider this case; this is not very important because most of the test specimens had higher values of L^2/Rh and probably buckled elastically. Figure 7.19(b) for $L^2/Rh = 200$ shows definite peaks for $a_0 < 0{\cdot}1$ and similar results were obtained for higher values of L^2/Rh. Figure 7.20 shows by circles the peak values of $(S/G)(L^2/h^2)$ which were thus found for various values of L^2/Rh and a_0, and in dashed lines the following empirical formula which fits

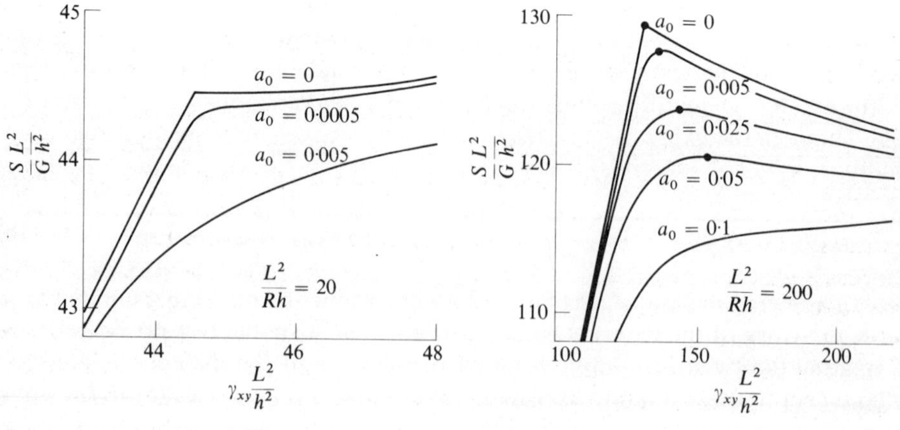

FIGURE 7.19 (from T. T. Loo, 'Effects of large deflec. and imperfections on elastic buckling of cyl. under torsion and axial compression', *Proc. 2nd U.S. Nat. Congress of Appl. Mech.*, p. 345, 1955, courtesy of ASME)

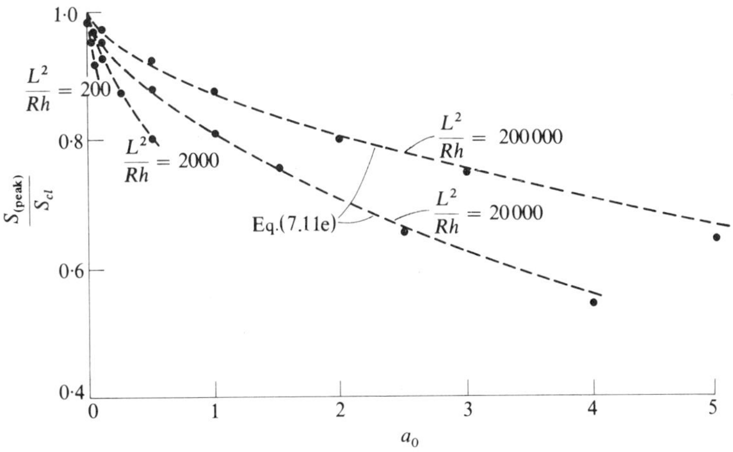

Figure 7.20 (from T. T. Loo, 'Effects of large deflec. and imperfections on elastic buckling of cyl. under torsion and axial compression', *Proc. 2nd U.S. Nat. Congress of Appl. Mech.*, p. 345, 1955, courtesy of ASME)

these results closely, $S_{cl} = Eh^2A/(1-\nu^2)L^2$ being related to L^2/Rh by Fig. 7.17(a):

$$1 - \frac{S}{S_{cl}} = 1{\cdot}14a_0^{0{\cdot}6}(L^2/Rh)^{-0{\cdot}18} \tag{7.11e}$$

Unfortunately these results can not be compared directly to those which were obtained in the last two articles for the axial compression and external pressure cases, because Loo did not use the formula (7.5b) for a_0 which was used in the other cases, using instead a formula of the type:

$$a_0 = U_0\left(\frac{l_x l_y}{(l_x + l_y)h}\right)^2 \tag{7.11f}$$

However, some comparison can be obtained from the fact that Loo also studied the axial compression case, using $\lambda_0 = 0$ in Eq. (7.11a) and the same methods which he used for the torsion case described above. For the axial compression case he obtained with $U_0 = 0{\cdot}00005$ a curve closely corresponding to that obtained from Fig. 7.10 for $U = 0{\cdot}0001$, and with the same value $U_0 = 0{\cdot}00005$ he obtained the curve shown in Fig. 7.21 for the torsion buckling case. If this curve can also be considered to correspond to $U = 0{\cdot}0001$ it can be seen that Loo's study indicates that somewhat smaller magnitudes of initial defects are needed to explain the test results in the torsion case than in the other cases. Whatever the cause of this difference in theoretical results may be, it is evident that the general methods of studying the effects of initial defects which have been described explain the trend of the experimental results in all of these cases very well, with assumptions for the size of the defects which are at least of the same order of magnitude.

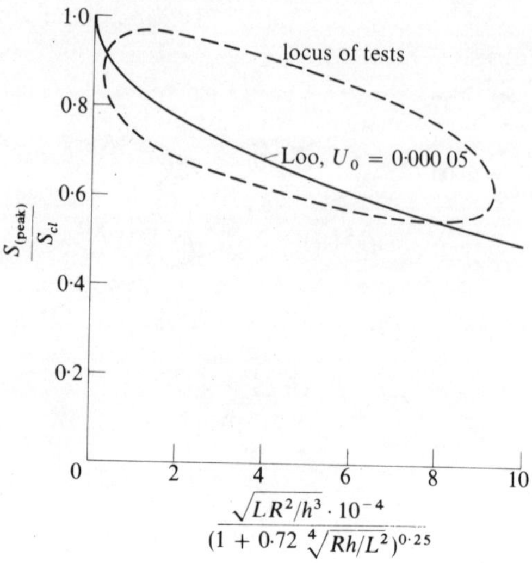

FIGURE 7.21 (from T. T. Loo, 'Effects of large deflec. and imperfections on elastic buckling of cyl. under torsion and axial compression', *Proc. 2nd U.S. Nat. Congress of Appl. Mech.*, p. 345, 1955, courtesy of ASME)

Buckling of Cylinders Under Combined Loading

The methods of making classical stability studies, as well as studies of buckling of real cylinders with defects, which have been presented in the last three sections can be applied to the buckling of cylinders under any combination of loading. One example was given in Sec. 7.3, in which, for $\alpha = 1$, a combination of circumferential and a certain amount of axial compression was studied. In that case it was known from tests that the buckling shape would be similar to that for circumferential compression alone; for the more general case it would of course be necessary to consider the possibility that buckling in more than one wave in the axial direction could occur at a lower loading. As in all theoretical studies of buckling, the investigator must be sure to consider all the buckling modes which may permit buckling at the lowest loading, and the case of combination loading is complicated by the fact that the easiest buckling mode may change (as in the case just mentioned) as the proportion of one type of load to others changes.*

The help of tests as a guide to, as well as confirmation of, theoretical studies is especially important here. The two methods of study, experimental and theoretical, each have their uses and limitations, their strengths and weaknesses, in all fields of science. Since their strengths and weaknesses usually lie in different areas they supplement each other, and this reinforcement is especially important

* These points are brought out in the comprehensive but approximate energy-method study: L. H. Donnell, 'Stability of isotropic or orthotropic cylinders or flat or curved panels, between and across stiffeners, with any edge conditions between hinged and fixed, under any combination of compression and shear', *NACA Tech. Note* No. 918, 1943.

in complex fields such as buckling of shells. A testing machine which is capable of testing cylinders under any combination of axial compression, bending, torsion, and internal suction is described in the author's paper on torsion buckling.*

In spite of the complications discussed above, approximate but useful relations between the buckling resistance to combinations of two types of loading and the resistance to each type of loading separately can be obtained from surprisingly simple 'interaction' curves or formulas. As an example, let σ and S be compressive axial and shear stresses in a cylinder wall due to loads, and σ_0 and S_0 the value of these stresses when failure occurs under pure axial compression and under pure torsion respectively. Then by plotting σ/σ_0 against S/S_0 a curve can be obtained which shows when failure should be produced under all possible combinations of these two types of stresses. This curve will obviously pass through the points P, Q: $\sigma/\sigma_0 = 1$, $S/S_0 = 0$ and $\sigma/\sigma_0 = 0$, $S/S_0 = 1$, as shown in Fig. 7.22(a). Moreover we can tell by elementary reasoning something about the slope of the curve at these points. The curve must be symmetrical about the σ/σ_0 axis, since for a symmetrical structure such as a cylinder a change in the sign of the shear should make no difference. Hence the curve must be perpendicular to the σ/σ_0 axis where it crosses it, as indicated by the short horizontal line. On the other hand the curve will not be symmetrical about the S/S_0 axis but must cross it at a positive angle such as shown in the figure, since compression will obviously

* Loc. cit., Eq. (6.34).

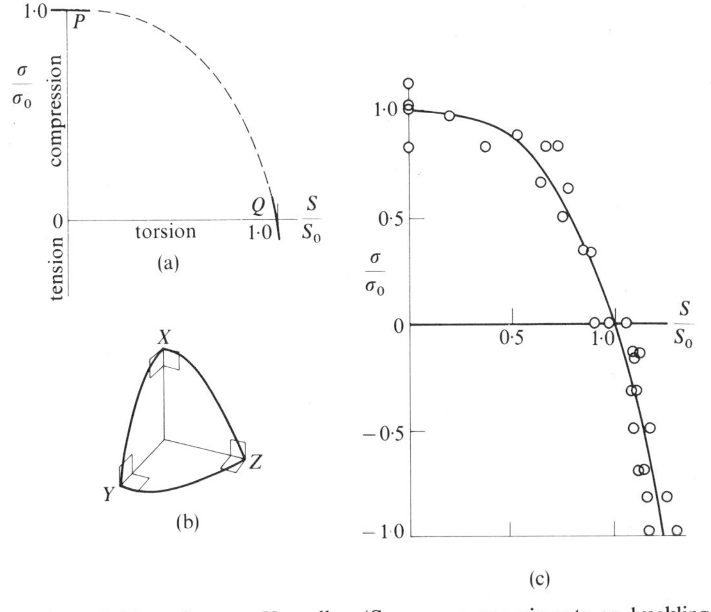

FIGURE 7.22 (from Bridget, Jerome, Vosseller, 'Some new experiments on buckling of thin wall construction', *Trans. ASME*, vol. 56, p. 569, 1934, courtesy of ASME)

decrease and tension increase the shear which the structure can take before buckling.

Some kind of curve must connect these two points, as suggested by the dotted line. It seems reasonable to assume that this curve will be continuous (even though its slope or more probably its curvature may be discontinuous where the mode of failure changes with the proportion of one loading to the other), and that it can be approximated by a simple formula of the type:

$$\left(\frac{\sigma}{\sigma_0}\right)^m + \left(\frac{S}{S_0}\right)^n = 1 \tag{7.12a}$$

which evidently passes through the points P, Q, Fig. 7.22(a). Since the effect of either loading should obviously increase with its magnitude m and n must evidently be positive, $m, n > 0$. In order for the curve to be symmetrical about the σ/σ_0 axis n would have to be an even integer, but this restriction is eliminated if we consider only the shape of the curve in the positive S/S_0 region. The requirement is then only that the slope be horizontal at the point P, which is satisfied if $n > 1$. The requirement that the curve pass through the point Q at an angle as shown in Fig. 7.22(a) is met if $0 < m \leqslant 1$.

Extending this argument to a combination of three loadings, Fig. 7.22(b) shows a reasonable interaction surface described by the formula:

$$X^m + Y^{m'} + Z^n = 1 \tag{7.12b}$$

where we may have, for example, $X = \sigma/\sigma_0$, $Z = S/S_0$, and Y is a similar ratio for circumferential stresses (or external pressures), in which case $n > 1$, $0 < (m, m') \leqslant 1$.

Figure 7.22(c) shows the results of tests on identically made cylinders, with dimensions R, L, h of about 1, 6, 0·002 inches respectively, which were tested in the machine mentioned above in a combination of axial compression and torsion.* The curve shown, which seems to fit the results as well as any, is a cubic parabola corresponding to $m = 1$, $n = 3$ in Eq. (7.12a).

7.5 APPLICATION OF ELASTICITY THEORY TO THICK CYLINDERS

Introduction

In the previous Secs. 7.1–7.4 we have considered small and large deflections and various ordinary and buckling loadings of thin-walled cylinders to which the classical shell theory based on the Love–Kirchhoff assumption applies. To complete our discussion of cylinder theory we must also say something about the cases for which this approximate assumption does not apply. There are a number

* F. J. Bridget, C. C. Jerome, and A. B. Vosseller, 'Some new experiments on buckling of thin wall construction', *ASME Trans.*, vol. 56, 1934, p. 569.

of solutions for cylinders which satisfy elasticity theory and rigorous boundary conditions on the ends as well as on the inner and outer surfaces. These are largely confined to such simple loading as axial tension or compression, pure bending and torsion of a cylindrical bar, or, in the case of loading on the inner and outer surfaces to axisymmetric loadings, such as the solutions (5.79e), (5.80b,c) previously given. However, as in the cases of bars and plates, these as well as other much more general solutions are contained in series solutions in the loading functions. In the remaining discussion we will confine ourselves to the case of distributed loading normal to the inner and outer surfaces, which is by far the most important case for practical applications.

General Series Solution in Loading Functions for Thick Cylinders

C. C. Wu and C. W. Lee have succeeded in developing this difficult solution.* As in similar developments for bars and for plates presented in Secs. 3.3 and 5.2, they start with first terms given by classical bending theory, in this case Eq. (6.34) (although, as will be seen, they might possibly have facilitated their work a little by using the more exact Eq. (6.36)). The second terms represent the most important corrections to the classical theory and involve higher derivatives of the loading functions, and so on. All the terms are calculated to approach an exact three-dimensional elasticity solution in the limit, as well as to satisfy exact boundary conditions on the inner and outer surfaces.

Coordinates, Symbols Used

Figure 7.23 shows the standard axial, angular, and radial (positive outward) coordinates z, θ, r, which were defined in Fig. 3.5 and used in the development of the basic elasticity equations (3.9g) for these coordinates; this figure also shows

* 'A refined theory for circular cylindrical shells', *Proc. First Int. Conf. on Structural Mechanics in Reactor Technology*, Paper No. J1/5, 1971.

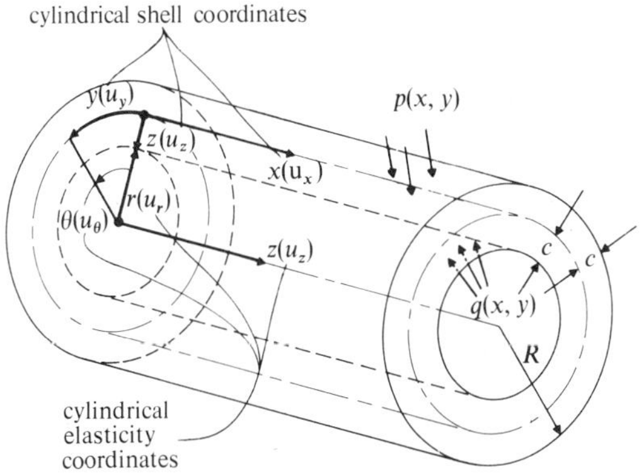

FIGURE 7.23

the shell coordinate system which we have been using and will use in the present development, which for cylindrical shells involves the axial, circumferential, and radial (positive inward) coordinates x, y, z. It is evident that to convert from the former system to the latter we must change z, θ, r, u_z, u_θ, u_r to x, y/R, $R-z$, u_x, u_y, $-u_z$ respectively, where R is the constant radius of the middle surface; the thickness is $h = 2c$, as shown.

Elasticity Conditions

To simplify the results, we will use the dimensionless symbols $\mathbf{x}$, $\mathbf{y}$, $\mathbf{z}$, $\mathbf{c} = x/R$, y/R, z/R, c/R respectively. Dividing $r = R - z$ into unity we obtain:

$$\frac{1}{r} = \frac{1}{R-z} = \frac{1}{R}(1 + \mathbf{z} + \mathbf{z}^2 + \cdots), \qquad \frac{1}{r^2} = \frac{1}{R^2}(1 + 2\mathbf{z} + 3\mathbf{z}^2 + \cdots), \text{ etc.} \qquad (7.13a)$$

Using the above equivalencies, the three basic equations of elasticity (3.9g) which the solution must satisfy can be expressed in the new coordinate and displacement symbols in the form:

$$\begin{aligned}
&\alpha \frac{\partial^2 u_z}{\partial \mathbf{z}^2} + \beta \frac{\partial^2 u_z}{\partial \mathbf{x}^2} + \frac{\partial u_x}{\partial \mathbf{z}\,\partial \mathbf{x}} + (1 + \mathbf{z} + \mathbf{z}^2 + \cdots)\left(\frac{\partial^2 u_y}{\partial \mathbf{y}\,\partial \mathbf{z}} - \alpha \frac{\partial u_z}{\partial \mathbf{z}}\right) \\
&\qquad + (1 + 2\mathbf{z} + 3\mathbf{z}^2 + \cdots)\left[\beta \frac{\partial^2 u_z}{\partial \mathbf{y}^2} - \alpha u_z + (\alpha + \beta)\frac{\partial u_y}{\partial \mathbf{y}}\right] = 0 \\
&\beta\left(\frac{\partial^2 u_y}{\partial \mathbf{z}^2} + \frac{\partial^2 u_y}{\partial \mathbf{x}^2}\right) + (1 + \mathbf{z} + \mathbf{z}^2 + \cdots)\left(\frac{\partial^2 u_z}{\partial \mathbf{y}\,\partial \mathbf{z}} - \beta \frac{\partial u_y}{\partial \mathbf{z}} + \frac{\partial^2 u_x}{\partial \mathbf{x}\,\partial \mathbf{y}}\right) \\
&\qquad + (1 + 2\mathbf{z} + 3\mathbf{z}^2 + \cdots)\left[\alpha \frac{\partial^2 u_y}{\partial \mathbf{y}^2} - \beta u_y - (\alpha + \beta)\frac{\partial u_z}{\partial \mathbf{y}}\right] = 0 \qquad (7.13b) \\
&\beta \frac{\partial^2 u_x}{\partial \mathbf{z}^2} + \alpha \frac{\partial^2 u_x}{\partial \mathbf{x}^2} + \frac{\partial^2 u_z}{\partial \mathbf{z}\,\partial \mathbf{x}} + (1 + \mathbf{z} + \mathbf{z}^2 + \cdots)\left(\frac{\partial^2 u_y}{\partial \mathbf{x}\,\partial_y} - \frac{\partial u_z}{\partial \mathbf{x}} - \beta \frac{\partial u_x}{\partial \mathbf{z}}\right) \\
&\qquad + (1 + 2\mathbf{z} + 3\mathbf{z}^2 + \cdots)\beta \frac{\partial^2 u_x}{\partial \mathbf{y}^2} = 0
\end{aligned}$$

where:

$$\alpha = 2(1-\nu) \approx 1{\cdot}4, \quad \beta = 1 - 2\nu \approx 0{\cdot}4. \qquad (7.13c)$$

Boundary Conditions

Taking the distributed normal pressure on the outside surface as $p(x, y)$ and on the inside surface as $q(x, y)$, as indicated in Fig. 7.23, the boundary conditions on these surfaces are $z = c, -c$; $\sigma_z = -q, -p$, σ_{yz}, $\sigma_{xz} = 0$. Using Eqs. (3.9h) and the above equivalencies, these conditions can be expressed in the new coordinates and displacement symbols in the following form. For $z = c, -c$, $\mathbf{z} = 1, -1$:

$$(1-\nu)\frac{\partial u_z}{\partial \mathbf{z}} + \nu \frac{\partial u_x}{\partial \mathbf{x}} + \nu(1 + \mathbf{z} + \mathbf{z}^2 + \cdots)\left(\frac{\partial u_y}{\partial \mathbf{y}} - u_z\right) = -(1+\nu)\beta R\left(\frac{q}{E}, \frac{p}{E}\right) \qquad (7.13d)$$

$$\frac{\partial u_y}{\partial \mathbf{z}}+(1+\mathbf{z}+\mathbf{z}^2+\cdots)\left(\frac{\partial u_z}{\partial \mathbf{y}}+u_y\right)=0, \qquad \frac{\partial u_x}{\partial \mathbf{z}}+\frac{\partial u_z}{\partial \mathbf{x}}=0$$

Series Solution

We define the new functions $\mathbf{d}(x, y)$, $\mathbf{s}(x, y)$ by the governing relations:

$$\Delta\mathbf{d} = d = p - q, \qquad \Delta\mathbf{s} = s = p + q \tag{7.13e}$$

where Δ represents the operator:

$$\Delta = \frac{\mathbf{c}^3}{3(1-\nu^2)}\left[\nabla^8+\nabla^2\left(-\gamma\frac{\partial^4}{\partial \mathbf{x}^4}+6\frac{\partial^4}{\partial \mathbf{x}^2\,\partial \mathbf{y}^2}+2\frac{\partial^4}{\partial \mathbf{y}^4}\right)+\gamma\frac{\partial^4}{\partial \mathbf{x}^4}+4\frac{\partial^4}{\partial \mathbf{x}^2\,\partial \mathbf{y}^2}+\frac{\partial^4}{\partial \mathbf{y}^4}\right]+\mathbf{c}\frac{\partial^4}{\partial \mathbf{x}^4},$$

$$\text{and } \gamma = \frac{12(1-\nu^2)}{5} \approx 2{\cdot}18 \tag{7.13f}$$

and where:

$$\nabla^2 = \frac{\partial^2}{\partial \mathbf{x}^2}+\frac{\partial^2}{\partial \mathbf{y}^2} = R^2\left(\frac{\partial^2}{\partial x^2}+\frac{\partial^2}{\partial y^2}\right), \qquad \nabla^4 = \nabla^2\nabla^2, \text{ etc.} \tag{7.13g}$$

It was pointed out in Sec. 5.2 in the discussion of Eqs. (5.18a) (which are the governing relations for plates, corresponding to Eq. (7.13e) for cylinders) that those equations were the same as the transverse equilibrium equation (4.19) for thin plates, $D\nabla^4 w = p$, if w is replaced by $3(1-\nu^2)c(\mathbf{t}_z - \mathbf{b}_z)/2E$. It is interesting and important to note that Eqs. (7.13e) are similarly related to our most accurate transverse equilibrium equation (6.36) for thin cylinders. Comparison of Eq. (6.36) for the lateral loading case with the first of Eqs. (7.13e) shows that if w is replaced by $(R/2E)\nabla^4\mathbf{d}$ (a correspondence which is confirmed by the primary term of the expression for $u_{z(z=0)} = w$ given below) the two equations are the same except for the terms labelled c_1 and c_4 in Table 6.7, and a small difference in the c_2 and c_5 terms. As pointed out in the discussion of Table 6.7, the c_1 and c_4 terms, and the exact magnitude of the c_2 and $c_5 = c_2 - 2$ terms are of negligible importance in problems to which the classical theory based on the Love–Kirchhoff approximation applies (but which of course are presumably not negligible in thick cylinder problems which we are now considering).

While Eq. (4.19) for thin plates is universally accepted and uncontroversial, the correspondence between Eq. (6.36) and Eqs. (7.13e) (whose terms were developed step by step with no preconceived idea of the final result, by starting with the simple Eq. (6.34) and satisfying exact elasticity theory for each step) is a striking confirmation of the validity of Eq. (6.36) in the much more complex and controversial field of shell theory, as well as confirmation of Wu and Lee's work.

Starting with terms obtained from Eq. (6.34), expressions for the next terms are assumed with unknown coefficients and the order of differentiation of **d** and **s** and the power of **z** increased by two. Substituting these into the elasticity and boundary conditions, the coefficients can be determined except for those of terms containing still higher derivatives of **d** and **s**, which are taken care of in the next round. In this way the following series expressions for the displacements was obtained:

$$
\begin{aligned}
\frac{E}{R}u_x &= \left\{\frac{\partial^3\mathbf{d}}{\partial\mathbf{x}^3}\left(\frac{\nu}{2}+\frac{9\nu}{10}\mathbf{c}^2+\cdots\right)-\frac{\partial^3\mathbf{d}}{\partial\mathbf{x}\,\partial\mathbf{y}^2}\left(\frac{1}{2}-\frac{8+2\nu}{15\alpha}\mathbf{c}^2\cdots\right)\right\}\\
&+\mathbf{c}\left\{\frac{\partial^3\mathbf{s}}{\partial\mathbf{x}^3}(\nu\cdots)-\frac{\partial^3\mathbf{s}}{\partial\mathbf{x}\,\partial\mathbf{y}^2}\left(\frac{1}{2}\cdots\right)\right\}\\
&+\left\{-\nabla^4\frac{\partial\mathbf{d}}{\partial\mathbf{x}}\frac{\mathbf{z}}{2}-\nabla^2\frac{\partial^3\mathbf{d}}{\partial\mathbf{x}^3}\left(\frac{\nu}{4}\mathbf{z}^2+\frac{10+19\nu}{60}\mathbf{c}^2\right)+\nabla^2\frac{\partial^3\mathbf{d}}{\partial\mathbf{x}\,\partial\mathbf{y}^2}\left(\frac{17-7\nu}{15\alpha}\mathbf{c}^2\cdots\right)\right.\\
&-\frac{\partial^5\mathbf{d}}{\partial\mathbf{x}^5}\left[\left(\frac{1+\nu}{12}\mathbf{z}^3+\frac{3-15\nu}{20}\mathbf{c}^2\mathbf{z}\right)\cdots\right]+\frac{\partial^5\mathbf{d}}{\partial\mathbf{x}^3\,\partial\mathbf{y}^2}\left[\left(\frac{\nu^2}{2\alpha}\mathbf{z}^3+\frac{12+38\nu-115\nu^2}{30\alpha}\mathbf{c}^2\mathbf{z}\right)\cdots\right]\\
&\left.+\frac{\partial^5\mathbf{d}}{\partial\mathbf{x}\,\partial\mathbf{y}^4}\left[\left(\frac{2-\nu}{6\alpha}\mathbf{z}^3-\frac{8-3\nu}{10\alpha}\mathbf{c}^2\mathbf{z}\right)\cdots\right]\right\}\\
&+\mathbf{c}\left\{-\nabla^4\frac{\partial\mathbf{s}}{\partial\mathbf{x}}\frac{\mathbf{z}}{2}+\nabla^2\frac{\partial^3\mathbf{s}}{\partial\mathbf{x}^3}\frac{\nu}{2}\mathbf{z}+\frac{\partial^5\mathbf{s}}{\partial\mathbf{x}^5}\left[\left(\frac{1-\nu}{4}\mathbf{z}^2-\frac{5+11\nu}{20}\mathbf{c}^2\right)\cdots\right]\right.\\
&\left.-\frac{\partial^5\mathbf{s}}{\partial\mathbf{x}^3\,\partial\mathbf{y}^2}\left[\left(\frac{\nu}{4}\mathbf{z}^2-\frac{24+13\nu+13\nu^2}{30\alpha}\mathbf{c}^2\right)\cdots\right]+\frac{\partial^5\mathbf{s}}{\partial\mathbf{x}\,\partial\mathbf{y}^4}\left[\left(\frac{12+13\nu}{15\alpha}\mathbf{c}^2\right)\cdots\right]\right\}\\
&+\left\{\nabla^6\frac{\partial\mathbf{d}}{\partial\mathbf{x}}\left(\frac{2-\nu}{6\alpha}\mathbf{z}^3-\frac{2+3\nu}{10\alpha}\mathbf{c}^2\mathbf{z}\right)+\nabla^4\frac{\partial^3\mathbf{d}}{\partial\mathbf{x}^3}\left[\left(\frac{2+\nu-\nu^2}{24\alpha}\mathbf{z}^4\right.\right.\right.\\
&\left.\left.-\frac{10-19\nu+19\nu^2}{60\alpha}\mathbf{c}^2\mathbf{z}^2+\frac{490+789\nu-229\nu^2}{12600\alpha}\mathbf{c}^4\right)\cdots\right]\\
&\left.+\nabla^4\frac{\partial^3\mathbf{d}}{\partial\mathbf{x}\,\partial\mathbf{y}^2}\left[\left(\frac{7-5\nu}{24\alpha}\mathbf{z}^4-\frac{5-\nu}{12\alpha}\mathbf{c}^2\mathbf{z}^2+\frac{751+7999\nu}{12600\alpha}\mathbf{c}^4\right)\cdots\right]\cdots\right\}\\
&+\mathbf{c}\left\{\nabla^6\frac{\partial\mathbf{s}}{\partial\mathbf{x}}\frac{\nu}{3\alpha}\mathbf{c}^2+\nabla^4\frac{\partial^3\mathbf{s}}{\partial\mathbf{x}^3}\left[\left(\frac{2-\nu}{12}\mathbf{z}^3-\frac{2+3\nu}{20}\mathbf{c}^2\mathbf{z}\right)\cdots\right]\right.\\
&\left.+\nabla^4\frac{\partial^3\mathbf{s}}{\partial\mathbf{x}\,\partial\mathbf{y}^2}\left[\left(\frac{2-\nu}{6\alpha}\mathbf{z}^3-\frac{16-11\nu}{30\alpha}\mathbf{c}^2\mathbf{z}\right)\cdots\right]\cdots\right\}\\
&+\left\{-\nabla^8\frac{\partial\mathbf{d}}{\partial\mathbf{x}}\left[\left(\frac{3-\nu}{120\alpha}\mathbf{z}^5+\frac{1-3\nu}{60\alpha}\mathbf{c}^2\mathbf{z}^3-\frac{87-157\nu}{4200\alpha}\mathbf{c}^4\mathbf{z}\right)\cdots\right]\cdots\right\}\\
&+\mathbf{c}\left\{\nabla^8\frac{\partial\mathbf{s}}{\partial\mathbf{x}}\left[\left(\frac{\mathbf{c}^2\mathbf{z}^2}{12}-\frac{\mathbf{c}^4}{36}\right)\cdots\right]\cdots\right\}\cdots
\end{aligned}
$$

$$
\begin{aligned}
\frac{E}{R}u_y &= \left\{\frac{\partial^3\mathbf{d}}{\partial\mathbf{x}^2\,\partial\mathbf{y}}\left[\left(\frac{2+\nu}{2}(1-\mathbf{z})-\frac{8+14\nu}{15}\mathbf{c}^2\right)\cdots\right]\right.\\
&\left.+\frac{\partial^3\mathbf{d}}{\partial\mathbf{y}^3}\left[\frac{1-\mathbf{z}}{2}-\frac{\mathbf{c}^2}{10}\cdots\right]\right\}+\mathbf{c}\left\{\frac{\partial^3\mathbf{s}}{\partial\mathbf{x}^2\,\partial\mathbf{y}}(1-\mathbf{z}\cdots)+\frac{\partial^3\mathbf{s}}{\partial\mathbf{y}^3}\left(\frac{1-\mathbf{z}}{2}\cdots\right)\right\}\\
&+\left\{-\nabla^4\frac{\partial\mathbf{d}}{\partial\mathbf{y}}\left(\frac{\mathbf{z}}{2}+\frac{2}{5}\mathbf{c}^2\right)-\nabla^2\frac{\partial^3\mathbf{d}}{\partial\mathbf{x}^2\,\partial\mathbf{y}}\left(\frac{\nu}{4}\mathbf{z}^2+\frac{44-5\nu-19\nu^2}{30\alpha}\mathbf{c}^2\right)\right.\\
&+\frac{\partial^5\mathbf{d}}{\partial\mathbf{x}^4\,\partial\mathbf{y}}\left[\left(\frac{5+2\nu}{12}\mathbf{z}^3-\frac{31-55\nu+4\nu^2}{30\alpha}\mathbf{c}^2\mathbf{z}\right)\cdots\right]\\
&\left.+\frac{\partial^5\mathbf{d}}{\partial\mathbf{x}^2\,\partial\mathbf{y}^3}\left[\left(\frac{2-\nu}{2\alpha}\mathbf{z}^3+\frac{14+15\nu-74\nu^2}{30\alpha}\mathbf{c}^2\mathbf{z}\right)\cdots\right]+\frac{\partial^5\mathbf{d}}{\partial\mathbf{y}^5}\left[\left(\frac{2-\nu}{6\alpha}\mathbf{z}^3-\frac{\nu}{2\alpha}\mathbf{c}^2\mathbf{z}\right)\cdots\right]\right\}
\end{aligned}
\tag{7.14}
$$

$$+\mathbf{c}\left\{-\nabla^4\frac{\partial \mathbf{s}}{\partial \mathbf{y}}\frac{\mathbf{z}}{2}+\nabla^2\frac{\partial^3\mathbf{s}}{\partial \mathbf{x}^2\,\partial \mathbf{y}}\frac{\nu}{2}\mathbf{z}+\frac{\partial^5\mathbf{s}}{\partial \mathbf{x}^2\,\partial \mathbf{y}}\left[\left(\frac{1-\nu}{4}\mathbf{z}^2-\frac{73-5\nu}{60}\mathbf{c}^2\right)\cdots\right]\right.$$

$$\left.-\frac{\partial^5\mathbf{s}}{\partial \mathbf{x}^2\,\partial \mathbf{y}^3}\left[\left(\frac{\nu}{4}\mathbf{z}^2+\frac{72-107\nu+25\nu^2}{30\alpha}\mathbf{c}^2\right)\cdots\right]-\frac{\partial^5\mathbf{s}}{\partial \mathbf{y}^5}\left[\left(\frac{7-12\nu}{15\alpha}\mathbf{c}^2\right)\cdots\right]\right\}$$

$$+\left\{\nabla^6\frac{\partial \mathbf{d}}{\partial \mathbf{y}}\left(\frac{2-\nu}{6\alpha}\mathbf{z}^3-\frac{2+3\nu}{10\alpha}\mathbf{c}^2\mathbf{z}\right)+\nabla^4\frac{\partial^3\mathbf{d}}{\partial \mathbf{x}^2\,\partial \mathbf{y}}\left[\left(\frac{4+3\nu-\nu^2}{24\alpha}\mathbf{z}^4\right.\right.\right.$$

$$\left.\left.-\frac{\nu+19\nu^2}{60\alpha}\mathbf{c}^2\mathbf{z}^2+\frac{108-229\nu-229\nu^2}{12\,600\alpha}\mathbf{c}^4\right)\cdots\right]$$

$$\left.+\nabla^4\frac{\partial^3\mathbf{d}}{\partial \mathbf{y}^3}\left[\left(\frac{3-\nu}{8\alpha}\mathbf{z}^4-\frac{1+\nu}{4\alpha}\mathbf{c}^2\mathbf{z}^2+\frac{123+2327\nu}{4200\alpha}\mathbf{c}^4\right)\cdots\right]\cdots\right\}$$

$$+\mathbf{c}\left\{\nabla^6\frac{\partial \mathbf{s}}{\partial \mathbf{y}}\frac{\nu}{3\alpha}\mathbf{c}^2+\nabla^4\frac{\partial^3\mathbf{s}}{\partial \mathbf{x}^2\,\partial \mathbf{y}}\left[\left(\frac{2-\nu}{12}\mathbf{z}^3-\frac{6+13\nu-9\nu^2}{30\alpha}\mathbf{c}^2\mathbf{z}\right)\cdots\right]\right.$$

$$\left.+\nabla^4\frac{\partial^3\mathbf{s}}{\partial \mathbf{y}^3}\left[\left(\frac{2-\nu}{6\alpha}\mathbf{z}^3-\frac{16-\nu}{30\alpha}\mathbf{c}^2\mathbf{z}\right)\cdots\right]\cdots\right\}$$

$$+\left\{-\nabla^8\frac{\partial \mathbf{d}}{\partial \mathbf{y}}\left[\left(\frac{3-\nu}{120\alpha}\mathbf{z}^5+\frac{1-3\nu}{60\alpha}\mathbf{c}^2\mathbf{z}^3-\frac{87-157\nu}{4200\alpha}\mathbf{c}^4\mathbf{z}\right)\cdots\right]\cdots\right\}$$

$$+\mathbf{c}\left\{\nabla^8\frac{\partial \mathbf{s}}{\partial \mathbf{y}}\left[\left(\frac{\mathbf{c}^2\mathbf{z}^2}{12}-\frac{\mathbf{c}^4}{36}\right)\cdots\right]\cdots\right\}\cdots$$

$$\frac{E}{R}u_z=\left\{\nabla^4\mathbf{d}\frac{1}{2}+\nabla^2\frac{\partial^2\mathbf{d}}{\partial \mathbf{x}^2}\frac{\nu}{2}\mathbf{z}+\frac{\partial^4\mathbf{d}}{\partial \mathbf{x}^4}\left[\left(\frac{1+\nu}{4}\mathbf{z}^2+\frac{3-15\nu}{20}\mathbf{c}^2\right)\right.\right.$$

$$\left.+\left(\frac{1+\nu}{4}\mathbf{z}^3-\frac{15-3\nu}{20}\mathbf{c}^2\mathbf{z}\right)\cdots\right]+\frac{\partial^4\mathbf{d}}{\partial \mathbf{x}^2\,\partial \mathbf{y}^2}\left[\left(\frac{2\nu+\nu^2}{2\alpha}\mathbf{z}^2\right.\right.$$

$$\left.\left.-\frac{12+68\nu-55\nu^2}{30\alpha}\mathbf{c}^2\right)+\left(\frac{2+3\nu+\nu^2}{6\alpha}\mathbf{z}^3-\frac{30-19\nu-\nu^2}{30\alpha}\mathbf{c}^2\mathbf{z}\right)\cdots\right]$$

$$\left.+\frac{\partial^4\mathbf{d}}{\partial \mathbf{y}^4}\left[\left(\frac{\nu}{2\alpha}\mathbf{z}^2-\frac{2+3\nu}{10\alpha}\mathbf{c}^2\right)+\left(\frac{1+\nu}{6\alpha}\mathbf{z}^3-\frac{1}{4}\mathbf{c}^2\mathbf{z}\right)\cdots\right]\right\}$$

$$+\mathbf{c}\left\{\nabla^4\mathbf{s}\frac{1}{2}-\nabla^2\frac{\partial^2\mathbf{s}}{\partial \mathbf{x}^2}\frac{\nu}{2}(1-\mathbf{z})-\frac{\partial^4\mathbf{s}}{\partial \mathbf{x}^4}\left[\left(\frac{\mathbf{z}}{2}\right)\cdots\right]+\left[\left(2\frac{\partial^4\mathbf{s}}{\partial \mathbf{x}^2\,\partial \mathbf{y}^2}+\frac{\partial^4\mathbf{s}}{\partial \mathbf{y}^4}\right)\frac{\nu}{2\alpha}\mathbf{z}^2\cdots\right]\right\}$$

$$+\left\{\nabla^6\mathbf{d}\left(\frac{\nu}{2\alpha}\mathbf{z}^2-\frac{8-3\nu}{10\alpha}\mathbf{c}^2\right)+\nabla^4\frac{\partial^2\mathbf{d}}{\partial \mathbf{x}^2}\left(\frac{\nu+\nu^2}{6\alpha}\mathbf{z}^3-\frac{29\nu-19\nu^2}{30\alpha}\mathbf{c}^2\mathbf{z}\cdots\right)\right.$$

$$+\nabla^4\frac{\partial^2\mathbf{d}}{\partial \mathbf{y}^2}\left(\frac{1+\nu}{6\alpha}\mathbf{z}^3-\frac{\mathbf{c}^2\mathbf{z}}{4}\right)+\nabla^2\frac{\partial^4\mathbf{d}}{\partial \mathbf{x}^4}\left(\frac{5\nu+2\nu^2}{24\alpha}\mathbf{z}^4-\frac{24+11\nu+26\nu^2}{60\alpha}\mathbf{c}^2\mathbf{z}^2\cdots\right)$$

$$\left.+\nabla^2\frac{\partial^4\mathbf{d}}{\partial \mathbf{x}^2\,\partial \mathbf{y}^2}\left(\frac{(1+\nu)^2}{12\alpha}\mathbf{z}^4+\frac{25-30\nu-37\nu^2}{30\alpha}\mathbf{c}^2\mathbf{z}^2\right)+\nabla^2\frac{\partial^4\mathbf{d}}{\partial \mathbf{y}^4}\left(\frac{4+\nu}{24\alpha}\mathbf{z}^4-\frac{\nu}{4\alpha}\mathbf{c}^2\mathbf{z}^2\right)\cdots\right\}$$

$$+\mathbf{c}\left\{\nabla^6\mathbf{s}\left(\frac{\nu}{2\alpha}\mathbf{z}^2-\frac{14+11\nu}{30\alpha}\mathbf{c}^2\right)-\nabla^4\frac{\partial^2\mathbf{s}}{\partial \mathbf{x}^2}\left(\frac{\nu^2}{2\alpha}\mathbf{z}^2+\frac{10-44\nu+9\nu^2}{30\alpha}\mathbf{c}^2\right)\right.$$

$$+\nabla^4\frac{\partial^2\mathbf{s}}{\partial \mathbf{y}^2}\left(\frac{1+\nu}{6\alpha}\mathbf{z}^3\right)+\nabla^2\frac{\partial^1\mathbf{s}}{\partial \mathbf{x}^4}\left(\frac{\alpha}{5}\mathbf{c}^2\mathbf{z}\right)$$

$$\left.-\nabla^2\frac{\partial^4\mathbf{s}}{\partial \mathbf{x}^2\,\partial \mathbf{y}^2}\left(\frac{75-71\nu+16\nu^2}{30\alpha}\mathbf{c}^2\mathbf{z}\right)-\nabla^2\frac{\partial^4\mathbf{s}}{\partial \mathbf{y}^4}\left(\frac{7}{12}\mathbf{c}^2\mathbf{z}\right)\cdots\right\}$$

$$-\left\{\nabla^8\mathbf{d}\left(\frac{1+\nu}{24\alpha}\mathbf{z}^4-\frac{5-3\nu}{20\alpha}\mathbf{c}^2\mathbf{z}^2+\frac{227-157\nu}{4200\alpha}\mathbf{c}^4\right)\right.$$

$$+\nabla^6\frac{\partial^2\mathbf{d}}{\partial\mathbf{x}^2}\left(\frac{2+3\nu+\nu^2}{120\alpha}\mathbf{z}^5-\frac{20+\nu-19\nu^2}{180\alpha}\mathbf{c}^2\mathbf{z}^3+\frac{3150-3869\nu+229\nu^2}{12600\alpha}\mathbf{c}^4\mathbf{z}\right)$$

$$\left.+\nabla^6\frac{\partial^2\mathbf{d}}{\partial\mathbf{y}^2}\left(\frac{2+\nu}{20\alpha}\mathbf{z}^5-\frac{11-4\nu}{30\alpha}\mathbf{c}^2\mathbf{z}^3+\frac{252-217\nu}{420\alpha}\mathbf{c}^4\mathbf{z}\right)\cdots\right\} \qquad (7.14\text{ cont.})$$

$$-\mathbf{c}\left\{\nabla^8\mathbf{s}\left[\frac{\mathbf{c}^2\mathbf{z}}{6}+\left(\frac{1+\nu}{24\alpha}\mathbf{z}^4-\frac{5+\nu}{60\alpha}\mathbf{c}^2\mathbf{z}^2\right)\cdots\right]-\nabla^6\frac{\partial^2\mathbf{s}}{\partial\mathbf{x}^2}\left(\frac{\nu+\nu^2}{24\alpha}\mathbf{z}^4-\frac{5\nu-3\nu^2}{20\alpha}\mathbf{c}^2\mathbf{z}^2\right)\cdots\right\}$$

$$+\left\{\nabla^{10}\mathbf{d}\left[\left(\frac{2+\nu}{720\alpha}\mathbf{z}^6-\frac{2-3\nu}{240\alpha}\mathbf{c}^2\mathbf{z}^4+\frac{70-157\nu}{8400\alpha}\mathbf{c}^2\mathbf{z}^2\right)\cdots\right]\cdots\right\}\cdots$$

$$-\mathbf{c}\left\{\nabla^{10}\mathbf{s}\left[\frac{\nu}{18\alpha}(\mathbf{c}^2\mathbf{z}^3-\mathbf{c}^4\mathbf{z})\cdots\right]\cdots\right\}\cdots$$

Wu and Lee also calculated the series expressions for the six stresses* (their symbols φ, ψ, s represent $-R^4\mathbf{d}/2$, $R^4\mathbf{s}/2$, y respectively, and their z, u_z are taken positive outward and so have opposite signs from those used here). These expressions for the stresses are too long to repeat here and they can of course be obtained without difficulty from the displacements by using Eqs. (3.9f, h), which become, using the above equivalencies:

$$R\mathrm{e}=\frac{\partial u_z}{\partial\mathbf{z}}+(1+\mathbf{z}+\mathbf{z}^2+\cdots)\left(\frac{\partial u_y}{\partial\mathbf{y}}-u_z\right)+\frac{\partial u_x}{\partial\mathbf{x}}$$

$$R(1+\nu)\sigma_x/E=\frac{\partial u_x}{\partial\mathbf{x}}+\frac{\nu}{\beta}R\mathrm{e},\qquad R(1+\nu)\sigma_z/E=\frac{\partial u_z}{\partial\mathbf{z}}+\frac{\nu}{\beta}R\mathrm{e}$$

$$R(1+\nu)\sigma_y/E=(1+\mathbf{z}+\mathbf{z}^2+\cdots)\left(\frac{\partial u_y}{\partial\mathbf{y}}-u_z\right)+\frac{\nu}{\beta}R\mathrm{e} \qquad (7.15\text{a})$$

$$2R(1+\nu)\sigma_{xy}/E=(1+\mathbf{z}+\mathbf{z}^2+\cdots)\frac{\partial u_x}{\partial\mathbf{y}}+\frac{\partial u_y}{\partial\mathbf{x}}$$

The expressions for σ_{yz}, σ_{xz} are given by the left-hand sides of the last two Eqs. (7.13d) divided by $-2R(1+\nu)/E$ and $2R(1+\nu)/E$ respectively.

As might be expected, the series solution in loading functions for cylinders (7.14) is somewhat more complex than the corresponding solutions (3.28) for beams and (5.19) for plates. This is especially so because of the use of the series expression (7.13a) for $1/r = 1/(R-z)$, which makes infinite series of individual terms as well as the expressions as a whole. It might seem that the final expressions could have been simplified a great deal if the elasticity conditions (3.9g) had been multiplied through by r^2, and the first two of the boundary conditions obtained from Eqs. (3.9h) by r, to clear of fractions and eliminate the necessity for using Eqs. (7.13a), before the development of the series solution was started. This possibility was tentatively explored but no solution was found in the time available. In any case this development demonstrates that the very useful general method of series solutions in loading functions for the improvement and

* Loc. cit., p. 733.

extension of the classical theories is applicable to shells as well as to bars and plates, and the number of terms given in expressions (7.14) should give good approximations for reasonably thick cylinders (as shown, for example, by Fig. 7.24, which is discussed later).

Convergence of Eqs. (7.14)

The limitations and the various types of solutions contained in Eqs. (7.14) are similar to those pointed out in the discussions of Eqs. (3.28) and (5.19) for beam and plates respectively. Like those earlier series the convergence of Eqs. (7.14) should be rapid for loadings with half wavelengths which are large compared to the thickness but the last term given, involving the tenth derivatives of **d** or s and higher derivatives not given, will converge poorly or diverge for components of the loading which have wavelengths of the order of or less than the thickness.

Edge Boundary Conditions

As in the case of the beam and plate series solutions, means of satisfying edge boundary conditions are contained in Eqs. (7.14), and are given as solutions of the homogeneous equation obtained by setting p, q equal to zero in Eqs. (7.13e):

$$\Delta\mathbf{d} = 0, \qquad \Delta\mathbf{s} = 0 \tag{7.15b}$$

Solutions will be similar to those for Eq. (7.3a), which were discussed in Sec. 7.1. As has been pointed out previously, these represent relations which are escalated with respect to their physical origins, and in spite of their eighth order cannot be expected to satisfy more than gross boundary conditions. Satisfaction of more complete or complete boundary conditions would require superposition of supplementary local stress fields obtained from three-dimensional elasticity theory. The methods discussed in Sec. 5.5 for thick plates can, as was stated there, be applied with fair approximation to thick cylinders and other shells by ignoring the curvature (as discussed in Sec. 7.1 this applies especially to membrane boundary conditions). Some local stress field solutions have been found for thick cylinders, for example for the axisymmetric case,* and an approximate solution for solid cylinders** which could be adapted to hollow cylinders, or applied to them by subtracting solutions for the gross cylinder and for the hole.

Applications; Uniform Loading

Because of its multiple series nature, Eqs. (7.14), unlike the corresponding beam and plate theories, do not contain exact closed solutions such as the known

* C. W. Lee, 'Analysis of thick-walled cylinders under axisymmetric edge loads', *Proc. First Int. Conf. on Pressure Vessels Technology, ASME*, Part I, Design and Analysis, p. 369, 1969.

** G. Horvay and J. A. Mirabal, 'The end problem of cylinders', *J. Appl. Mech.*, Dec. 1958, p. 561.

Lamé solution, Eqs. (5.79e), for a disc or thick cylinder under uniformly distributed inside and outside pressure. However, we can of course obtain a series solution for this case from Eqs. (7.14), and it is interesting to compare this with the Lamé solution as a check upon its convergence and the accuracy which can be obtained from the limited number of terms which are given.

It will be sufficient for this purpose to compare the expressions for the radial displacement which we call here u_z. Taking $\mathbf{d} = Dx^4$, $\mathbf{s} = Sx^4$ and substituting these into Eqs. (7.13e) in which we take $p(x, y) = p_0$, $q(x, y) = p_i$, we find:

$$D = \frac{p_0 - p_i}{24\left(\frac{4}{5}c^3R + cR^3\right)}, \quad S = \frac{p_0 + p_i}{24\left(\frac{4}{5}c^3R + cR^3\right)} \tag{7.15c}$$

Substituting these expressions for **d**, **s** into the expression for u_z in Eq. (7.14), using all the terms given and $1/[(4c^3R/5) + cR^3] = (1/cR^3)/[1 + (4c^2/5R^2)] = (1/cR^3)[1 - (4c^2/5R^2) + (4c^2/5R^2)^2 \cdots]$ the resulting expression for u_z, with as many terms as are fully determined is:

$$\begin{aligned} \frac{E}{R}u_z = \frac{R}{c}\Bigg\{\Bigg[&\left(\frac{1}{2} + \frac{1-\nu}{2}\frac{c}{R} - \frac{1+3\nu}{4}\frac{c^2}{R^2}\cdots\right) \\ &+ \left(\frac{\nu}{2} - \frac{1-\nu}{2}\frac{c}{R} - \frac{3+\nu}{4}\frac{c^2}{R^2}\cdots\right)\frac{z}{R} + \frac{1+\nu}{4}\left(\frac{z^2}{R^2} + \frac{z^3}{R^3} + \cdots\right)\Bigg]p_0 \\ &- \Bigg[\left(\frac{1}{2} - \frac{1-\nu}{2}\frac{c}{R} - \frac{1+3\nu}{4}\frac{c^2}{R^2}\cdots\right) \\ &+ \left(\frac{\nu}{2} + \frac{1-\nu}{2}\frac{c}{R} - \frac{3+\nu}{4}\frac{c^2}{R^2}\cdots\right)\frac{z}{R} + \frac{1+\nu}{4}\left(\frac{z^2}{R^2} + \frac{z^3}{R^3} + \cdots\right)\Bigg]p_i\Bigg\} \end{aligned} \tag{7.15d}$$

For comparison the expression for $u_r (= -u_z$ above) in Eqs. (5.79e) can also be put in series form by using expression (7.13a) for $1/r$ and $r = R - z$, $a = R - c$, $b = R + c$. Carrying out this operation, we find that the Lamé solution in series form yields exactly the same Eq. (7.15d). We can then readily compare the values obtained using the number of terms shown above with the exact values obtainable from the original closed Lamé solution. Figure 7.24 shows the ratio between the values obtained for $w = u_{z(z=0)}$ from the truncated series expression

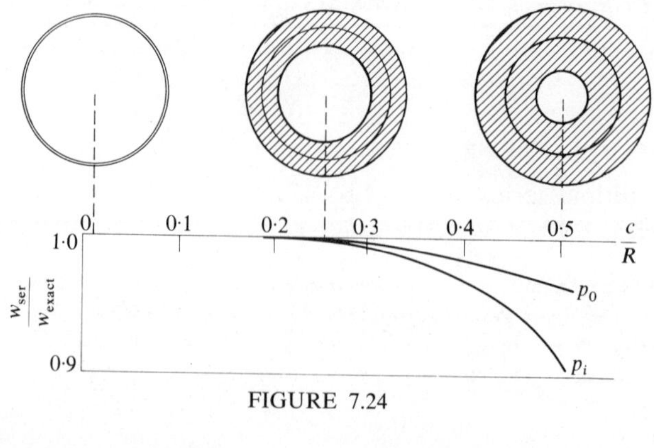

FIGURE 7.24

and the exact expression. This result indicates that the number of terms in the series expressions (7.14) can give a reasonable approximation for values of c/R even in the very thick walled range.

Harmonically Distributed Loads

For a more general application of Eqs. (7.14) let:

$$p = \sum_m \sum_n P_{mn} \sin\frac{mx}{R} \sin\frac{ny}{R}, \qquad q = \sum_m \sum_n Q_{mn} \sin\frac{mx}{R} \sin\frac{ny}{R}$$
$$\mathbf{d} = \sum_m \sum_n D_{mn} \sin\frac{mx}{R} \sin\frac{ny}{R}, \qquad \mathbf{s} = \sum_m \sum_n S_{mn} \sin\frac{m}{R} \sin\frac{ny}{R} \tag{7.15e}$$

where the n's must be integers for continuity if a whole cylinder is being considered, but the m's need not be. We will not consider boundary conditions in this study; the results would apply to the middle part of a long cylinder in any case, and at least the most important hinged-edge conditions could evidently be satisfied if desired by proper choice of the origin of coordinates and m (and of n if a panel is being considered).

Substituting expressions (7.15e) into Eqs. (7.13e) and equating the coefficients of the individual harmonic terms we find:

$$D_{mn} = \frac{P_{mn} - Q_{mn}}{A_{mn}}, \qquad S_{mn} = \frac{P_{mn} + Q_{mn}}{A_{mn}} \tag{7.15f}$$

where A_{mn} equals:

$$\frac{c^3}{3R^3(1-\nu^2)}[(m^2+n^2)^4 + (m^2+n^2)(\gamma m^4 - 6m^2n^2 - 2n^4) + \gamma m^4 + 4m^2n^2 + n^4] + \frac{c}{R}m^4 \tag{7.15g}$$

These values can be substituted into the expressions (7.14) for displacements or the expressions for stresses derived from them. For example, the maximum stress is likely to be σ_y at the inside surface $z = c$, and at peak points of the loading such as at $x = \pi R/(2m)$, $y = \pi R/(2n)$. The first terms of the expression for σ_y, with these values of x, y, z are:

$$\begin{aligned}
\sigma_y = -\frac{1}{2}\sum_m \sum_n \Bigg\{ \Bigg[& m^4\left(1 + \mathbf{c} + \frac{9}{5}\mathbf{c}^2 \cdots\right) + m^2n^2\left(\frac{2+\nu}{1-\nu^2}\mathbf{c} + \frac{8+2\nu}{3(1-\nu^2)}\mathbf{c}^2 \cdots\right) \\
& + n^4\left(\frac{\mathbf{c}+\mathbf{c}^2}{1-\nu^2} \cdots\right) - m^6\left(\frac{\nu}{1-\nu^2}\mathbf{c} - \frac{12-17\nu}{15(1-\nu)}\mathbf{c}^2 \cdots\right) \\
& - m^4n^2\left(\frac{1+2\nu}{1-\nu^2}\mathbf{c} + \frac{13+15\nu+17\nu^2}{15(1-\nu^2)}\mathbf{c}^2 \cdots\right) \\
& - m^2n^4\left(\frac{2+\nu}{1-\nu^2}\mathbf{c} + \frac{8+2\nu}{3(1-\nu^2)}\mathbf{c}^2 \cdots\right) - n^6\left(\frac{1}{1-\nu^2}\mathbf{c} + \frac{1}{1-\nu^2}\mathbf{c}^2 \cdots\right) \cdots \Bigg] D_{mn} \\
& + \Bigg[m^4(\mathbf{c} \cdots) + m^2n^2\left(\frac{2}{1-\nu^2}\mathbf{c}^2 \cdots\right) + n^4\left(\frac{1}{1-\nu^2}\mathbf{c}^2\right) - m^6\left(\frac{\nu}{1+\nu}\mathbf{c}^2\right) \\
& - m^4n^2\left(\frac{1+\nu-\nu^2}{1-\nu^2}\mathbf{c}^2 \cdots\right) - m^2n^4\left(\frac{2}{1-\nu^2}\mathbf{c}^2 \cdots\right) - n^6\left(\frac{1}{1-\nu^2}\mathbf{c}^2 \cdots\right) \cdots \Bigg] S_{mn} \Bigg\}
\end{aligned} \tag{7.15h}$$

7.6 THICK CYLINDERS—CORRECTIONS TO CLASSICAL DEFLECTIONS

As discussed in Secs. 3.5 and 5.6 for beams and plates, it is not necessary to use the rather complex elasticity-based methods developed in the last section when all that is needed is a correction to the deflections predicted by classical theory to allow for the effects of transverse strains. The method originated by Timoshenko, of taking the total deflection w_t as the sum of the purely flexural deflection w_f predicted by classical theory, and a corrective deflection w_s due to transverse strains (chiefly shear) alone, can be applied also to shallow shells. Since for this case we can take the transverse shear forces F_{xz} and F_{yz} as given by Eqs. (6.31g), and these expressions are the same (for m_x, $m_y = 0$) as for flat plates as given by Eqs. (4.15) or (5.84b), all of the development represented by Eqs. (5.84c, d, e) applies to shallow shells as well as to plates.

We can therefore find the total deflection w_t by solving Eq. (5.84e) simultaneously with Eq. (6.34) in which the w in the first term (which represents the flexural resistance to the loads) is replaced by w_f and the w in the second term (which represents the membrane resistance to the loads) is replaced by w_t. Or we can obtain a direct relation between the loading and the w_t in the same way as in Eqs. (5.84g, h), which gives, instead of Eq. (6.34):

$$\frac{2Ec^3}{3(1-\nu^2)}\left[1+\frac{(8-3\nu)c^2}{11(1-\nu)}\nabla^2\right]\nabla^8 w_t+\frac{2Ec}{R^2}\frac{\partial^4 w_t}{\partial x^4}=\nabla^4 p \tag{7.16a}$$

for the lateral load case. This can be modified for sandwich construction in the same way as discussed under Eq. (5.86a), and other types of loads added to the right-hand side as in Eq. (6.34) or (5.85).

Alternative Solution

Like Eq. (3.67a) for beams and (5.84i) for plates, we can also obtain a relation between the total deflection and the loading from the series expression for u_z in Eqs. (7.14). Taking $z = 0$, $\mathbf{s} = 0$ and $u_{z(z=0)} = w_t$ this gives:

$$\frac{E}{R}w_t=\frac{1}{2}\nabla^4\mathbf{d}+\mathbf{c}^2\left[\frac{3-15\nu}{20}\frac{\partial^4\mathbf{d}}{\partial\mathbf{x}^4}-\frac{12+68\nu-55\nu^2}{60(1-\nu)}\frac{\partial^4\mathbf{d}}{\partial\mathbf{x}^2\,\partial\mathbf{y}^2}\right.$$
$$\left.-\frac{2+3\nu}{20(1-\nu)}\frac{\partial^4\mathbf{d}}{\partial\mathbf{y}^4}-\frac{8-3\nu}{20(1-\nu)}\nabla^6\mathbf{d}\right]-\mathbf{c}^4\left[\frac{227-157\nu}{8400(1-\nu)}\nabla^8\mathbf{d}\cdots\right] \tag{7.16b}$$

where $\Delta\mathbf{d} = p$ and Δ is defined by Eq. (7.13f). This relation is more complex than Eq. (7.16a) but it is not limited to the shallow shell case.

Figure 7.25(b) and (c) show deflections predicted by Eqs. (7.16a) and (7.16b) respectively, as their ratio to those predicted by the simple shallow shell classical theory, Eq. (6.34); these ratios are given for the case of harmonically distributed loading p, as given by Eq. (7.15e) with $m = n = 2, 3, 8$, and for various values of c/R. We take $q, \mathbf{s} = 0$ and the total deflection $w_t = \Sigma\Sigma\, W_{mn} \sin(mx/R)\sin(ny/R)$.

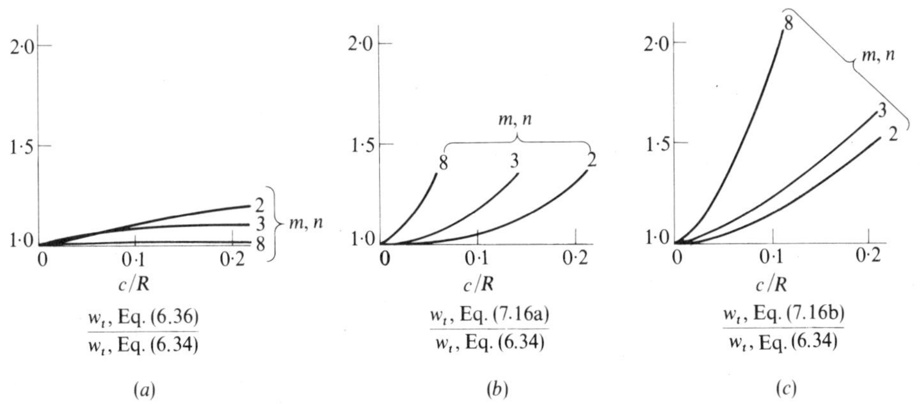

FIGURE 7.25

Also shown in Fig. 7.25(a) are the similar ratios obtained with the more exact classical theory Eq. (6.36). The results at (a) show the corrections for the error in the classical theory due to using shallow shell simplifications, (b) shows corrections for the error due to neglecting transverse strains, while (c) gives corrections for *both* errors.

Range of Application of Various Approximations

As in the discussion at the ends of Chaps. 3 and 5 for beams and plates, we can summarize the range of application of the various classical shell theories, and the improvements upon them which have been presented, by the relations between certain quantities.*

h = thickness
R = radius of curvature
l = 'primary half wavelength of the deflection' or distance between points of inflection.
w_{max} = maximum deflection.

'Classical' theories, which ignore transverse strains, are generally applicable when $h \ll l, R$. It is not easy to set definite upper limits for h/l and h/R for different types of classical theories because of the complexities involved. However, the chart shown in Fig. 7.26 represents an attempt in this direction. The lines shown represent the values of h/l and h/R for which the error in the deflection w or the stress σ_y is about 5 percent, as calculated by comparing results obtained by using the classical theories (6.34) and (6.36) and the relations associated with them, with the results obtained by using the more accurate Eqs. (7.15h) and (7.16a, b), for harmonic loading of a cylinder, as given by the first of Eqs. (7.15e) with $m = n$.

* Such comparisons can be made from various viewpoints. Compare for example 'On two-dimensional equations of the general linear theory of thin elastic shells', by A. L. Gol'denveizer, *Problems of Hydrodynamics and Continuum Mechanics.*

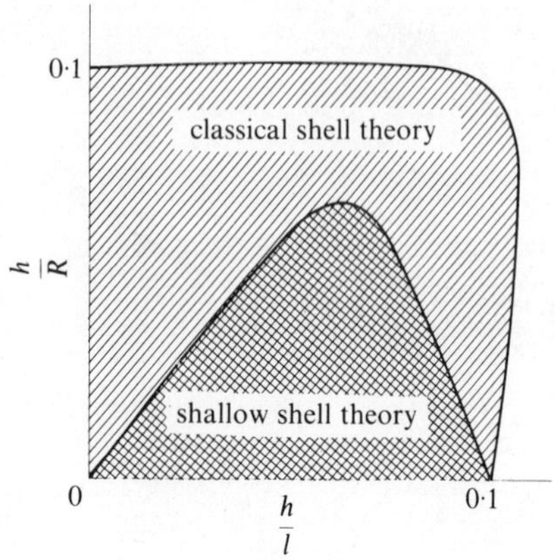

FIGURE 7.26

According to this chart, shallow shell classical theories should be applicable for values h/l and h/R in the double cross-hatched areas, while complete classical theories should be used for values in the single cross-hatched area, and a more exact elasticity-based theory such as the series solution (7.14) should be used for values outside these areas. Obviously such a study cannot give a definitive answer for all cases, but it should give a rough idea of the limitations of the classical theories.

As in the cases of beams and plates, it is probably safe to say that linear theories ignoring nonlinear membrane strains are applicable when $w_{max}/h < 0{\cdot}2$. For deflections up to about $w_{max}/h = 10$ it is usually unnecessary to use more than the squares of the slopes, $(\partial w/\partial\alpha)^2$ and $(\partial w/\partial\beta)^2$, in the expressions for membrane strains in the classical theories. For still larger deflections the full expressions (6.18) should be used; these do not include any nonlinear effects in the flexural strains, and such effects are probably never required in practical applications of classical theories. Nonlinear, non-classical theories, that is theories which consider both the effects of finite deformations and transverse strains, would be needed only in problems such as large deflections of thick-walled shells. Such deflections could only occur in the elastic range if the material is rubber-like, in which case simple linear stress–strain relations would probably also not apply; such cases are outside the range of this book.

AUTHOR INDEX

SUBJECT INDEX